W9-AEK-676

SECOND EDITION

EDITORS:

Claude Bouchard, PhD

Peter T. Katzmarzyk, PhD

Pennington Biomedical Research Center
Baton Rouge, Louisiana

Human Kinetics

Library of Congress Cataloging-in-Publication Data

Physical activity and obesity / Claude Bouchard, Peter T. Katzmarzyk, editors. -- 2nd ed.
p. ; cm.
Includes bibliographical references and index.
ISBN-13: 978-0-7360-7635-7 (hard cover)
ISBN-10: 0-7360-7635-2 (hard cover)
1. Obesity. 2. Exercise--Physiological aspects. I. Bouchard, Claude. II. Katzmarzyk, Peter T., 1968-
[DNLM: 1. Obesity--prevention & control. 2. Exercise--physiology. WD 210 P578 2010]
RC628.P496 2010
616.3'98062--dc22

2009047894

ISBN-10: 0-7360-7635-2 (print)
ISBN-13: 978-0-7360-7635-7 (print)

The Web addresses cited in this text were current as of June 2009, unless otherwise noted.

Acquisitions Editor: Michael S. Bahrke, PhD; **Managing Editor:** Melissa J. Zavala; **Assistant Editors:** Christine Bryant Cohen, Casey A. Gentis, and Kathy Bennett; **Copyeditor:** Joyce Sexton; **Indexer:** Craig Brown; **Permission Manager:** Dalene Reeder; **Graphic Designer:** Bob Reuther; **Graphic Artist:** Angela K. Snyder; **Cover Designer:** Bob Reuther; **Visual Production Assistant:** Jason Allen; **Art Manager:** Kelly Hendren; **Associate Art Manager:** Alan L. Wilborn; **Illustrator:** TwoJay!; **Printer:** Thomson-Shore, Inc.

Printed in the United States of America 10 9 8 7 6 5 4 3 2

The paper in this book is certified under a sustainable forestry program.

Human Kinetics
Web site: www.HumanKinetics.com

United States: Human Kinetics
P.O. Box 5076
Champaign, IL 61825-5076
800-747-4457
e-mail: humank@hkusa.com

Canada: Human Kinetics
475 Devonshire Road Unit 100
Windsor, ON N8Y 2L5
800-465-7301 (in Canada only)
e-mail: info@hkcanada.com

Europe: Human Kinetics
107 Bradford Road
Stanningley
Leeds LS28 6AT, United Kingdom
+44 (0) 113 255 5665
e-mail: hk@hkeurope.com

Australia: Human Kinetics
57A Price Avenue
Lower Mitcham, South Australia 5062
08 8372 0999
e-mail: info@hkaustralia.com

New Zealand: Human Kinetics
P.O. Box 80
Torrens Park, South Australia 5062
0800 222 062
e-mail: info@hknewzealand.com

E4628

Contents

Part II: Prevalence and Cost Issues 39

Part III: Determinants of Physical Activity Levels 69

Part IV: Physical Activity and Risk of Obesity 97

Part V: Physical Activity and Biological Determinants of Obesity . . . 129

Part VIII: Clinical Implications 263

Part IX: Policy and Research Issues 335

Contributors

Barbara E. Ainsworth, PhD, MPH
Department of Exercise and Wellness
School of Applied Arts and Sciences
Arizona State University
Phoenix, AZ

Lars Bo Andersen, MD
Department of Sports Medicine
Norwegian School of Sport Sciences
Oslo, Norway
Research in Childhood Health
University of Southern Denmark
Odense, Denmark

Ross Andersen, PhD
Canada Research Chair
Professor
Department of Kinesiology and Physical Education
McGill University
Montreal, PQ

Timothy Armstrong, PhD
Coordinator, Surveillance and Population Based Prevention Unit
Chronic Diseases and Health Promotion Department
World Health Organization
Geneva, Switzerland

Tom Baranowski, PhD
Professor of Pediatrics (Behavioral Nutrition & Physical Activity)
Children's Nutrition Research Center
Department of Pediatrics
Baylor College of Medicine
Houston, TX

David R. Bassett Jr., PhD
Obesity Research Center
Department of Exercise, Sport, and Leisure Studies
University of Tennessee
Knoxville, TN

Adrian Bauman, PhD, MB, MPH
Centre for Physical Activity, Nutrition and Obesity Research
School of Public Health
University of Sydney
New South Wales, Australia

Louise A. Baur, MBBS (Hons), BSc (Med), PhD
Discipline of Paediatrics & Child Health
University of Sydney
Clinical School, The Children's Hospital at Westmead Australia

Steven N. Blair, PED
Department of Exercise Science
Department of Epidemiology and Biostatistics
Arnold School of Public Health
University of South Carolina
Columbia, SC

Elisabet Børsheim, PhD
Department of Surgery, Metabolism Unit
University of Texas Medical Branch/Shriners Hospitals for Children at Galveston
Galveston, TX

George A. Bray, MD
Pennington Biomedical Research Center
Baton Rouge, LA

Wendy J. Brown, PhD
School of Human Movement Studies
University of Queensland
Australia

Mercedes R. Carnethon, PhD
Department of Preventive Medicine
The Feinberg School of Medicine
Northwestern University
Chicago, IL

Timothy Church, MD, MPH, PhD
Professor
John S. McIlhenny Endowed Chair of Health Wisdom
Laboratory of Preventive Medicine Research
Pennington Biomedical Research Center
Baton Rouge, LA

Kerry S. Courneya, PhD
University of Alberta
Edmonton, AB

Christopher A. DeSouza, PhD
Integrative Vascular Biology Laboratory
Department of Integrative Physiology
University of Colorado
Boulder, CO

Rod K. Dishman, PhD
Professor
Exercise Science
Department of Kinesiology
The University of Georgia
Athens, GA

Joseph E. Donnelly, EDD
Professor/Director, Energy Balance Laboratory & The Center for Physical Activity and Weight Management
The Schiefelbusch Institute for Lifespan Studies
University of Kansas
Lawrence, KS

David J. Dyck, PhD
Department of Human Health & Nutritional Sciences
University of Guelph
Guelph, ON

Conrad P. Earnest, PhD
Director, Exercise Biology Lab
Pennington Biomedical Research Center
Baton Rouge, LA

Janice Eng, PhD
Professor
Department of Physical Therapy and Graduate Program in Rehabilitation Sciences
University of British Columbia
Rehab Research Lab, GF Strong Rehab Centre
Vancouver, BC

Johan G. Eriksson, MD, PhD
Department of General Practice and Primary Healthcare
University of Helsinki
Finland and National Public Health Institute
Helsinki, Finland

Mikael Fogelholm, ScD
Adjunct Professor
Director, Academy of Finland, Health Research Unit
Helsinki, Finland

Kevin R. Fontaine, PhD
Division of Rheumatology
Johns Hopkins University School of Medicine
Baltimore, MD

Lise Gauvin, PhD
Department of Social & Preventive Medicine
Centre de recherche Léa-Roback sur les inégalités sociales de santé de Montréal
Groupe de recherche interdisciplinaire en santé
CR-CHUM (Centre de recherche du Centre Hospitalier de l'Université de Montréal)
Université de Montréal
Montréal, PQ

Isabelle de Glisezinski, PhD
INSERM U858, I2MR, Université Paul Sabatier, 31403 Toulouse cedex, France
Service d'Exploration de la Fonction Respiratoire et de Médecine du Sport, Hôpital Larrey, TSA 30030, 31059 Toulouse cedex 9, France

Bret Goodpaster, PhD
Division of Endocrinology and Metabolism
Department of Medicine
University of Pittsburgh
Pittsburgh, PA

Steven L. Gortmaker, PhD
Department of Society, Human Development, and Health and Nutrition
Harvard School of Public Health
Boston, MA

Frank L. Greenway, MD
Medical Director and Professor
Pennington Biomedical Research Center
Baton Rouge, LA

Bernard Gutin, PhD
Department of Nutrition
School of Public Health
University of North Carolina
Chapel Hill, NC

Anthony C. Hackney, PhD, DSc
Professor of Exercise Physiology & Nutrition
Department of Exercise & Sport Science
Department of Nutrition
University of North Carolina
Chapel Hill, NC

James M. Hagberg, PhD
Department of Kinesiology
School of Public Health
University of Maryland
College Park, MD

Mark Hamer, PhD
Department of Epidemiology and Public Health
University College London
London, U.K.

Steven B. Heymsfield, MD
Global Center for Scientific Affairs
Merck & Company
Rahway, NJ

John H. Himes, PhD, MPH
Division of Epidemiology and Community Health
University of Minnesota School of Public Health
Minneapolis, MN

David A. Hood, PhD
School of Kinesiology and Health Science
Muscle Health Research Centre
York University
Toronto, ON

Jennifer M. Hootman, PhD
Epidemiologist, Arthritis Program
Division of Adult and Community Health
Centers for Disease Control and Prevention
Atlanta, GA

Edward S. Horton, MD
Joslin Diabetes Center
Harvard Medical School
Boston, MA

John M. Jakicic, PhD
Department of Health and Physical Activity
Physical Activity and Weight Management Research Center
University of Pittsburgh
Pittsburgh, PA

W. Philip T. James, CBE, MD, DSc
London School of Hygiene and Tropical Medicine
International Obesity Task Force
London, U.K.

Ian Janssen, PhD
School of Kinesiology and Health Studies
Department of Community Health and Epidemiology
Queen's University
Kingston, ON

Victor Katch, PhD
Professor, Movement Science
School of Kinesiology
Associate Professor, Pediatrics, Section of Pediatric Cardiology
School of Medicine
University of Michigan
Ann Arbor, MI

Wendy M. Kohrt, PhD
Professor of Medicine, Division of Geriatric Medicine
University of Colorado Denver
Aurora, CO

Catherine M. Kotz, PhD
Veterans Affairs Medical Center
Minnesota Obesity Center
Minneapolis, MN
Department of Food Science and Nutrition
University of Minnesota
Saint Paul, MN

Markku Laakso, MD, PhD
Department of Medicine
University of Kuopio
Kuopio, Finland

I-Min Lee, MD, ScD
Brigham and Women's Hospital
Harvard Medical School
Boston, MA

Teresa Liu-Ambrose, PhD
Assistant Professor
Department of Physical Therapy and Graduate Program in Rehabilitation Sciences
University of British Columbia
Centre for Hip Health, Vancouver Coastal Research Institute
Vancouver, BC

Tim Lobstein, PhD
International Obesity Task Force, International Association for the Study of Obesity
London, U.K.
Science Policy Research Unit
University of Sussex
Brighton, U.K.

Ruth J.F. Loos, PhD
Medical Research Council Epidemiology Unit
Institute of Metabolic Science
Cambridge, U.K.

Cheryl Lovelady, PhD, MPH, RD
Professor of Nutrition
University of North Carolina
Greensboro, NC

Ian A. Macdonald, PhD
School of Biomedical Sciences
University of Nottingham
Nottingham, U.K.

Robert M. Malina, PhD
Professor Emeritus, Department of Kinesiology and Health Education
University of Texas at Austin
Research Professor, Department of Health and Physical Education
Tarleton State University
Stephenville, TX

Christopher D. Morrison, PhD
Pennington Biomedical Research Center
Baton Rouge, LA

Michelle F. Mottola, PhD
Director, R. Samuel McLaughlin Foundation–Exercise and Pregnancy Lab
Associate Professor, School of Kinesiology, Faculty of Health Sciences, Department of Anatomy & Cell Biology
Schulich School of Medicine & Dentistry
University of Western Ontario, London
Associate Scientist, Child Health Research Institute, Lawson Health Research Institute
London, ON

W. Kerry Mummery, PhD
Centre for Social Science Research
Associate Dean
College of Health and Human Services
Faculty of Sciences, Engineering & Health
Central Queensland University
Queensland, Australia

Peter W. Nathanielsz, MD, PhD, ScD
Center for Pregnancy and Newborn Research
University of Texas Health Sciences Center
San Antonio, TX
Department of Comparative Medicine
Southwest Foundation for Biomedical Research
San Antonio, TX

Robert L. Newton Jr., PhD
Assistant Professor
Pennington Biomedical Research Center
Baton Rouge, LA

Paul E. O'Brien, MD
Director, Centre for Obesity Research and Education
Monash University
Melbourne, Australia

Neville Owen, PhD
Cancer Prevention Research Centre
School of Population Health
University of Queensland
Brisbane, Australia

Russell R. Pate, PhD
Department of Exercise
University of South Carolina
Columbia, SC

John C. Peters, PhD
The America on the Move Foundation
UCD, Center for Human Nutrition
Denver, CO

Kenneth E. Powell, MD, MPH
Epidemiologic and Public Health Consultant
Atlanta, GA

Tuomo Rankinen, PhD
Pennington Biomedical Research Center
Human Genomics Laboratory
Baton Rouge, LA

Denis Richard, PhD
Laval University Institute of Cardiology and Pulmonology
Merck Frosst/CIHR Research Chair on Obesity
Laval Hospital Research Center
Québec, PQ

Chris Riddoch, PhD
Professor of Sport and Exercise Science
School for Health
University of Bath
Bath, U.K.

Robert Ross, PhD
School of Kinesiology and Health Studies/Department of Medicine
Division of Endocrinology and Metabolism
Queen's University
Kingston, ON

Larissa Roux, MD, PhD
The Copeman Healthcare Centre
Vancouver, BC

James F. Sallis, PhD
Professor of Psychology
San Diego State University
San Diego, CA

Art Salmon, EdD
Sport & Recreation Branch
Ontario Ministry of Health Promotion
Toronto, ON

Yves Schutz, PhD, MPH
Department of Physiology
Faculty of Medicine
University of Lausanne
Lausanne, Switzerland

Roy J. Shephard, MD, PhD
Faculty of Physical Education & Health
Department of Public Health Sciences, Faculty of Medicine
University of Toronto
Toronto, ON

Chantal Simon, MD, PhD
University Lyon 1, Lyon, France
University of Strasbourg, Strasbourg, France

Steven R. Smith, MD
Pennington Biomedical Research Center
Baton Rouge, LA

Adrian H. Taylor, PhD
School of Sport & Health Sciences
University of Exeter, U.K.

Angelo Tremblay, PhD
Division of Kinesiology
Department of Social and Preventive Medicine
Faculty of Medicine
Université Laval
Québec City, PQ

Mark S. Tremblay, PhD
Director of Healthy Active Living and Obesity Research
Children's Hospital of Eastern Ontario Research Institute
Ottawa, ON

Margarita S. Treuth, PhD
Associate Professor
Department of Physical Therapy, School of Health Professions
University of Maryland Eastern Shore
Princess Anne, MD

Richard P. Troiano, PhD
National Cancer Institute
National Institutes of Health
Bethesda, MD

François Trudeau, PhD
Department of Physical Activity Sciences
Université du Québec à Trois-Rivières
Trois-Rivières, PQ

Catrine Tudor-Locke, PhD
Associate Professor
Walking Behavior Laboratory
Pennington Biomedical Research Center
Baton Rouge, LA

Mark H. Vickers, PhD
Liggins Institute and The National Research Centre for Growth and Development
University of Auckland
Auckland, New Zealand

Youfa Wang, MD, PhD
Associate Professor
Center for Human Nutrition, Department of International Health
Bloomberg School of Public Health
Johns Hopkins University
Baltimore, MD

Darren Warburton, PhD
Cardiovascular Physiology and Rehabilitation Laboratory
Experimental Medicine Program, Department of Medicine
University of British Columbia
International Collaboration on Repair Discoveries

Rena R. Wing, PhD
Professor of Psychiatry & Human Behavior
The Warren Alpert Medical School of Brown University
Director, Weight Control & Diabetes Research Center
The Miriam Hospital
Providence, RI

Stephen C. Woods, PhD
Department of Psychiatry
Director, Obesity Research Center
University of Cincinnati
Cincinnati, OH

Shawn D. Youngstedt, PhD
Department of Exercise Science, Norman J. Arnold School of Public Health
University of South Carolina
Department of Psychiatry, Dorn VA Medical Center
Columbia, SC

Preface

Physical Activity and Obesity was published in 2000 and was the result of a perceived need for such a book in the exercise science and sports medicine communities. The book included 19 chapters written by leading authorities in the field. Several years later, discussions began on whether the field had progressed sufficiently to warrant a second edition of the book. The conclusion was that the field of physical activity and obesity was burgeoning with new research and that more than a second edition was required. Indeed, it was the view of the editors that a new book highlighting these recent developments was needed. *Physical Activity and Obesity, Second Edition* was born out of these discussions.

This book has been written in collaboration with leading scientists from many fields. We have aimed to provide short, authoritative chapters written by leading scientists in the field rather than the typical longer chapters found in many academic texts. We hope that the experiment will be successful and that the reader finds this a useful approach. We would like to thank the contributing authors for adhering to the strict formatting guidelines and tight timelines for this project.

We would like to thank a number of people. Ms. Lisa Landry and Ms. Nina Laidlaw at the Pennington Biomedical Research Center have provided able assistance to the editors. In particular, their assistance with the editing and formatting of such a large number of chapters and their facilitation of communication between the editors and contributing authors have been extremely valuable to us. At Human Kinetics, Michael Bahrke, acquisitions editor, Elaine Mustain, developmental editor, and Melissa Zavala, managing editor, have guided the book through all stages and into print. They provided the framework for the excellent writing and critical concepts from each of the authors and their colleagues that form the heart of this book.

part I

Definition and Assessment of Physical Activity and Obesity

The opening section of this book sets the stage for the sections that follow. Chapter 1 summarizes the problem and provides an introduction to some of the key research issues in the field of physical activity and obesity. Chapter 2 provides a perspective on the continuum of physical activity and exercise behavior and introduces several key concepts and terms that are carried forward throughout the book. The characterization of sedentary behavior and its potential effects on health are emerging as important areas of research. Chapter 3 provides an overview of the methods used to assess the level of sedentary behavior among humans. On the other hand, chapters 4 and 5 describe the methods used to assess physical activity levels in adults and children, respectively. Chapters 6 and 7 provide the necessary background on the most common methods currently used to assess obesity in adults and children, respectively. Given the widespread use of self-reported data for the assessment of both physical activity and obesity, this section concludes with chapter 8, which provides a discussion of the limitations and pitfalls of relying on self-reported data.

1

Introduction

Claude Bouchard, PhD, and Peter T. Katzmarzyk, PhD

The first incarnation of this book was published in 2000 (1). It was motivated by the fact that in the view of many, the prevalence of obesity around the world had reached epidemic proportions. A decade later, the situation has deteriorated further: The World Health Organization (WHO) estimates that the number of overweight and obese adults around the globe is well in excess of 1 billion. The global picture has also worsened considerably for children and adolescents, as is reviewed later in this book. A major concern is that the obesity pandemic will translate into a growing burden of disease worldwide (2). There is already evidence for this in population data showing a strong and growing increase in the prevalence of type 2 diabetes around the globe. The fear is that the prevalence of diabetes, cardiovascular disease, and other morbidities associated with excess adiposity and ensuing metabolic dysfunction will increase to an extent that will make it impossible to cope even for the most developed economies of the world. In this regard, the WHO reported recently that 1.9 million premature deaths result each year from sedentary lifestyles (3).

In this context, we originally thought that physical activity had the potential to attenuate the rate of growth of the obesity epidemic and contribute to efforts to bring it under control. One decade later, this concept remains valid and is more important than ever. Over the last decade, a whole body of data has provided support for the notion that it is necessary to quantify not only the level of physical activity but also the time spent in sedentary pursuits. It turns out that the level of physical activity is at best only moderately correlated with the level of inactivity, such that the two metrics provide independent information. This is particularly important for obesity and its related morbidities. Indeed, recognition is now almost universal that it is essential for us to consider both metrics when discussing prevention of weight gain, treatment of overweight or obesity, weight maintenance after weight loss, or improvement of obesity-associated metabolic dysfunctions.

A General Model

The epidemic of excess weight is driven by widespread energy imbalance favoring storage of the energy surplus not expended. We have proposed a model in which the potential contributors to this epidemic are grouped under four major headings: built environment, social environment, behavior, and biology (4). Figure 1.1 depicts a hierarchical model of these four classes of determinants. Factors in the built environment (e.g., reliance on the automobile, building design, lack of safe sidewalks) and the social environment (e.g., advertising, pressure to consume) are such that the global environment has become "obesogenic" not only in developed countries but also increasingly in most developing areas of the world. An obesogenic environment favors the adoption of obesogenic behavior (e.g., consumption of large portion size meals; high-fat diets; high sugar intake; many hours spent watching TV, playing video games, or sitting at the computer). We hypothesize that obesogenic environments and behavior are fueling the rise in the prevalence of overweight and obesity that the world is currently experiencing.

However, one should not omit biology from the discussion. Several lines of evidence support the contention that there are individual differences in the predisposition to gain weight and that genetic variation has much to do with the risk of becoming obese, particularly of becoming severely obese. Reviewing the arguments in favor of this concept is beyond the scope of this short introductory chapter; but the evidence is compelling, as discussed recently by O'Rahilly and Farooqi (5) and as reviewed recently in a supplement of *Obesity* (6). This is schematically depicted in figure 1.1 by the

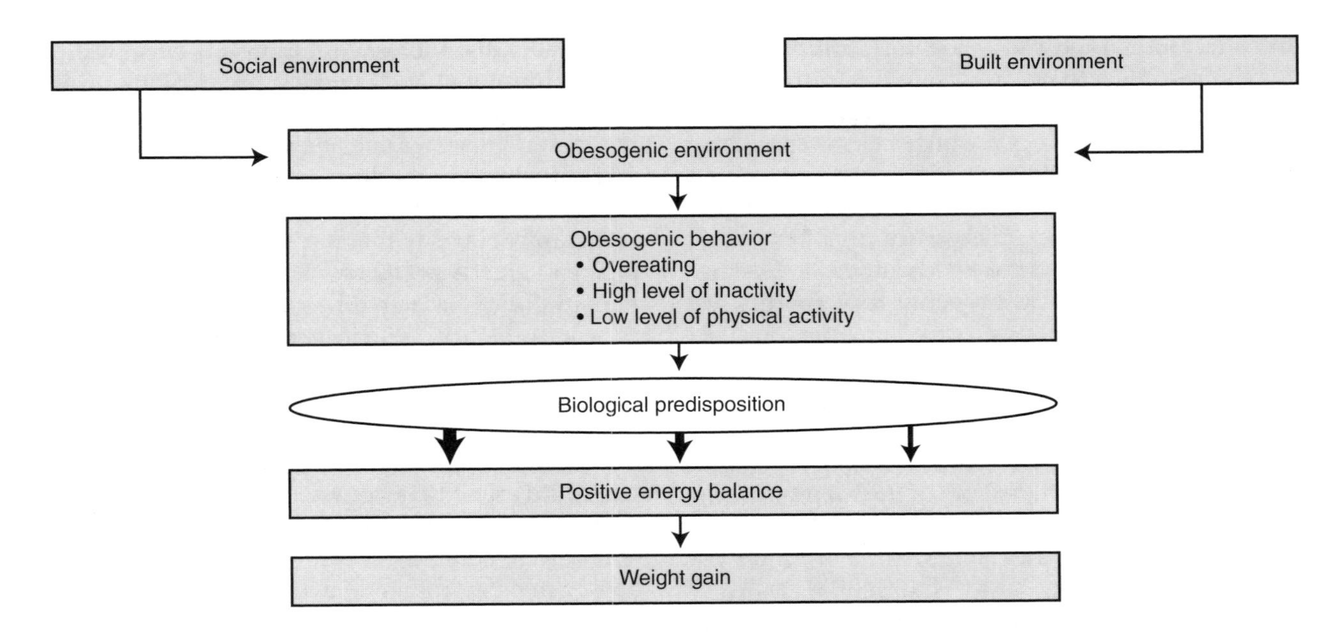

Figure 1.1 The obesogenic environment is defined by many characteristics in the social and built environment. An obesogenic environment favors the adoption and maintenance of obesogenic behaviors. The net effect of an obesogenic mode of life is modulated by biological traits that are highly prevalent in the population.

Adapted, by permission, from C. Bouchard, 2007, "The biological predisposition to obesity: Beyond the thrifty gene scenario (commentary)," *International Journal of Obesity* 31: 1337-1339.

three arrows linking biological predisposition to positive energy balance. The thin arrow represents a low biological predisposition, while the intermediate and the thickest arrows stand for increasing biological susceptibility to being in positive energy balance for long periods of time (4).

Gene–Physical Activity Interaction Effects

We often hear that since the human gene pool could not have changed rapidly enough to explain the current rise in the prevalence of obesity, this increase is undoubtedly caused by changes in our environment and our way of life. It is not uncommon to read that biology has nothing to do with the current epidemic of excess weight. But the converse is also true; there are biologists who hold the view that the epidemic is caused entirely by vulnerabilities in our genome. As is generally the case when such diametrically opposed views are upheld, the truth lies somewhere in the middle.

Hundreds of candidate genes have been studied for their potential role in the predisposition to obesity. Many of them may turn out to be true contributors as obesity polygenes; but unfortunately, the results available thus far on these genes come from studies that are seriously underpowered. More recently, genome-wide association studies based on hundreds of thousands of genomic markers have led to exciting new gene discoveries for several common diseases. A few of these reports have dealt with obesity and have generated one gene with a substantial effect size plus a few others with rather minor contributions. A number of variants in the "fat mass and obesity associated" gene *(FTO)* showed a very strong association with body mass index (BMI), obesity, or fat mass in several independent studies (7). On average, homozygotes for the risk allele at the *FTO* gene weigh 3 to 4 kg (6.6 to 8.8 lb) more and have a 1.67-fold increased risk of obesity compared to homozygotes for the nonrisk allele. Each risk allele increases the risk of obesity by more than 30%. Similar risk levels have been observed in children and adolescents, particularly for total adiposity. The risk allele is quite prevalent among Caucasians, as about 16% of them are homozygotes. The population-attributable risk for overweight has been estimated at 20%.

Gene–physical activity interaction refers to a situation in which the response or the adaptation to a change in physical activity behavior is conditional on the genotype of the individual. Interactions between the *FTO* genotype and physical activity level have been recently reported in two independent studies, one performed on 17,508 Danish adults and the other on 704 Old Order Amish adults (8, 9).

In both cases, it is only the sedentary homozygotes for the risk allele who are heavier than the other two genotypes. The homozygotes for the risk allele are normal weight if they are physically active. This interaction effect is illustrated in figure 1.2.

An important issue is whether one could take advantage of this gene–physical activity interaction effect, and of others yet to be identified, to design more effective obesity prevention or treatment programs.

Progress Over the Past Decade

Even though significant progress has been made since the 2000 edition of *Physical Activity and Obesity* was published, the advances have been only incremental and represent a series of small steps forward in the quest for a better understanding of the true causes of obesity at the individual and population levels, effective preventive and treatment strategies, and efficacious management of obesity-associated morbidities. There has been no quantum leap in obesity science in the last decade. However, useful advances have been made in the definition of the central and peripheral regulation of energy balance, particularly in terms of the regulation of appetite and satiety, the biology of adipose tissue, and its role in hormonal and cytokine production and in inflammation. We have also seen a paradigm shift in the study of genes and alleles contributing to the predisposition to obesity. During the same period, we have witnessed major setbacks in the pharmacotherapy of obesity that were preceded by a short period of great expectations. However, most experts remain optimistic regarding the future of obesity pharmacotherapy as reviewed recently in a volume of *Handbook of Obesity* (10). Moderate progress has been achieved regarding dietary or activity behavioral changes designed to prevent or treat obesity, but no major breakthrough has occurred. In fact, one could argue that the most important advances of the decade have been observed on two fronts: (a) in our understanding of the powerful effects of the obesogenic environment as defined by social and built-environment determinants, and (b) in bariatric surgery as a treatment for excess adiposity and metabolic dysfunction.

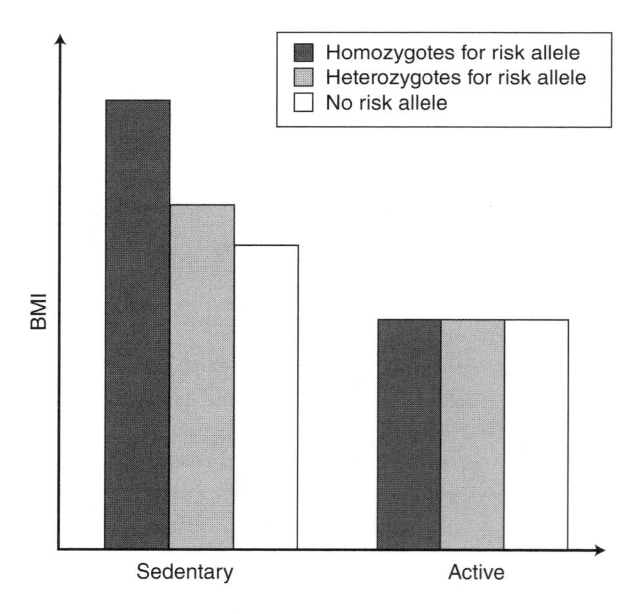

Figure 1.2 Homozygotes for the risk allele at the *FTO* gene are heavier when subjects are sedentary. The effect of the risk allele is progressively abolished with increasing physical activity level.

Data from C.H. Andreasen et al., 2008, "Low physical activity accentuates the effect of the FTO rs9939609 polymorphism on body fat accumulation," *Diabetes* 57: 95-101; and E. Rampersaud et al., 2008, "Physical activity and the association of common FTO gene variants with body mass index and obesity," *Archives of Internal Medicine* 168: 1791-1797.

Important Differences Between Overweight and Obesity

We believe that the distinction between overweight and obesity is highly justified in terms of both the etiology of the conditions and the levels of risk for morbidity and mortality. The distinction between the two conditions is also useful for an understanding of the contributions and limits of regular physical activity.

The main difference between the two conditions resides in the greater excess of weight and in the higher percentage of the body as fat in the obese. A second difference is that positive energy balance has been more pronounced and has been sustained for a longer period of time in the obese than in the overweight. A third difference pertains to energy expenditure: Obese individuals have a higher resting metabolic rate as a result of a greater respiring tissue mass and a higher total energy expenditure above resting energy expenditure than normal-weight people. The latter is caused by the fact that more energy is required to move a bigger mass.

One implication of these differences is that a sedentary lifestyle or a low level of habitual activity (or the two in combination) has the potential to account for a large proportion of the adult overweight cases. For instance, a nonresting daily energy expenditure depressed by about 300 kcal will generate chronic positive energy balance and translate into a surplus of calories consumed of more than 100,000 kcal over a year. Assuming an average efficiency of tissue deposition, this would result in a gain of 6 to 8 kg (about 13 to 18 lb) of body weight over a year. The weight gain would be progressively less with time as the resting metabolic rate and the energy cost of moving the body increase with body mass accretion. After a while, energy balance would be restored but at a new body mass level, one that may now be in the overweight range. One should note that the path to overweight under conditions of sedentarism does not require any increase in energy intake. Thus, it is possible to become overweight without being truly hyperphagic in comparison to other people of the same sex, age, and body build.

In contrast, obesity, particularly the more severe types (i.e., BMIs of 40 and more), requires positive energy balance conditions that are sustained for longer periods of time. Such conditions will generally be achieved only when energy intake is increased and energy expenditure is depressed, and if energy intake continues to increase after new steady state body weight plateaus are achieved. Thus, one could speculate that a sedentary mode of life plays an important role in both overweight and obesity but is, under certain circumstances, a condition sufficient to lead to an overweight state.

A Major Challenge

Throughout the evolutionary journey of *Homo sapiens,* the struggle to free human beings from muscular work and physical exertion was a constant feature. It made very impressive gains around the period of the industrial revolution and, even more, in the last century with the technological progress achieved in industrialized countries. Thus, the amount of energy expended by individuals to ensure sustained food supply, decent housing under a variety of climatic conditions, safe and rapid transportation, personal and collective security, and diversified and abundant leisure activities has decreased during the last century. The decrease in the levels of energy expenditure associated with physical work continues unabated with the growth around the world in the number of automobiles, central heating and cooling systems, elevators, escalators, television sets, computers, cell phones, motorized heavy construction equipment, clothes washers and dryers, dishwashers, vacuum cleaners, powered lawn mowers, snowblowers, and other labor-saving and muscular work–saving technologies. At the same time, data from many countries suggest that there has been no change in the time spent in leisure-time physical activity. However, it is now widely recognized that the reduction in the amount of physical work may have gone too far. The benefits of a physically active lifestyle have been compared with those of an inactive mode of life; and, while not all the evidence is in, it seems that human beings are better off when they maintain a physically active lifestyle (1).

Nowhere are the consequences of the present level of low energy expenditures more obvious than in energy balance and body weight regulation. While all of these trends in the reduction of muscular work are taking place, caloric intake appears to be increasing slightly or at best remains constant. Evolution has endowed *Homo sapiens* with complex regulatory systems of appetite and satiety as well as with physiological and metabolic characteristics determining basal metabolic rates and food-, exercise-, or cold-induced thermogenesis. The recent past in affluent societies reveals that these biological systems cannot cope well in an environment in which palatable foods are abundant and energy expenditure of activity is low. In particular, the lesson from the last few decades is that it seems to be impossible for a large fraction of sedentary individuals to regulate food and caloric intake so as to be in balance at low levels of daily energy expenditure and to get rid of the caloric surplus consumed in the form of heat dissipation. The energy expenditure from physical activity is thus too low for most people to be able to eat normally without having to be on caloric restriction diets from time to time or having to be constantly restraining their food intake. Energy wastage mechanisms are simply unable to cope, and positive energy balance is the net result.

These conditions provide a fertile environment for the expansion of the obesity epidemic. Because the health consequences of excess body fat do not become immediately manifest, the epidemic of obesity in children, adolescents, and young adults will translate later into an unprecedented number of cases of type 2 diabetes, hypertension, cardiovascular disease, gallbladder disease, postmenopausal breast cancer, colon and other cancers, osteoarthritis, back pain, and physical and mental disabilities. The global burden of obesity-associated disabilities and health care costs is already very high but will become staggering in the coming decades.

Even though individuals bear responsibility for maintaining healthy weights, it is obvious that this approach is not sufficient: It has not succeeded in containing the present obesity epidemic. What is needed is a series of major policies aimed at transforming our environment and the way we live. Indeed, nothing short of a paradigm shift has any chance of success in the efforts to curtail the increase in the number of people who are chronically in positive energy balance. Changes throughout the food chain and in food labeling, portion sizes, caloric density, advertising, city planning, building codes, mass transit systems, car use, foot and cycling paths, pedestrian areas in city centers, and school schedules and programs are among those that need to be implemented if we are to attenuate the impact of the current obesogenic environment.

Content of the Book

This book has been written with the collaboration of the most prominent scientists and clinicians in the field. It is organized around nine parts with a total of 89 chapters. Part I provides key definitions and discusses several methods for the assessment of obesity and physical activity levels. Part II includes an overview of the current epidemics of physical inactivity and obesity along with the associated economic costs. Part III provides chapters on the determinants of physical activity levels. Part IV examines the issue of physical activity as a risk factor for weight gain and the development of obesity. Part V covers topics related to physical activity and the putative biological determinants of obesity, while part VI covers physical activity and behavioral and environmental determinants of obesity. Part VII examines the role of physical activity in the prevention and treatment of obesity; and following that, part VIII explores the clinical implications of physical inactivity and obesity. Finally, part IX provides an overview of some of the policy and research issues facing the field of physical activity and obesity.

Covering this wide range of topics with 89 chapters could have resulted in a much larger volume. However, an effort was made to focus each chapter on only the key issues. Moreover, a size limit was imposed on the manuscript of each contributing author. Each author was asked to provide a very concise and authoritative review of the evidence pertaining to his or her topic. Only the most critical references were to be cited, and citing relevant review papers was one strategy that was helpful in meeting the rule imposed on the number of references.

2

The Physical Activity and Exercise Continuum

Darren Warburton, PhD

As documented throughout this book, there is considerable evidence supporting the important role of physical activity in the primary and secondary prevention of obesity and obesity-related diseases (1, 2). Physical activity is thought to be of benefit for over 25 chronic conditions—in particular, cardiovascular disease, breast and colon cancers, type 2 diabetes, and osteoporosis (1, 3).

When considering the health benefits of physical activity, it is essential to have a clear understanding of the various components of physical activity and exercise. In particular, it is important to recognize that there is a continuum of behavior ranging from being inactive to being very physically active. Moreover, physical activity (including its subcomponents) is distinct from exercise. Accordingly, the primary purpose of this chapter is to briefly describe the various components of physical activity and exercise across a continuum. A secondary purpose is to evaluate briefly the commonly used exercise prescription and energy expenditure methodologies for physical activity and exercise promotion.

Definition of Health, Physical Activity, and Exercise

Owing to the extensive discussions (throughout this book) regarding the roles that physical activity and exercise play in the determination of health status, it is essential that we first clearly define each variable. In this context, it is important to highlight that *health* does not represent merely the absence of disease but rather reflects the physical, social, and psychological well-being of an individual. Moreover, owing to the numerous and varied interrelationships between the determinants of health status, the assessment of health status is not straightforward. For instance, according to Bouchard and Shephard (4), at least five aspects of health status need to be considered in its evaluation: (1) genetics; (2) biochemical, physiological, and morphological conditions that determine the onset of illness, disease, impairment, or disability; (3) functional well-being; (4) psychological well-being associated with mood and cognitive processes; and (5) health potential relating to longevity and functional potential.

Classically defined, *physical activity* refers to all leisure and nonleisure body movements resulting in an increased energy output from rest (4). However, it is important to note that there are four broad physical activity domains; these include occupational (work related), domestic (housework, yard work, child care, chores), transportation (bicycling or walking), and leisure time (discretionary or recreational time for physical activity, sport, exercise, and hobbies).

Traditional health promotion programs have often focused on increasing leisure-time physical activity to reduce the risk for chronic disease and increase overall health status. However, more recent efforts have identified the health benefits of physical activities from all four domains (1, 2). This is particularly salient given the finding that most individuals only have 3 to 4 h per day available for leisure-time physical activities. Moreover, assessments of leisure-time physical activity have demonstrated a relative stability in the percentage of adults who meet the most common recommendations for leisure-time physical activity (i.e., 30 min of moderate-intensity exercise on a daily basis) (5). In comparison, industrialization has been associated with a marked decrease in nonleisure-time physical activity, as well as increased participation in sedentary behaviors (sitting, watching television, playing video games), decreased manual labor, reduced active transportation (i.e., walking or bicycling to work), and increased access to labor-saving devices (5, 6). Collectively, these changes place the general population at increased risk for obesity and obesity-related disease and highlight the importance of developing interventions that target decreasing sedentary behaviors and enhancing nonleisure-time physical activity levels.

The accurate assessment of physical activity is of paramount importance in discussions of the health benefits of physical activity (7). Physical activity is often assessed via questionnaire (survey), direct observation, physical activity diary, direct measurement (e.g., pedometers, accelerometers, global positioning systems, and heart rate monitors), or some combination of these. As outlined elsewhere in this book (see chapters 3 through 5), there are various limitations to the currently available assessment techniques for physical activity, with most experts advocating the direct assessment of physical activity using objective methods over a prolonged period of time (7). Moreover, evaluation of the level of sedentarism (see chapter 3) appears to be essential in providing important insights into the physical activity–inactivity profile of an individual (7).

Classically defined, *exercise* refers to structured and repetitive leisure-time physical activity whose main objective is to maintain or improve physical fitness, exercise performance, health status, or more than one of these (2, 8). Within the exercise literature, numerous studies have evaluated the relationship between health status and one or more components of health-related physical fitness (1, 2). *Health-related physical fitness* consists of the components of cardiorespiratory fitness, motor fitness, musculoskeletal fitness, body composition, and metabolism (4). As outlined throughout this book, compelling evidence indicates that substantial health benefits may be derived from improvements in health-related physical fitness (2). Moreover, individuals at the upper end of the physical activity continuum consistently have the lowest risk for chronic disease and premature mortality (1, 2).

The Continuum of Physical Activity Behavior

The continuum of physical activity behavior ranges from being inactive to being very physically active. There are numerous stages within the continuum of physical activity behavior. Sedentary behaviors (e.g., watching television, sitting, reading, working at a desk) are associated with the lower end of the physical activity continuum. Researchers often interchangeably use the terms sedentarism and physical inactivity. However, it is important to recognize that most people engage in some level of physical activity throughout the day (9). Therefore, often a person who is physically inactive is not completely sedentary unless confined to bed rest or dependent upon others (9).

Generally speaking, individuals may move across the physical activity continuum from a relatively inactive state by increasing their physical activity levels or reducing their sedentary behaviors or doing both. However, it is important to highlight that in many situations, sedentary behaviors are independent from (not correlated with) physical activity behaviors. For instance, an endurance athlete who participates in 3 to 4 h of daily regimented vigorous activity can also spend the majority of his or her day engaging in inactive pursuits (such as sitting while working at a computer) without being classified as physically inactive or sedentary. In fact, this individual may be classified at the upper end of the physical activity continuum despite engaging in sedentary behaviors for the majority of the day. This dichotomy highlights the importance of assessing the complete activity–inactivity profile of the individual across the day. This may be particularly important for the prevention or treatment of obesity; in these areas there is still much to be learned regarding the minimal and optimal levels of physical activity and the role of sedentary behavior for effective weight management (1). This point has been raised clearly by the eloquent work of Hamilton and collaborators (10). These authors indicate that relatively little is known about the effects that changes in sedentary behaviors (particularly sitting for prolonged periods) and nonexercise physical activities (such as brief yet frequent muscle contractions completed while standing) will have on health status.

It is apparent that daily living involves a very high volume of intermittent nonexercise physical activity (even in individuals who would be traditionally classified as sedentary). In fact, Hamilton and colleagues (10) demonstrate that nonexercise activity makes up a much larger component of the daily total energy expenditure than exercise. They advocate maintaining or increasing nonexercise activity throughout the day to ameliorate the risks associated with sedentary behaviors. Hamilton and colleagues also note that sedentary behaviors and nonexercise physical activity are very distinct behaviors that do not necessarily reflect the lower end of the physical activity continuum. In other words, some individuals can be considered physically active if they are meeting the current recommendations for moderate-to-vigorous physical activity, yet they may also be very sedentary throughout the rest of their day. In fact, the authors argue that people must be made aware of the potential health hazards associated with sitting too much and the potential health benefits of maintaining nonexercise activity throughout the day.

Estimates and Indices of Physical Activity

In the promotion of health benefits of physical activity, many practitioners attempt to classify the intensity (table 2.1) or energy expenditure (or both) of specific exercises and activities (table 2.2). Exercise prescription generally involves a characterization of the dose, including the frequency, relative intensity, type, and duration of activity. Objective physiological markers (such as heart rate) are often used for prescriptive purposes during exercise training programs. Heart rate reserve (see table 2.1) is being used increasingly in effective exercise prescription for health in both asymptomatic and symptomatic populations (8). However, subjective indicators of the relative intensity of effort are also commonly used (e.g., rating of perceived exertion [RPE] and perceptions regarding breathing and body temperature); these are easy to understand and implement with the general population (8). The use of objective or subjective indices (or both) of relative effort allows for the development of an individualized exercise program that is specifically tailored to the needs of the client. This form of exercise prescription has been extensively used in the prevention and treatment of obesity (1). Current evidence indicates that moderate-intensity (40% to 59% of heart rate reserve) exercise lasting 45 to 60 min per day is likely required to prevent weight gain, and 60 to 90 min of moderate-intensity exercise is likely required to sustain long-term weight loss (1).

Table 2.1

Relative Intensities for Aerobic Exercise Prescription (for Activities Lasting up to 60 min)

	Intensity	%HRR	%HRmax	RPE (6-20)	RPE (0-10)	Breathing rate	Body temperature	Sample activity
	Very light effort	<20	<50	<10	<2	Normal	Normal	Dusting
Range required for health	Light effort	20-39	50-63	10-11	2-3	Slight increase	Start to feel warm	Light gardening
Range required for health	Moderate effort	40-59	64-76	12-13	4-6	Greater increase	Warmer	Brisk walking
Range required for health	Vigorous effort	60-84	77-93	14-16	7-8	More out of breath	Quite warm	Jogging
	Very hard effort	>84	>93	17-19	9	Greater increase	Hot	Running fast
	Maximal effort	100	100	20	10	Completely out of breath	Very hot/ perspiring heavily	Sprinting all-out

HRR = heart rate reserve; HRmax = maximal heart rate; RPE = rating of perceived exertion.

Adapted from D.E. Warburton, C. Nicol, S.S. Bredin, 2006, "Prescribing exercise as preventive therapy," *Canadian Medical Association Journal* 174: 961-974.

Table 2.2

Estimated Energy Expenditures for a Variety of Common Leisure-Time Physical Activities and Activities of Daily Living as a Function of Time

Physical activity	METs	EE (kcal · kg^{-1} · min^{-1})	ENERGY EXPENDITURE (KCAL) IN A GIVEN TIME					
			10 min	20 min	30 min	40 min	50 min	60 min
LEISURE-TIME ACTIVITIES								
Backpacking	7.0	0.12	82	163	245	327	408	490
Basketball, game	8.0	0.13	93	187	280	373	467	560
Basketball, shooting baskets	4.5	0.08	53	105	158	210	263	315
Bicycling, general stationary	7.0	0.12	82	163	245	327	408	490
Bicycling, 10.0-11.9 mph (light)	6.0	0.10	70	140	210	280	350	420
Bicycling, 12.0-13.9 mph (moderate)	8.0	0.13	93	187	280	373	467	560
Bicycling, 14.0-15.9 mph (vigorous)	10.0	0.17	117	233	350	467	583	700
Bowling	3.0	0.05	35	70	105	140	175	210
Calisthenics (light/moderate)	3.5	0.06	41	82	123	163	204	245
Calisthenics (vigorous)	8.0	0.13	93	187	280	373	467	560
Dancing, general aerobic	6.5	0.11	76	152	228	303	379	455
Dancing, social or ballroom (fast)	4.5	0.08	53	105	158	210	263	315
Fishing, from a boat (sitting)	2.5	0.04	29	58	88	117	146	175
Fishing, in a stream (waders)	6.0	0.10	70	140	210	280	350	420
Frisbee playing	3.0	0.05	35	70	105	140	175	210
Golfing, using a power cart	3.5	0.06	41	82	123	163	204	245
Golfing, walking and carrying clubs	4.5	0.08	53	105	158	210	263	315
Golfing, walking and pulling clubs	4.3	0.07	50	100	151	201	251	301
Hiking, cross-country	6.0	0.10	70	140	210	280	350	420
Ice hockey	8.0	0.13	93	187	280	373	467	560
Jogging, general	7.0	0.12	82	163	245	327	408	490
Playing catch, football or baseball	2.5	0.04	29	58	88	117	146	175
In-line skating	12.5	0.21	146	292	438	583	729	875
Running, 5.0 mph (12 min/mile)	8.0	0.13	93	187	280	373	467	560
Running, 7.5 mph (8 min/mile)	12.5	0.21	146	292	438	583	729	875
Running, 10.9 mph (5.5 min/mile)	18.0	0.30	210	420	630	840	1050	1260
Skiing, cross-country (4.0-4.9 mph, moderate)	8.0	0.13	93	187	280	373	467	560
Skiing, water	6.0	0.10	70	140	210	280	350	420

Skiing, downhill	6.0	0.10	70	140	210	280	350	420
Snowmobiling	3.5	0.06	41	82	123	163	204	245
Softball, general	5.0	0.08	58	117	175	233	292	350
Swimming, general leisure	6.0	0.10	70	140	210	280	350	420
Swimming laps, freestyle (vigorous)	10.0	0.17	117	233	350	467	583	700
Tennis, general	7.0	0.12	82	163	245	327	408	490
Walking, 2.0 mph	2.5	0.04	29	58	88	117	146	175
Walking, 3.5 mph	3.8	0.06	44	89	133	177	222	266
Walking, 5.0 mph	8.0	0.13	93	187	280	373	467	560
ACTIVITIES OF DAILY LIVING								
Carrying small children	3.0	0.05	35	70	105	140	175	210
Chopping wood	6.0	0.10	70	140	210	280	350	420
Cleaning house, general	3.0	0.05	35	70	105	140	175	210
Groceries, carrying without shopping cart	2.5	0.04	29	58	88	117	146	175
Groceries, carrying up stairs	7.5	0.13	88	175	263	350	438	525
Ironing	2.3	0.04	27	54	81	107	134	161
Mopping	3.5	0.06	41	82	123	163	204	245
Mowing lawn, general	5.5	0.09	64	128	193	257	321	385
Raking lawn	4.3	0.07	50	100	151	201	251	301
Shoveling snow, manually	6.0	0.10	70	140	210	280	350	420
Sweeping floors or carpet	3.3	0.06	39	77	116	154	193	231
Sweeping sidewalk	4.0	0.07	47	93	140	187	233	280
Vacuuming	3.5	0.06	41	82	123	163	204	245
Walking the dog	3.0	0.05	35	70	105	140	175	210
Walking, household	2.0	0.03	23	47	70	93	117	140
Walking, pushing or pulling a stroller with child	2.5	0.04	29	58	88	117	146	175
Washing dishes	2.3	0.04	27	54	81	107	134	161
Watering household plants	2.5	0.04	29	58	88	117	146	175
Watering lawn or garden	1.5	0.03	18	35	53	70	88	105
Weeding garden	4.5	0.08	53	105	158	210	263	315

Data provided are based on an individual weighing 70 kg.

Adapted from D.E. Warburton, C. Nicol, S.S. Bredin, 2006, "Prescribing exercise as preventive therapy," *Canadian Medical Association Journal* 174: 961-974.

Many health and fitness professionals use estimates of energy expenditures for various physical activities (8). Standardized activity and energy tables using metabolic equivalents (METs; $1\ \text{MET} = 3.5\ \text{ml} \cdot \text{kg}^{-1} \cdot \text{min}^{-1} = 1\ \text{kcal} \cdot \text{kg}^{-1} \cdot \text{h}^{-1}$) are commonly employed to provide an objective and easy-to-understand means of estimating energy expenditure and the absolute intensity of various activities (8) (see table 2.2). Although standardized energy expenditure tables (based on METs) are often used in weight management programs, they have several limitations, including lack of consideration for inter-individual differences in baseline fitness levels, skill, coordination, and exercise economy, the effects of environmental factors (e.g., cold, wind, heat, altitude), and differences in the exercise intensity of effort during particular physical activities (8). These tables appear to be particularly limited in middle-aged and elderly people (8). Moreover, these tables were generally developed based on data from normal-weight individuals; thus their use with the overweight and obese is somewhat limited (6). Obese individuals exhibit a higher energy expenditure than lean individuals during both weight-bearing and non-weight-bearing activities (6). This has important consequences when absolute MET values are used to estimate energy expenditure, resulting in an overestimation during non-weight-bearing activities (owing to reduced energy expenditure per kilogram body mass during these activities) and an underestimation of energy expenditure during weight-bearing activities in the obese (6).

Many agencies and individuals also quantify daily energy expenditure levels using multiples of basal metabolic rate via the Physical Activity Level (PAL) index (equivalent to total energy expenditure [assessed by doubly labeled water] divided by 24 h basal metabolic rate) (6). A PAL of 2.0 indicates a twofold increase (in comparison to resting metabolic rate) in daily energy expenditure. The use of PAL provides a simple means of categorizing and standardizing daily energy requirements (6). The PAL is affected by body size and age owing to their effects on basal metabolic rate. A single number is used to categorize energy requirements, with a higher PAL representing a greater level of daily physical activity (5). For instance, sitting at rest equates to a PAL of approximately 1.2; a sedentary office worker exhibits a PAL of 1.4; and a PAL of 1.75 is seen in an individual who engages in regular physical activity. With the use of PAL (5), people are often placed in one of four physical activity categories: inactive/sedentary = 1.0 to 1.39; low active = 1.4 to 1.59; active = 1.6 to 1.89; and very active = 1.9 to 2.5. A PAL of ≥1.70 is currently recommended to prevent overweight or obesity (6); this equates to approximately 60 min of additional physical activity for most inactive individuals (5).

Summary

The health benefits of habitual physical activity are irrefutable. However, a clear understanding of the various domains of physical activity is required. It is important to recognize that physical activity has many domains, including occupational, domestic, active transportation, and leisure-time (discretionary) activities. Current evidence indicates that the greatest reduction in physical activity with industrialization has occurred in nondiscretionary physical activities. Moreover, only a small portion of the day is spent in the pursuit of discretionary leisure-time physical activities. It is also important to recognize that there is a continuum of behavior ranging from being inactive to being very physically active. Reducing sedentary behaviors and increasing discretionary and nondiscretionary physical activity levels throughout the day provide enormous opportunities for improving health status on the individual and population level.

3

Assessing the Level of Sedentarism

Mark S. Tremblay, PhD

Sedentarism has often been conceptualized as reflecting the low end of the physical activity continuum (1). However, individuals can achieve high levels of both physical activity and sedentary behaviors, and one type of behavior does not automatically displace the other. Evidence suggests that sedentarism has independent effects on obesity and health and consequently should be treated as a separate and distinct construct (2-5). For the purpose of this chapter, sedentarism is defined as purposeful and extended engagement in behaviors characterized by minimal movement, low energy expenditure, and rest.

Conceptualizing sedentarism as distinct from physical activity is important for a variety of reasons. Approaches to reduce sedentarism may be different than those designed to increase physical activity. For example, Tremblay and colleagues (6) illustrated that a reduction in sedentarism can be achieved through almost limitless micro-intervention opportunities designed to promote energy expenditure. This approach has been demonstrated to have beneficial health effects and has germinated the field of inactivity physiology research (3). For some people (especially those who have not and will not embrace an organized or structured program of physical activity), reducing sedentarism may be a more achievable and viable approach as a proximal goal for increasing movement and energy expenditure and improving weight management and maintenance. Furthermore, for those with identified sociodemographic challenges (and associated increased risk of overweight and obesity [1]), a reduction in sedentary behavior can be achieved with minimal resource requirements (e.g., registration fees, transportation, equipment). Finally, methodologies for the assessment of sedentarism require a focus on indicators different than simply low levels of physical activity.

Relationship Between Sedentarism and Health

The relationships among sedentarism, obesity, and comorbidities are relatively understudied; therefore the importance of sedentarism to health or obesity has not been clearly delineated either in comparison to or in addition to physical activity. Figure 3.1 provides a conceptual illustration of the relationships among physical activity, sedentarism, and health (including obesity). The best scenario for the promotion of healthy body weight and the prevention of chronic disease is certainly one that places an individual in the lower right quadrant (high physical activity, low sedentarism). However, the efficacy

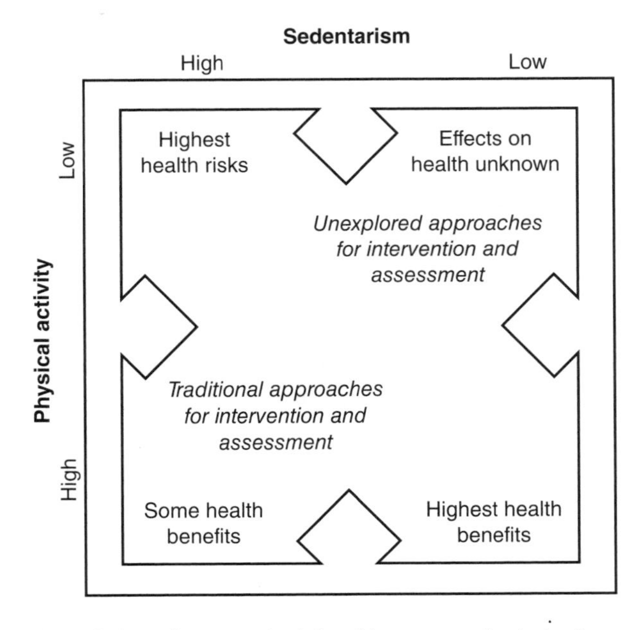

Figure 3.1 Conceptual relationships among physical activity, sedentarism, and health (including obesity).

of a behavior pattern that results in placement in the top right quadrant (low physical activity, low sedentarism) is relatively unknown. Recent research has demonstrated that sedentarism, as assessed by television time, is associated with indicators of overweight in children (2) and adults (4). Using data from the Youth Risk Behavior Survey, Eisenmann and colleagues (2) observed that the odds of being overweight were no different across physical activity levels provided that youth had low levels of television viewing. Healy and coworkers (4) reported that among adults who met public health guidelines for physical activity, television viewing time was positively associated (in a dose–response fashion) with a number of metabolic disease indicators, including waist circumference, systolic blood pressure, and blood lipid and blood glucose levels. These studies support the need to assess both physical activity and sedentary behavior indicators when one is studying the effects on health.

Hamilton and colleagues (3) have demonstrated plausible biological mechanisms to explain how changes in sedentary behavior and incidental movement could provoke changes in chronic disease risk. Through a series of "inactivity physiology" studies, the authors examined the role of sedentary behavior on mortality, cardiovascular disease, type 2 diabetes, metabolic syndrome risk factors, and obesity. In addition to finding strong evidence of an association between sitting time and each of these conditions, they observed that lipoprotein lipase (key enzyme associated with cardiovascular and metabolic disease risk factors) was significantly influenced by changes in sedentary behaviors. The change in lipoprotein lipase was more pronounced in response to an experimentally induced increase in sedentary behavior when compared to the addition of vigorous exercise training to normal activity levels.

Behavior Compensation

The emerging evidence that sedentary behavior has independent effects on health and may also mediate the relationship between physical activity behavior and health elevates the importance of carefully assessing whether the introduction of one behavior affects the other. For example, the introduction of a structured physical activity program may successfully increase the minutes of moderate and vigorous physical activity; however, compensatory behavioral adjustments throughout the day (or week) may result in an overall reduction in daily movement or energy expenditure. This behavior compensation, whereby total energy expenditure is reduced or sedentarism is increased (or both changes occur) in response to a physical activity intervention, complicates or contaminates the study of the relationship between physical activity and health.

Methods to Assess Sedentarism

To understand the relationships illustrated in figure 3.1 and to examine the consequences of behavior compensation require careful, robust, context-appropriate assessments of sedentarism (and physical activity; see chapters 4 and 5). Incomplete, improper, or inappropriate assessment procedures will serve to mask or exaggerate true relationships.

Clues for assessment methods are drawn from sedentary behaviors. Whereas a piece of exercise equipment is a cue for a method of physical activity assessment (e.g., frequency, intensity, and duration of use), a piece of equipment that promotes sedentary behavior is a cue for a method of sedentarism assessment. For example, as already discussed, televisions, computers, and video game consoles promote extended sitting and idle behavior, so screen time (frequency × duration) is an appropriate indicator of sedentarism. Indeed, screen time (or some derivative—television time, computer time, video game time) has become a common measure of sedentariness (1, 2, 4, 7, 8). The 2008 Active Healthy Kids Canada Report Card on Physical Activity for Children and Youth (www.activehealthykids.ca) identified excessive screen time as a key opportunity to reverse the physical deactivation of our children. The Report Card indicates that Canadian children and youth aged 10 to 16 years are averaging 6 h of leisure-time screen time per day based on the 2005-2006 Health Behavior in School-aged Children Survey. Such data suggest that this form of sedentary behavior is both common and extensive. Pediatric groups have established screen time guidelines (9) to inform children and their parents about healthy screen time usage. These guidelines can be used to assess the proportion of the population meeting or exceeding the guidelines at a point in time, or to monitor trends over time (7, 8).

The same methodologies we use to assess physical activity can guide our assessment of sedentarism. Table 3.1 provides a series of potential assessment methods in comparison with traditional measures of physical activity (and inactivity). Methods are categorized as direct (objective), reported (subjective), and global assessments (for a discussion on the advantages and disadvantages of the various methodologies see chapter 8). The table contrasts

■ Table 3.1 ■

Examples of Potential Assessment Procedures and Indicators of Physical Activity and Sedentarism

PHYSICAL ACTIVITY (PA) MEASURE		SEDENTARY BEHAVIOR (SB) MEASURE	
Assessment procedure	**Sample indicators**	**Assessment procedure**	**Sample indicators**
Direct (objective) methods		**Direct (objective) methods**	
Direct observation	Minutes of PA, types of PA	Direct observation	Minutes of SB, types of SB
Portable indirect calorimetry	Energy expenditure, $\dot{V}O_2$	Portable indirect calorimetry	Minutes at resting metabolic rate
Doubly labeled water	Energy expenditure	Doubly labeled water	Energy expenditure
Accelerometry	Minutes above thresholds	Accelerometry	Minutes below thresholds
Pedometry	Steps per day	Pedometry	Steps per day
Heart rate monitoring	Minutes above thresholds	Heart rate monitoring	Minutes below thresholds or at rest
Respiration rate	Minutes above thresholds	Respiration rate	Minutes below thresholds or at rest
Reported (subjective) methods		**Reported (subjective) methods**	
PA questionnaire or interview	Energy expenditure, minutes of PA	SB questionnaire or interview	Minutes of SB, chair time, screen time
PA activity diary	Energy expenditure, minutes of PA	SB activity diary	Minutes of SB, chair time, screen time
PA log	Frequency of PA, types of PA	SB log	Frequency of SB, types of SB
Exercise equipment usage recall	Frequency, duration, intensity	Labor-saving device usage recall	Frequency, duration
Active transportation recall	Type, frequency, duration, distance	Automobile usage recall	Frequency, duration, distance, car time
Stairs climbed recall	Floors per day	Elevator and escalator usage recall	Floors per day, frequency of use
Global assessments		**Global assessments**	
Occupational classification	MET value, energy expenditure	Occupational classification	Chair time, screen time, car time
PA comparisons to peers	Rating (e.g., higher, same, lower)	SB comparisons to peers	Rating (e.g., higher, same, lower)
Connectedness with nature	Outdoor time	Household cocooning	Indoor time, screen time

the different contexts for the measurement of physical activity versus sedentary behavior in an effort to provoke the conceptualization of additional assessment methodologies in these two domains. Note how physical activity measures like minutes of physical activity, minutes above thresholds, active transportation distance, energy expenditure, and outdoor time contrast with sedentarism measures like minutes of sedentary behavior, minutes below thresholds, car time, screen time, chair (sitting) time, and indoor time.

Using pedometers to assess step counts is one method to assess sedentarism. Based on extensive literature reviews (10, 11), it has been recommended that a zone-based hierarchy of step counts be used to classify physical activity behaviors among adults and children. Using this classification system, adults are categorized as sedentary if they achieve <5000 steps per day (10, 11), a threshold supported by evidence that the likelihood of being obese is increased in those categorized as sedentary using this cut point (10). Suggested thresholds for children aged 6 to 12 years are <7000 steps per day for girls and <10,000 for boys, although it is acknowledged that additional prospective, longitudinal, criterion-referenced studies are required to refine these targets.

Reilly and colleagues recently reviewed the use of direct measurements of sedentary behaviors via accelerometers (12). They concluded that objective measures of sedentary behavior using accelerometers are consistent regardless of the measurement epoch selected or the age of the child participant. Esliger and Tremblay provide an example of a comprehensive assessment of sedentary behavior derived from minute-by-minute, week-long accelerometry measurements (13). This detailed profile includes assessments of weekday-specific, weekend day–specific, and average-day sleep and sedentary behavior, as well as when (time of day) and how (sporadic, short bouts, long bouts) the sedentary behavior was accumulated. Detailed profiles of physical activity patterns are increasingly common, but equivalent profiles for sedentary behavior are rare in the research literature.

Typically, many of the indicators in table 3.1 are used to assess population-level adherence to established guidelines for physical activity (e.g., prevalence meeting physical activity guidelines; see chapters 9 and 10); although far less common, the same assessment can be made for sedentarism (e.g., prevalence meeting screen time guidelines [9]). It is worth noting that the proliferation of "screens" (television, computer, video game, cell phone, portable DVD player, etc.), diversification of uses (active video-based gaming), and simultaneous multiscreen usage has and will continue to complicate the measurement of screen time.

Further research is required to explore the validity of using surrogate inhibiting or enabling measures for the assessment of sedentarism. For example, information on park space, playground availability, ownership of exercise equipment, and gym membership is often used to assess or inform physical activity levels. In a similar manner, measures such as televisions or screens per household, Internet access in the home, cable or satellite television access, and automobiles per household member may serve as useful indices of sedentarism.

Examples of Sedentarism Assessment

Rosenberg and colleagues (5) used accelerometers to investigate the ability of the International Physical Activity Questionnaire (IPAQ) to assess sedentary behavior via sitting-time questions. Correlations between sitting time and accelerometer counts <100 (indicating minimal movement) were 0.33 for the long-form and 0.34 for the short-form questionnaires. The authors further observed no relationship between sitting time and likelihood of being classified as physically inactive. They concluded that sedentary behavior should be explicitly measured in population surveillance and research instead of being defined by a lack of physical activity.

Shields and Tremblay recently reported on the sedentary behaviors of Canadian adults using a nationally representative sample (7, 8). They examined television viewing and computer use separately. Twenty-nine percent of adults aged 20 years or older reported watching television 2 h or more per day, and 15% reported leisure-time computer usage of more than 10 h per week (8). The pattern of usage varied significantly by age group, with younger adults far more likely to accumulate screen time on the computer compared to older Canadians. The opposite was true of television viewing. Differences in sociodemographic characteristics were evident but varied between television and computer usage. For example, recent immigrants were less likely than people born in Canada to be frequent television viewers but more likely to be frequent computer users.

It is likely that not all sedentary pursuits are equal contributors to obesity. Shields and Tremblay recently explored this possibility (7). In a nationally representative sample of Canadian adults (n = 42,612), they observed that among men and women the odds of being obese increased with hours of weekly television viewing, and that this relationship was independent of leisure-time physical activity and diet. When the effects of age and other confounding variables were controlled, only a modest association was observed between frequent computer use and obesity. In contrast, reading time was not associated with obesity for either sex.

Implications

Improvements in movement and energy expenditure measurement techniques will lead to a more comprehensive understanding of sedentary behavior and its relative contributions to the prevention and management of obesity, and the maintenance of healthy body weights. Following this, it is likely that existing relationships between physical activity and health will need to be reexamined. Furthermore, as evidence of the importance of assessing sedentary behavior mounts, there will be a need to adjust leadership training curricula (e.g., for physical and health education teachers, health practitioners, recreation leaders) to include the assessment and management of sedentary behavior separate and distinct from physical activity. More detailed assessment of sedentary behavior will increase awareness and will likely debunk the legitimacy of the most cited reason for not being physically active—time.

Summary

Careful, robust, context-appropriate assessments of sedentarism, in addition to assessments of physical activity, are required. A variety of methods for assessing sedentarism are available that draw on the experience from physical activity assessment. A recent international conference summary report calls for further research on sedentary behavior and weight gain, as well as compensatory physical activity behavior, and the development of guidelines for sedentarism (1). Further exploration of the relationships among sedentarism, physical activity, and obesity (and other indicators of health) is required to inform future interventions.

Assessing the Level of Physical Activity in Adults

Barbara E. Ainsworth, PhD, MPH

Recommendations for the dose of health-enhancing physical activity and energy expenditure in adults are well known. In 1996, *Physical Activity and Health: A Report of the Surgeon General* identified an energy expenditure of 150 kcal per day or 1000 kcal per week in moderate- or vigorous-intensity activity (or both) as effective in reducing an adult's chronic disease risk (1). In 2007, the American College of Sports Medicine and the American Heart Association recommended that adults engage in a minimum of 30 min per day of moderate-intensity physical activity on five or more days per week, in addition to their daily physical activity patterns, to minimize their chronic disease risks and enhance health status (2). These recommendations arose from epidemiological and clinical studies evaluating various combinations of frequency, duration, and intensity of physical activity and energy expenditures that are associated with the lowest risks for chronic disease morbidity and mortality and that optimize health and well-being. The studies used various methods to quantify and track physical activity patterns and their associated energy costs. This review provides an overview of methods available to assess physical activity and energy expenditure in free-living adults.

Movement Construct

Methods used to measure physical activity and energy expenditure vary in complexity, precision, ease of use, and feasibility. Figure 4.1 presents a

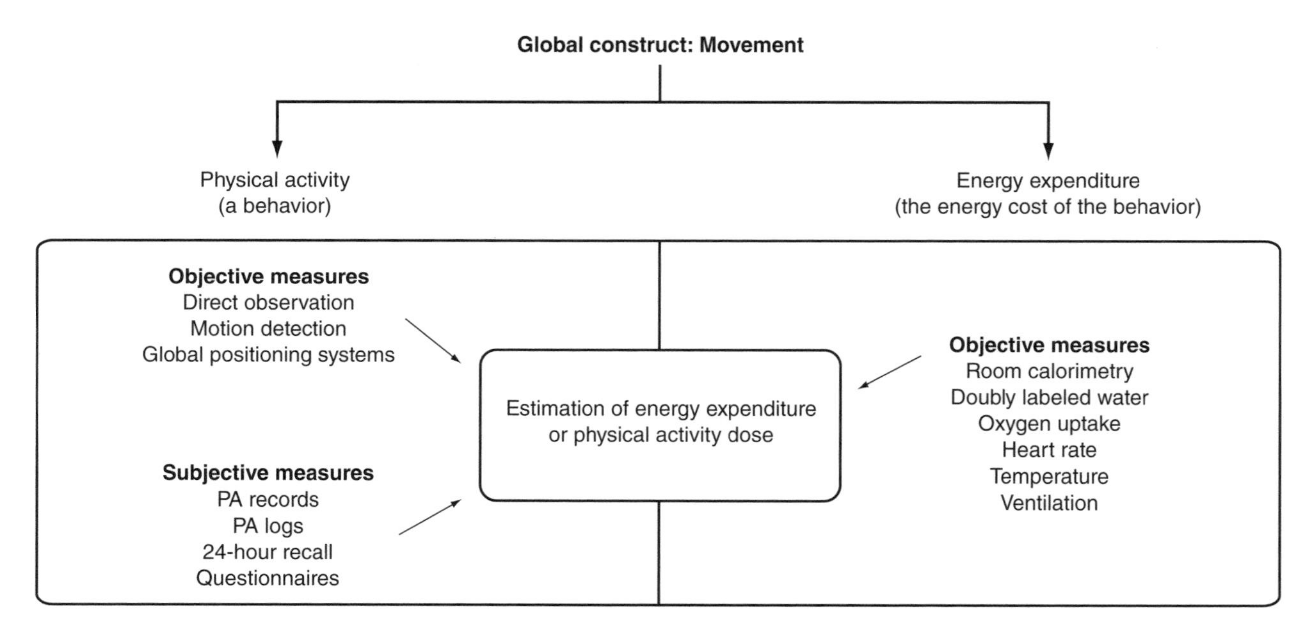

Figure 4.1 A conceptual framework for the measurement of a global construct of movement.

conceptual framework for the measurement of a global construct—movement (3). The construct has two dimensions, physical activity (a behavior) and energy expenditure (the energy cost of the behavior), that can be measured using objective and subjective methods.

The measurement of physical activity via objective methods involves objective recording of behaviors through direct observation of another person's movement, through use of motion detectors worn by individuals, or through use of remote sensing devices. Subjective methods include the use of physical activity records or logs and questionnaires that identify and quantify one's physical activity patterns performed during the past day, week, month, or year. Questionnaires may be self- or interviewer-administered. Objective methods of measuring energy expenditure require laboratory settings or biochemical processes to determine caloric energy output. Subjective methods use physiological measurements to track increases in the components of movement metabolism, oxygen utilization, heart rate, body temperature, or ventilation during physical activity from which energy expenditure can be estimated. Emerging within the field of physical activity assessment are devices that measure aspects of physical activity and energy expenditure simultaneously. Two examples are accelerometers, which provide an objective measure of physical activity duration and intensity, and heart rate monitors, which provide an objective measure to estimate energy expenditure. For all methods, it is possible to compute a summary score to determine associations between energy expenditure or physical activity and health outcomes of interest. Table 4.1 provides an overview of physical activity and energy expenditure measurement methods.

■ Table 4.1 ■

An Overview of Physical Activity and Energy Expenditure Measurement Methods

	Dimension measured	Units	Technical or subject burden
OBJECTIVE MEASURES			
Direct observation	PA	F, D, I, M	Moderate
Room calorimetry	EE	Kcal from heat production	High
Doubly labeled water	EE	Kcal from O_2 consumption	High
Indirect calorimetry	EE	Kcal, METs from CO_2 production	High
Heart rate	EE	Kcal, METs from VO_2 relationship	Moderate
Accelerometer	PA EE	F, D, I Kcal from METs regression	Moderate Moderate
Pedometer	PA	Steps	Low
Global positioning system	PA	Distance	High
SUBJECTIVE MEASURES			
PA records	PA EE	F, D, M Kcal, METs from Compendium	High
24 h recall	PA EE	F, D, M Kcal, METs from Compendium	Moderate
PA log	PA EE	F, D, M Kcal, METs from Compendium	Low
PA questionnaire	PA EE	F, D, M Kcal, METs from Compendium	Low

PA = physical activity; EE = energy expenditure; F = frequency; D = duration; I = intensity; M = mode; Kcal = kilocalories; MET = metabolic equivalent.

Measuring Physical Activity Behaviors

Physical activity behaviors are measured to describe the type of activities performed and to quantify the amount of movement individuals are doing. Physical activity measures often reflect what type of activity people are doing (mode or type); how long they perform an activity in minutes or hours (duration); how often they do the activity in days, months, or years (frequency); and how hard the activity feels or how much the activity increases the resting metabolic rate (intensity). Intensity is measured in terms of METs, shorthand for metabolic equivalents. One MET reflects the kilocalorie energy requirement of sitting quietly in a chair. MET levels are classified as light (<3 METs), moderate (3.0 to 5.9 METs), and vigorous in intensity (≥ 6.0 METs) (4). To estimate kilocalorie energy expenditure, one multiplies the activity duration in hours by body weight in kilograms by MET level. Physical activity is often measured using objective and subjective methods (e.g., accelerometer and questionnaire) simultaneously to allow for maximum precision in the assessment of frequency, duration, and intensity and to obtain information about the mode of physical activity performed.

Objective measures of physical activity involve the use of direct observation, motion detectors, and remote sensing systems. *Direct observation* is used to characterize physical activity behaviors without input from those being observed. Direct observation systems include the following characteristics: a well-defined observation strategy to sample activities per unit of time; a list of activity categories for coding movement types; a list of associated variables that may influence behavior (e.g., context, others' behavior, environmental settings); supplemental methods to record concurrent levels of energy expenditure; data entry procedures (e.g., pencil and paper, computer, Palm Pilot); and detailed scoring schemes used to summarize the data. The reader is referred to McKenzie for a detailed description of the various methods available (5).

Motion detectors are mechanical and electronic devices worn on the body to provide an objective measure of physical activity. Their use in measuring free-living physical activity in research and practice settings has become increasingly popular. Accelerometers are small, battery-operated electronic motion sensors designed to measure the rate and magnitude of bodily movement. Worn on a waist belt or on the arm or ankle, accelerometers record the duration of movement at varying intensity levels each minute. Regression equations are used to identify the time spent in sedentary or light-, moderate-, or vigorous-intensity activities and to estimate energy expenditure. The reader is referred to Ward and colleagues (6) for more information about accelerometers. Pedometers are small electronic devices attached to the belt that indicate the number of steps taken. Pedometers have grown increasingly popular as a behavioral intervention tool to inform adults and motivate them to increase walking and other forms of daily activity. Tudor-Locke and Bassett (7) provide a system for classifying the level of physical activity based on daily steps taken. *Global positioning systems (GPS)* are a part of newer technologies that are emerging for objective assessment of physical activity behaviors. Remote sensing with portable GPS built into a wearable device (e.g., wristwatch or shoulder bag) records the location and distance traveled. As long as the GPS units are detectable by the satellite systems that track the signal, spatial patterns for individual movement can be recorded. Such recordings have value in the study of how individuals interact with their environments during physical activity.

Subjective measures of physical activity involve the use of physical activity records and logs, 24 h recalls, and recall questionnaires. *Physical activity records* provide a detailed account of activities done within a given period of time. Completed by the individual doing the activity, they are useful for identifying the type (e.g., watching TV, occupational activity, walking), duration, and frequency of activities and may take the form of a written diary, a record book, or dictation into a tape recorder. The records are scored using the Compendium of Physical Activities, which is a detailed list of various types of physical activities and their associated energy costs (i.e., METs) from which intensity levels can be identified (8). The *24 h physical activity recall* is a modified physical activity record that uses trained facilitators to prompt respondents in the recall of all physical activities performed on the previous day, from rising from bed to retiring at night. The level of detail obtained is similar to that derived from physical activity records. An advantage of using 24 h recalls is that it is possible to collect detailed information about physical activity patterns with minimal respondent burden. *Physical activity logs* are a modified form of the physical activity record that allows respondents to check the types and duration of activities performed during intermittent periods of the day or at the end of a single day. Physical activity logs can be flexible in design, with activities listed in the log tailored for a research or practice setting. For example, a walking log may focus on walking performed during different periods of the day. A transportation log may include activities specific only to the transit mode.

Physical activity questionnaires are the most common instruments used to assess physical activity behaviors in public health surveillance systems, in large-scale epidemiological studies of health-related outcomes, and in some intervention research settings. *Global questionnaires* are short instruments designed to provide a general classification of one's physical activity status (e.g., active vs. inactive). Very short in length, global questionnaires often reflect participation in structured exercise, occupation, or transportation settings and provide little detail on specific patterns and types of physical activity performed. *Recall questionnaires* assess the frequency and duration of specific activities within categories or types of physical activity. The recall period is generally one to two weeks. Most recall instruments have 7 to 12 questions; these instruments are often used in surveillance to assess the proportion of the population that meets physical activity recommendations and in research settings to assess the dose (frequency × duration) of activity at varying intensity levels (i.e., light, moderate, or vigorous). *Quantitative history* questionnaires are designed to obtain the frequency and duration of multiple types of physical activity over a prolonged period in the past. The questionnaires generally have 30 or more items with a recall frame of one year to a lifetime. Quantitative history instruments are frequently used in studies designed to identify detailed patterns of physical activity habits with disease or positive health status endpoints.

Measuring Energy Expenditure

Energy expenditure is defined as the outcome of physical activity behaviors and reflects the heat produced from skeletal muscle contraction or volume of oxygen utilized ($\dot{V}O_2$) and carbon dioxide produced ($\dot{V}CO_2$) during movement. Here we discuss three common field methods. *Doubly labeled water* is the most common objective field method of measuring energy expenditure. Doubly labeled water consists of the stable water isotopes ^2H_2O and $H_2^{18}O$ and is administered according to body size. Labeled hydrogen (^2H_2O) is excreted as water alone, while labeled oxygen ($H_2^{18}O$) is lost as water and CO_2 ($C^{18}O_2$). The difference in the isotope turnover rate provides a measure of metabolic expired carbon dioxide, a result of energy expenditure. Energy expenditure can be estimated from oxygen consumption and the heart rate (HR) response to movement. *Oxygen consumption* reflects the rate at which oxygen is used by the tissues. With 1 L of oxygen utilization equal to about 5 kcal of energy expenditure, higher levels of oxygen consumption during an activity reflect higher intensities of activity. The procedure used to measure oxygen consumption is called indirect calorimetry. The *heart rate response* to movement is used to estimate activity-related energy expenditure based on an assumed linear relation between the HR and $\dot{V}O_2$ responses to activity. During moderate-intensity activity, the HR and $\dot{V}O_2$ increase linearly; however, during low- and very high-intensity physical activity, the HR and $\dot{V}O_2$ are nonlinear depending on individual characteristics (e.g., age, fitness levels). As there is considerable between-person variation in the HR-$\dot{V}O_2$ relationship, there are correction methods, such as the FLEX HR, that establish individual calibration curves for the HR-$\dot{V}O_2$ association. For more information, readers are referred to a detailed overview of physical activity and energy expenditure assessment methods presented by Pettee and colleagues (9).

Newer integrated systems that include GPS, accelerometers, pedometers, and HR sensors in a single unit are being developed to assess a broad range of physical activity behaviors and related energy expenditure (10). Applications for these techniques include the use of neural networks to predict movement types and patterns from accelerometer recordings, to provide feedback about activity patterns in rehabilitation settings, to assist with weight loss and chronic disease risk reduction programs, and to gain a better understanding of how people interact with the environment for leisure and transportation purposes.

Summary

Many objective and subjective methods are available to measure physical activity and its related energy expenditure under free-living conditions. These measures exist along a continuum of precision, cost, and subject and administrative burden. The simultaneous use of objective and subjective measures provides the best combination for determining the duration, frequency, intensity, and mode of physical activity performed in order to assess mortality, morbidity, and positive health outcomes.

5

Assessing the Level of Physical Activity in Children

Russell R. Pate, PhD

A pandemic of childhood obesity has focused considerable attention on the physical activity levels of young people around the world. Numerous studies have concluded that most children and adolescents in the developed countries fail to meet current public health guidelines for physical activity. Hence, there is great interest in monitoring and promoting physical activity in this population in a wide range of settings. Accurate assessment of physical activity levels and evaluation of physical activity interventions require application of valid measures of physical activity. Such measures are needed for use in a wide range of professional and scientific settings, and there is a need for measures that vary widely in cost and burden. Accordingly, a great deal of recent research has focused on developing measures of physical activity that are valid, reliable, and cost-effective when applied to young people at all developmental levels, from early childhood to late adolescence. This chapter provides a summary of the methods currently available for measuring physical activity in children and youth and an overview of emerging methodologies.

Established Measures of Physical Activity

Physical activity has been defined as bodily movement that is produced by the contraction of skeletal muscle and that substantially increases energy expenditure, and it includes several elements that are of interest to researchers and health professionals. These are frequency of activity participation, duration of activity bouts, intensity of activity, type of activity, time spent in specified intensity zones, and settings in which activity occurs. Available instruments typically address a few but not all of these components.

Doubly Labeled Water

Doubly labeled water (DLW) is sometimes considered the "gold standard" of physical activity measures. The DLW technique is an objective measure of total energy expenditure over a period of several days. If paired with a measure of resting energy expenditure, DLW provides an accurate estimate of daily physical activity energy expenditure. However, DLW provides no information about the manner in which physical activity occurred. Further, while the ingestion of the stable isotopes is perfectly safe for children, this method is expensive; and the oral administration of the isotopes may not be well received by young children, making it impractical for most large-scale studies.

Heart Rate Monitoring

Heart rate monitoring as a measure of physical activity is based on the linear relationships among physical activity, energy expenditure, and heart rate. Most heart rate monitors utilize electrodes embedded in a chest strap that transmit the electrocardiographic signal to a watch or other receiver. Minute-by-minute average heart rates are computed and can be stored over several days to allow for longer data collection periods. Heart rate monitoring can provide information regarding the frequency, intensity, and duration of activity. When

Acknowledgements
The author appreciates the contributions of Jonathan Mitchell, Kerry McIver, and Gaye Groover Christmus to this chapter.

determining activity intensity, researchers often use periods of time above a certain heart rate threshold as minutes of activity at specified intensities.

Heart rate devices impose moderate subject burden but can provide information regarding physical activity over several days. One of the problems with heart rate as a measure of physical activity is that it can be increased by stimuli other than physical activity (1). An additional issue in children is that resting heart rate decreases with progression into adolescence, and this decrease must be considered in longitudinal analyses of children's physical activity. The use of individual calibration curves for the exercise intensity–heart rate relationship can greatly enhance the validity of this method, although the burden of individual calibration reduces its feasibility for large-scale studies.

Accelerometry

Accelerometers are instruments that provide objective information about overall physical activity and intensity of physical activity. The most common types are uniaxial accelerometers, which measure movement in one plane, typically the vertical. The raw data output, "activity counts," can be translated into minutes of sedentary and light-, moderate-, and vigorous-intensity activity through the use of calibration equations or can be used to estimate energy expenditure in terms of metabolic equivalents (METs) or kilocalories. Several device-specific and age-specific calibration equations are reported in the literature. When using accelerometers, researchers must consider the time-sampling interval, particularly with young children, who tend to have sporadic physical activity patterns. While a 1 min time-sampling interval is appropriate in studies of adults, 30 s or 15 s intervals may be more appropriate in studies of elementary school and preschool children, respectively.

Accelerometers can provide an objective assessment of children's physical activity, including information about activity frequency, intensity, and duration; but they do not provide information about the context or setting in which an activity occurs. Additionally, several research questions remain regarding the use of accelerometers, including whether they measure activity counts differently in children of different weights and with different gaits.

Direct Observation

Direct observation involves trained individuals who use a structured system to record levels of physical activity during specified observation time intervals. Related contextual factors may be recorded as well. Direct observation can provide highly detailed descriptions of physical activity behaviors and is particularly useful when one is studying young children. Using direct observation, researchers can record not only frequency, intensity, duration, and type of physical activity, but also the social (e.g., playing with peers or interacting with adults) and physical (e.g., at school or at home, indoors or outdoors) environmental contexts in which activity occurs. Understanding the contexts in which children are active or sedentary is important in developing physical activity interventions for children and families. Direct observation requires extensive and continuing observer training to ensure validity and interobserver reliability (2). It is burdensome and costly to researchers but provides information on a wide range of activity components and related factors. Participant burden is low, and reactivity is minimal, particularly in young children.

Pedometers

Pedometers provide an objective measure of the number of walking or running steps an individual takes during an observation period. Accumulated step count can be used as an indicator of total activity. Pedometers are useful for measuring physical activity in large-group studies due to their low cost and ease of use. Step counts can be used in analyses without an extensive data reduction process, and parents, teachers, or older children can record step counts. No calibration equations are needed to interpret pedometer data. Despite their ease of use, however, the utility of pedometers in research studies is limited. Step counts provide no information about physical activity intensity; for example, steps accumulated during running are not differentiated from those accumulated during slow walking. In addition, the gait speed fluctuations seen with the sporadic physical activity patterns of children may affect the validity of step count as a marker of overall physical activity. Children also may react to pedometers, opening and closing the devices to check their step counts.

Self-Report Instruments

Large-scale and epidemiologic studies often utilize self-reported physical activity because self-report instruments are a cost-effective approach to reaching large numbers of children. The burden on children and parents is not excessive, but accurate recall is not a realistic expectation in children below the age of 10. Self-reports can include questionnaires, diaries, proxy reports, surveys, and interviews. All of these instruments require children or their parents to recall or document physical activity behaviors over a period of time, such as one day, three days, a week, the previous month, or the previous year. There are more than 30 survey instruments that can be self-administered, interviewer-administered, or completed by proxy, usually by a parent, guardian, or teacher (1). These instruments often include questions regarding the frequency, intensity, and duration of activity and types of activities performed. Self-report data often are converted into descriptions of activity that can include rating (low, moderate, or high active), arbitrary activity scores, time spent in certain intensities of activity, summary scores, or calculation of calories expended (3).

Self-reports of physical activity can provide rich descriptions of activity behavior but subject burden can be high, especially if a large number of activity components are assessed. In addition, considerable bias can be introduced when individuals are asked to recall their behaviors. Researchers have to rely on individuals both to accurately remember behaviors and to record them free of any biases. Young children (those less than 10 years of age) probably cannot accurately recall and report their participation in physical activity; for this age group, proxy reports are often necessary. The validity of these reports also may be questionable given that parents, guardians, or teachers do not observe all of a child's activities on a particular day. An additional concern about self-reports by young children is that these youngsters may not be able to understand the differences between physical activity and exercise, sports, and games.

Emerging Methods in Physical Activity Assessment

Because of the limitations of established methods, new techniques are being developed at a rapid pace. These emerging methods, which make use of new technologies, are expected to enhance the validity and comprehensiveness of physical activity measurement.

The SenseWear Pro Armband (Body Media, Pittsburgh, PA) is a physical activity monitor that incorporates accelerometry and a variety of physiological parameters, including heat flux, galvanic skin response, skin temperature, and near-body temperature, in producing an estimate of energy expenditure (EE). This device is worn over the right triceps and can detect most movements and physiological responses to those movements. While the armband shows promise in providing estimates of EE, a calibration study showed that it overestimates EE in children (4). Also, the SenseWear armband is not ideal for use in young children because of its size. The equations for estimating EE are proprietary; and the device, as with other accelerometers, does not provide contextual information regarding the physical or social environments in which activities are performed.

The Intelligent Device for Energy Expenditure and Activity (IDEEA; MiniSun, Fresno, CA) constantly monitors body and limb motions through sensors attached to the chest, thighs, and feet (5). The device assesses daily physical activity by monitoring limb movements, postures, transitory motion, and gaits. It has been shown to be a reliable tool for identifying specific types of physical activities, but this function has not been evaluated in children (6). The IDEEA monitor can determine physical activity frequency, intensity, duration, and type. The multiple sensors, however, are a limitation to its use in children. In addition, the device cannot store activity data for periods of more than two to three days due to memory limitations.

In order to study the physical environment and its association with physical activity, several researchers have utilized a combination of global positioning systems (GPS) and accelerometers (7). Global positioning system data loggers facilitate collection of second-by-second position and speed-of-motion data for each second that an individual is in motion. These devices can provide information on the frequency, duration, and location of activity, which can be used to identify the environments in which physical activity is most common. A GPS system can provide valuable information, but there may be concerns about individual privacy, and these devices have not yet been evaluated in children.

Conclusion

Methods for measurement of physical activity in children and youth have been the focus of much research over the past decade. As a result of this work, both researchers and practitioners now have access to methods that are much improved over those available prior to the mid-1990s. However, the expanding interest in prevention of childhood obesity and the consequent growing interest in promoting physical activity suggest that there will be a continuing effort to refine existing methods and to develop new methods for measurement of physical activity in young persons.

Objective measures of physical activity, such as accelerometry, hold great promise. However, research is needed to minimize or overcome the limitations of accelerometry as it has been used to date. Also, it appears likely that there will be a continuing need for self-report measures for use in surveillance systems and large-group epidemiologic studies. But research is needed to enhance the validity of such measures. Therefore, it seems certain that methodological research on measurement of physical activity in children and youth will remain a high priority for the foreseeable future.

6

Evaluation of the Overweight Patient

George A. Bray, MD

Weight loss is the big American game. At any one time, millions of people are trying to lose weight. One approach to quantifying this craze was through the Behavioral Risk Factor Surveillance System; state health departments in collaboration with the U.S. Centers for Disease Control and Prevention contacted a random sample of American homes via a telephone survey. In that survey, published in 2000, 46% of women and 33% of men were trying to lose weight. Women reported trying to lose weight at a lower body mass index (BMI) than did men. Among the women who admitted to being overweight (BMI 25-29.9 kg/m^2), 60% were trying to lose weight; but the percentage of men trying to lose weight did not reach this level until BMI exceeded 30 kg/m^2. The likelihood of trying to lose weight increased with the amount of education the respondent had. Nearly 20% of the men and women reported taking in fewer calories and attempting to exercise more than 150 min per week as their strategy to accomplish their goal. With this degree of interest in the problem of losing weight, it is important for the physician to have a plan for how to approach the issues raised by prevention and treatment of overweight.

Evaluation

The first step for the physician and the overweight patient—before deciding on diet, behavior modification, or any other therapy—is to evaluate the risk of overweight to this individual. Selection of treatment can then be made using a risk–benefit assessment. A summary of this strategy, using BMI as a guideline, is presented in table 6.1. Once the evaluation is complete and both the physician and patient wish to engage in an effort to lose weight, the potential treatment options can be discussed. The choice of therapy depends on several factors, including the degree of risk associated with the overweight, patient preference, and whether the patient is ready to make the changes needed to lose weight.

A Clinical Perspective on Setting Weight Loss Goals

From a medical perspective, a successful patient will achieve a weight loss of more than 5% of initial weight. This is sufficient to reduce significantly the risk of developing diabetes in individuals with

Table 6.1

Use of the Body Mass Index to Select Appropriate Treatments

BODY MASS INDEX CATEGORY (KG/M^2)					
Treatment	25-26.9	27-29.9	30-34.9	35-39.9	≥40
Diet, exercise, lifestyle	+	+	+	+	+
Pharmacotherapy		With comorbidities	+	+	+
Surgery				With comorbidities	+

impaired glucose tolerance. A weight loss of 5% to 15% will reduce most of the risk factors associated with overweight, such as dyslipidemia (except total cholesterol), hypertension, and diabetes mellitus. In the Diabetes Prevention Program, a multicenter trial including participants with impaired glucose tolerance, weight loss of 7% reduced the rate of progression from impaired glucose tolerance to diabetes by 58%. Similar results have been reported with other programs aimed at preventing progression from impaired glucose tolerance to diabetes using diet and exercise or medications.

One feature of modest weight loss is that body weight may remain lower after months to years. This can be emphasized to the patient during the discussion of goals for the program. In a study of weight change in young and middle-aged women who were participating in the Nurses' Health Study, Field and colleagues found two important things. First, they noted that women who lost ≥5% of their weight over a two-year period (1989-1991) gained less weight than did their peers between 1989 and 1995. Second, the participants who engaged in vigorous physical activity gained approximately 0.5 kg (about 1 lb) less than did their inactive peers. Thus, encouraging weight loss to help slow later gain and encouraging physical activity are two important lessons from our patients.

Clinical Evaluation of the Overweight Patient

The basic components involved in the evaluation of any overweight or obese patient are a medical examination and a laboratory assessment. These should include a record of the historical events associated with the patient's weight problem, a physical examination for pertinent information, and appropriate laboratory evaluation. I will use the criteria recommended by the U.S. Preventive Services Task Force (1) and also take into account the reports from the National Heart, Lung, and Blood Institute (NLHBI) (2) and the World Health Organization (3). The importance of evaluating overweight individuals has increased as the epidemic of overweight has worsened and the number of potential patients needing treatment has increased.

Clinical History

Among the important elements of the clinical history to identify is whether there are specific events associated with the increase in body weight. Has there been a sudden increase in weight, or has body weight been rising steadily over a long period of time? Weight gain is associated with an increased risk to health. Three categories of weight gain are identified: <5 kg (<11 lb), 5 to 10 kg (11 to 22 lb), and >10 kg (>22 lb). In addition to total weight gain, you need to consider the rate of weight gain after age 20 when deciding on the degree of risk for a given patient. The more rapidly the patient is gaining weight, the more concerned you should be.

Etiologic factors that cause overweight should be identified, if possible (4). If there are clear-cut factors, such as drugs that produce weight gain, or cessation of smoking, these should be noted during the clinical evaluation.

Successful and unsuccessful weight loss programs that the patient has undertaken should also be identified. For example, a sedentary lifestyle increases the risk of early death, and individuals with no regular physical activity are at higher risk than individuals with modest levels of physical activity. Thus, even in the absence of successful weight loss, patients should be encouraged to maintain adequate levels of physical activity.

Family History

It is important to determine whether the patient comes from a family in which overweight is common—the usual setting—or whether she or he has become overweight in a family where few people are overweight. The latter setting suggests a need to search for environmental factors that may be contributing to weight gain. Recent studies have shown that among children and adolescents with a BMI above 30 kg/m^2, alteration in the melanocortin-4 receptor occurs in 2.5% to 5.5% of these individuals. Among genetic defects associated with any chronic disease, this is one of the most common; and evaluation of whether this defect is present may become important in the treatment of overweight people.

Physical Examination

A physical examination of the patient is important to determine the degree of overweight and obesity, and to immediately identify any overt health concerns. Minimally, a physical examination should consist of measurements of BMI, waist circumference, blood pressure, and other visible signs of obesity-related complications such as Acanthosis nigricans.

1. *Determining BMI and waist circumference—the vital signs associated with overweight.* I will use the algorithm in figure 6.1 from the NHLBI (2). The BMI provides the first assessment of risk. Individuals with a BMI below 25 kg/m^2 are at very low risk, but nonetheless, nearly half of those in this category at ages 20 to 25 will become overweight by age 60 to 69. Thus, a large group of pre-overweight individuals need preventive strategies. Risk rises with a BMI above 25 kg/m^2. The presence of complicating factors further increases this risk. Thus, an attempt at a quantitative estimate of these complicating factors is important.

The first step in clinical examination of the overweight patient (5) is to determine vital signs, which include BMI and waist circumference as well as pulse and blood pressure. Accurate measurement of height and weight is the initial step in the clinical assessment (6), since these are needed to determine the BMI. The BMI is calculated as the body weight (kg) divided by the stature (height [m]) squared: kg/ht^2. Body mass index has a reasonable correlation with body fat and is relatively unaffected by height. The height, weight, BMI, and other relevant clinical and laboratory data should be recorded during this evaluation. This helps categorize the

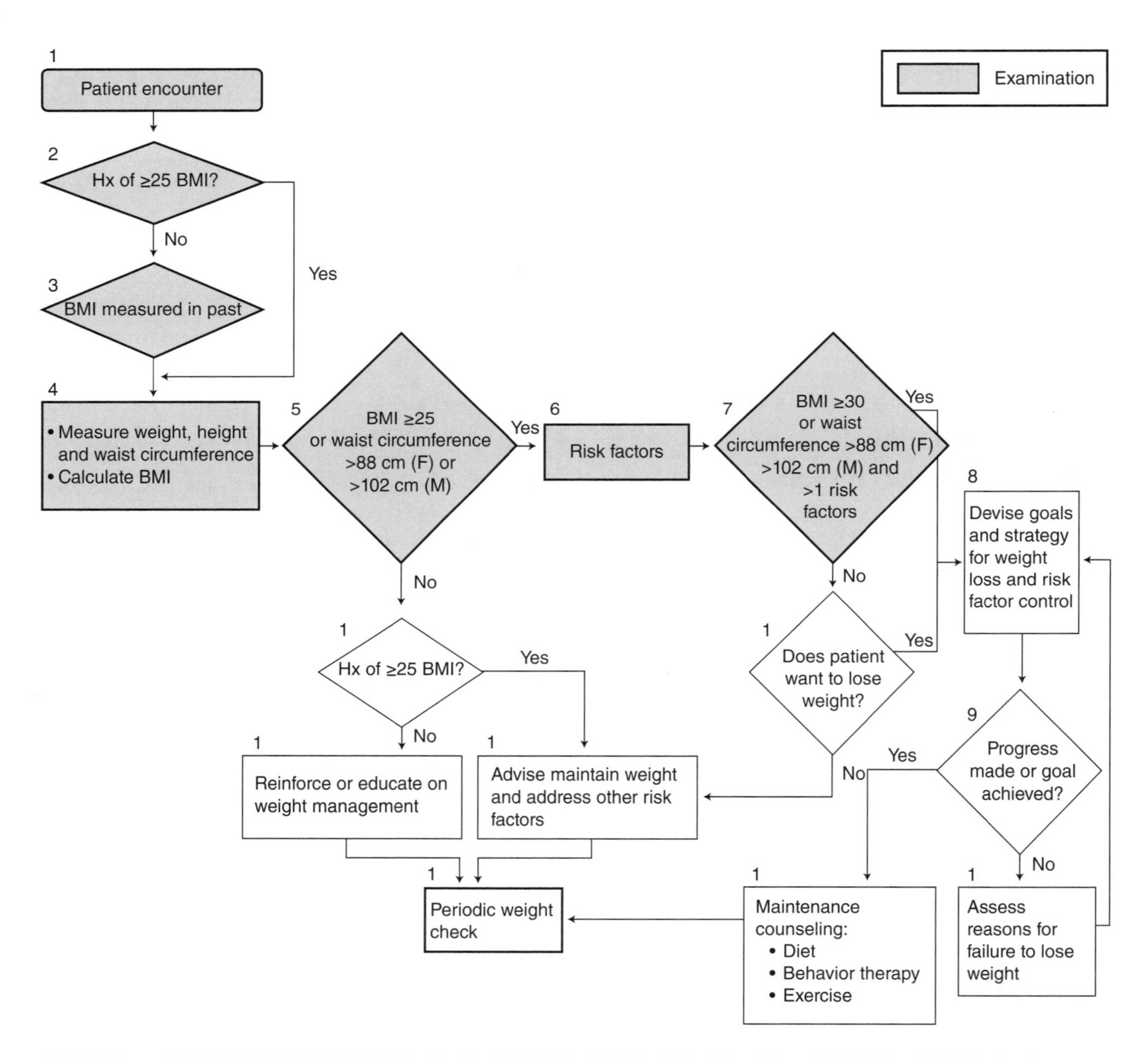

Figure 6.1 Algorithm for evaluating risk from overweight and approaches to treatment (from NHLBI). The initial step involves determining height and weight for establishing BMI and then proceeding along the appropriate lines of this algorithm.

patient as pre-overweight or overweight, and with or without clinical complications. The BMI needs to be interpreted in an ethnically specific context. An Asian conference selected lower levels of BMI to define overweight (BMI >23 kg/m^2) and obesity (BMI >25 kg/m^2).

Body mass index has a curvilinear relationship to risk. Several levels of risk can be identified. These cut points are derived from data collected on Caucasians. It is now clear, however, that different ethnic groups have different percentages of body fat for the same BMI. Thus, the same BMI presumably carries a different level of risk in each of these populations. One needs to take these differences into consideration when making clinical judgments about the degree of risk for the individual patient. During treatment for weight loss, the body weight is more useful than the BMI, since the height is not changing, and the inclusion of the squared function of height makes the BMI more difficult for physician and patient to evaluate.

2. *Waist circumference.* Waist circumference is the second vital sign in the evaluation of the overweight individual. The waist circumference is the most appropriate measurement for calculating central adiposity. It is determined using a metal or a nondistensible plastic tape. The two most common locations for measurement are at the level of the umbilicus and at the midpoint between the lower rib and the suprailiac crest. Although visceral fat can be measured more precisely with computed tomography (CT) or magnetic resonance imaging (MRI), these are expensive procedures; and clinical studies show that the waist circumference is essentially as good an indicator of visceral fat and much less difficult to obtain.

Measuring the change in waist circumference is a good strategy for following the clinical progress of weight loss. It is particularly valuable when patients become more physically active. Physical activity may slow loss of muscle mass and thus slow weight loss, while fat continues to be mobilized. Waist circumference can help in making this distinction. The relationship of central fat to risk factors for health varies among populations as well as within them. Japanese Americans and Indians from South Asia have relatively more visceral fat and are thus at higher risk for a given BMI or total body fat than are Caucasians. Even though the BMI is below 25 kg/m^2, central fat may be increased particularly in Asian populations. Thus central adiposity is important, especially with BMI between 22 and 29 kg/m^2.

3. *Blood pressure.* Careful measurement of blood pressure is important. Hypertension is amenable to improvement with diet and is an important criterion for a diagnosis of the metabolic syndrome. Having the patient sit quietly for 5 min before measuring the blood pressure with a calibrated instrument will help stabilize it. The blood pressure criteria from the Seventh Joint National Commission recommendations should be followed.

4. *Other items for the physical examination.* Acanthosis nigricans deserves a comment. This is a clinical condition with increased pigmentation in the folds of the neck, along the exterior surface of the distal extremities, and over the knuckles. It may signify increased insulin resistance or malignancy and should be evaluated.

Laboratory Evaluation

One strategy for refining the meaning of the BMI and the waist circumference is with laboratory measurements of lipids, glucose, and C-reactive protein (CRP). An increased fasting glucose, low high-density lipoprotein (HDL) cholesterol, and high triacylglycerol values are atherogenic components of the metabolic syndrome. Along with elevated blood pressure, it is possible to categorize the patient as having metabolic syndrome by one of several sets of criteria. In the International Diabetes Foundation (IDF) criteria, presence of increased central adiposity is required with abnormalities in two of the other four categories needed for making the diagnosis of metabolic syndrome. I prefer the IDF criteria, since they focus on the importance of central adiposity. In addition to the lipids that are determined as part of the assessment of the metabolic syndrome, a patient should have a measurement of low-density lipoprotein (LDL) cholesterol, which is a key risk factor. Also important is a measurement of highly sensitive C-reactive protein (hs-CRP). It is now clear that risk for heart disease can be predicted from both the LDL-cholesterol and hs-CRP.

The final issue is the plateau in body weight. Although not part of the evaluation of the overweight individual, it needs to be discussed with patients before they begin any program. After a period of time, weight loss slows and then stops. A value of 10% below starting weight would be a reasonable estimate of where most people will stop. Patients need to recognize this and need to know that when they maintain this lower weight for a time, they can then resume weight loss before another plateau occurs.

7

Assessment of Obesity in Children

John H. Himes, PhD, MPH

Accurate identification of the obese child is critical in a number of settings, each with different purposes. The specific purposes of the assessments should dictate the most appropriate methods and criteria to be used, and they demonstrate various applications of data concerning childhood obesity. After a discussion of definitions and nomenclature associated with childhood obesity, this chapter addresses assessment of obesity in children relative to clinical, public health, and research applications. Finally, high-priority needs for improving assessment of obesity in children are outlined.

Definitions and Nomenclature

Obesity refers to excess total adipose tissue in the body, usually expressed as fat weight or percentage of the body weight that is fat. Because girls have more total body fat than boys from before birth onward, and because body fat changes systematically as the child grows, definitions of obesity should consider chronological age and gender. Obesity is meant to be a status related to health and health risks, not cosmetic concerns. It is not the excess fat per se that is a concern but rather the current and subsequent health-related sequelae associated with the excess fat.

As easy and uncontroversial as this general definition seems, it has been difficult to find wide agreement on firm criteria for defining exactly what constitutes excess body fat and obesity in children based on health risk. Many of the obesity-associated health risks in children relate to risk factors for subsequent obesity and its adverse sequelae in adulthood, so large longitudinal studies are required. At the same time, the best estimates of total body fat usually require laboratory methods that complicate the execution of large population-based studies. Ironically, the emergence and popularity of an easily measured alternative indicator of total body fat, the body mass index (BMI), have probably distracted investigators from developing the longitudinal and population-based data sets and the scientific consensus required to standardize definitions of obesity in childhood based on the most valid estimates of total body fat. Although useful in many applications, BMI remains an imperfect measure of total body fat in children and adolescents (1).

The most commonly used definitions of obesity in children are actually couched in terms of statistical distributions of BMI rather than total body fat. In the United States, children 2 years of age and older with BMI ≥85th percentile and <95th percentile for age and gender (2) are considered overweight. Children with BMI ≥95th percentile or ≥30 kg/m^2 (whichever is less) are considered obese (3). Children less than 2 years with BMI ≥95th percentile are considered overweight; the term obese is reserved for older ages when predictability of subsequent status and health risks is stronger. This recent recommendation supplants the terms "at risk of overweight" and "overweight" previously used for children meeting the same BMI percentile criteria. Similar percentile cutoffs and language are used by investigators in several other countries who have developed their own BMI reference curves.

Two different sets of BMI reference curves have been recommended for international use. The International Task Force on Obesity developed reference curves for identifying child overweight and obesity from a pooled sample of BMI data from six national studies (including the United States) (4). The curves identifying overweight and obesity pass through the points of the BMI distributions at each age that correspond to the adult BMI values equivalent to 25 kg/m^2 and 30 kg/m^2. These childhood values approximate BMI percentiles 82 to 84 and 96 to 97 relative to the Centers for Disease Control and Prevention (CDC) Growth Charts (2), so prevalences of overweight and obesity for a sample do not markedly differ with use of the two sets of curves.

The World Health Organization (WHO) released a set of BMI curves as standards for all children from birth to 5 years of age based on uniformly collected data from six countries, including the United States (5). The samples included only very healthy, high-socioeconomic mothers who did not smoke during pregnancy and who breast-fed their term infants. The WHO upper percentiles of BMI are somewhat lower than corresponding values on the CDC Growth Charts at a given age (2), so prevalences of child overweight and obesity will be systematically larger with use of the WHO criteria for a given sample. Data are insufficient to determine whether these WHO BMI cutoffs differentially identify subsequent health risk compared with those from the CDC percentiles.

Clinical Applications

Clinicians may diagnose obesity, rule out other conditions, provide guidance, or monitor progress of interventions. For most clinical applications, the BMI criteria outlined earlier have been recommended (3). There are no obvious approaches to evaluating the severity of obesity for children with BMI levels substantially above the 95th percentile except the use of z-scores for BMI, and these are often unfamiliar to clinicians. For monitoring changes in BMI status of severely obese children, changes in the actual BMI value may be the most appropriate metric and can be useful for communicating with older children and parents, given appropriate explanation.

Assessments using alternative measures of subcutaneous fatness (e.g., skinfold thickness) or measures of fat distribution (e.g., waist circumference) are not recommended for routine clinical assessments of obesity because they provide little information for identifying the fattest children beyond that obtained with BMI. A comprehensive review of medical issues related to assessment of overweight and obesity in children, including recommendations for history, physical, and laboratory determinations, has been recently published (3).

In the past, concern has been expressed about the potential misclassification of early-maturing adolescents as obese with use of a BMI reference based on chronological age. The concern arises because BMI is positively associated with indicators of sexual maturation, and the maturation-related BMI effects are primarily hormonally mediated rather than due to long-term energy balance per se. Nevertheless, new analyses indicate that at high BMI levels, for example 85th or 95th percentiles, where the prevalence of early-maturing children is high, the BMI differences between children maturing at early and average rates are small, so the actual maturation-related effects are generally $<\pm 1$ kg/m^2 and probably clinically unimportant (6).

Public Health Applications

In public health applications, childhood overweight and obesity may be a focus in screening, in surveillance, and in the evaluation of programs. The BMI criteria usually are appropriate for these purposes. The measurements of stature and weight are straightforward and inexpensive, and appropriate equipment is portable so that measurements can be collected in many different settings. Multiple observers, of course, require suitable training and standardization.

For some more specialized public health applications, measures of fat distribution may be included in addition to BMI. For example, screening of adolescents to identify those at risk for cardiovascular disease, insulin resistance, or metabolic syndrome may include waist circumference as an indicator of subcutaneous abdominal fat and visceral fat deposition. Also, waist circumference may be expressed as a ratio relative to hip circumference or to stature for some applications.

Research Applications

In research, specific measures of fatness and even definitions of obesity may vary greatly and depend on the particular research questions and the samples used. Often, children considered obese in research are those defined as the fattest by some fatness measure, for example the upper quintile of the sample distribution of total body fat using dual-energy X-ray absorptiometry (DXA). Caution should be exercised on several counts, however, with this approach. Investigators should realize that such definitions are sample specific and that the actual prevalence of obesity defined in terms of recommended BMI criteria may vary considerably from that in other samples. Also, the specific individuals identified as the fattest children in a group may actually differ considerably based on the fatness measure used (7).

A research setting opens the door to many methods of assessment of body fat in children; selected measurement approaches with general strengths and weaknesses are listed in table 7.1. Each of the methods includes its own set of assumptions, which should be understood before findings are interpreted or compared across methods. Details and references concerning these and related methods of estimating body composition are reviewed elsewhere (8). When possible, investigators should include research findings in terms of common measures, for example BMI or skinfolds, as well as those for the particular fatness measures that are the main focus of the research. This will allow inferences to be drawn broadly and still not stifle innovation.

In interpreting results, one should carefully distinguish between associations with continuously distributed measures of fatness and with overweight or obesity status defined as categories. Often, it is difficult to directly apply the findings from fatness-related research to clinical or public health practice because results are not cast in terms of commonly used categories.

Needs for Improved Obesity Assessment in Children

1. Consensus recommendations from an international expert committee or other prominent body regarding definitions of overweight and obesity in terms of total body fat measured with valid methods and based on concurrent and subsequent health outcomes. The overall goal would be to determine how much body fat constitutes "excess" related to health. This process would require a comprehensive synthesis of a very large body of scientific literature and would probably also uncover substantial gaps in knowledge. Nevertheless, such recommendations, even if preliminary, would have enormous impact on the field in standardizing language and criteria and in providing a common basis of reference for practice and research applications.

■ Table 7.1 ■

Body Composition Methods for Estimating Total and Regional Body Fat in Children

Method	Strengths	Weaknesses
Stature, weight, body mass index	a, b, c, e, f	n
Skinfold thickness (caliper)	a, b, d, e	j
Subcutaneous fat thickness (ultrasound)	a, d, e	j
Body circumferences	a, b, d, e, f	j, n
Body density (water displacement)	c, f	g, h, i, l
Body density (air displacement plethysmography)	c, f	g, h, i, m
Isotope dilution (tritium)	c, e, f	g, h, k, l, m
Isotope dilution (deuterium)	c, e, f	g, h, l, m
Isotope dilution (doubly labeled water)	c, e, f	g, h, l, m
Total-body electrical conductivity (TOBEC)	c, f	g, h, i, m
Bioelectrical impedance analysis (BIA)	a, b, c, e, f	j, l
Potassium isotope counting (^{40}K)	c, f	g, h, i, m
Dual-energy X-ray absorptiometry (DXA)	c, f	g, h, i, k, l, m
Computed tomography (CT)	d, f	g, h, i, j, k, l, m
Magnetic resonance imaging (MRI)	d, f	g, h, i, j, l, m

a = relatively easy to use; b = relatively inexpensive to use; c = usually estimates total body fat; d = usually estimates regional body fat; e = fairly portable for field applications; f = high measurement accuracy; g = difficult to use; h = relatively expensive or very expensive to use; i = not usually portable for field applications; j = may require development of separate equations to estimate total body fat; k = involves ionizing radiation; l = may not be appropriate for younger children; m = may be available only at certain laboratories; n = not specific to body fat.

2. Development of high-quality reference curves for total adiposity for important populations using valid measures of total body fat and constructed using the most appropriate statistical methodologies. There are precious few population reference data for any measures of fatness, especially measures of total body fat (9). All levels of assessment of overweight and obesity in children require an appropriate basis for comparison. Ideally, the basis of comparison should be total body fat as the criterion measure. In some respects, high-quality reference curves are an important component for the development of the consensus definitions specified in the preceding paragraph.

3. Development of large long-term cohort studies of children using the best methods possible for assessment of total body fat to determine the associations with concurrent and subsequent risk factors and morbidity, as well as optimum criteria and thresholds for anthropometric and laboratory measures of adiposity and obesity status. Our current stage of understanding of the health implications of obesity requires us to take the next step beyond associations with surrogate measures of fatness like BMI toward better understanding of the health consequences of total body fatness in populations. Such studies will be expensive and complicated undertakings, but they are crucial for better defining assessment criteria in terms of health outcomes. Thoughtful study designs could yield important information with spans of 10 to 15 years in duration.

4. Development of valid, widely applicable, and fairly portable research methods to assess total body fat that are suitable for epidemiological field studies of child obesity, including children as young as 4 to 5 years of age. Current potentially useful methods such as foot-to-foot bioelectrical impedance analysis have been hindered as manufacturers have withheld prediction equations and research output (e.g., resistance and reactance) on proprietary grounds and have opted for popular market sales over important research applications (10). Investigators and commercial partners should work together to make available innovative technical approaches to estimate total body fat that researchers can apply to a wide variety of field settings and research questions. In the long run, such work will also contribute to new products for the benefit of both commercial endeavors and the research community.

Conclusions

The public and many professionals often take for granted the accurate identification of children as overweight or obese, supposing that the diagnosis is straightforward and without controversy. Yet there remain many unanswered questions regarding the most accurate assessment of obesity in children. Improved understanding of assessment and defining it in terms of total body fat and health risks are important to help us better understand the etiology, prevention, and management of obesity and what it means for health to be an obese child.

Limitations of Self-Report in Physical Activity and Obesity Research

Richard P. Troiano, PhD

Due to relatively low cost of administration and ease of mass distribution, questionnaires that collect self-reported height and weight and physical activity are major sources of data on weight status and physical activity. Self-reports are frequently used in epidemiological studies and population surveillance. In clinical or intervention settings, height and weight may be measured, but physical activity assessment is likely to be based on self-report. This chapter describes concerns about agreement between self-reports and objective measures of weight status and physical activity, as well as implications of lack of agreement.

Methods of Reported Weight Status

Unlike many other factors assessed by self-report, height and weight are usually captured with a standardized assessment. Across many studies and surveys, the most likely questions are minor variations on these two:

- How tall are you without shoes?
- How much do you weigh without shoes?

In the research context, the most common use of responses to these questions is the calculation of body mass index (BMI), which may then be used to classify an individual's weight status as underweight, normal weight, overweight, or obese. Such data are used to track the national prevalence of overweight and obesity and to evaluate associations between weight status and outcomes or predictive factors.

Methods of Reported Physical Activity

Self-report approaches to assess physical activity vary widely. The assessment can range from a single question that asks respondents to rate their level of activity as inactive, somewhat active, or very active; to one or two questions that ask respondents to quantify the frequency and duration of their moderate- and vigorous-intensity activity from recreation or from all sources; to a battery of questions that ask respondents to quantify frequency and duration of activities in multiple contexts. These more extensive approaches may include separate questions that address transportation, occupation, household tasks, recreational or leisure activities, sedentary pursuits, and strengthening and flexibility exercises in addition to cardiorespiratory activities. The most comprehensive self-report would be a multiday physical activity log or diary.

The responses to the simplest single-question assessment can be used to classify individuals based on their self-assessed level of activity. More extensive and quantitative questions can be used to estimate time spent at particular levels of activity (e.g., sedentary or moderate to vigorous intensity) or an integrated measure of time and intensity, such as metabolic equivalent (MET)-minutes per day or week, to determine compliance with recommended levels of activity or even to estimate energy expenditure.

Evaluating Agreement Between Self-Reports and Objective Measures

Concerns about the limitations of self-reported information are based on comparisons with objective measures. For weight status, self-reports are compared to measured height and weight. For physical activity, reports can be compared to objective measures from devices such as pedometers, accelerometers, or heart rate monitors, or to measures of energy expenditure obtained from indirect calorimeters or doubly labeled water.

Agreement Between Reported and Measured Weight Status

In general, height is overreported and weight is underreported (1). Women underreport weight to a greater extent than men, and the degree of underreporting for both men and women increases with increasing BMI. If BMI is calculated, underestimated weight and overestimated height will both lead to underestimated BMI and potential misclassification of weight status categories.

Proposed explanations for the observed misreporting include recall errors and social desirability bias, whereby respondents report weighing what they once weighed, what they wish they weighed, or what they know they are supposed to weigh. An additional explanation for the overestimation of height, particularly among older adults, is the loss of height with age. The reported height may reflect a measure from decades earlier that has persisted in memory or from a driver's license, while the individual's actual height has decreased due to vertebral compression.

An illustration of the effect of reported weight status misclassification can be observed in estimates of adult obesity prevalence in the United States in 2004, based on three nationally representative surveys that differ in the method of data collection for height and weight (see table 8.1).

Each of the surveys provides an essentially concurrent nationally representative measure of obesity prevalence, so the three should agree. However, underestimated BMI due to self-report leads to obesity prevalence that is 7 percentage points lower in males and about 10 percentage points lower in females compared with measured height and weight.

Agreement Between Reported and Measured Physical Activity

Physical activity is generally overreported. The overreporting is particularly evident for vigorous-intensity physical activity (5). Analysis of accelerometer data collected in the National Health and Nutrition Examination Survey (NHANES) 2003-2004 showed that adult population prevalence of adherence to the current recommendation to obtain at least 30 min of moderate or greater intensity on five or more days per week was less than 4%, compared with estimates of 25% to 50% for surveys that use reported physical activity (6). The NHANES 2003-2004 accelerometer data indicated that the mean time in vigorous activity was less than 2 min per day among adults, even when activity bouts of

Table 8.1

U.S. Prevalence of Obesity From Three National Surveys

		OBESITY PREVALENCE (%)	
Survey*	Data source	Males	Females
BRFSS 2004	Telephone self-report	23.6	22.5
NHIS 2004	Face-to-face self-report	23.9	23.7
NHANES 2003-2004	Measured height and weight	31.1	33.2

*BRFSS = Behavioral Risk Factor Surveillance System (2); NHIS = National Health Interview Survey (3); NHANES = National Health and Nutrition Examination Survey (4).

Data from BRFSS Nationwide (States and DC) Prevalence data for 2004: Overweight and obesity. http://apps.nccd.cdc.gov/brfss/sex.asp?cat=OB&yr=2004&qkey=4409&state=UB. Accessed 4/30/08; M. Lethbridge-Çejku, D. Rose, and J. Vickerie, 2006, "2004 summary health statistics for U.S. adults: National Health Interview Survey," *Vital Health Statistics* 10(228). National Center for Health Statistics: Hyattsville, MD, p 81; and C.L. Ogden et al. 2006, "Prevalence of overweight and obesity in the United States, 1999-2004," *Journal of American Medical Association* 295(13): 1549-55.

less than 10 min were included (6). Similar results based on accelerometer data have been reported for Sweden (7).

Potential explanations for misreports in physical activity are more complex than for height and weight. Errors in recall and social desirability are still factors. However, accurately and quantitatively reporting physical activity can be a challenging task for respondents. This challenge was increased as physical activity recommendations evolved from a fitness paradigm that focused on sustained bouts of exercise to a public health model that includes concepts like accumulation of moderate-intensity activity through making active choices in routine tasks.

Reporting exercise is relatively easy in that exercise tends to be regular or programmed (e.g., the lunchtime run or twice-weekly class at the health club) and well defined as people put on special clothes or go to a specific location. Distinct clothing and locations provide cues that make exercise events easier to remember, and the regularity of the activities is likely to facilitate duration estimates. In contrast, recalling and reporting activities that are consistent with more recent recommendations can be difficult. Respondents are asked to recall accumulated activities that occur in short bouts or that result from active choices such as taking the stairs instead of the elevator, parking farther away, or walking for short-trip errands. For many, even the encouraged iconic brisk walks are less likely than most exercise to be regular and well defined by memory cues.

An additional challenge is presented when the questionnaire inquires about frequency (times or days) and duration (per time or per day) of moderate-intensity activity from multiple and varied sources, which is not uncommon. The active health-conscious respondent who walks briskly for 10 min each way to and from the bus stop on her daily commute, rides her bike after dinner twice a week for 45 min, and takes a 3 h hike on Saturday is faced with a dilemma. Does she report the frequency and duration of the most common event or perhaps the longest event? Or should she attempt to do the calculus to total up the minutes of accumulation of all the activities ($10 \times 10 + 2 \times 45 + 1 \times 180 = 370$) and divide by the total number of events in the week ($10 + 2 + 1 = 13$) to arrive at an average duration ($370 / 13 = 28.5$) of 30 min and a frequency of 13 times per week, or perhaps an hour per day on six days per week? It might be easier to just guess at a number. Particularly if one is not as health conscious and active as in our example, guessing 30 min on five days may seem like a good option because one has frequently heard those numbers. Few respondents will perform the algebra involved.

Another factor likely to contribute to overestimated reports of physical activity is the difficulty of estimating intensity. Most researchers are considering absolute intensity when including a question in a survey or analyzing questionnaire results. For example, for the researcher, moderate-intensity activity is defined as a level of 3.0 to 5.9 METs. But most respondents have no idea about MET levels, so the researcher provides either physiologic cues (somewhat elevated heart rate, "You can talk, but not sing") or examples (brisk walking, doubles tennis, scrubbing floors). However, the physiologic cues are dependent upon fitness and therefore convert the question to one based on relative intensity that changes with age and fitness. For many inactive or older individuals, activities are likely to be reported as moderate (relative) that the researcher would call light intensity (absolute). Similarly, moderate-intensity activities from the researcher's perspective may be reported as vigorous.

Beyond Bias

Most of the discussion so far has addressed over- or underestimation that represents bias. Bias is a problem if the absolute level of obesity or adherence to physical activity recommendations is desired. However, in some applications, such as epidemiological studies, the absolute level is considered less important than relative ranking of individuals. Systematic bias, which does not affect ranking, will not preclude finding relationships between exposures and outcomes. If the misreporting affects rankings but can be statistically modeled, it might be possible to correct the reported estimates to get an estimate similar to what would be obtained from objective measures.

The misreporting of both weight status and physical activity appears to be nonsystematic and not amenable to correction factors. Plankey and colleagues (8) analyzed reported and measured height and weight in NHANES II data. They confirmed the frequent observation that underreporting of BMI varied systematically with measured BMI. However, the pattern of reporting error was only weakly associated with self-reported BMI. These investigators concluded that it was not possible to determine a regression model that could eliminate the reporting error to predict measured BMI from self-reports (8).

Taking a different approach, Troiano and Dodd (9) demonstrated complex and severe misclassification for physical activity by examining cross-classification of physical activity by self-report and accelerometer measures in NHANES 2003-2004. An example for men ages 20 to 59 years is shown in table 8.2. The table classifies individuals by bouts of moderate- or greater-intensity activity recorded by accelerometer (Actigraph threshold = 2020 counts/min) or reported on the questionnaire. For each data source, there is a category for zero time (measured or reported), labeled 0. Nonzero measures were divided into quintiles, labeled 1 to 5. The numbers in the cells represent percentage of the U.S. population of men ages 20 to 59 years.

The row and column totals indicate that 37.5% of the men had no measured bouts of moderate- or greater-intensity activity, whereas only 10% reported no activity. The sum of the diagonals (18.3%) provides the proportion of the population for which the ranking agrees. If cells within plus or minus one category are included, the agreement increases to 49%. Thus, over half of the men would be misclassified by two or more categories. Another indication of the severe self-report misclassification is the low variation in percentages across report categories within the low-activity accelerometer categories (e.g., rows 0 to 2). Contrast these rows with accelerometer category 5, where the values increase monotonically, and the value in report category 5 is almost 10-fold greater than that in report category 0. Misclassification is also evident in the typically low to moderate correlations between self-reports and objective measures of physical activity.

Implications of Reporting Error

The complex misreporting in self-reported weight status and physical activity can affect surveillance, epidemiologic, and intervention research. The magnitude of risk factors such as obesity or physical inactivity can be dramatically underestimated, and trends over time could be masked by changes in social desirability. The resulting misclassification can attenuate relative risk estimates in epidemiological studies. Baseline physical activity level of participants in intervention research may be misclassified, and reporting error could make it difficult to determine compliance with interventions.

Self-report is still the best way to determine the type of physical activity performed by individuals and the context in which it is performed (e.g., transportation, occupation, or recreation). However, quantifications of weight status and physical activity by self-report have demonstrated limitations. Wherever possible, height and weight should be measured and physical activity should be assessed by objective measures if quantification is desired. In situations in which objective measures are not feasible, researchers are encouraged to remind themselves and readers of the limitations of self-report by referring to findings with terms like "reported obesity" rather than "obesity" or "reported physical activity" rather than "physical activity."

■ Table 8.2 ■

Weighted Cross-Classification Percentages for Categories of Activity Measured by Accelerometer and Self-Report in Men Ages 20 to 59 Years

CATEGORY BASED ON ACCELEROMETER[1]	CATEGORY BASED ON SELF-REPORT 0	1	2	3	4	5	Total
0	***4.75***	8.70	8.20	4.93	6.42	4.51	37.50
1	1.97	***2.03***	2.36	2.49	2.20	1.38	12.42
2	1.28	2.09	***2.43***	2.21	1.69	2.87	12.56
3	0.91	2.26	1.80	***2.37***	2.39	2.74	12.47
4	0.68	1.74	1.88	3.62	***2.30***	2.25	12.48
5	0.52	0.86	1.47	2.40	2.86	***4.46***	12.56
Total	10.11	17.68	18.15	18.01	17.85	18.20	100.0

[1]For each measure, category 0 corresponds to no activity, and categories 1-5 correspond to quintiles of nonzero activity.

Data from R.P. Troiano and K.W. Dodd, 2008, "Differences between objective and self-report measures of physical activity. What do they mean?" *Journal of Korean Social Measurements and Evaluation*. 10(2): 31-42.

part II

Prevalence and Cost Issues

The second section of this book takes a bird's-eye view of the problem of physical inactivity and obesity by highlighting the most current information on global prevalences and costs. Chapters 9 and 10 describe the worldwide prevalence of physical inactivity in adults and children, respectively, while chapter 11 deals with the global prevalence of adult obesity. Chapter 12 describes the economic consequences of the global obesity epidemic and provides examples of the burden of obesity in several countries. Chapter 13 describes the level of childhood obesity across the regions of the world where data exist. Chapter 14 presents an estimate of the economic burden associated with physical inactivity, and the potential interactions between physical inactivity and obesity on health care costs. This section concludes with an examination of issues related to estimating the cost-effectiveness of physical activity interventions (chapter 15).

Global Prevalence of Adult Physical Inactivity

Wendy J. Brown, PhD

Comparing levels of inactivity around the world is problematic because there is no universally agreed-upon definition of "inactivity" (or "activity"), and different jurisdictions use very different methods for assessing population levels of these constructs. Prior to the turn of the last century, the focus of most national surveys (where they existed) was on participation in *leisure-time activity* (recreation and sports), and individuals with the lowest participation rates were assumed to be *inactive.* However, with the accumulation of evidence about the benefits of more moderate-intensity activity, attention turned to the measurement of physical activity (PA) in multiple domains, including work (occupational and domestic) and transport, as well as during discretionary or leisure time. In terms of international comparisons, this is important because there is wide variation in levels of activity in each of these domains across countries. For example, in countries where occupations involve physical labor (e.g., in agriculture, heavy industry), where domestic activities involve growing food or carrying water, or where transport is largely on foot or by bicycle, there may be little involvement in leisure activities. Yet the people in these countries will have higher daily energy expenditure than those in many more developed countries where there is ample access to active leisure opportunities but limited activity in transport or in occupational or domestic work. Again, countries that have low levels of activity in all these domains would be categorized as most inactive.

Comparisons From National Population Surveillance Instruments

Many countries, including the United States, Canada, New Zealand, and several European nations, have established population monitoring surveys that are used to derive national prevalence estimates for inactivity and activity from generic questions about participation in moderate-intensity and vigorous activity, as well as walking. The results cannot, however, be directly compared. A 2004 study compared participant responses to the U.S. Behavioral Risk Factor Survey (BRFSS), the Active Australia survey (AA), and the (then) new International Physical Activity Questionnaire (IPAQ) (1). The results showed that subtle differences among these three surveys (e.g., in the reference period used to report frequency ["last" or "usual" week] and in the inclusion of different examples of moderate and vigorous activities from the leisure, transport, occupational, and domestic settings) resulted in wide variation in the total PA minutes reported by people who completed all three surveys with reference to the same time period. In brief, the BRFSS and the IPAQ captured 144 and 500 min (respectively) more activity time per week than the AA survey, while prevalence estimates for "meeting guidelines" were 10% higher from BRFSS and 26% higher from IPAQ than from the AA survey (56% "active") when the same threshold for "active" was used (1). The conclusion was that the more detailed the questions, the more PA is recorded—and that comparisons should not be made from one survey to another.

Acknowledgements
My thanks to Dr. Tim Armstrong and Professor Adrian Bauman for their helpful comments on an earlier draft of this chapter.

Making comparisons across surveys is also problematic when there is no consensus on what constitutes inactivity and activity. Inactivity is often defined as doing little or no PA in any of the leisure, occupational, or transport domains. The word "inactive" is also used to describe people who do not reach a defined threshold for meeting guidelines. Conversely, those who do reach the threshold are defined as active. But there are differences between surveys in where this threshold is set. Consensus seems to be emerging that "active" (also referred to as "sufficiently active") is commensurate with at least 150 min of moderate-intensity PA per week *or* 60 min of vigorous PA each week (or a combination of moderate and vigorous activities with equivalent energy expenditure) (2). Recently, researchers have suggested that the 150 min threshold should be in addition to a basal amount of moderate activity each day as captured by instruments like the IPAQ that address PA in multiple domains (3). If international comparisons of PA are to be made, then it is important that the survey used, the mode of administration, the data reduction methods, and the definitions of inactive and active be consistent.

"International" Surveys for Measuring Physical (In)Activity

In recent years, progress has been made toward the development of instruments that can be used to make internationally comparable estimates of physical (in)activity in populations with very different patterns of energy expenditure in work, transport, and leisure. The IPAQ (www.ipaq.ki.se) was developed by researchers from 12 countries as a basis for international surveillance, and the Global Physical Activity Questionnaire (GPAQ) was developed by the World Health Organization (WHO) as part of its STEPwise approach to chronic disease risk factor surveillance (4). Both have acceptable measurement properties, and the long-form (31-item) IPAQ and the 16-item GPAQ allow differentiation of data derived from the key domains of PA, which is important for developing targeted interventions in countries with vastly contrasting patterns of PA. To date, however, the only published results from these surveys are from studies that have used the short-form (9-item) IPAQ.

International Comparisons in Europe

The first international comparison study was performed by a group of researchers from Spain and Ireland who conducted face-to-face interviews with about 1000 adults from each of the (then) 15 member states of the European Union (EU) in 1997 (5). The focus of the interviews was on participation in sports (e.g., football, golf), recreational activities (e.g., aerobics, walking), and outdoor pursuits (e.g., hill walking, canoeing). Metabolic equivalents were assigned to each activity, and the sum of MET-hours per week for all reported activities was calculated for each country. The study demonstrated a "north–south" gradient in leisure-time activity, with Sweden, Finland, the Netherlands, Ireland, and Denmark showing much higher leisure-time PA than Greece, Italy, Spain, and Portugal (which could therefore be described as more inactive). Exceptions were Austria, which had the third-highest PA levels, and Belgium, which had the second-lowest levels after Portugal. More than 90% of Swedes and Finns reported some leisure activity, compared with 40% of the Portuguese sample (5).

In 2002, the IPAQ was incorporated into the Eurobarometer survey of the same 16 member states of the EU. The IPAQ questions covered weekly frequency and duration of walking and moderate and vigorous activities; total time was weighted by generic MET values (walking, 3.3; moderate, 4; vigorous, 8). In their 2004 study, Rutten and Abu-Omar (6) reported median MET-hours per week values; these were significantly higher (overall median, 24 MET-hours per week) than those reported in the earlier European study (15 MET-hours per week) (5), which would not have been unexpected given that the IPAQ assesses activities in the four domains of leisure, work, transport, and home rather than only leisure-time activities. The highest PA was in the Netherlands, which, given the culture and infrastructure around cycling for transport in that country, was not surprising (see figure 9.1). Walking was highest in Spain and Denmark, and the lowest PA was seen in Northern Ireland, which recorded a median of only 11.6 MET-hours per week (6). The north–south gradient seen in the earlier leisure activity survey (5) was no longer apparent.

Using data from the same Eurobarometer study, members of the original IPAQ development group derived prevalence estimates for "health-enhancing activity" using a relatively high cut point of 3000 MET-minutes over seven days, or 1500 MET-minutes of vigorous activity accumulated over three days or more (3). This was designed to represent 5 × 30 min of moderate-intensity activity or 3 × 20 min of vigorous activity *on top of* a basal 60 min of moderate activity each day. The average prevalence of this health-enhancing PA level was 31%, suggesting that more than two-thirds of the adult populations of the EU countries are insufficiently

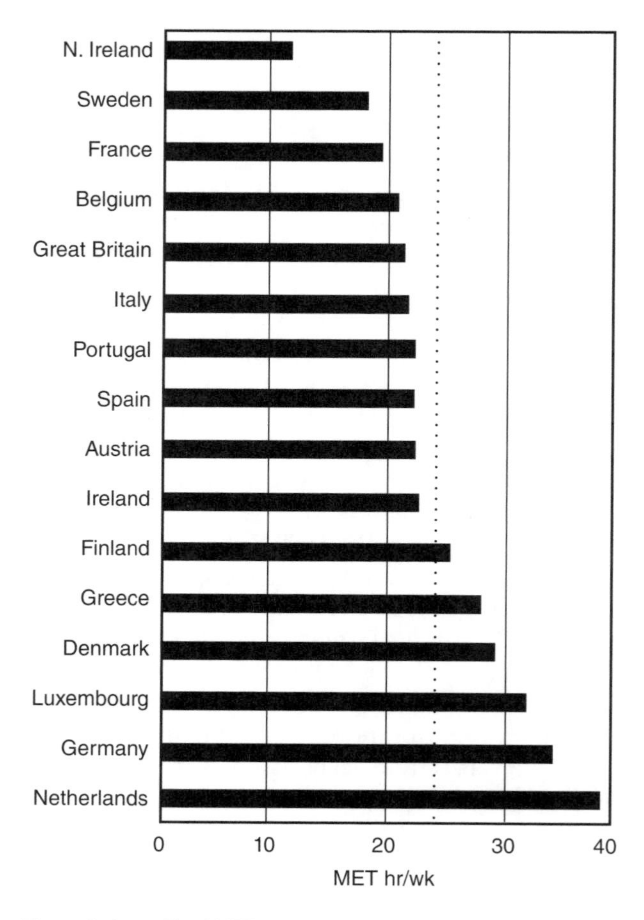

Figure 9.1 Total MET-hours per week reported on the short International Physical Activity Questionnaire in 16 European Union nations. Dashed line indicates median value of 24 MET-hours per week.

Data from A. Rütten and K. Abu-Omar, 2004, "Prevalence of physical activity in the European Union," *Sozial-und Präventivmed* 49: 281-289.

active for health benefits (or are inactive). Almost half of the insufficiently active (again 31% of the total sample) did not report enough activity to meet a lower 600 MET-minutes per week criterion; they were referred to as sedentary in this study. Belgium (40%) and France (43%) had the highest proportions of sedentary people, and the Netherlands (19%) and Denmark (22%) had the lowest (3).

Global Physical Activity

The IPAQ was also used in a 2008 study that included data from 51 mainly low-middle–income countries participating in the WHO World Health Survey (7). Using cut points of 480 (vigorous) and 600 (combined) MET-minutes per week, the researchers found that only 18% of the pooled sample were inactive (did not reach the threshold). This is much lower than the 31% estimate in the 2006 EU study (3). Levels of inactivity were consistently low (<10%) in parts of Africa (e.g., Ethiopia, Zambia, Malawi), Eastern Europe (e.g., Hungary, Croatia, Estonia), and Southeast Asia (e.g., India, Nepal, Bangladesh) and in the Western Pacific Region (e.g., China, Vietnam, Laos). People living in urban areas were more likely to be inactive than their counterparts in rural areas, highlighting the detrimental effect on activity levels of the rapid urbanization and technological changes that are occurring in some of these countries (7).

This issue is nicely illustrated by the results of a small IPAQ survey of Old Order Amish people living in southeastern Ontario, Canada. Amish people follow rules that ban the use of motorized transportation, electricity, and modern conveniences; most work on farms where they till the soil with horses and grow vegetables in family gardens (8). All of the 98 men and women who participated in this study met the U.S. guidelines, and PA levels were far greater than any others in the published literature. The men recorded an average of 299 MET-hours per week of activity and the women 207, compared with a median for the EU countries, using the same survey, of 24 MET-hours per week (see figure 9.1). This represents about 50 h of walking and moderate activity and several hours of vigorous activity each week (8).

As part of the WHO work on the global burden of disease in 2004, Bull and colleagues attempted to overcome the challenges of comparing data from different surveys (9). The researchers sourced data from several countries in each WHO region, including Africa (Ethiopia and South Africa), three regions of the Americas (A: Canada and United States; B: Argentina and Brazil; C: Peru), the Eastern Mediterranean (Saudi Arabia and Egypt), Europe (A and B: Western Europe; C: Estonia, Latvia, Lithuania, the Russian Federation), Southeast Asia (Bangladesh and India), and two regions of the Western Pacific (A: Australia, New Zealand, and Japan; B: China), to create prevalence estimates for each region. Some countries had data from national surveillance instruments with sample sizes in excess of 5000 (e.g., Canada, United States, the Russian Federation, New Zealand, and China) while others had only regional surveys with smaller samples. Almost all used different surveys and derived different summary scores even with use of the same survey. Some of the data came from samples in large cities (e.g., Egypt), and some countries had only leisure-time data (e.g., South America, Saudi Arabia, and Japan). There were few data from Southeast Asia. A detailed comparison of all the surveys can be found in the WHO report (9). The prevalence estimates for inactivity (little or no activity at work or in transport or leisure) in men and women are shown in figure 9.2.

Detailed examination of these data highlights the importance of including assessment of PA in the domains of work and transport as well as in leisure or discretionary time. For example, activity for transport was highest in Africa, the Eastern Mediterranean, China, and Southeast Asia and lowest in North America. Work-related activity was also low in North America, Western Pacific region A (Australia, New Zealand, and Japan), and Western Europe. These three regions, however, had the highest levels of discretionary activity. Overall, when activity in all domains was considered, the highest levels of inactivity were in Eastern Europe and the Americas. The least inactive regions were the Western Pacific and Africa (9).

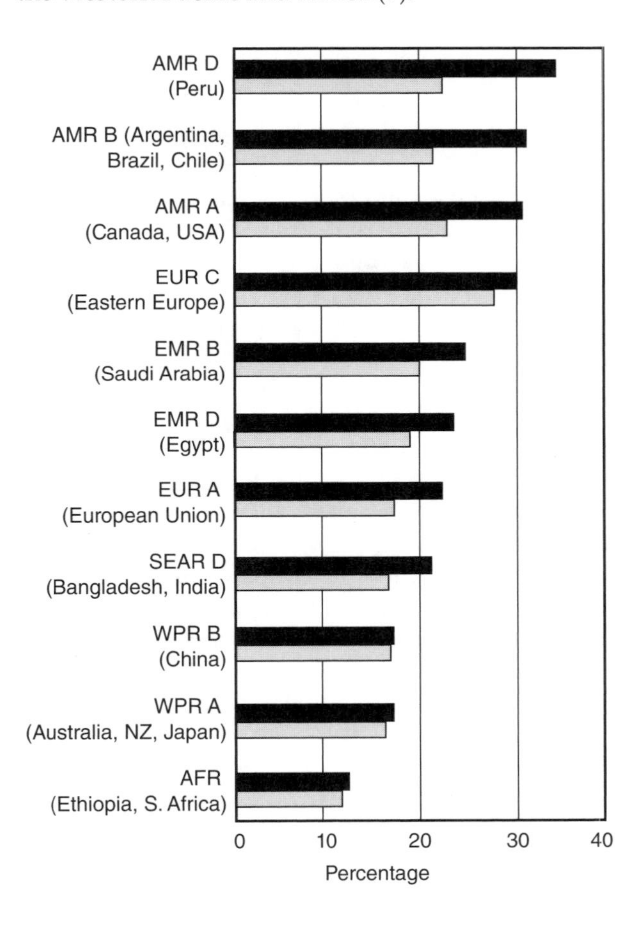

Figure 9.2 Prevalence estimates for inactivity (little or no PA in the domains of work, home, transport, or leisure) in men (pale bars) and women (dark bars) for selected WHO regions (examples of countries from which data were available are in parentheses). (AMR = America; EUR = Europe; EMR = Eastern Mediterranean; SEAR = Southeast Asian Region; WPR = Western Pacific; AFR = Africa).

Data from F. Bull et al., 2004, Physical inactivity. In *Comparative quantification of health risks: Global and regional burden of disease attributable to selected major risk factors*, edited by M. Ezzati et al. (Geneva, Switzerland: World Health Organization), pp 729-881.

Minority Populations

While none of the international comparisons described here directly address racial and ethnic group differences within multicultural societies, several national surveys have considered this issue. In the United States, the prevalence of regular PA is assessed using the Behavioral Risk Factor Surveillance System, a state-based survey that includes questions similar to but subtly different from the IPAQ questions. In 2005, the proportion of people meeting the U.S. guideline of at least 30 min a day of moderate activity on five or more days a week, *or* at least 20 min of vigorous-intensity activity on three or more days, was 49.7% for men and 46.7% for women (10). Prevalence estimates were lower among Hispanic (men 41.9%, women 40.5%) and non-Hispanic black people (men 45.3%, women 36.1%) than among white men and women (men 52.3%, women 49.6%). In Canada, the Community Health Survey showed that the prevalence of meeting the Canadian guideline (≥1.5 kcal · kg^{-1} · day^{-1}) was about the same in North American Aboriginal people (47%) as in white Canadians (49%) (11). National surveys in New Zealand have also shown very little difference in the proportions of Maori people (men 60%, women 51%) and European New Zealanders (men 58%, women 50%) who meet the New Zealand guideline of 150 min of activity per week (12).

Conclusions

Comparisons of the prevalence of physical inactivity in countries across the world are constrained by the use of different measurement instruments and, when the same survey is used, by variations in the definition of and threshold for inactivity. Even when the same survey is used, there may be cultural differences in understanding of the terms used; and in some countries, PA behavior varies with seasonal change. The available data suggest that when PA is considered in the domains of work, transport, and leisure, inactivity reflects urbanization and technological development and is therefore more prevalent in wealthier countries. Exceptions do exist; in developed countries where the bicycle is still a major form of transport, PA levels remain relatively high. If meaningful international comparisons of physical inactivity are to be made, it is important that efforts be continued to create truly "international" surveys, with standardized methods of data reduction and reporting.

10

The Prevalence of Children's Physical Activity

Chris Riddoch, PhD

Children's physical activity behavior is thought to be a major contributor to the increasing prevalence of overweight and obesity in children. Current recommended levels of physical activity for children are 60 min per day of moderate- or vigorous-intensity physical activity (MVPA). This recommendation was originally based on self-reported physical activity and was for general health, not specifically for avoiding obesity. The recommendation was formulated on the basis of well-established physiological principles and expert consensus rather than scientific data on physical activity–health associations.

Until recently, our knowledge of children's physical activity levels and patterns, and in particular our understanding of associations between physical activity and health, has been severely restricted by the lack of an accurate measurement tool. However, recent technological advances in the development of objective methods of assessing physical activity have enabled us to overcome children's limited ability to recall and quantify physical activity; hence we can now measure children's physical activity levels and patterns with far higher levels of precision. For the first time, we can identify and quantify levels of physical activity with relative precision and more confidently assess associations with health outcomes such as obesity.

Large-scale surveillance systems exist that routinely include assessment of physical activity (e.g., Youth Risk Behavior Surveillance System in the United States, Health Behavior in School-Aged Children in Europe). These large surveys routinely utilize self-report methods to measure physical activity, and thus the data are necessarily prone to error. These data have recently been reviewed (1) and in general suggest disappointingly low numbers of children reporting sufficient physical activity. A major limitation of such studies is that they have used different physical activity questionnaires for the assessment of self-reported physical activity. Only the Health Behavior in School-aged Children study has applied a common methodology across different countries. While such methods can be useful for monitoring trends over time, more accurate data are required if health associations are to be investigated (2).

This review brings together for the first time more recent data on children's physical activity that have been gathered with the use of objective methods (accelerometry) in large samples of children. This focus on objective measurement data will give us more accurate and meaningful insight into children's physical activity levels and patterns.

Accelerometer Measures of Physical Activity

The use of accelerometers enables us to record every movement of the child, at all intensities, not just the more memorable bouts of physical activity that the child may recall. Such methods provide us with two main physical activity variables. First, total physical activity can be assessed that reflects all movement performed by the child over a defined time period, usually one week. Data are usually reported as accelerometer counts per minute (cpm) averaged over the defined measurement period. This variable is theoretically related to the total amount of energy expenditure achieved through physical activity as it includes all movement. It is therefore logical to presume that this variable might be more important when we are assessing relationships with obesity risk. The second important variable is the number of minutes (usually per day) of MVPA. This variable is important because children's physical activity recommendations are stated in terms of MVPA. It is believed to matter for general health reasons, especially for metabolic or cardiovascular health, as this intensity of physical activity places greater physiological stress on the child's body. This review addresses both variables but focuses on MVPA because this is the variable used in framing children's physical activity recommendations. Somewhat surprisingly, MVPA has also been shown to be more strongly associated with obesity than has total physical activity (3).

Levels of Physical Activity

This brief review focuses specifically on four of the largest studies that utilize accurate measurement and that we might consider representative of large and diverse populations of children across six countries. These studies on large and well-characterized samples constitute the best evidence on levels and patterns of physical activity in contemporary child and adolescent populations. The results are broadly comparable, as the various studies have used very similar methods. They encompass the full child and adolescent age range and include data from six countries. These studies are the National Health and Nutrition Examination Survey (NHANES; United States) (4), the European Youth Heart Study (EYHS; Denmark, Portugal [Madeira], Estonia, Norway) (5), the Avon Longitudinal Study of Parents and Children (ALSPAC; England) (6), and the Trial of Activity for Adolescent Girls (TAAG; United States) (7). The results, in terms of levels of physical activity, are shown in table 10.1.

It can be seen that despite considerable consistency in instrumentation and protocols reported among the studies, data reduction procedures have varied. Comparisons between studies therefore remain a little problematical. In particular, cut points for the lower threshold of MVPA vary within the range 906 to 3600 cpm. While some discrepancies are to be expected according to the age, gender, and size of the children, such large discrepancies must also be due to inconsistencies in the calibration studies that produce them. Calibration studies have been performed in different settings (laboratory or free-living), have used different activities, and have focused on children of different ages. Both NHANES and EYHS adopt relatively low thresholds for MVPA (906-2020 cpm), whereas TAAG and ALSPAC use 3000 to 3600 cpm. All studies are comparable in terms of total physical activity data, as the counts per minute variable is not dependent on any threshold. Despite these inconsistencies, the data as a whole do nevertheless constitute the most accurate and representative available data on how active or inactive contemporary children are.

Table 10.1 shows that children's total physical activity (counts per minute) is broadly consistent across the three studies reporting this variable, demonstrating marked reductions in physical activity levels with increasing age and consistent gender differences, with girls around 20% less active than boys. These differences are mirrored in the MVPA data, however defined. The data are broadly consistent across countries, suggesting that in these Western industrialized countries, children are generally consistent in their physical activity behavior. It is not possible to compare these data with data from self-report studies, as the two methods measure different dimensions of physical activity and, importantly, do so with large error differentials. In effect, data produced by self-report and objective methods refer to different physical activity constructs and do so with considerably different levels of precision.

From these data we can now gain greater insight regarding whether children are "active enough." The crucial question is whether these levels of physical activity are low enough to be contributing to—or driving—the childhood obesity epidemic. This issue is dealt with in more detail in a later chapter, but two further analyses from the studies reported in this review are of particular importance as they provide crucial information on whether children are sufficiently active to achieve optimal health. First, a cross-sectional analysis of the EYHS data showed that 116 min per day and 88 min per day of MVPA (defined as >2000 cpm) were necessary in 9- and 15-year-olds, respectively, to avoid increased risk of a cluster of cardiovascular disease risk factors, including obesity (8). Secondly, in a cross-sectional analysis of 11-year-old children in the ALSPAC study, 35 min per day (boys) and 23 min per day (girls) of MVPA (defined as >3600 cpm) were insufficient to avoid increased risk of obesity compared to that in children who participated in MVPA 55 min per day (boys) and 37 min per day (girls) (3). Given that these studies applied different criteria for the MVPA threshold, the data are broadly in agreement. It appears, therefore, that only children currently in the top quintile of MVPA, however defined, are

Table 10.1

Physical Activity Levels Measured by Accelerometer of Boys and Girls in Four Recent Studies

Study	Location	Participants	Age or grade	Total physical activity (cpm)		MVPA (min/day)		Def. of MVPA (METs)	MVPA cut point (cpm)	Reference
				Boys	Girls	Boys	Girls			
NHANES*	United States, nationally representative	1778 boys and girls	6-11 years	647 (21)	568 (12)	95 (5)	75 (2)	4	1400-2020 (age dependent, 6-18 years)	(4)
			12-15 years	521 (24)	328 (14)	45 (3)	25 (2)			
			16-19 years	429 (11)	328 (12)	33 (2)	20 (2)			
EYHS**	Denmark, Portugal (Madeira), Estonia, Norway	2185 boys and girls	9 years	784 (282)	649 (204)	192 (66)	160 (54)	3	9 years, 906	(5)
			15 years	615 (228)	491 (163)	99 (45)	73 (32)		15 years, 1706	
ALSPAC***	Southwest England	5595 boys and girls	11 years	644 (528-772)	529 (444-638)	25 (16-38)	16 (10-25)	4	3600	(6)
TAAG****	36 schools across United States	**3x repeated cross-sectional samples:**						4.6	3000	(7, 9)
		1603	6th grade				23.7			
		3085	8th grade				22.4			
		3378	8th grade				20.8			

*National Health and Nutrition Examination. Data are means (SEM).

**European Youth Heart Study. Data are means (SD).

***Avon Longitudinal Study of Parents and Children. Data are medians (IQR). MVPA = moderate-to-vigorous physical activity.

****Trial of Activity in Adolescent Girls. Baseline data from intervention study (6th grade) and two independent sets of control group data from 8th grade. Data are means.

likely to avoid obesity and other elevated cardiovascular disease risk factors. In other words, up to 80% of today's children may be engaging in insufficient MVPA and are at increased risk of chronic disease.

Future Directions

We are still in the early stages of understanding children's physical activity behavior using this more advanced technology. One avenue of future investigation is to ascertain whether particular patterns of physical activity are particularly obesogenic. For example, children with the same physical activity level might accrue their physical activity in very different combinations of intensity, duration, frequency, and mode. A second important need is for longitudinal data to confirm the cross-sectional data currently reported from these studies. There is also developing interest in the concept of "sedentariness." It is now increasingly accepted that sedentary behavior is not just "low levels of physical activity"—rather, it is a different set of behaviors (television or computer use, reading, homework, etc.) that in themselves constitute a potential risk to health irrespective of physical activity level. Sedentary behavior may have a different set of determinants and even a different physiology from physical activity. It is quite possible for a child to have high levels of *both* physical activity and sedentariness. At this point, we have only limited objective data on sedentary behavior, although accelerometry holds possibilities for more accurately quantifying time spent in these pursuits. However, at this point the data are too limited to review.

Conclusions

Obesity levels in children are increasing at an alarming rate. Increasing numbers of children are obviously exceeding the point at which energy intake chronically exceeds energy expenditure, resulting in fat accumulation. It is impossible to state with certainty that reduced physical activity levels are the major driver of these trends, as we have no trend data for physical activity over the 30 years during which obesity levels have been rising. It is equally difficult to ascertain the role of dietary changes over this period, again because of a lack of accurate trend data on children's diet. What is clear is that increasing numbers of children are now in a situation in which over time they fail to burn off the calories they eat. The evidence from these four large, representative studies demonstrates with some clarity that large numbers of children are insufficiently active to achieve optimal health, including maintenance of a healthy weight.

11

Global Prevalence of Adult Obesity

W. Philip T. James, CBE, MD, DSc

The cutoff point for specifying obesity has been almost universally accepted as a body mass index (BMI) of ≥30 following a series of World Health Organization (WHO) meetings, although in Japan, obesity is specified with BMIs ≥25 and as "obesity disease" when, in addition, there is an increase in visceral fat of ≥100 cm^2. In China, the criterion is a BMI ≥28, but most countries have now accepted the WHO specification despite the recognition finally agreed upon in a WHO meeting that the susceptibility of adults to the impact of weight gain varies substantially in Asians (1). The reason for these different definitions varies, but the WHO cutoff point arose from a pre-World War II decision by U.S. insurance companies to classify subjects as obese when they were 20% or more above the upper limit of a normal weight for height, this limit being originally based on the estimated level at which mortality rates began to rise. This upper normal limit was later crudely rounded to a BMI of 25, which then meant that a value of 30 was taken for obesity.

When we consider prevalences, it is useful to remember that obesity rates may show surprisingly rapid increases because the distribution of adult BMIs, although approximately Gaussian when a population's average BMI is 20 to 21 (the optimum average for minimum detrimental health effects), soon becomes skewed as the average BMI increases, with the more extreme BMIs going up proportionally faster. This is analogous to the skewing of blood pressure or blood cholesterol and in part explains why in the United States there are so many extraordinarily obese, physically impaired adults when the average BMI of the population does not seem much greater than elsewhere. When considered on a population or group basis, the relationship between the average BMI of a group or population and the prevalence of obesity is so consistent that one can readily estimate the proportion of obese people from knowing the average BMI for the group (2).

The significance of a BMI of 30 may vary not only between ethnic groups but also within a single population. Thus men involved in manual work or athletes involved in sports requiring muscular power may be very muscular, with BMIs above 30, but not be obese—their high BMI in practice signifying a work-induced increase in muscle mass and a comparatively high lean body mass and low fat mass. With aging also, for a given BMI there is a progressive increase in the proportion of body fat in most societies, which predominantly reflects the impact of declining physical activity with age in both sexes. Further, women tend to have a greater proportion of fat from the age of puberty than men.

Secular Increases in Obesity

The secular increase in adult obesity rates was not well documented until recently because obesity was simply regarded as an individual manifestation of personal habits that medically might lead to problems. Furthermore, in public health terms, obesity was seen as a risk factor only for people with higher blood cholesterol levels and blood pressure. Historically, human obesity has existed for millennia, but a high population prevalence has become apparent only in the last three to four decades. The previously low prevalences reflect the poor conditions that most societies endured, the demands of physically taxing jobs, and the impact of wars and famines. This in turn led to a view of obesity as a sign of success in postwar German men; currently, obese men and women in West Africa are considered both successful and attractive.

However, by the early 1980s, obesity was being recognized as a major public health problem in the United States and the United Kingdom. Since then there have been only a few exceptions to the progressive increase in national obesity rates (see figure 11.1). Cuba saw a major reduction in obesity rates when, following the collapse of the Soviet Union, the substantial annual financial support for Cuba suddenly stopped and food was in very short supply. In Japanese women <50 years old, there has also been a marked, even, and progressive secular decline in the whole distribution of BMIs, which implies a major societal change in their circumstances (3). This might well reflect the fact that younger women have been emancipated and permitted to go out to work rather than being confined to the home as was traditionally the case, thereby allowing them far greater spontaneous physical activity. In Finland, despite the marked industrial changes, with a substantial reduction in those physically working on farms, obesity rates have stopped increasing; this probably reflects both the fall in the average fat content of the national diet together with the substantial public health emphasis now being given to leisure-time activity.

Figure 11.1 illustrates the increasing prevalences of obesity in different parts of the world. Data of this type are difficult to obtain because only in the last decade have national governments realized that they should be undertaking national surveys on a representative basis so that their public health problem can be assessed. Data are often available from areas adjacent to research centers or from selected regions in a country, but these supposed national figures often fail to reflect the sometimes marked difference in obesity prevalence between urban and rural communities. Data emerging from lower-income countries, for example India and China, reveal average increases in BMI of five units, with the proportion of people who have BMIs ≥25 going up fourfold and, depending on the average BMIs, corresponding marked increases in obesity. These weight increases are paralleled by dramatic rises in type 2 diabetes and hypertension rates, affecting 20% to 30% of the adult urban population. This sudden surge in body weight is linked to a reduction in employment-related physical work, which can drop to a third of that demanded in the countryside, and a fivefold decrease in those who are classified as sedentary. There is also evidence that once the men become sufficiently affluent to drive their own car to work, as in China, this change is associated with a further increase in obesity rates (4).

On a global basis, the increase in BMI is progressive as average incomes rise, reflecting the change in work patterns as an increasing proportion of subsistence farmers transfer to lighter but better-paid jobs in the towns. On a global basis it can be

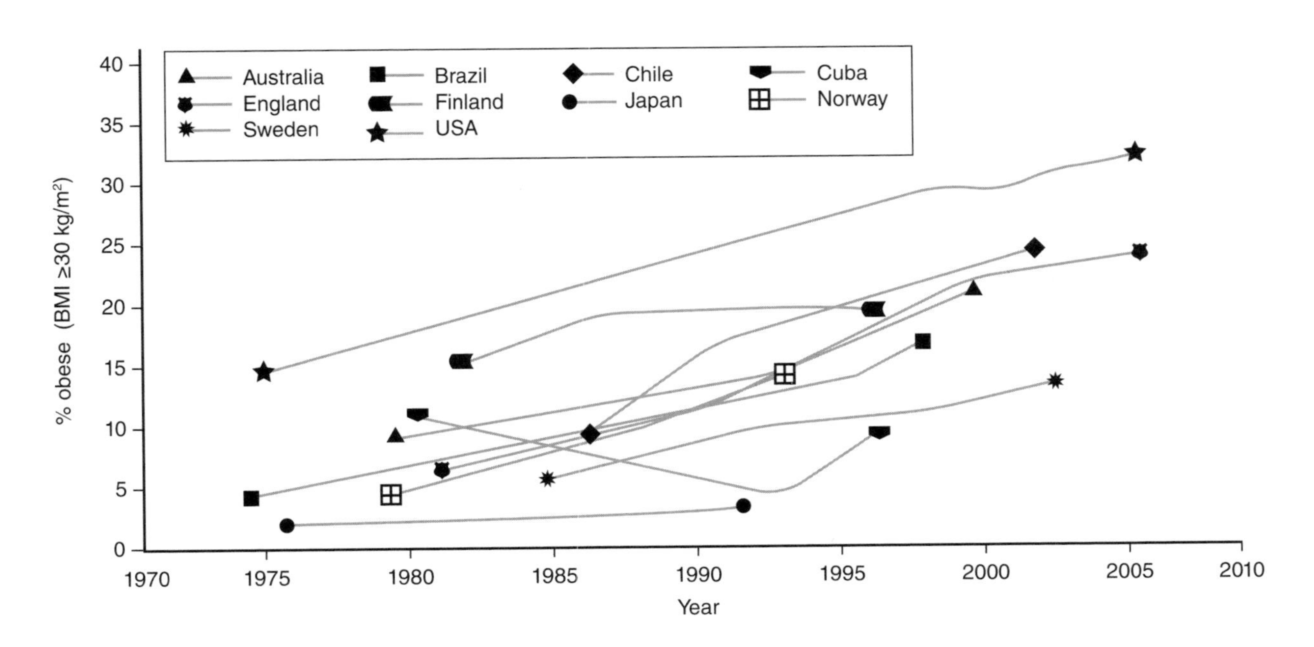

Figure 11.1 Measured nationally representative (unless otherwise specified) sequential surveys showing the prevalences of obesity in various countries.

Data from W.P. James, 2008, "The epidemiology of obesity: the size of the problem," *Journal of Internal Medicine* Apr. 263(4): 336-52.

shown that national average BMIs rise progressively until annual income is about $10,000 per capita per annum, and in parallel with these increases in BMI are increases in the average blood cholesterol concentrations (5).

Overall Global Prevalences

Currently the International Obesity Task Force estimates that there are >520 million obese adults in the world; an additional 1.0 billion are overweight with use of the classic WHO criteria, the overweight then being specified as having BMIs of 25 to 29.9. China, however, considers ≤24 the upper limit for a "normal" or "acceptable" BMI for adults in the 1.7 billion Chinese population; in contrast, a BMI of ≤23 is considered the limit of normality for the adults in the nearly 2 billion Indian, Japanese, Indonesian, and Thai populations. In China, ≥28 is taken as the obesity cutoff; so, in a 2002 national survey, instead of 2.9% as the percentage of the adult population considered obese (using a BMI ≥ 30), 7.1% was the Chinese prevalence figure for BMIs of ≥28 (6). Given the increasing evidence that the majority of the world's populations are likely more susceptible to the comorbid complications of excess weight gain than Northern Americans or Northern Europeans, it is clear that probably well over 2 billion adults are at risk from excess weight gain.

Regional Differences

Figure 11.2 shows the overall prevalences of obesity in men and women separately in different regions of the world, expressed in an age-matched mode to allow for demographic differences in population structures. By standardizing the data to conform with an international United Nations population structure, one makes allowances for the fact that high birth rate populations, for example in the Middle East and Africa, have a greater proportion of children. As these individuals grow into adult life, the overall adult population will escalate markedly; thus the absolute numbers of overweight and obese people in the world will also increase very substantially as the younger overweight and obese children enter adult life in these regions. The data in figure 11.2, however, give comparable age-adjusted data for the whole world rather than absolute numbers.

It is evident that there is a progressive increase in adult obesity, but additional sex differences as well as marked regional variations are also seen. Thus the Middle East, on a regional basis, has the highest prevalence of female obesity. This in part is explained by the women's multiple pregnancies, but the effects are dominated by dietary changes and marked physical inactivity. There has been a substantial change in these populations' diets, but this might be expected to have had a comparable effect on the men, a group in which obesity rates are appreciably lower. One of the overriding features of the life pattern of the women in this region is, however, the remarkable constraint on their ability to engage in physical activity. Thus in some regions (e.g., Saudi Arabia), women are unable to move outside their home unless accompanied by their husband or a male relative; they are unable to drive anywhere on their own to engage in exercise and usually conform to the societal demand to stay within the confines of the home. Thus many women say that their only opportunity for activity is to deliberately walk up and down stairs at home in an attempt to follow medical advice to undertake more physical activity.

Asia has the lowest prevalence of obesity and overweight; Africa has slightly higher rates. Latin America, which includes Mexico and the Caribbean and has obesity rates almost comparable to those in the United States, has prevalence figures nearly as high as those for the women in the European region, which include the more obese Southern and Eastern European women.

Sex Differences

In almost all countries, a greater proportion of women than men are obese, whereas usually there is a greater proportion of men who are overweight than women. These differences in sex-related prevalences for overweight and obesity are evident in all the regions included in figure 11.2, with women consistently having a higher prevalence of obesity than men but with men having higher prevalences of overweight. Small dietary differences between men and women have repeatedly been described, but all the traditional data on physical activity patterns suggest that in most countries men tend to be more physically active (7). In addition, men, when in positive energy balance, in general deposit more of their excess energy as protein-rich and metabolically active lean tissue, which then limits the further accumulation of excess weight; women, on the other hand, deposit proportionally less excess energy as lean tissue and therefore increase their overall weight to compensate for the same degree of energy imbalance. This may in part explain the earlier development of overweight and obesity in middle-aged women as countries come out of

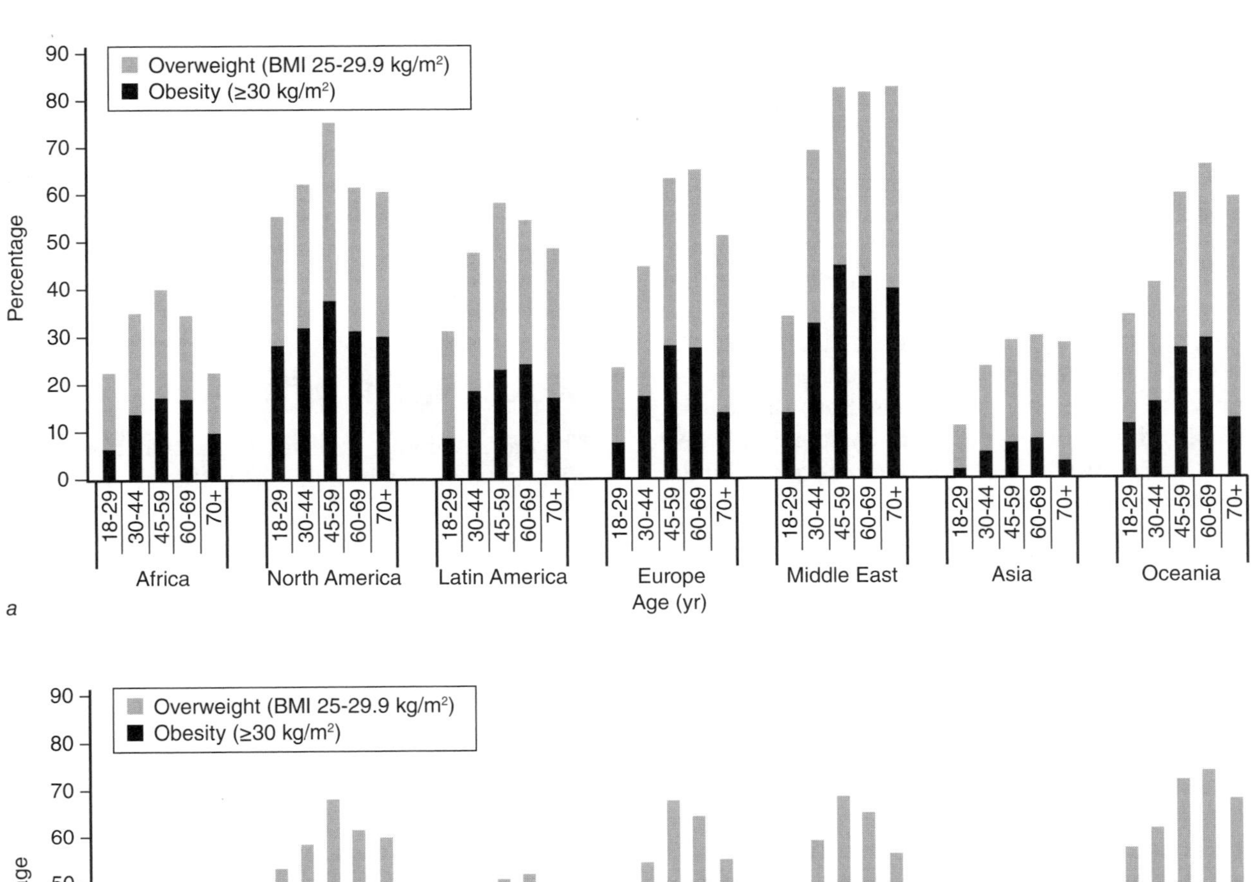

Figure 11.2 The regional prevalences of overweight and obesity by age in *(a)* adult women and *(b)* adult men, taking account of population size differences within each region and using standard criteria of overweight as BMIs of 25 to 29.9 and obesity as BMIs of 30+.

poverty. Subsequently the men begin to increase weight, and children's obesity rates increase later.

Age, Period, and Cohort Effects

In each region it is clear that both women and men are showing a progressive increase in overweight and obesity prevalences starting in early adult life and continuing until about the age of 60 years. Only in the 70+-year group is there a suggestion of a fall in the prevalence rates. In view of the data presented in figure 11.1, it is clear that the higher overweight and obesity rates in older adults evident regionally in figure 11.2 are not a cohort effect whereby older adults throughout their lives have been intrinsically heavier than the younger cohorts. In fact, the opposite is true, in that the new cohorts of young adults are starting their adult lives heavier than their predecessors; thus both a cohort and a secular effect apply, and one can expect an acceleration of the upward trend for each age category in the years to come. Detailed analyses from three sequential

Australian surveys (8) used a new mathematical age-period-cohort Poisson regression model to dissect out the age-related, secular (period), and finally birth cohort effects on the prevalence rates of obese and overall overweight in Australia between 1990 and 2000. Across the life span, the prevalence of obesity increased for each cohort of adults born after 1926: With each successively born cohort there was an increase in the prevalence of obesity at any given age. Thus the rate of obesity for males and females born in 1976 through 1980, seen in the 1990 to 2000 surveys, was double that of those born 10 years earlier; for the middle-aged, the percentage increase in rate of obesity was less. Age demonstrated a highly significant effect, with both the period, or secular, and cohort effects also having an evident impact on the prevalence of overall overweight (i.e., BMI ≥25) in the total population. When stratified by sex, the age and period effects were statistically significant in men as in women, but the observed birth cohort effect did not reach statistical significance in the men.

What this means is that the recognized progressive fall in physical activity during adult life can explain much of the observed increase seen in figure 11.2; but in addition, secular changes lead to period effects that probably involve progressive changes in both diet and physical inactivity. The birth cohort effect in part probably relates to increasing birth weights with increasingly heavier women entering pregnancy. These fetal programming and epigenetic effects seem exceptionally important in poorer societies and may explain the remarkable susceptibility to comorbidities as adults in these societies, born 20 to 70 years ago in difficult circumstances, now begin to put on weight (9).

Socioeconomic Differences

The increase in prevalence of obesity in poor countries is particularly evident in the more affluent sections of the community; but in Western societies, and increasingly as the average incomes of other countries rise, there is a reversal in the socioeconomic gradient. Thus, in most affluent societies, obesity rates progressively increase as the relative purchasing power of the population declines. The higher obesity rates in the poorer sections of most middle- and high-income countries partly relate to the relative cost of low-energy-dense, nutrient-rich foods, but the differing opportunities to engage in leisure-time activity may also contribute to these socioeconomic differences.

12

The Economic Cost of Obesity

Peter T. Katzmarzyk, PhD

Obese individuals have an elevated risk of developing several chronic conditions as well as dying prematurely. Given the high and increasing prevalence of obesity in many countries, this suggests that obesity is taking a mounting toll on society. There are several ways to estimate the societal burden of obesity. Estimates of the number of cases of chronic diseases attributable to obesity, the number of obesity-related deaths, lost work productivity attributable to obesity, and the economic costs to the health care system have all been used to document the impact of obesity on society. The focus of this chapter is on the economic burden associated with obesity. Studies on the economic burden of illness are important because they quantify the burden of a risk factor in terms that the general public, the media, and elected officials can readily understand (national currency) and thus have the potential to inform and influence public health policy.

Obesity has the potential to affect the economy in two main ways. First, the treatment of obesity and obesity-attributable morbidities imposes direct medical costs on the health care system. The direct medical costs associated with obesity have been estimated for a number of countries and regions. Second, the indirect costs of obesity to society can be estimated through the loss of productivity due to such things as work absenteeism, disability, and premature mortality. The indirect costs of obesity are more difficult to estimate than the direct medical costs; however, several studies have addressed this issue as well.

Methods to Estimate Costs

Two main approaches have been used to study the economic burden of obesity. The first approach is prevalence based and relies on the prevalence of obesity, as well as the association between obesity and treatable chronic diseases, to estimate obesity-attributable medical costs. The second approach is to link individual-level data on body weight status with directly measured health care expenditures or work performance.

Prevalence-Based Methods

Prevalence-based studies generally estimate the economic costs of obesity by using population-attributable risk percentage (PAR%) methodology. The PAR% is computed from the population prevalence (P) of obesity and the relative risk (RR) of a given disease associated with obesity: $PAR\% = [P(RR - 1)] / [1 + P(RR - 1)]$. For example, if the relative risk of postmenopausal breast cancer associated with obesity is 1.47 and the prevalence of obesity in Canada is 14.7%, the PAR% would be $[0.147(1.47 - 1)] / [1 + 0.147(1.47 - 1)] = 6.5\%$. In other words, 6.5% of postmenopausal breast cancer cases in Canada are considered to be directly attributable to obesity (1). With this information, the PAR% can then be applied to the total medical expenditures for postmenopausal breast cancer to arrive at the obesity-attributable costs. When this methodology is applied to all obesity-related diseases and the estimates are summed, the total health care costs attributable to obesity can be estimated for a country or a region.

Several assumptions and limitations need to be considered with use of a prevalence-based approach to estimating the economic costs attributable to obesity. First, it must be kept in mind that the PAR% is a *theoretical* estimate of the attributable risk associated with a risk factor at the population level. It is assumed that if the prevalence of obesity is reduced through intervention efforts, then the burden associated with it will also decrease. Second, the assumption is made that the RR associated with obesity in the population of interest is the same as that from the sample used to derive the RR estimate. However, this may not be the case. Relative risk estimates are often obtained from single large studies in the published literature or from meta-analyses of existing studies; they are rarely obtained from the same population that was used to estimate the prevalence of obesity. Thus, there may be differences in the classification and measurement of obesity between the two aspects of the study. Another assumption is that if a given percentage of a disease is attributable to obesity, then that same percentage of the economic cost of that disease is also attributable to obesity. This may not always be the case, as access to and usage of health care services may not be uniform in obese versus nonobese people. Finally, most studiesof attributable risk dichotomize body weight status into obese and nonobese groups rather than estimating the risk across a continuum of body mass index (BMI). The precision of the cost estimates is likely reduced in these cases, and the overall costs associated with unhealthy body weights may be underrepresented.

Individual-Level Methods

In contrast to prevalence-based methodologies in which data are aggregated at the level of the population, another approach is to link an individual's directly measured health care utilization, expenditures, or work performance to his or her body weight status. This allows for a more direct association between obesity, health care utilization, and the expenditure of health care dollars. The procedure in these studies is normally to use BMI data from a survey and to link these data with health care data via a medical insurance number or other identifying code.

In addition to measuring health care expenditures, individual-level methods are also used to estimate differences in work productivity and absenteeism associated with obesity. Height and weight data can be collected from employees, and this information is linked to employment records of attendance, disability, and performance.

Direct Costs of Obesity

Within the context of the economic burden of obesity, direct costs generally refer to medical expenses directly related to the treatment of obesity and its comorbidities. It is difficult to compare direct costs of obesity across countries because of cross-national differences in public and private health care plans, standards of care, modes of treatment, and so on. However, standardizing the costs as a percentage of total health care dollars spent allows for at least some standard metric for comparison.

Table 12.1 provides an overview of the results from some recent studies of the economic cost of obesity from different regions of the world. The selected studies were conducted in Australia (2), Canada (1), the United Kingdom (3), the United States (4), and China (5). The threshold used to define obesity in these studies was a BMI ≥30 kg/m^2, with the exception of the China (28 kg/m^2) and United States (29 kg/m^2) studies. In order to facilitate comparisons, the currencies have been converted to U.S. dollars using exchange rates for the year the study was conducted, and the results have also been expressed as a percentage of total health care expenditures for the country where available. Overall, the expenditures range from 1.4% to 5.7% of direct medical expenditures across the countries. China expends the least amount on obesity, and the United States spends the greatest amount as a proportion of total health care dollars.

Although the focus of this chapter is on the economic costs of obesity per se, there are health risks and economic consequences for overweight individuals as well. With a focus on the economic burden of obesity only, the total economic burden associated with excess body weight is likely underestimated. Given the high prevalence of overweight, the economic consequences for this group may be substantial and may be even greater than those associated with obesity. For example, the study from mainland China estimated that the economic cost of obesity (BMI ≥28 kg/m^2) was 8.07 billion yuan, whereas the direct economic cost attributable to overweight (BMI 24 to 27.9 kg/m^2) in the same study was 13.0 billion yuan (5).

In addition to findings from the population-level prevalence-based studies, there is some evidence from studies using individual-level data that obese individuals incur greater direct medical expenses than those who are not obese. For example, an analysis of obese and nonobese Kaiser Permanente of Colorado members determined that medical resources utilization was significantly higher in obese versus nonobese patients (figure 12.1) (6). Further, direct medical costs were higher in obese ($585.44) versus nonobese ($333.24) health plan members, and the costs increased by 2.3% for each

Table 12.1

Direct Medical Costs of Obesity in Various Countries

Country	Year	Obesity classification	Direct cost components	Direct costs**, reference
Australia	2005	BMI ≥30 kg/m^2	Hospital admissions, hospital days, medical consultations, drugs, allied health practitioners	1.08 billion AUS, Yates and Murphy (2) ($845 million USD), $554 per obese adult
Canada	2001	BMI ≥30 kg/m^2	Hospital care, drugs, physician care, other medical institutions, other health professionals	1.6 billion CAN, Katzmarzyk and Janssen (1) ($1.07 billion USD), 1.5% of total direct health expenditures
United Kingdom	2002	BMI ≥30 kg/m^2	Physician consultations, admissions, day cases, outpatient attendances, drugs for treating obesity and its consequences	990-1225 million GBP, House of Commons Health Committee (3) ($1431-1770 million USD), 2.3-2.6% of National Health Service expenditure
United States	1995	BMI ≥29 kg/m^2	Physician and nursing care, hospital care, drugs	$51.6 billion, Wolf and Colditz (4), 5.7% of direct national health expenditures
China	2003	BMI ≥28 kg/m^2	Direct costs of clinic visits and in-patient stays for treatment of obesity-related hypertension, diabetes, coronary heart disease, and stroke	8.1 billion yuan, Zhao et al. (5) ($972 million USD), 1.4% of total national medical expenditures

**Currencies have all been converted to USD for comparison (standardized to the year of the study).

increase in BMI unit (6). Similarly, among 27,977 employees from several large U.S. companies, total annual health care costs were highest among the obese ($2686) versus overweight ($2384) and normal-weight ($2198) employees (7). Self-reported height and weight data collected during the 1995-1996 National Population Health Survey in Canada have also been linked with the medical insurance registry in Ontario to obtain the expenditures for physicians' services (for the year prior to the survey) associated with various levels of BMI (8). The results indicate that annual per capita physician costs increase by $8.90 for each unit increase in BMI. These data support the notion that obese individuals incur higher overall medical expenses than people who maintain a normal body weight.

Indirect Costs of Obesity

With some assumptions, the direct medical costs of obesity can be estimated fairly well, as described in the preceding section. Estimating the indirect costs involves more assumptions regarding the costs associated with lost productivity. For example, the economic losses associated with absenteeism, disability, or premature mortality must be assigned a dollar value. Studies of indirect costs and lost productivity due to obesity may be more relevant to employers, particularly in countries or states where universal health care plans take care of the majority of the medical expenses.

A recent review has summarized the evidence on the indirect costs of obesity (9). The authors reviewed 31 relevant studies published between 1992 and 2007 and identified five classes of indirect costs. These included absenteeism, disability, premature mortality, presenteeism (loss of productivity from workers who are present but cannot perform all or some of their assigned duties), and workers' compensation (9). There is fairly good evidence that obese individuals have higher levels of absenteeism due to sickness, injury, or disability, but the results for presenteeism and workers' compensation are mixed (9). Although the available evidence suggests

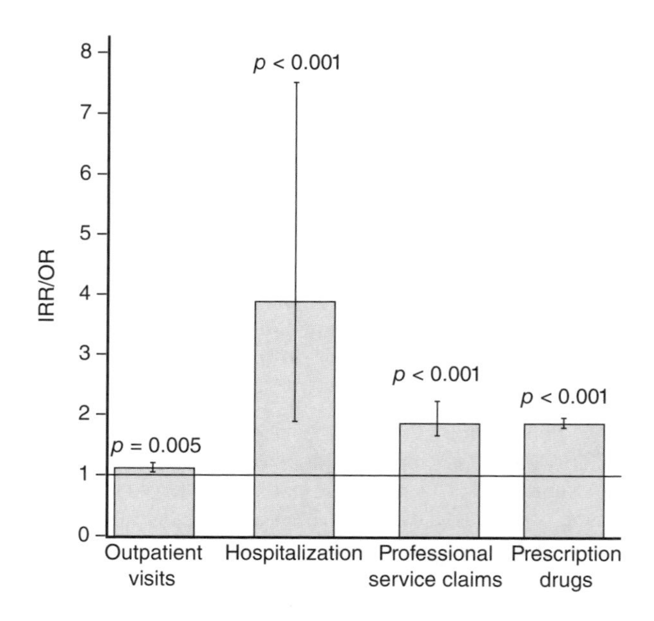

Figure 12.1 Risk of excess medical resources utilization in obese versus nonobese patients enrolled in the Kaiser Permanente of Colorado medical plan (n = 539 obese; 1225 nonobese). IRR: incidence rate ratio for outpatient visits, professional service claims, and prescription drugs; OR: odds ratio for hospitalization. Error bars represent 95% confidence intervals.

Data from M.A. Raebel et al., 2004, "Health services use and health care costs of obese and nonobese individuals," *Archives of Internal Medicine* 164: 2135-2140.

that obesity incurs high indirect costs through several pathways, there is a need for more research to further delineate the total spectrum of indirect costs of obesity to society.

Lifetime Health Care Costs

Most papers concerned with the economic burden of obesity take a snapshot approach by using data at a single point in time. However, there is also interest in determining the lifetime accrual of expenditures in obese versus nonobese individuals. Given that obese individuals have higher overall death rates compared to normal-weight individuals, they are less likely to live to older ages when health care costs increase dramatically. Allison and colleagues (10) modeled the economic costs of obesity while taking into account the higher mortality rates among obese individuals, and the results indicated that health care costs associated with obesity were approximately 25% lower when the differential in mortality was taken into account. A more recent study from the Netherlands also modeled the lifetime medical costs in obese and nonobese adults. During their lifetime, obese individuals had higher medical expenditures; however, due to their lower life expectancy, the overall lifetime medical costs were lower than among the nonobese (11). The fact that the nonobese live longer on average allows for the accumulation of more medical expenditures for diseases unrelated to obesity, which are higher in the later ages of life. This is an interesting perspective on direct medical expenditures; however, these models do not account for the indirect costs of obesity, such as lost work productivity due to obesity-related illnesses, premature mortality, or quality of life issues among the obese population.

Summary

The increasing prevalence of obesity in many countries is taking a mounting toll on public health expenditures. Studies of the direct medical costs associated with obesity suggest that between 1.4% and 5.7% of national health expenditures in several countries is attributable to obesity. The indirect costs are also substantial; however, more research is required to achieve a better understanding of the spectrum of indirect cost components affected by obesity.

13

Global Prevalence of Childhood Obesity

Tim Lobstein, PhD

Although several different methods and approaches are used to measure obesity in children, all the available data have one thing in common: They show a substantial and rapid increase in the numbers of children affected, in most regions of the world. In more developed economies, child obesity prevalence levels have doubled in the last two decades (1) and have risen particularly strongly among lower-income households and minority ethnic groups. In less developed economies, child obesity levels are also rising, especially in urban areas and among better-off households (2).

Definitions of Child Overweight and Obesity

For young children, it is common practice to use "weight-for-height" rather than body mass index (BMI) to indicate nutritional status. This practice is based on existing definitions used to assess underweight and stunting, whereby a child's weight-for-age, height-for-age, and weight-for-height are compared with standard growth curves taken from a reference population.

In recent years, BMI has been increasingly accepted as a valid indirect measure of adiposity in older children and adolescents for survey purposes (1, 3), leading to various approaches to selecting appropriate BMI cutoff values to take account of the fluctuations in BMI during normal growth. A number of different BMI-for-age reference charts have been developed, such as those used by the U.S. National Center for Health Statistics and those developed by other national authorities. These provide a set of cutoffs to define overweight and obesity among children of each gender at every age.

It should be noted that the definitions of overweight and obesity are good for making comparisons between different population groups and comparing a population over time; but for the clinical assessment of individual children, more careful examination of the child is needed to monitor individual growth trends and to ensure that, for example, a high BMI is not due to extra muscle mass or to stunted linear growth.

Confusion over the use of different reference curves led to the establishment of an expert panel, convened by the International Obesity Task Force (IOTF), which proposed a set of BMI cutoffs based on pooled data collected from Brazil, Britain, Hong Kong, Singapore, the Netherlands, and the United States. The panel agreed that overweight and obesity would be defined in children according to the BMI centile curves that passed through the adult cutoff points of BMI 25 and 30 at age 18. The resulting set of age- and gender-specific BMI cutoff points for children was published in 2000 (4).

Although the World Health Organization (WHO) previously recommended using a set of cutoffs based on a reference population derived from the United States, it has recently reviewed its recommendations. The U.S. data had included large numbers of formula-fed infants with growth patterns that differed from those of breast-fed infants, and this was likely to result in underestimation of the true extent of overweight in younger children. The World Health Organization has now published a new set of standard growth charts based on data from healthy breast-fed babies and infants aged 0 to 5 years and has extended these to provide a set of reference charts for children aged 5 to 19 years (5). WHO recommends that cutoffs for overweight and obesity for children aged 0-5 years should be set at

+2 s.d. and +3 s.d. respectively, while for children aged 15-19 years they should be set at +1 s.d. and +2 s.d. respectively.

As a result of these different approaches, one should take care when looking at published prevalence figures for overweight and obesity, and the prevalence levels based on one set of cutoffs or reference curves should not be compared directly with those based on another. Furthermore, the use of the cutoffs may differ; that is, some reports give the prevalence value for all "overweight" children including those who are obese, while others give the prevalence excluding those who are obese. Readers should also note that prevalence levels using reference curves from the United States sometimes refer to "at risk of overweight" and "overweight" for the top two tiers of adiposity, and sometimes to "overweight" and "obese."

In the present chapter, the prevalence levels are based on the IOTF international classification scheme, as most available survey evidence has been obtained with use of this approach and the results tend to be more conservative than with some other approaches (1).

Prevalence Levels in Childhood and Adolescence

Policy makers attempting to tackle the rise in child obesity are beset not only by the problem of comparing prevalence figures based on different definitions of child obesity but also by a dearth of comparable representative data from different countries. Even if figures are available, they need to be examined carefully. Firstly, did the survey collect weights and heights from real measurements or from a parent's report? Was the survey nationally representative or from a subpopulation, for example a subpopulation living in the more accessible urban areas? And, when two surveys are compared across a period of time, are they properly comparable in terms of the children's ages and in terms of their ethnic and sociodemographic mix, which may itself have changed over time?

The figures presented here are based on the latest and most reliable data available, some of which were previously published by Wang and Lobstein (2).

Worldwide Prevalence Levels

Estimates for the global prevalence of overweight and obesity among school-age children were made in 2004, when it was concluded that approximately 10% of school-age children (aged 5 to 17) were overweight, including some 2% to 3% who were obese. This global average covers a wide range of prevalence levels in different regions and countries, with the prevalence of overweight in Africa and Asia averaging well below 5% and in the Americas and Europe above 20%. Projections to the year 2010 are shown in table 13.1.

Americas

Comprehensive and comparable national representative data on trends in the prevalence of obesity are available for the United States, where surveys have been undertaken since the 1960s. Data for 2003–2004 show that 36% of children aged 6 to 17 years were overweight, including 13% who were obese. In Canada, 26% of younger children and 29% of older children were found to be overweight in a 2004 survey, almost exactly double the prevalence levels found among children 25 years earlier.

In Mexico, a survey of children aged 11 to 14 years in 1998–1999 showed 33% of both males and females to be overweight or obese, based on U.S. definitions. The levels were higher among better-off families and in urban areas. The prevalence of overweight among school-age children in Brazil was 14% in 1997 (compared with just 4% in 1974), and levels were higher in urban populations. In Chile in 2000, the prevalence of overweight among schoolchildren was 26%.

Table 13.1

Estimated Prevalence of Excess Body Weight in School-Age Children in 2010

Region*	Obese	Overweight (including obese)
Americas	15%	46%
Mideast and North Africa	12%	42%
Europe and former Soviet Union	10%	38%
West Pacific	7%	27%
South East Asia	5%	23%
Africa	<1%	<5%

*Countries in each region are according to the World Health Organization.

Data from Y. Wang and T. Lobstein, 2006, "Worldwide trends in childhood overweight and obesity," *International Journal of Pediatric Obesity* 1: 11-25.

Data available for schoolchildren in most other South and Central American countries are sparse, but some data have been collected for preschool children. In Bolivia, the prevalence of overweight (defined as one standard deviation above a reference mean) was 23% in 1997, and in the Dominican Republic it was 15% in 1996. In a few countries in the region, obesity prevalence has fallen: In Colombia it fell from 5% to 3% between 1986 and 1995.

Europe

The highest child obesity prevalence levels in Europe are found in several Southern European countries. A survey in 2001 showed that 36% of 9-year-olds in central Italy were overweight, including 12% obese. In Northern Greece in 2000, 26% of boys and 19% of girls aged 6 to 17 years were overweight or obese, while data from Crete in 2002 showed 44% of boys aged 15 to be overweight or obese. In Spain, 35% of boys and 32% of girls aged 13 to 14 years were overweight or obese in a survey in 2000.

Northern European countries tend to have lower prevalence levels. In Sweden in 2000-2001, the prevalence of overweight and obesity combined was 18% for children aged 10 years. In the Netherlands the figures are particularly low, with only 10% of children aged 5 to 17 overweight, including only 2% obese, according to a 1997 survey. In France the figures are a bit higher, at 15% overweight including 3% obese in a northern French survey in 2000, and in England higher still, with 29% overweight, including 10% obese, in a 2004 survey.

The reasons for a north–south gradient are not clear. Genetic factors are unlikely to be the explanation, as all countries in Western Europe have shown a marked increase in prevalence in recent decades. The child's household or family income may be a relevant variable, possibly mediated through income-related dietary factors such as maternal nutrition during pregnancy or breast- or bottle-feeding in infancy, as well as the quality of the diet during childhood.

In Eastern Europe, prevalence of child obesity tends to be lower, but the region is catching up with its western neighbors. Surveys in 2000 and 2001 show a prevalence of overweight and obesity combined of 13% of schoolchildren in the Czech Republic, 17% in Serbia, and 18% in Poland.

Africa, Eastern Mediterranean, and Middle East

Studies from the late 1970s through the 1990s demonstrate that several countries in this region appear to be showing high levels of childhood obesity. In Egypt, for example, the prevalence of overweight (based on standard deviation greater than one) was over 25% in preschool children. Among children aged 11 to 14 years, the prevalence of overweight including obesity was 14% for boys and 21% for girls (using U.S. cutoff definitions).

Similar figures are found in other parts of the region. Some 20% of adolescents aged 15 to 16 years in Saudi Arabia are overweight (based on BMI >120% reference median value). In the city of Riyadh, overweight among boys aged 6 to 14 years rose from 3% in 1988 to 25% in 2005. In Bahrain in 2002, 30% of boys and 42% of girls aged 12 to 17 were overweight, including over 15% obese among both genders.

Asia and Pacific

Based on surveys in the 1980s and 1990s, the prevalence of obesity among preschool children is around 1% or less in many countries in Asia and the Pacific Region, such as Bangladesh (1.1%), the Philippines (0.8%), Vietnam (0.7%), and Nepal (0.3%). In recent years, the prevalence of overweight has markedly increased in some areas, such as Ho Chi Minh City in Vietnam, where some 30% of preschoolers are overweight, although the figure remains low (some 3%) among older schoolchildren.

It should also be noted that no data are available for some countries in the region (e.g., the Pacific islands) where adult obesity prevalence rates are known to be high.

In more economically developed countries, the prevalence figures for preschool and school-age children are considerably higher. Among Australian children and adolescents aged 7 to 15 years, the prevalence of overweight (including obesity) doubled from 11% to 21% between 1985 and 1995, and was found to be 27% in a regional survey of 4- to 12-year-olds in 2003 to 2004.

In mainland China, whose population accounts for one-fifth of the global population, the prevalence of obesity has been rising quickly in both adults and children during the past two decades. A survey in 1997 showed the prevalence of overweight, including obesity, among schoolchildren to be 7.7%, but this represented an average of urban and rural populations; in urban areas the level was 12.4%. Perhaps of much greater concern was the finding that among preschool children the prevalence was 29%.

Sub-Saharan Africa

There are very few surveys from African countries that can provide prevalence figures for childhood obesity, as most public health nutrition programs have focused on undernutrition and food safety problems. The prevalence of childhood obesity remains very low in this region, although it appears to be rising in several countries. In South Africa, where obesity has become prevalent in adults, childhood obesity is also rising: The prevalence of overweight (including obesity) among young people aged 13 to 19 years is over 17%, with boys less commonly overweight (7%) than girls (25%). Prevalence was highest (over 20% for both boys and girls) in white and Indian population groups.

Secular Trends and Demographic Differences

As noted already, the last two decades have seen unprecedented increases in the prevalence of child obesity. North America and some countries in Europe have shown particularly high yearly increases in prevalence. Data from other regions (e.g., Latin America) indicate that similar rates of increase may be occurring in less developed countries.

In contrast, several countries are showing only modest increases. China has shown a small rise in the prevalence of overweight among rural children, but a more marked increase among urban children. The rapid rise in the prevalence of overweight is seen in most developed economies; but an interesting exception is Russia, where the economic downturn in the early 1990s may have been linked to a decline in the prevalence of overweight children seen during the period.

Ethnic and Racial Factors

In the United States, obesity prevalence levels for children differed between ethnic groups in surveys across the 1980s and 1990s. Hispanic and Afro-Caribbean youth have been more likely to be overweight than white youth: In 2000, some 12% of white children aged 6 to 11 years were obese (defined by U.S. cutoffs) while the figure rose to 20% among Afro-Caribbean children and 24% among Mexican Americans.

South African figures, as noted earlier, have shown strong differences by racial group. In the United Kingdom, Afro-Caribbean girls were more likely to be overweight than girls in the general population. Indian and Pakistani boys were more likely to be overweight than other ethnic groups. A survey of over 2000 adolescents aged 11 to 14 in East London reported high levels of overweight and obesity among all ethnic groups examined (white British, Indian, and Bangladeshi), with the highest levels among Indian males (36% overweight) and black African females (40% overweight). This was unlikely to be due to economic differences, as no association between BMI and measures of socioeconomic status was found, although the group as a whole was relatively deprived compared with the U.K. population (37% were from homes with no employed parent, 48% from homes eligible for social benefits such as free school meals).

Socioeconomic Factors

Examination of differences in the distribution of overweight and obesity among children from various social classes (defined by family income levels or educational levels of the main income earner) shows a complex pattern. In the more economically developed, industrialized countries, there is a general tendency for children in lower socioeconomic groups to show higher prevalence levels of overweight and obesity. In European countries, there is evidence that these differences are increasing.

In contrast, in countries that are not economically developed or are undergoing economic development, overweight and obesity levels tend to be highest among families with the highest incomes or educational attainment. These opposing patterns are further complicated by urban–rural differences, which themselves are likely to be confounded by income and educational factors. In Brazil in 1997, 20% of children in higher-income families were overweight or obese, compared with 13% of children in middle-income families and only 6% of children in lower-income families. In China there is a similar association between child overweight and family income level and educational level.

Conclusion

The best available evidence indicates that the prevalence of childhood obesity continues to increase in many areas of the globe. The collection of national-level data on childhood obesity remains a research priority. Efforts should be made to express prevalences of overweight and obesity using the IOTF criteria to enable comparisons across countries.

14

The Economic Cost of Physical Inactivity

Ian Janssen, PhD

This chapter presents information on the economic burden of physical inactivity. As will be discussed, several different methodological approaches have been employed to examine different types of costs attributable to physical inactivity. The specific purpose of most cost-of-illness studies in the physical activity literature has been to quantify, in economic terms, the impact of physical inactivity on society. A larger goal of these studies has been to highlight the extent of the economic burden of physical inactivity and the stress that high inactivity rates place on overburdened medical care systems. By quantifying the financial resources devoted to treating the diseases associated with physical inactivity, these studies have helped to convince policy makers of the need to provide more funding for prevention initiatives that focus on increasing physical activity levels in the population.

Direct Versus Indirect Costs

Before discussing the economic burden of physical inactivity, it is necessary to distinguish between the different types of costs that can be examined. First and foremost are *direct costs*. Direct costs can be defined as the values of goods and services for which payment was made and resources used in treatment, care, and rehabilitation related to illness or injury (1). Direct costs include, but are not limited to, expenditures for care in hospitals and other institutions, drug costs, salaries of health care workers, and health administration. Although the method of payment would differ, direct costs are incurred in both publicly funded (e.g., United Kingdom, Canada) and insurance-based (e.g., United States) medical care systems. For studies in the physical activity literature, direct health care costs reflect the money required to treat the diseases that are directly attributable to physical inactivity such as cardiovascular disease, diabetes, colon and breast cancer, type 2 diabetes, and osteoporosis.

Two American studies are highlighted here to illustrate the effect of physical inactivity on direct health care costs. Wang and colleagues examined a group of 23,490 young and middle-aged American employees who were enrolled in a medical insurance plan (2). After covariates were controlled for, direct health care costs were $2586 annually in inactive employees, a value that was $285 and $221 higher than in moderately active (one or two sessions a week) and very active (three or more sessions a week) employees, respectively. In a group of 42,520 Medicare retirees in the United States, direct medical care costs were $12,450 in sedentary individuals. This was $1690 higher than in moderately active (one to three sessions a week) retirees and $2686 higher than in very active (four or more sessions a week) retirees (3). Thus, independent of age, two large studies of American adults have shown that direct health care costs are about 15% higher in inactive individuals by comparison to active individuals. Interestingly, within these two studies, physical inactivity affected different components of direct health care costs, including inpatient costs, outpatient costs, and drug expenditures (2, 3). Other authors have noted that savings in direct costs for regularly active persons are observed within both genders, adults of all ages, smokers and nonsmokers, and those with and without disability (4).

Three studies have examined the effect of physical activity on health care spending according to adiposity status as measured by the body mass index (2, 3, 5). All three of these studies showed independent effects of physical activity and adiposity on direct health care costs. Thus, active obese persons had lower direct costs than inactive obese persons and higher direct costs than active lean persons. An example of this effect within a sample of older adults is shown in figure 14.1. These findings highlight the importance of physical activity in health care spending within obese persons, and are consistent with the observation that physically active obese persons have a lower morbidity and mortality risk than inactive obese persons (6). Another recent study was able to demonstrate that 26% of the $1.35 billion (U.S. dollars) in direct health care costs (inpatient + outpatient costs) attributable to physical inactivity in China in the year 2000 was accounted for by the beneficial effects of physical activity on adiposity status (7). Thus, while physical activity has effects on direct health care costs that are independent of obesity, a part of the effect is due to the negative relation between physical activity and obesity.

Studies on the economic burden of illness can also consider *indirect costs,* which are a reflection of the value society places on health and life (1). Indirect costs are not as straightforward as direct costs, as they ultimately rely on assignment of a monetary value to the worth of individuals to society. In most situations, indirect costs are calculated using the human capital approach, which considers the value of lost productivity as a result of disability and premature death using lost earnings as a surrogate. In some cases, indirect costs are calculated using the willingness-to-pay approach, which attempts to recognize pain and suffering and the psychosocial consequences of illness.

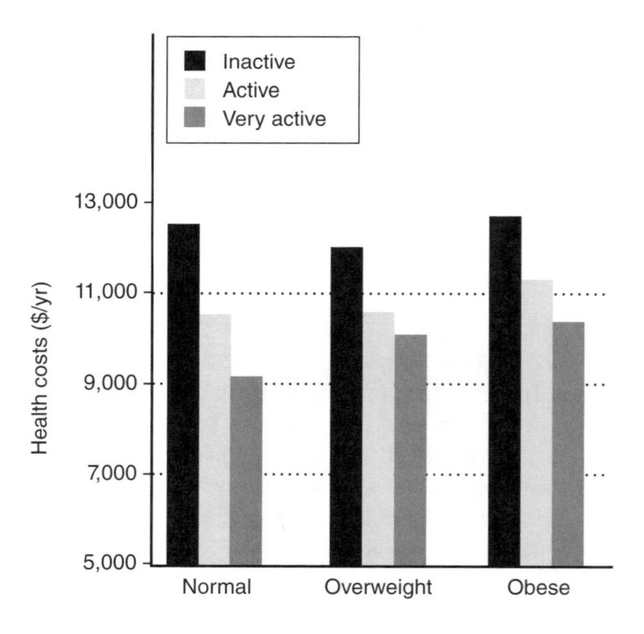

Figure 14.1 Direct health care costs (dollars per year) according to adiposity and physical activity in older adults.

Data from F. Wang et al., 2005, "BMI, physical activity, and health care utilization/costs among Medicare retirees," *Obesity Research* 13: 1450-1457.

In an examination of the adult population in Canada, Katzmarzyk and Janssen (8) reported that the indirect cost of physical inactivity, calculated using the human capital approach, was $3.7 billion per year. Interestingly, the indirect costs were far greater than the direct health care costs associated with physical inactivity ($3.7 vs. $1.6 billion per year). This is the only study that has examined the indirect costs of physical inactivity at the population level. Given the greater effect of physical activity on indirect costs observed by Katzmarzyk and Janssen (8), and the lack of consideration of indirect costs in other physical activity cost-of-illness analyses, it is possible that the economic impact of physical inactivity has been underestimated based on the available literature as a whole.

Population-Level Versus Individual-Level Costs

Most cost-of-illness studies employ a top-down approach whereby the health care costs are estimated at the *population level.* Typically, these estimates are generated by multiplying the population-attributable risk fraction (PAR%, percentage of disease caused by inactivity) to the health care costs associated with each of the diseases. The PAR% are calculated from the prevalence of physical inactivity in the country and the relative risk of the diseases associated with physical inactivity. These relative risks used in the PAR% calculations can be obtained either from meta-analyses or from risk estimates generated in prospective studies of representative samples of the population.

An example from Canada demonstrates how the top-down approach can be applied (8). Within Canada, the prevalence of physical inactivity in adults in the year 2001 based on national surveys was 53.5%. A meta-analysis of the existing literature on the relation between physical activity and coronary heart disease suggested that the relative risk of coronary heart disease in an inactive adult relative to an active adult is 1.45 (45% higher risk). Based on this information, the PAR% of coronary heart disease in Canada in 2001 was 19.4% (PAR% = [prevalence (relative risk – 1)] / [1 + (relative risk – 1)]). In other words, 19.4% of coronary heart diseases cases were directly attributable to physical inactivity. The Economic Burden of Illness report for Canada indicated that the direct health care cost of coronary heart disease in the country in 2001 was $2.43 billion. Thus, the direct health care cost of coronary heart disease in the year 2001 within Canada that was attributable to physical inactivity was estimated to be $471 million ($2.43 billion × 19.4%). When this was added to the costs incurred by the other major diseases associated with physical inactivity (hypertension, stroke, colon cancer, breast cancer, osteoporosis), the total direct costs of physical inactivity in Canada were estimated to be $1.62 billion, which represented 1.5% of the total direct health care costs in the country.

While the results from these top-down population-level studies are informative for both researchers and policy makers, it is essential to recognize that they are based on several assumptions and provide only theoretical estimates of the economic burden of physical inactivity at the *population* level. An alternative approach in cost-of-illness studies is to directly link physical activity data to health care expenditures in a large cohort, thereby providing concrete estimates of the medical costs of physical inactivity at the *individual* level.

Referring in detail to a recent study by Anderson and colleagues (5) will highlight the utility of the latter approach. The participants in this study were 4674 individuals aged 40 and older who were enrolled in a health plan. Data on direct health care costs, including expenses for health care professionals and hospital claims, were extracted for each participant from administrative health care claim data for a four-year period extending from 1996 to 1999. Subjects completed a survey regarding their physical activity and other lifestyle habits, anthropometry (height and weight), age, sex, and chronic disease status. Survey data were then linked to the administrative data, and regression techniques were used to predict health care spending according to physical activity level, adiposity status, and other characteristics. Average health care costs, adjusted for the other covariates, increased from $4240 in the most active subjects to $4966 in the moderately active subjects and to $5783 in the least active subjects. The corresponding values for normal-weight, overweight, and obese adiposity categories were $3994, $5239, and $6146, respectively. When considering the prevalences of inactivity and obesity, which were very high, the authors of this study calculated that 23.5% of the direct health costs within this cohort were attributable to physical inactivity, overweight, and obesity!

Current Costs Versus Lifetime Costs

Two approaches to establish a time frame are used in cost-of-illness studies: prevalence and incidence (1). *Prevalence-based* studies examine costs incurred during a given time period, regardless of the time of onset of the disease, or in the case of physical activity, regardless of the onset of physical inactivity. The time period examined is typically a year, although it can extend over multiple years (with costs usually presented in yearly averages). All of the studies discussed thus far in this chapter examined the costs of physical inactivity using a prevalence-based approach.

The effort in *incidence-based* studies is to estimate all of the future costs associated with the onset of illness in the baseline year. This approach is typically used for illness or diseases per se, rather than lifestyle behaviors such as physical inactivity that precede the diseases on which the costs are based. In fact, no studies on the economic burden of physical inactivity have employed an incidence-based approach. To complete such a study one would need to collect primary data on the incidence of physical inactivity, would need to know the likely course of disease development for all of the diseases associated with physical inactivity (e.g., time course for the development of stroke in a typical newly inactive person vs. a typical active person), and would need information on the likely course and duration of those diseases. Much of the information required for incidence-based studies can only be estimated. Thus the lifetime costs of physical inactivity could only be predicted, and the predictions would be based on several assumptions.

A common assumption by researchers and policy makers alike is that increasing physical activity at the population level would equate to lower health care spending. This assumption is based on evidence from *prevalence-based* cost-of-illness studies. However, to fully understand the economic burden of physical inactivity, one needs to consider the prolonged life associated with improving physical activity in addition to the elevated health care costs observed in inactive persons in a given year. The annual health care savings observed in a physically active individual may be offset, in part or whole, by an increase in long-term spending related to prolonged survival. An incidence-based approach would allow one to estimate the lifetime health care costs according to physical activity. Although studies are unavailable for physical activity, two studies in the area of obesity used an *incidence-based* methodology and so were able to address this issue. While one of these studies, conducted in the United States, showed that obesity is associated with higher lifetime direct costs (9), the second, conducted in the Netherlands, indicated that obesity is associated with lower lifetime direct costs (10).

Conclusions

As suggested in the preceding section, no studies worldwide have used an incidence-based approach to estimate the lifetime costs associated with physical inactivity. Thus, the effects of physical inactivity on health care spending over the lifetime of an individual, which may have important policy implications, are unknown. Furthermore, there is no published information on the direct or indirect economic costs attributable to physical inactivity in the pediatric age group. Since most of the health outcomes of physical inactivity in adults are consequent to chronic diseases of old age, childhood physical inactivity may not be associated with meaningful health care expenditures. Nonetheless, this warrants exploration. Given that the focus of this textbook is on physical inactivity and obesity, a fitting way to close this chapter is to highlight the observation that physical inactivity and obesity have independent and additive effects on health care spending in adults of all ages.

15

The Cost-Benefit Relationship of Physical Activity Interventions for Obesity

Larissa Roux, MD, PhD

Can physical activity (PA) promotion prevent and treat obesity and improve health outcomes? With so many competing priorities, is investment in PA promotion worthwhile? On the surface, the answers to these questions may seem obvious. We know that the questions are important: The prevalence of obesity has risen sharply over the past four decades (1), and cost-of-illness studies have shown that Americans spend $51.6 billion or 5.7% of their national health expenditure on the direct costs of obesity (2). It also makes sense that promotion of PA, our main means of energy expenditure, should be one of the main lines of attack in confronting the obesity epidemic. Furthermore, ecological and cross-sectional studies have suggested that greater participation in PA is associated with lower prevalence of obesity (3).

So, is the answer to these questions an unequivocal yes? Before answering, it is worthwhile to consider the implications of the answer. Turning the tide in the epidemics of chronic disease will require sustained and comprehensive efforts to modify deeply entrenched behaviors and lifestyles, and therefore great and indefinite investment of resources. In the case of obesity, the potential rewards of such investment are high; but in an environment of scarce resources and competing health priorities, the consequences of incorrect assumptions, incomplete knowledge, and support of inefficient interventions are great. For example, understanding what constitutes appropriate duration and intensity of PA to achieve certain goals, or determining the relative merits of different PA programs including outcomes and costs, could make the difference between effective and ineffective or disastrous policy. Besides wasting valuable resources, investment in an ineffective strategy could jeopardize future policy development in this area.

The most rational and comprehensive approach to developing strategy for an urgent public health issue under conditions of uncertainty and scarcity would carefully define the problem, acquire detailed information on the success of interventions in achieving outcomes of interest, and measure the costs and short- and long-term benefits of implementing the best interventions. In recent years, cost-effectiveness analysis (CEA) has emerged as a powerful tool that brings efficacy data from interventional studies and cost data to life in the development of informed health policy. In this chapter, it will be argued that CEA creates a framework for identifying and exploring gaps in our knowledge of obesity, and offers a strategy for making the best policy decisions in urgent and uncertain times.

Cost-Effectiveness Analyses in Obesity

The effectiveness of PA promotion in the prevention or treatment of obesity can be measured by its impact on relevant outcomes such as change in body mass index, reduction of comorbidities, or improvement in survival. The Panel on Cost-Effectiveness in Health and Medicine has adopted quality-adjusted life-years as a standard measure of the effectiveness of an intervention. This measure accounts for impact on both quantity and quality of life by summing the products of time spent in specific health states and individuals' preferences for those health states (4). (Preferences are measured using utilities; CEAs that use this metric for quality of life are more specifically referred to as cost utility analyses.) Interventions that improve quality of life and increase survival will have the most favorable quality-adjusted life-years. Because some of the benefits of PA promotion do not become immediately evident (and are therefore difficult to measure directly), one can assess interventions by converting existing (often short-term) data on their impact on PA levels and the health benefits of increased PA into probabilities that can be used to model health implications of PA across a lifetime.

Determining Cost

Cost-effectiveness studies extend traditional cost-of-illness studies (which summarize all costs associated with a health condition) by accounting not only for cost of illness but also for cost of interventions and cost savings if illnesses are prevented or reduced in severity. Recognizing the tremendous importance of intervention- and outcome-related costs, many interventional studies have begun collecting costs alongside efficacy data; prospective, detailed measurement of costs can greatly refine our estimates of cost-effectiveness.

Putting It Together in the Real World

Measures of effectiveness of interventions and associated costs are combined in cost-effectiveness ratios, which may be used to broadly compare interventions and inform policy. Although the cost-effectiveness of PA promotion in and of itself in the prevention and treatment of obesity has not been studied, a few economic evaluations have examined multimodal interventions that included a PA component.

For example, in an early effort in this area, our group conducted an evaluation of outpatient weight loss strategies in overweight and obese adult U.S. women (5). Evaluated interventions included routine primary care and varying combinations of diet, exercise, behavior modification, or pharmacotherapy. Efficacy data were obtained from clinical trials, population-based surveys, and other published literature, while information on costs and quality of life required primary collection. This study showed that a three-component intervention of diet, exercise, and behavior modification cost U.S.$12,600 per quality-adjusted life-year gained compared with routine care, and suggested that comprehensive care of obesity compares favorably with other established health care interventions. These results were found to be somewhat dependent on our measurement of obesity-related quality of life and the probabilities of weight loss maintenance, which highlighted the need for accurate measurements of these phenomena in future studies.

Investigators in the Assessing Cost-Effectiveness in Obesity project used cost-effectiveness methods in a thoughtful and systematic approach to informing policy directed at preventing unhealthy weight gain in Australian children and adolescents (6). Part of their analysis focused on modeling the impact of a school-based PA program, the Walking School Bus (WSB), on changes in body mass index and subsequently on disability-adjusted life-years. The study showed that the cost of the WSB program was $760,000 per disability-adjusted life-year saved—an amount that was not considered to be highly cost-effective. Before making policy recommendations based on these findings, the investigators subjected this cost-effectiveness estimate to a second stage of analysis, which considered factors such as strength of evidence, equity, feasibility, and sustainability to more fully assess the attractiveness of the program.

In reaching their estimates of cost-effectiveness, the authors in both of these studies grappled with methodological challenges. Creation of models that aim to completely describe the course of obesity and the effects of interventions immediately identifies critical data gaps. For instance, if an intervention is successful in reducing body mass index, what is the likelihood that this reduction will be sustained over time? In the absence of good probability data, the Assessing Cost-Effectiveness in Obesity model assumed that body mass index reductions were sustained. This assumption can be tested in sensitivity analyses, whereby imprecise estimates are varied to show what effect they have on the ultimate cost-effectiveness ratios. Although data gaps may influence the accuracy of our CEA estimates, they also point the way to research priorities—as data accumulate, CEAs will get more and more refined. This continuous evolution of CEAs in response to new data means that many CEAs may be out of date before they are published! Although this is a problem, it does mean that CEAs are continuously getting better and perhaps more useful. Obesity models must also balance considerations of completeness and parsimony: Inclusion of more data sources may allow us to model reality more closely, but also creates the risk of distorting our estimates if data quality is poor. Furthermore, as models expand, their inputs are less likely to be understood by a single investigator. This means that these studies will become increasingly multidisciplinary—a phenomenon that, while adding sophistication, makes accurate modeling highly dependent on close cooperation and communication.

Apart from these issues, models focusing on disease prevention face other, unique challenges. In particular, prevention models may need to account for a large number of diseases. How many diseases are affected by PA? The decision about which diseases to model may depend on their relative importance, interdependence, considerations of parsimony, or data availability and quality. Other complexities in prevention models include how to interpret and account for causality (for instance, does exercise increase quality of life, or do people with a high quality of life exercise more?). Attributable risk is also a complex matter to interpret. While modeled risks of disease are often considered to be independent, in reality they may be related; when someone increases PA, do dietary and smoking habits remain constant or do they change in response?

Addressing Challenges Through the Project MOVE Approach

Although still subject to many of the challenges of CEAs, our recent study of the cost-effectiveness of PA promotion strategies (Measurement of the Value of Exercise—MOVE) (7) made progress in addressing some of the problems noted in the previous section. This study evaluated PA promotion strategies strongly recommended in the Centers for Disease Control and Prevention's Guide to Community Preventive Services (8), which summarizes the conclusions of an exhaustive review of available epidemiologic and clinical literature on PA promotion. Starting with the Community Guide recommendations placed our subsequent modeling efforts on a solid foundation of evidence for the derivation of effect size and cost estimates, and helped to at least partially bridge the data gap that has limited previous analyses. An effort was made to develop the model parsimoniously, using as few high-quality outcome and cost data sets as possible. The research team included diverse multi-institutional and cross-disciplinary partnerships with a well-defined strategy for communication and quality control. In this way, a parsimonious model based on the best possible (and often specialized) estimates of cost and benefit was created. The model was also constructed so that it would remain responsive to new data and so that it is transferable to other contexts.

It should be pointed out that because of the complex relationship between PA and obesity, the status of obesity as an intermediate variable along the pathways to disease, and the potential to double-count costs, obesity was not included as a disease outcome in this model. However, the model cohort is based on the 2003 Behavioral Risk Factor Surveillance Survey (BRFSS) and is representative of the population prevalence of overweight (37.9%) and obesity (25.2%) (9). Again, basic assumptions and choice of data inputs were thought to favor parsimony without necessarily oversimplifying the model's ability to account for disease progression.

This study showed that all of the evaluated PA interventions, applied to the adult U.S. population, offered good value for money. All the interventions examined appeared to reduce disease incidence and improve quality of life at costs comparable to those of many well-accepted public health strategies. Cost-effectiveness ratios ranged between $14,000 and $69,000 per quality-adjusted life-year gained, relative to no intervention. Additionally, reductions in disease incidence ranged from 5 cases per 100,000 people for colon cancer to nearly 500 cases per 100,000 people for coronary heart disease. Estimated cost-effectiveness was most influenced by intervention cost and effect size. However, varying parameter estimates across a wide range of uncertainty through multiple sensitivity analyses still resulted in cost-effectiveness ratios below widely accepted thresholds for public health value.

The findings of this analysis have several important implications for research and public health practice. First, the results from this study support using any of the evaluated interventions as part of public health efforts to promote PA. Second, the study helps to demonstrate that it is possible to carry out complex prevention modeling of community-based interventions using decision modeling approaches that are often focused on clinical analyses. Third, the modeling approach employed is a useful adjunct to the rigorous evidence-based review carried out by the Task Force on Community Preventive Services to identify recommended interventions for the Community Guide. Last, applied in context with information on program reach, effectiveness, feasibility, and community priorities and resources, such a CEA can also be a powerful decision-making tool for public health practitioners and advocates.

Future Directions

The rapid expansion of knowledge about the pathogenesis of obesity and its associated costs has created exciting circumstances for economic evaluation. Cost-effectiveness analyses have changed the landscape of obesity research. They make it increasingly possible for us to see the benefits of our programs relative to competing priorities for funding and, where these are weak, to create a broad framework to identify data gaps and opportunities for further research.

Future CEAs in obesity highlighting the role for PA will benefit from higher-quality data inputs and will likely play an increasing role in the formulation of health policy as estimates of cost-effectiveness become increasingly accurate and precise.

part III

Determinants of Physical Activity Levels

This section of the book describes the available evidence for several key determinants of physical activity levels. Chapter 16 examines biological determinants of physical activity, while chapter 17 focuses more specifically on genetic determinants and genetic markers associated with physical activity levels. Chapter 18 tackles the emerging science related to epigenetics and fetal programming as they relate to physical activity levels. The second half of this section examines sociodemographic correlates of physical activity levels, focusing on socioeconomic position (chapter 19), ethnicity (chapter 20), psychological factors (chapter 21), and the built environment (chapter 22). The evidence presented in this section provides some important potential targets for physical activity interventions focused on the prevention and treatment of obesity.

16

Biological Regulation of Physical Activity Level

Catherine M. Kotz, PhD

Physical activity is all activity that requires muscular work, and can be separated into chosen (volitional) physical activity or exercise and spontaneous (nonchosen, or nonvolitional) physical activity. Chosen physical activity, such as exercise, is well defined, and the neuromuscular signaling required to perform such work is well understood (1). The motor cortex controls the mechanical aspects of exercise, whereas higher cortical centers regulate the initiation of this form of physical activity. Although we do not fully understand all the biological inputs involved in making choices to be physically active, factors such as conditioning (i.e., Skinnerian) to cues in our environment and our daily habits play an important role. These factors are discussed in other chapters. The choice to perform physical activity is portrayed as free choice, involving our cognition; but, as with the regulatory mechanisms underlying eating behavior, biological cues influence cognitive choice behavior, and thus it is likely that biological factors underlie physical activity behavior. Regulatory mechanisms underlying the initiation and control of spontaneous physical activity are less well described. Shared neural systems regulate levels of both spontaneous physical activity and exercise, and spontaneous physical activity levels in humans reflect daily chosen activities (e.g., exercise) (2). An important contrast between the two forms of physical activity, however, is that spontaneous physical activity may not be chosen and therefore may not be necessarily regulated by higher cortex. Need for motor control output still implies a brain site of regulation for spontaneous physical activity, but spontaneous activity may originate in more autonomic brain sites such as the hypothalamus.

Spontaneous Physical Activity

As spontaneous physical activity is all physical activity considered nonvolitional, or not "by choice," this excludes formal exercise and thus leads to the distinction between the two forms of physical activity. Although the interpretation of movement "by choice" can vary, spontaneous physical activity may include activities such as fidgeting, tremors, and restless behavior. All spontaneous physical activities, large and small, require energy in the form of adenosine triphosphate (ATP) and therefore result in the expenditure of energy. Spontaneous physical activity levels in humans vary, but estimates on the daily contribution to energy expenditure range from 100 to 900 kcal/day (2, 3). This variation in energy expenditure, on a daily basis, could contribute to the large differences in body weight regulation among humans. In rodents, high levels of spontaneous physical activity predict a lean body weight (4), whereas comparatively low spontaneous physical activity levels are associated with the cause and maintenance of obesity.

Physical Activity Regulation

Perhaps the clearest evidence that physical activity is biologically regulated is the demonstration that physical activity, both volitional and nonvolitional, clusters within animal strains and likewise within families (5, 6). The animal strain variation implies that genes and their associated action proteins confer heritability of physical activity level. Selective breeding of animals has yielded rat and mouse strains with consistently divergent levels of physical activity (7). Thus, just like other behaviors such as eating and social activities, physical activity behavior may be inheritable and under biological control. The level to which this occurs is relatively understudied. When differences in physical activity are noted, they are usually associated with a marker of disease or some other process. Certainly, changes in physical activity levels are associated with disease; most forms of malaise present with low physical activity levels. Thus it can be difficult to dissociate between some forms of illness and biologically controlled low activity levels.

Similarly, animal lines of low physical activity have been used to represent models of depression (e.g., Wistar-Kyoto rats). Although these animal lines have additional physiological markers associated with depression, it is also plausible that these models could characterize inherited low physical activity behavior. Based on the multiple determinants important to body weight regulation, high levels of activity do not necessarily translate into low body weight, since they can be more than offset by elevated eating behavior. Likewise, low levels of activity do not automatically imply the obese state, since food intake can be reduced to offset low activity levels. For instance, the c57 mouse line has comparatively high levels of physical activity, yet strains can become obese with age. These mice also have food intake elevated enough to offset any advantage gained by the energy expenditure associated with their inherently high physical activity levels.

Multiple biological inputs, including hormonal, circadian, and neuronal, determine physical activity behavior. Based on the historic use of physical activity as a marker of behavior or disease states rather than as a causative factor in obese or obesity-resistant states, it is clear that many of these inputs remain undescribed or unacknowledged. Hormones and neuropeptides identified that may be important to physical activity include (but are not limited to) orexin (hypocretin), agouti-related protein, ghrelin, neuromedin U, neuropeptide Y, cholecystokinin, corticotropin-releasing hormone, and estrogen. Recently, brain-specific homeobox transcription factor (Bsx) was shown to be important to locomotor behavior, as knockout of this gene affects physical activity levels (8). These regulatory mechanisms are described in other chapters.

Physical Activity and Body Weight

By virtue of the energy expenditure associated with all forms of physical activity, there is an impact on body weight regulation. A large amount of animal evidence shows the effects of spontaneous physical activity on body weight; this is supported by work in the human literature, which shows the importance of physical activity–associated energy expenditure (2, 3, 9). A clear illustration of how biologically regulated physical activity can affect body weight regulation comes from studies with obesity-prone and -resistant rats. Obesity-resistant rats have significantly greater physical activity than obesity-prone rats, and their physical activity predicts body weight gain (4). The energetic cost of activity varies according to type of activity; and while this energetic efficiency may also be regulated, the energetic consequences of all physical activity are clearly important in determining propensity for obesity.

Whether physical activity behavior in animal models is representative of that in humans is unclear, but obesity-resistant rats move significantly more than obesity-prone rats primarily during the active, or dark, phase (4). This suggests that obesity-resistant rats are more physically active than obesity-prone rats during normal activity times. Rat behavior chamber measurements indicate that obesity-resistant rats have more bouts of activity and that the bouts last significantly longer than those in obesity-prone rats. Velocity calculations for obesity-prone and -resistant rats are similar, suggesting that speed of movement is not an important contributor to the "work" associated with the amount of physical activity performed. The differences in spontaneous physical activity levels between obesity-resistant and -prone rats are a repeated and consistent observation, suggesting that physical activity behavior is a trait characteristic of these rats (4).

The thermogenic consequences of physical activity can be difficult to assess in humans and in animal models (10). Measures of daily energy expenditure by indirect calorimetry, which yields the sum of all energy expenditure (physical activity, basal metabolic rate, and thermic effect of food), reveal similar daily energy output between obesity-prone and -resistant rats (4). Interpretation of this potentially counterintuitive finding is that, despite the large differences in physical activity behavior between these rat strains, there are no differences in physical activity–derived energy expenditure; rather, the strain differences in body weight are consequent to variations in caloric intake. One must consider, however, the larger body mass that is supported by the already obese, obesity-prone rat. The comparatively larger mass of obesity-prone rats should—based on rules of energetic costs of tissue substrate oxidation and the cost of moving a greater body mass—increase energy expenditure. Yet, the fact that no differences in energy expenditure are observed indicates that obesity-resistant rats expend more energy than would be predicted from their body mass. Although there is no perfect method of correcting for differences in body mass (10), multiple correction methods indicate that the caloric expenditure in obesity-resistant rats is significantly greater than that of obesity-prone rats (4). Further, the relationship between energy expenditure during physical activity and total daily energy expenditure is positive and significant, implying that activity contributes importantly to total energy expenditure. Together, these observations suggest that elevated physical activity behavior in obesity-resistant rats, which is observed as early as 6 weeks of age, contributes to their enhanced energy expenditure and resistance to obesity.

Potential Mechanisms

Several brain mechanisms exist that may be involved in controlling physical activity (as covered in other chapters) and the thermogenesis generated. An example of a biological input regulating physical activity is that of the neuropeptide orexin A. Orexin A is a multifunctional neuromodulator important to energy expenditure, and strain differences in orexin stimulation pathways for physical activity and the associated energy expenditure track with the propensity for obesity (4). In humans, lack of brain orexin is associated with elevated body mass index. In genetically manipulated animals, loss of brain orexin results in reduced locomotor activity and increased body weight. Orexin A produces pronounced spontaneous physical activity after central administration. Stimulation efficacy for orexin-mediated physical activity is not equal across brain sites, suggesting brain site-specific regulation of physical activity. Orexin produces both large motor (e.g., ambulatory) and small motor (e.g., grooming) activities. Although all physical activity has energetic consequences, ambulatory activity has, by virtue of the required muscular effort, the largest effect on heat production and thus energy expenditure. Starting at a very young age (6 weeks), before differences in body weight exist between obesity-prone and -resistant rats, obesity-resistant rats are highly responsive to the physical activity–promoting effects of orexin A and have elevated brain expression of orexin receptors. Together with our data showing baseline differences in physical activity between obesity-resistant and -prone rats, this suggests that both baseline physical activity levels and sensitivity to brain mechanisms that enhance physical activity behavior are biological traits of obesity-resistant rats and are important to resisting obesity.

Conclusion

In conclusion, animal and human studies imply biological and inheritable regulation of physical activity, which confers variable obesity resistance. Several hormones and neuropeptides have been shown to regulate physical activity, thermogenesis, and therefore the propensity for obesity. Like other behaviors, physical activity behavior is regulated by a complex network of circulating and neuronal signals for physical activity. There may be an association between chosen physical activity such as exercise and spontaneous physical activity. Whether activity is volitional or nonvolitional may be subject to interpretation, but the thermogenic consequences of either chosen or spontaneous physical activity behavior affect obesity. The biological inputs of these behaviors are beginning to be described, and future research in this area should yield promising new targets for enhancing physical activity.

17

Genetics and Physical Activity Level

Tuomo Rankinen, PhD

The beneficial effects of regular physical activity on primary and secondary prevention of several common chronic diseases, including obesity and its comorbidities, have been well established; and reduction of sedentarism is one of the main goals of public health initiatives. The main challenge for implementation of these recommendations is the poor compliance to physical activity interventions. Research on physical activity as a behavior has mainly focused on psychological, social, and environmental factors that contribute to levels of physical activity. The biological basis of activity behavior has only recently become a topic of interest, and advances in techniques of molecular genetic research have opened new avenues to test for the role of specific genes and mutations in activity-related traits. The purpose of this review is to summarize the recent advances in genetic epidemiology and molecular genetic studies on physical activity levels in humans.

Evidence From Genetic Epidemiology Studies

Studies on the genetics of physical activity level are not extensive, but available evidence from twin and family studies suggests that genetic factors contribute significantly to the propensity toward being sedentary or physically active. Several twin studies have investigated various domains of physical activity behavior (see [1] for details). Phenotypic similarity in activity behavior, quantified as intrapair correlation coefficient, is consistently greater in monozygotic (MZ) than in dizygotic (DZ) twins. This is indicative of a contribution of genetic factors to the trait variance, because MZ twins share 100% of their genes identical-by-descent, while DZ twins share only 50% of their genes identical-by-descent.

The most comprehensive twin study on physical activity was published only recently. Investigators representing seven large twin studies from Australia, Denmark, Finland, Norway, the Netherlands, Sweden, and the United Kingdom created a cohort of 37,051 twin pairs: 13,676 MZ pairs, 17,340 same-sex DZ pairs, and 6035 opposite-sex DZ pairs (2). Information on exercise participation was derived from questionnaires, and the final outcome variable was dichotomized as exercisers and nonexercisers: The exercisers were defined as individuals who reported at least 60 min of weekly activity with a minimum intensity of 4 metabolic equivalents. The mean prevalence of exercise participation was 44% in males and 35% in females. The lowest participation rates were found among Swedish twins (37% in males and 23% in females), and the highest in twins from Australia (64% in males, 56% in females).

Tetrachoric correlations showed that the intrapair resemblance in exercise participation was significantly higher in MZ twins than in DZ twins (2). Furthermore, correlations among same-sex DZ twins tended to be greater than in opposite-sex DZ pairs. The most parsimonious model from structural equation model fitting revealed that variance in exercise participation was explained by additive genetic and unique environmental effects in all but one country-by-sex subgroup (figure 17.1). The only exception was Norwegian males, among whom shared environment also contributed significantly to the variance in exercise participation. Heritability estimates ranged from 27% (Norwegian males) to 71% (U.K. females); the median heritability across all groups was 62% (2). These estimates are in line with observations from previous studies with smaller numbers of twins (1).

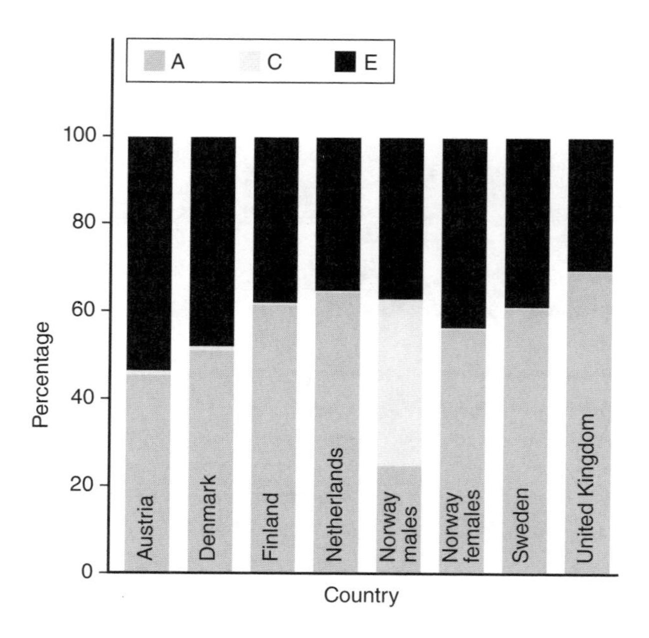

Figure 17.1 Contribution of additive genetic (A), shared environmental (C), and unique environmental (E) effects to the variance in exercise participation among 37,051 twin pairs aged 19 to 40 years.

Data from J.H. Stubbe et al., 2006, "Genetic influences on exercise participation in 37,051 twin pairs from seven countries," *PLoS ONE* 20: 1:e22.

The most recent family study on physical activity level was published in 2006. Cai and colleagues reported results of heritability analyses in 1030 Hispanic children from 319 families of the Viva La Familia Study (3). Families were identified through an overweight proband, and all of his or her siblings between 4 and 19 years of age were recruited. Free-living physical activity was measured using accelerometers. Children wore the accelerometer for three days, and the activity level was quantified as total activity, as well as percentage of awake time spent in sedentary, light, moderate, and vigorous activities. The children spent the majority of their awake time in sedentary (37.8%) and light (52.7%) activities, while the proportion of time used for moderate and vigorous activities was 9.4% and 0.3%, respectively. Total activity count was significantly higher in boys than in girls, but percentage of time spent in different activity categories did not differ between the genders. After adjustment for age and sex, the maximal heritability estimate for total physical activity reached 55%. Estimates for sedentary, light, and moderate activities ranged from 46% to 57%. However, the heritability of vigorous activity was estimated to be considerably lower (18%), which most likely reflects the low level of such activities among these children (3).

Results of the twin studies consistently support the hypothesis that genetic factors contribute to the propensity toward being sedentary or physically active. Although the heritability estimates from the first family studies were moderate at best, more recent data based on slightly larger sample sizes and, in particular, more objective measures of physical activity level provide heritability estimates that are very similar to those reported in twin studies. These findings provide a good justification to search for genes and DNA sequence variants contributing to physical activity behavior in humans.

Evidence From Molecular Studies

The latest update of the human gene map for performance and health-related fitness presented six candidate genes and 12 quantitative trait loci for physical activity phenotypes (4). The candidate genes with positive findings include dopamine D2 receptor *(DRD2)*, angiotensin-converting enzyme *(ACE)*, leptin receptor *(LEPR)*, melanocortin 4 receptor *(MC4R)*, calcium-sensing receptor *(CASR)*, and aromatase *(CYP19A1)*. The first four genes were investigated with an a priori hypothesis on the association between physical activity and DNA sequence variation.

Using data from the Quebec Family Study and the HERITAGE Family Study, Simonen and coworkers reported associations between a C/T transition in codon 313 of the *DRD2* gene and physical activity level. Caucasian women who were homozygous for the T-allele were significantly less active than the other genotypes in both studies (5). Also in the Quebec Family Study cohort, Loos and colleagues reported significant associations between a C/T polymorphism located in the 5'-region of the *MC4R* gene and physical activity phenotypes (6). Homozygotes for the T-allele had significantly lower moderate-to-strenuous physical activity levels and higher inactivity scores than the other genotypes (figure 17.2).

A glutamine (Gln) to arginine (Arg) substitution in codon 223 of the *LEPR* gene was associated with total physical activity level, calculated via dividing 24 h energy expenditure by sleeping energy expenditure measured in a respiratory chamber, in Pima Indians. The Arg233Arg homozygotes showed a 5% lower physical activity level than the Gln223Gln homozygotes (7). In a group of never-treated stage I hypertensives (8), the ACE I/D genotype was associated with physical activity status assessed by a questionnaire. The frequency of the D/D genotype was significantly higher in the sedentary group than

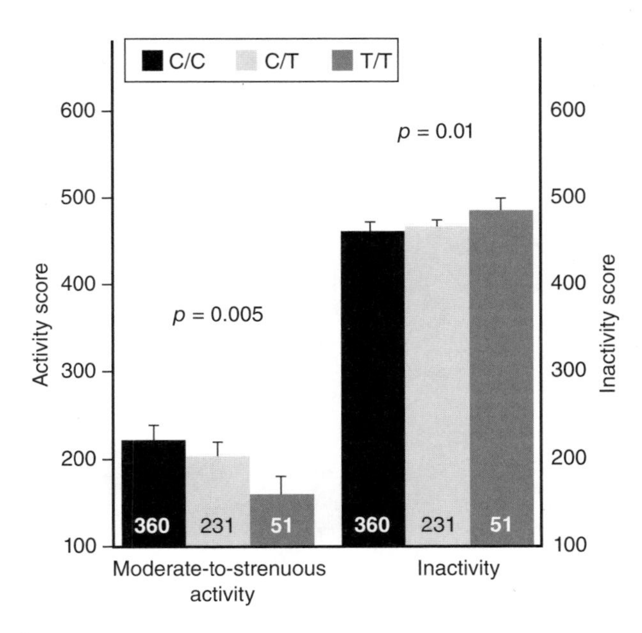

Figure 17.2 Moderate-to-strenuous physical activity and physical inactivity levels according to a melanocortin 4 receptor *(MC4R)* genotype in the Quebec Family Study cohort. The polymorphism is a C/T transition located 2745 base pairs upstream of the *MC4R* gene start codon.

Data from R.J. Loos, 2005, "Melanocortin-4 receptor gene and physical activity in the Quebec Family Study," *International Journal of Obesity* 29: 420-428.

among physically active subjects. Approximately 76% of the D/D homozygotes were sedentary, whereas the corresponding frequency in the I-allele homozygotes was 48% (8).

Four studies have reported genome-wide linkage scans for physical activity traits. In the Quebec Family Study, the strongest evidence of linkage was observed on chromosome 2p22-p16 for the physical inactivity phenotype (9). Suggestive linkages were found on 13q22 with total activity and moderate to strenuous activity and on 7p11 with both inactivity and moderate to strenuous activity (9). De Moor and colleagues reported genome-wide linkage scans for participation in competitive sports in 700 female DZ twins (1946 markers) and for exercise participation in Dutch sibling pairs (361 markers). Suggestive quantitative trait loci were identified on chromosomes 3q22-q24, 4q31-q34, and 19p13.3 (10, 11).

The latest evidence of linkage for physical activity traits comes from the Viva La Familia Study (3). A genome-wide linkage scan with a 10-centimorgan microsatellite panel revealed quantitative trait loci on chromosome 18q12.2-q21.1 for sedentary and light activities with logarithm of odds scores of 4.07 and 2.79, respectively. Maximum logarithm of odds scores of 2.28 and 2.2 for total and moderate activities, respectively, were detected about 20 centimorgans downstream at 18q21.32, near the *MC4R* locus (3).

Future Directions

The future for genetic studies of physical activity behavior looks promising. Several stumbling blocks that have slowed the progress in the field have been either solved or significantly alleviated. Once these improvements can be combined with the most recent advancements in molecular genetic techniques, our understanding of the genetics of physical activity behavior will improve drastically. The power of genome-wide association studies has been already demonstrated with several complex phenotypes, and publication of the first genome-wide association studies for physical activity traits is only a matter of time. Likewise, laboratory methods to measure DNA methylation and other epigenetic modifications have greatly improved, and they will open new opportunities to study the biology of physical activity levels.

Summary

The studies summarized in this review indicate that physical activity–related traits are influenced by genetic factors with maximal heritability estimates ranging from 20% to 70%. The majority of the data available are based on genetic epidemiology studies, but the first molecular genetic studies support the notion that it is possible to detect biologically relevant genetic effects on habitual physical activity at the molecular level. The major limitation at the moment is the paucity of studies addressing the genetics of physical activity behavior in humans. However, it seems that the number of investigators focusing on physical activity as a major endpoint rather than a confounding factor is increasing. For example, two large studies investigating genetics of physical activity were published in late 2006 (2, 3).

Increase in the critical mass of genetic studies on physical activity would be particularly important for replication of the positive association studies. Currently, most of the findings are based on single reports, although in some cases the associations were found in multiple cohorts (e.g., [5]). Larger studies would also allow us to investigate gene–gene interactions as well as potential pleiotropic effects on physical activity and health outcomes. A recent study showed that the phenotypic association between exercise participation and self-rated health status observed in over 5000 adult twins and their siblings was fully accounted for by additive genetic correlation (12). This observation raises an interesting question—whether activity behavior and health-related phenotypes are affected at least to some extent by the same gene(s). Another limitation has been the measurement of physical activity level. Random variation in activity level estimates obtained from questionnaires and diaries can be quite high, which seriously limits the statistical power to detect small or moderate genetic associations. However, improvements in objective activity level recording technologies will at least alleviate if not solve the problems.

18

Epigenetic or Programming Effects on Physical Activity Level

Mark H. Vickers, PhD

The prevalence of obesity and related metabolic sequelae has reached epidemic proportions throughout the world and poses a significant health and economic burden for both developed and developing societies. Furthermore, the marked increases in childhood obesity and accompanying changes in levels of physical activity will translate to a further increase in adult obesity in the near future, thus obesity in children has been ranked as a critical public health threat. It is a widely held view that the development of an obesogenic environment, due to ease of access to highly calorific food and reduced energy expenditure in work and leisure activities, is the primary cause of obesity in the general population.

Multiple systems regulate energy homeostasis, and there is strong evidence for a genetic component to human obesity with the identification of a number of associated genes. However, the genetic component of this condition cannot account for the dramatic increase in the prevalence of obesity in recent years. In this context, an increasing number of epidemiological and experimental studies have highlighted a relationship between the periconceptual, fetal, and early-infant phases of life and the subsequent development of adult obesity and metabolic disease. A model of this relationship, referred to as the "developmental origins of health and disease" (DOHaD) model, speculates that the fetus makes predictive adaptations in response to adverse environmental cues in utero that result in permanent readjustments in homeostatic systems to aid immediate survival and improve success in an adverse postnatal environment. However, when there is a mismatch between the prenatal predictions and postnatal environment, these adaptations, known as predictive adaptive responses (1), may ultimately be disadvantageous in postnatal life, leading to an increased risk of chronic noncommunicable disease in adulthood, the inheritance of risk factors and a cycle of disease transmission across generations, or both.

To date, most of the evidence relating to the DOHaD hypothesis has to do with metabolic and cardiovascular disease; relatively little attention has been paid to indices related to physical activity. In addition, the studies that have examined activity have primarily focused on stress- and anxiety-related behaviors and not directly on physical activity per se. Furthermore, clinical and epidemiological evidence of changes in physical activity resultant from a poor fetal environment are limited, primarily due to two factors. First, lifestyle influence obscures linkages between metabolic predisposition and maturity-onset behavioral patterns; second, subject diaries related to perceived activity levels in clinical cohorts are required, and these have inherent errors in precision of reporting.

Evidence From Animal Models and Epidemiology

A number of recent reports in animal models suggest that aspects of locomotor activity are determined by factors operating in early life. Developmental programming in rats utilizing a variety of maternal manipulations has shown that voluntary locomotor behavior is significantly altered in offspring in postnatal life (figure 18.1). Importantly, in a model of maternal undernutrition, reduced physical activity has been shown to precede the development of an obese phenotype; programming-induced alterations in physical activity level were observed as early as postnatal day 35 (pubertal age) (2).

A maternal low-protein (LP) diet has been shown to result in a significant reduction in voluntary locomotor activity in male and female rat offspring. Gender-specific differences have been demonstrated, and the changes in locomotor activity are dependent upon the window of prenatal LP nutrition. Recent work in the mouse has also shown that a maternal LP diet fed exclusively during oocyte maturation leads to behavioral abnormalities in the offspring, including reduced locomotor activity (3). However, in contrast to findings from models

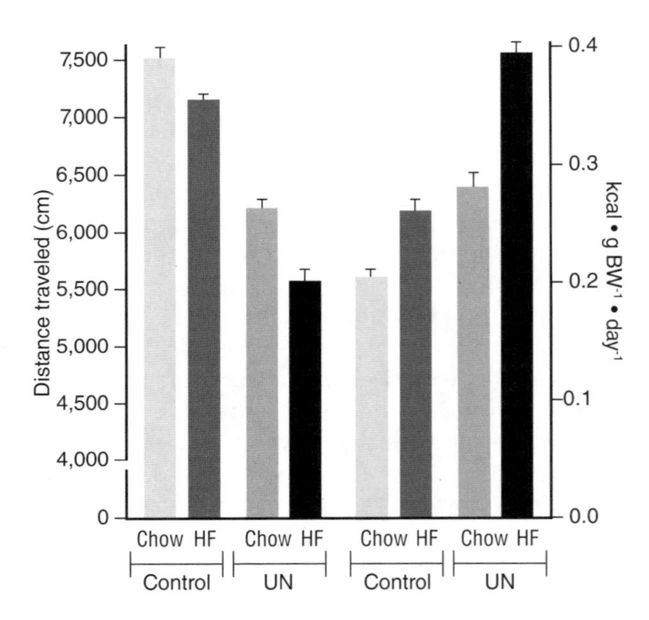

Figure 18.1 Voluntary locomotor activity and caloric intake in adult male rat offspring of mothers that were undernourished throughout pregnancy (UN) compared to offspring of normal pregnancies (CONTROL) that were both fed either a chow or high-fat (HF) diet postweaning. Offspring of UN mothers are significantly less active as adults and also display altered feeding behaviors (i.e., hyperphagia). $p < 0.05$ for effect of developmental programming and postnatal diet. Data are mean ± SEM. Food intake is expressed as calories consumed per gram body weight per day.

Adapted from M.H. Vickers et al, 2003, "Sedentary behavior during postnatal life is determined by the prenatal environment and exacerbated by postnatal hypercaloric nutrition," *American Journal of Physiology-Regulatory Integrative and Comparative Physiology* 285(1): R271-R273.

of global undernutrition in the rat (2), offspring of LP-fed dams did not exhibit increased adiposity in parallel to reduced physical activity levels when fed a standard diet postnatally. The lower levels of locomotor activity observed in the LP-exposed rats may provide sufficient disturbance of energy balance to promote obesity only when coupled to increased energy intake. These differences may simply be explained by the degree and severity of the dietary intervention in pregnancy.

In addition to models of altered maternal nutrition, gestational dexamethasone (DEX) exposure in the rat has been shown to elicit gender-specific alterations in locomotor activity in parallel with changes in learning and memory. This indicates that the presence or absence of somatic effects may not be an adequate predictor of postnatal behavioral changes following an adverse prenatal environment. Daily maternal DEX treatment in rats can result in postnatal hypoactivity in male offspring, although, conversely, maternal DEX treatment in late gestation has been shown to result in male rat offspring that display significantly increased locomotor activity in adult life. Multiple courses of betamethasone to pregnant guinea pigs have also been shown to elicit marked sex-specific effects, with female offspring exhibiting significantly increased locomotor activity in an open field test whereas no differences were observed in males (4). Prenatal cytokine exposure (using tumor necrosis factor-alpha), in addition to resulting in obesity in offspring, resulted in gender-specific programming effects including reduced locomotor activity in males but not females. Of note, male offspring of dams exposed to interleukin-6 and DEX in this study did not manifest any changes in locomotor activity while females showed an increase in physical activity in postnatal life (5).

Initial epidemiological studies indicated that fetal growth restriction correlated with later disease, implying that fetal nutritional deprivation was a strong programming stimulus. This prompted the development of experimental animal models using controlled maternal calorie, protein, or macronutrient deficiency during key periods of gestation. However, in many societies, maternal and postnatal nutrition are either sufficient or excessive, and obesity is one of the most common nutritional problems complicating pregnancy in developed countries. Evidence now suggests a "U"-shaped phenomenon whereby offspring exposed to either early-life nutritional deprivation or to an early environment overly rich in nutrients appear to be at risk. Thus, there is increasing interest in the potentially detrimental influence of maternal obesity and excess maternal nutrition on the risk of disease in childhood and beyond. However, as with models of maternal undernutrition, most attention has focused on obesity and related metabolic disorders, and little attention has been paid to indices related to physical activity. Maternal exposure to a high-fat diet in the rat has been shown to result in reductions in voluntary locomotor activity in offspring, although the results are not consistent. In the mouse, maternal diet–induced obesity leads to a reduction in physical activity in offspring, although a high polyunsaturated fat diet fed throughout pregnancy resulted in offspring with increased locomotor activity in a swim test. Thus it may be that the fatty acid composition of the maternal diet is crucial in programming activity levels and energy expenditure. Clinical evidence on developmental programming of physical activity level is scarce. Epidemiological studies of survivors of the Dutch famine of 1944-1945 have recently shown that prenatal exposure to famine resulted not only in food preferences toward unhealthy diets but in trends toward reduced physical activity in adulthood.

Developmental programming utilizing manipulations of maternal nutrition appears to provide consistent data on reductions in physical activity in offspring. These effects can be significantly compounded in the presence of a postnatal obesogenic dietary environment. The implications of such findings are profound: If reduced physical activity is determined during prenatal development, this may explain why public health attempts to improve exercise behavior in subjects with the metabolic syndrome are largely ineffective.

Possible Mechanisms

In human and in experimental animal models, hyperactivity has been linked to altered dopamine (DA) signaling in the nucleus accumbens and the medial prefrontal cortex. Hyperactive rats display increased DA levels and a reduction in DA transporter activity in conjunction with changes in hippocampal synaptic function (6). This may suggest that hyperactivity in female offspring following antenatal betamethasone exposure results from glucocortocoid actions on the developing DA signaling system.

Some changes may relate to altered muscle physiology of developmentally programmed offspring, and maternal manipulations that influence myogenesis in utero may affect the capacity for adult physical activity. Muscle fiber type is a key determinant of metabolic flexibility, and the capacity to adapt metabolism to variations in nutrient supply is greater in lean than in obese individuals. It has been demonstrated that developmental programming using maternal undernutrition leads to a shift in muscle fiber composition. Maternal undernutrition also results in reduced muscle mass in adulthood compared to that in controls. Thus reduced physical activity in programmed offspring may also stem from physical adaptations to changes in exercise capacity.

Leptin, an adipocyte-derived hormone that acts on hypothalamic neurons located in the arcuate nucleus of the hypothalamus to regulate energy homeostasis, has well-characterized effects on locomotor activity. Infusion of leptin in rodents lacking endogenous leptin promotes physical activity and energy expenditure and improves insulin signaling, whereas hyperleptinemia is associated with physical inactivity and insulin resistance. It has been reported in the rat that reductions in voluntary locomotor activity in offspring following maternal undernutrition can be reversed by treatment with leptin during the neonatal period (7).

Epigenetic Basis for Reduced Physical Activity

Any plausible molecular mechanism for developmental programming of changes in physical activity must explain how very early environmental disruption can set in place persistent molecular changes that will lead to damaging effects in later life. Epigenetics is the study of heritable changes in gene expression that are not mediated by DNA sequence alterations. Because of their inherent malleability, epigenetic mechanisms are susceptible to environmental influences; and this environmental susceptibility is expected to be enhanced during early development. Accordingly, nutritional perturbation of epigenetic gene regulation is a likely link between early nutrition and later metabolism and chronic disease susceptibility (8). It has been demonstrated that mammalian phenotype can be persistently altered via nutritional influences on the establishment or maintenance (or both) of epigenetic gene regulatory mechanisms (8, 9). For example, early methyl donor malnutrition could effectively lead to premature "epigenetic aging," thereby contributing to an enhanced susceptibility to chronic disease in later life (8).

Data relating epigenetic changes to altered physical activity level are limited, although studies on the influence of maternal care have identified profound changes in promoter methylation linked to altered behavioral characteristics in offspring (6). Meaney and Szyf found that the level of maternal care during the suckling period permanently changed offspring physiology and behavior affecting the ontogeny of methylation at specific CpG sites in the glucocorticoid receptor promoter (6). More recently, dietary protein restriction of pregnant rats has been shown to induce DNA hypomethylation and increased expression of glucocorticoid receptor and peroxisome proliferator-activated receptors-alpha in livers of offspring, the effects of which are reversed with maternal folate supplementation (10). Dietary choline deficiency has also been associated with hypomethylation of genes within the mouse hypothalamus, correlating with increased expression of genes associated with cell cycle inhibition. However, neither of these studies determined whether the epigenetic alterations arising from alterations in maternal diet persisted into adulthood and produced the subsequent metabolic phenotype that ensued (8).

Conclusions

Data linking developmental programming with altered physical activity level are scarce. However, experimental paradigms utilizing altered maternal nutrition appear to generate a phenotype characterized by reduced locomotor activity, a feature common to both genders. It is now evident that even brief periods of undernutrition at the earliest stages of embryonic development can exert lifelong effects upon food intake and locomotor activity in offspring. It is sometimes difficult, however, to separate behaviors directly related to physical activity level from those that are stress mediated. For example, many tests assessing locomotor activity require placement of the animal in a novel environment for a short period of time; this may elicit a stress response, so monitored behavior may not directly relate to physical activity level per se but rather to anxiety. These issues may be circumvented through the use of longer-term monitoring devices such as telemetry. Data from models utilizing glucocorticoids are less consistent and demonstrate marked sex-specific and dose-related effects. The possibility that feeding and other behaviors may be programmed *in utero* may thus be of considerable importance in the obesity field. These results suggest that both relative maternal under- and overnutrition can lead to the development of appetite disorders as well as diminished exercise behavior concomitant with the physiological features of the metabolic syndrome.

These observations raise the intriguing possibility that some behaviors and lifestyle choices that exacerbate the metabolic syndrome in humans may not be voluntary but may be an inherent part of the syndrome and may have a prenatal origin; thus the "couch potato" syndrome may have its origins during prenatal development. Our understanding of the specific mechanisms by which epigenetic gene regulation is first established during mammalian differentiation is fairly rudimentary (8). Thus data on epigenetics in the DOHaD paradigm are limited; but this is a rapidly emerging research field, with a current focus on the ability of early nutrition to influence DNA methylation to give rise to permanent changes in metabolism. This work has obvious major implications for public health policy; it may be better to expend health care funding on improving pregnancy care rather than waiting until metabolic and cardiovascular disorders manifest in offspring years or decades later.

19

Socioeconomic Position and Physical Activity Levels

Lise Gauvin, PhD

A substantial body of research shows that mortality, longevity, morbidity, and quality of life are associated with socioeconomic position (SEP), with more well-off persons and populations experiencing better health outcomes than less well-off individuals and groups (1). Socioeconomic position is an overarching concept that refers to "the socially derived economic factors that influence what positions individuals or groups hold within the multiple-stratified structure of a society" (2, p. 23). The existence of strong relationships between higher SEP and better health outcomes is now a matter of record (3), and underlying causal processes are currently being investigated (1). Furthermore, efforts are being deployed to reduce health inequalities related to SEP worldwide (see report from WHO Commission on the Social Determinants of Health [CSDH] released in 2008; www.who.int/social_determinants/thecommission/finalreport/en/index.html). Despite the interest in SEP as it relates to obesity and overall cardiovascular risk factors (1, 4, 5), efforts to more precisely quantify the direction and magnitude of associations between SEP and physical activity (PA) have a more limited history (6), as does incorporating knowledge about SEP–PA relationships in the context of interventions aimed at changing eating and PA (7). In this chapter, the concept of SEP is defined both conceptually and operationally. Then, research on the association of SEP with PA is reviewed, and the manner in which SEP–PA relationships have been addressed in intervention research is underscored. Throughout the chapter, gaps in the knowledge base are emphasized.

Conceptualization and Operationalization of SEP

The expressions socioeconomic position, socioeconomic status (SES), and social determinants of health are sometimes used interchangeably but have distinct meanings. Krieger (8) presents a pertinent glossary of terms, and Galobardes and colleagues (2) provide additional nuances and overview operationalizations. In keeping with the definition cited in the preceding paragraph, Krieger (8) offers additional clarity in distinguishing the concepts of SEP and SES. That is, she suggests that "SEP refers to an aggregate concept that includes both resource-based and prestige-based measures, as linked to both childhood and adult social class position" (p. 697). Resource-based aspects of the SEP concept designate material and social assets, whereas prestige-based aspects refer to individuals' rank or standing within a given social hierarchy. Socioeconomic status is more restrictive than SEP, as it designates a person's or a group's position within a larger whole without necessary regard for overall material and social assets. For instance, being in the top quintile of earners in a given profession or being in the lowest quintile of income in a low-income country represents a measure of status, whereas having sufficient income to pay for food, housing, and other basic needs or to have access to universal health care services or fitness resources refers to material or social assets and thus to SEP.

Although SES and SEP are often correlated, they do not represent isomorphic constructs. For example, in countries where there are extensive social policies and at least average economic development, a large proportion of the population may have sufficient income to pay for food and housing and to have access to a variety of universal health and social services including PA resources. In these same countries, prestige may be determined by occupation and educational attainment or leisure activities (e.g., golfing, skiing). As a result, the resource-based measure may not be correlated with status or position in that society. Conversely, in other nations where social policies are less extensive, resource-based measures (e.g., access to and quality of health care services, unemployment and other insurance benefits) may be highly related to occupational status and educational attainment.

In addition to distinctions between resource-based and prestige- or status-based conceptions of SEP, the notion of SEP must be specified along at least two other dimensions. First, SEP can be thought of as an individual or ecological attribute. Thus, resource-based and prestige-based indicators can be assessed at the level of individuals; families; social groups defined by location of residence or sharing of a cultural, ethnic, or racial attribute (e.g., residents of a specific neighborhood, immigrants, aboriginal populations); or populations governed by elected officials in a given jurisdiction (e.g., state, province, country, cluster of nations regrouped into an economic union). Examination of SEP–health and SEP–PA relationships therefore should be conducted at multiple levels, as associations at one level may or may not be paralleled by similar relationships at another level (i.e., ecologic and atomistic fallacies).

Second, SEP can change across time, as resources can become plentiful or scarce and the position of a person or group of persons can shift upward or downward across time. Thus SEP at one point in life may or may not be identical to SEP at another point in life, just as the SEP of a country on the world scene can fluctuate. These considerations are important, as some authors have hypothesized that deprivation may have more serious consequences for health if experienced at critical development periods (9). Similarly, there is evidence that deteriorating economic conditions in a given jurisdiction can negatively affect population health (3). Hence, understanding the scale and time at which SEP is assessed is critical to gaining a full understanding of its association with health and PA outcomes alike.

Overall, then, SEP conveys the idea that material and social conditions are unevenly distributed across persons and populations, thus situating people or population groups into different hierarchically structured positions. At the individual level, some of the most common measures of SEP include education, income, wealth (income and material belongings), occupation, employment status, quality of housing, gender, and ethnic or cultural or racial group (2). Familial-level SEP indicators can include family income, family composition, employment status of different members of the family, and housing. At the area level, researchers often use aggregates of individual-level measures (e.g., Townsend index, indices of Multiple Deprivation used in the United Kingdom) that characterize the population living in a certain territory: proportion of unemployed persons, proportion of persons of low income, proportion of immigrants, or proportion of seniors living in an area. Cross-national comparisons are often performed using OECD (Organization for Economic Co-operation and Development, www.oecd.org) indicators of national wealth, wealth, or quality of life.

Socioeconomic Position and Physical Activity Levels

At least one review of literature (6) that extended up until October 2004 examined the literature on SEP and adult involvement in leisure PA in industrialized nations. Following a thorough search strategy of the published literature in PubMed, Web of Knowledge, PsychInfo, and Sports Discus, a total of 28 cross-sectional studies and 5 longitudinal studies examining the association of SEP and PA at the individual level were identified. Across all indicators of SEP, an association was found with leisure-time PA, with higher SEP linked to greater likelihood or amount of PA. However, another review on coronary heart disease risk factors in children based on seven studies from the United Kingdom showed little evidence that physical inactivity was related to SEP (5). There were, however, several limitations in the research. First, the research focused mostly on the association of SEP to leisure-time PA. This is an important gap, since at least some emerging research shows that children (10) living in families with low income (a prestige-based indicator of SEP) are more likely to use walking as a mode of transportation at least in part because the family cannot afford a motor vehicle (a resource-based indicator of SEP). As a result, more information is required about transport PA (e.g., walking and cycling). Similarly, it comes as no surprise in industrialized nations that PA has been largely engineered out of occupational activities and is concentrated in lower-paying jobs. Associations of SEP to occupational PA merit further investigation also.

In addition, the search strategy in both reviews overtly limited papers to investigations conducted in industrialized nations. This appears as an important gap, since a recent review (4) shows that the association of individual-level SEP to indicators of overweight and obesity may differ across nations differing in levels of wealth. That is, in nations with higher levels of socioeconomic development, higher levels of individual SEP were associated with lower likelihood of overweight and obesity in both men and women. Conversely, in nations with low levels of socioeconomic development, the association was in the opposite direction, with higher levels of SEP associated with greater likelihood of overweight and obesity. Although there have been efforts to quantify the burden of chronic diseases and risk factors in countries with different levels of socioeconomic development, showing that the burden of physical inactivity is about the same across low-, middle-, and high-income countries (11), a more fine-grained analysis of the relationship between SEP and PA measured at the individual level across countries differing in economic wealth has not yet been conducted but could provide useful information.

Given the limited amount of research on SEP–PA relationships, it is not surprising to note that there are also limited data regarding the processes that lead to the existence of such disparities. In other words, causal explanations regarding SEP–PA relationships have yet to be formulated. In the area of SEP–health outcomes, two broad categories of explanations have been proposed and are referred to as psychosocial and neo-materialist (1). Both types of explanations postulate that the material and social conditions that accompany differing levels of SEP translate into different health outcomes through a variety of stress-related processes. The psychosocial explanations for SEP–health relationships assert that it is mainly through intraindividual psychological processes (e.g., beliefs about one's self-worth, perceptions of social support, perceptions of one's place in the social hierarchy based on relative position according to income) that material and social conditions related to SEP are internalized and transformed into deleterious physiological processes. Conversely, neo-materialist explanations for SEP–health associations posit that being higher up on the SEP scale affords one more and a greater variety of resources for overcoming or avoiding harmful material and social conditions. Therefore people higher up on the SEP scale can shield themselves from bad material and social conditions through the mobilization of more and a broader variety of resources, whereas people on the lower end of the scale lack resources and are thus exposed to deleterious risk conditions. A full explanation of the mechanisms underlying SEP–health relationships, and by extension SEP–PA relationships, has yet to be developed (1, 3).

Socioeconomic Position and Physical Activity Interventions

Despite the limited amount of literature on SEP–PA relationships, there is growing recognition that the presence of SEP disparities as they relate to both PA and health is a relevant consideration in health promotion or disease prevention programs. As a result, some researchers have begun to tailor PA interventions to low-SEP populations. Tailoring interventions, which often target individual psychosocial characteristics such as intentions, attitudes, and self-efficacy, to selected low-SEP groups represents one viable approach. Some data (7) show that interventions aimed at healthy eating and improved PA can be culturally appropriate and sometimes just as effective as interventions delivered to individuals higher up on the SEP scale. However, data also show that individually based interventions that include a strong educational component are not as effective with persons of low SEP (1).

Thus, an equally viable approach consists of implementing population-based interventions to entire populations that target public policies (e.g., implementing policies related to mandatory physical education in schools) and organizational and community environments (e.g., creation of PA resources such as fitness facilities or walking paths). Evaluation efforts would then focus on examining the extent to which these interventions are effective in increasing activity involvement across population strata differing in SEP, as well as on determining the impact of the intervention in changing population patterns of involvement in PA (i.e., central tendency and distribution). It is entirely plausible that population-based interventions could either increase or decrease disparities in PA involvement across persons or groups with low and high SEP to the extent that the implementation and uptake of policies and use of environmental resources are different across the better off and the less well off. Given the growing but still limited amount of research on policy and environmental determinants of PA, it is not surprising to note that there are also limited data pertaining to the impact of population-based PA interventions in influencing inequalities in PA associated with SEP.

Conclusion

Although it comes as no surprise that those persons and populations who have the most resources and who are higher up in any given social hierarchy have the most favorable health outcomes and health behavior profiles, these considerations have been given only limited attention in the understanding of the determinants of PA levels and in the crafting of intervention strategies. As outlined here, there is a small but compelling basis from which to extend the knowledge base on SEP–PA associations and from which to advance public policies and practices to increase PA involvement in entire populations throughout the world.

20

U.S. Ethnic Differences in Physical Activity and Sedentary Behavior

Robert L. Newton Jr., PhD

Ethnic differences in physical activity and sedentary behavior have been observed for several decades. These differences have been used to explain differences in the prevalence of obesity and its comorbidities. Both self-report and objective measurements have been utilized, though they have not always produced the same results. This chapter reviews data related to levels of physical activity and sedentary behavior in ethnic groups within the United States. Both self-report and objective data are discussed, with respect to both children and adults. It should be noted that due to the dearth of information available on certain ethnic groups (e.g., Asian Americans, Pacific Islanders, Native Americans), the chapter focuses on the data available for European, African, and Hispanic Americans.

Physical Activity in Adults

National surveillance studies of physical activity levels of adults have employed both self-report and objective monitoring strategies.

Self-Reported Activity

Levels of physical activity are provided by national epidemiological studies such as the National Health Interview Survey (NHIS) (1), the National Health and Nutrition Examination Survey (NHANES) (2), and the Behavioral Risk Factor Surveillance System (BRFSS) (3). In NHANES (2), being active was defined as engaging in vigorous physical activity three or more times per week. In BRFSS (3) and NHIS (1), moderate activity was defined as activity that occurred in 10 min bouts, five days per week, and vigorous activity was defined as activity that was greater than 50% of age- and gender-specific maximum cardiorespiratory capacity, occurring three days per week.

These national epidemiological studies show that ethnic differences in activity vary by gender. In NHANES (2), Hispanic American men had the lowest levels of activity, while African American men between the ages of 20 and 39 years and European American men aged 40 years and older actually had the highest rates of activity. Data from NHIS (1) show that African and European American men have higher levels of moderate and vigorous activity, respectively, and that Hispanic American men are the least active. In BRFSS (3), it was shown that 52.3% of European American men were regularly active, while less than 45% of African and Hispanic American men were. Across all three studies, European American women were more active than their ethnic counterparts.

These large-scale, national epidemiological studies show clear differences in women, but the results vary in men. However, all of these data are based upon self-reported physical activity, which is subject to bias. Use of objective measures of physical activity is a way to overcome the bias inherent in self-reported behavior. Currently, accelerometers provide an objective assessment of physical activity.

Objective Measurement

The NHANES (4) collected accelerometer data on over 3000 adults in 2003-2004. The data were analyzed in a number of different ways, including counts per minute, time in moderate and vigorous activity, time in 10 min bouts of moderate and vigorous activity, and whether or not the American College of Sports Medicine recommendations were met. Adults were classified into the age categories 20 to 59 years or 60 years and older.

These data show that Hispanic American men and women aged 20 to 59 years have greater counts per minute compared to both European and African Americans. When data were analyzed as bouts of 10 min or more, African American men and women aged 60 years and older spent fewer minutes in activity than Hispanic and European Americans. Furthermore, Hispanic American men spent more total time in moderate and vigorous activity, and African American women aged 60 years and older spent less total time in moderate and vigorous activity compared to European American women (see figure 20.1). Additional findings showed that most adults spent less than 20 min per day in moderate and vigorous activity. Furthermore, fewer than 5% of the population, regardless of ethnicity, met the current recommendation of the Centers for Disease Control and Prevention and the American College of Sports Medicine for physical activity. Therefore, in terms of ethnic differences in activity, the main conclusion from these data is that European Americans are not more active than other ethnic groups. This singular finding stands in direct contrast to the majority of previous self-reported data. In addition, differences between the ethnic groups were small, especially given that overall, American adults are not very active.

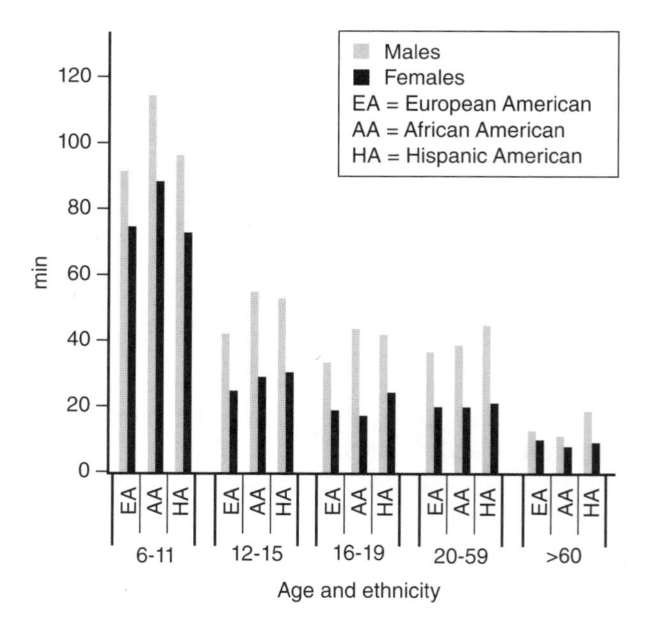

Figure 20.1 Average amount of time in moderate and vigorous activity per day by age, ethnicity, and gender.

Data from R.P. Troiano et al., 2008, "Physical activity in the United States measured by accelerometer," *Medicine and Science in Sports Exercise* 40: 181-188.

Sedentary Behavior in Adults

Ethnic differences have also been noted for sedentary behavior. The US Surgeon General (1), NHANES (2), and BRFSS (3) defined sedentary in terms of no leisure-time physical activity.

Self-Reported Behavior

The surgeon general (1) reported that African and Hispanic Americans had higher rates of inactivity compared to European Americans and that the most sedentary groups were African and Hispanic American women. The NHANES (2) and BRFSS (3) data indicate that European Americans have the lowest rates of no leisure-time physical activity (about 25%) and that Hispanic Americans have the highest rates (~35%). Individuals of a lower social class tend to be more active than those of a higher level. However, there are fewer ethnic differences within the same social class than when making broader comparisons including all groups (5).

When sedentary behavior was defined as TV viewing, those individuals who watched more than 2 h of TV per day had a greater likelihood of being overweight compared to those watching less than 2 h. In addition, overweight and obese individuals spent more time watching TV per day than normal-weight individuals. More importantly, a greater percentage of African and Hispanic Americans spent more than 4 h per day in sedentary pursuits, and African Americans were more likely to watch 2 h or more of TV per day in comparison to European American and Hispanic American adults (6). Therefore, epidemiological data strongly indicate that European Americans are the least sedentary ethnic group.

Objectively Measured Sedentary Behavior

The NHANES also reported on objectively measured sedentary behavior using accelerometers (7). The most surprising finding was that Hispanic Americans were less sedentary compared to European Americans across most age groups (see figure 20.2). In addition, there were few differences between African and European American adults. It is especially noteworthy that African and European American females spent similar amounts of time engaged in sedentary behavior despite the higher prevalence of both overweight and obesity in the former group. Overall, adults spent between 7 and 9 h per day engaged in sedentary behaviors.

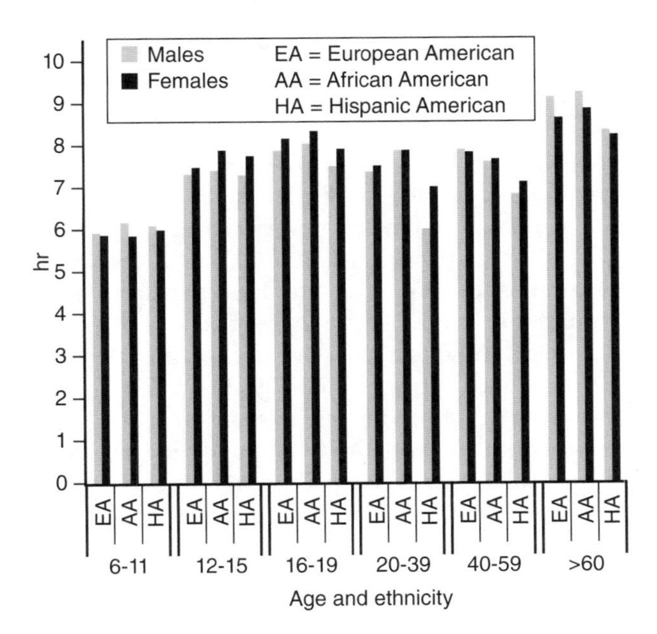

Figure 20.2 Average hours of sedentary behavior per day by age, ethnicity, and gender.

Data from C.E. Matthews et al., 2008, "Amount of time spent in sedentary behaviors in the United States, 2003-2004," *American Journal of Epidemiology* Apr 1, 167(7): 875-81.

Physical Activity in Children

Similar to studies among adults, both self-reported and objectively monitored physical activity levels are surveyed in children and youth.

Self-Reported Activity

As with adults, epidemiological data generally indicate that European American children engage in more activity than African American or Hispanic American children (2, 8-10). The surgeon general's report (2) indicated that Caucasian children engaged in more bouts of vigorous activity compared to African American or Hispanic American children. Similarly, NHANES III data (8) showed that European American boys and girls (87.9%, 77.1%) engage in more frequent bouts of vigorous activity compared to African American (77.6%, 69.4%) and Hispanic American (80.2%, 72.6%) children. More recent data on physical activity levels are provided by the Youth Media Campaign Longitudinal Survey (9). This survey showed that a smaller proportion of African and Hispanic American children (about 25%) reported involvement in organized sport compared to European American children (about 50%). However, differences in free-time activity were much smaller, as approximately 75% of African and Hispanic American and 80% of European American children reported participating within the previous week. Data from the National Longitudinal Study of Adolescent Health (10) showed that African and Hispanic American girls reported fewer bouts of moderate-intensity activity in comparison to Caucasian girls. However, there were no ethnic differences in boys.

Objectively Measured Activity

The NHANES accelerometry study (4) represents the largest study of accelerometers and youth to date. The children were divided into the age groups of 6 to 11 years, 12 to 15 years, and 16 to 19 years. In comparison to European American boys, African American boys had more counts per minute at the older and younger age groups, and Hispanic American boys had more counts at the older group. When expressed as time in moderate to vigorous physical activity (MVPA), African American boys were more active than European American boys at every age group and more active than Hispanic American boys at the younger age group. Hispanic American boys were more active than European American boys at the oldest age group. This replicates the counts per minute findings. The only ethnic difference in girls was found for counts per minute at the youngest age group; here African American girls were more active than Hispanic American girls. This exact same result was obtained for minutes in MVPA (see figure 20.1). However, there were no ethnic differences in the percentage of children meeting the American College of Sports Medicine recommendations (42%). These results are contradictory to self-reported activity in girls, which reflected lower levels of activity in African American girls.

Sedentary Behavior in Children

There are several indicators of sedentary behavior in children and youth. In many studies television viewing is used as a measure of sedentary behavior.

Self-Reported Behavior

Epidemiological data indicate that there are ethnic differences in sedentary behavior (8). African American boys and girls spent more hours on TV or video games per week in comparison to European and Hispanic Americans. African American boys and girls spent 29 and 27 h in front of the television and 6 and 2 h on video games, respectively. In comparison, European and Hispanic American boys and girls spent approximately 21 and 1.5 h on TV and games, respectively. These data indicate that African

American children spend more time in sedentary pursuits than children of other ethnic groups, which may predispose them to weight gain (8).

The association between physical activity, sedentary behavior, and body weight has been examined. Overall, both high levels of sedentary behavior and low levels of physical activity have been associated with greater odds of developing obesity. For boys, bouts of MVPA predicted weight gain one year later. Among girls, time in sedentary behavior predicted weight gain for European American girls, while MVPA predicted weight gain for Hispanic and African American girls (10). Therefore, MVPA appears to be a consistent predictor of short-term weight gain across ethnic groups.

Objectively Measured Behavior

As with physical activity data, there were no consistent differences between ethnic groups in the amount of time spent engaged in sedentary behavior according to the NHANES accelerometer substudy (7). The only difference was that African American girls aged 6 to 11 were less sedentary compared to European American and Hispanic American girls of the same age. There were no other significant differences in sedentary time for girls, and there were none for boys (see figure 20.2). It is important to note that self-report data is largely based on TV viewing, which only accounts for a portion of sedentary time. Objective data captures all sedentary activities. Therefore, it is possible for African American children to spend more time in front of the TV while also spending equal amounts of total sedentary time.

Conclusions

A major conclusion that can be drawn from this review is that ethnic differences in physical activity and sedentary behavior vary depending on whether subjective or objective methods are used. It is well known that self-reported data are subject to bias, in that individuals tend to overestimate the amount of time spent in physical activity and presumably underestimate the time spent being sedentary. Although objective data are not perfect, they provide a better estimate of activity. Therefore, objective data should be given more weight. Given this conclusion, there are few ethnic differences in physical activity and virtually no ethnic differences in sedentary behavior. The long-held belief that European Americans are more active and less sedentary than other groups may have to be questioned.

The lack of ethnic differences in adults and children may have health implications. African and Hispanic American women and girls still have higher rates of obesity in comparison to European American women. Objective data suggest that it is not that African American and Hispanic American females are less active but that all women are not active enough. However, it is unclear whether meeting the current physical activity recommendations would affect obesity in African and Hispanic females since they are just as active as European American females using this criteria. Given that sedentary behavior has more room for change, future projects may need to focus on reducing sedentary time while simultaneously increasing physical activity.

21

Psychological Factors and Physical Activity Level

Rod K. Dishman, PhD

The brain controls physical activity in different environments and social contexts. This control includes conscious choices that can be either planned or spontaneous and less conscious drives or urges. Inherent and acquired traits that influence these choices and drives are shaped by learning and memories of reinforcement history, by genes, and by various factors (developmental, social, environmental) that interact in ways that are not as yet known. Studies show that psychological, environmental, and genetic factors each explain a substantial amount of physical activity, but how they interact to determine levels and patterns of physical activity has received inadequate research attention. Nonetheless, most of the variation in leisure-time physical activity is ultimately determined by personal choices made in varying environments that facilitate or impede physical activity.

Moderators and Mediators of Physical Activity

Despite widespread intervention, physical activity among adults and youths in developed and developing nations remains below levels recommended for health promotion. The modest success of interventions in yielding sustainable increases in physical activity has led to renewed interest in identifying psychological *moderators* (variables that modify the relation or effect between an independent variable and physical activity) and *mediators* (variables that transmit the relation or effect of an independent variable on physical activity) that can guide successful interventions to increase physical activity levels (1). About 50 different behavioral correlates of physical activity have been reported among adults and youths worldwide. Of those correlates, putative moderators and mediators include personal factors that influence choices to be active, such as self-efficacy, perceived behavioral control, attitude, goals, intentions, enjoyment, social norms, and perceptions of social support and access to physical activity settings. Personality appears to be weakly associated with physical activity (2), but it could plausibly influence spontaneous physical activity or act as a moderator or mediator to help explain social–cognitive or gene–environment influences on exercise behavior. In addition, real and perceived features of social and physical environments operate at the levels of families, schools, places of employment, and neighborhoods, which are all located within communities, to influence physical activity. Figure 21.1 illustrates these theorized relations and interactions.

Social-cognitive variables (i.e., beliefs that are formed by social learning and reinforcement history) are influences on self-initiated change in health behaviors such as physical activity (3). Self-efficacy is a belief in personal capabilities to organize and execute the courses of action required to attain a behavioral goal. Like self-efficacy, perceived behavioral control includes efficacy beliefs about internal factors (e.g., skills, abilities, and self-motivation or willpower) and external factors (e.g., time, opportunity, obstacles, and dependence on other people) that are imposed on behavior. Each construct is distinguishable from outcome expectancy, which is

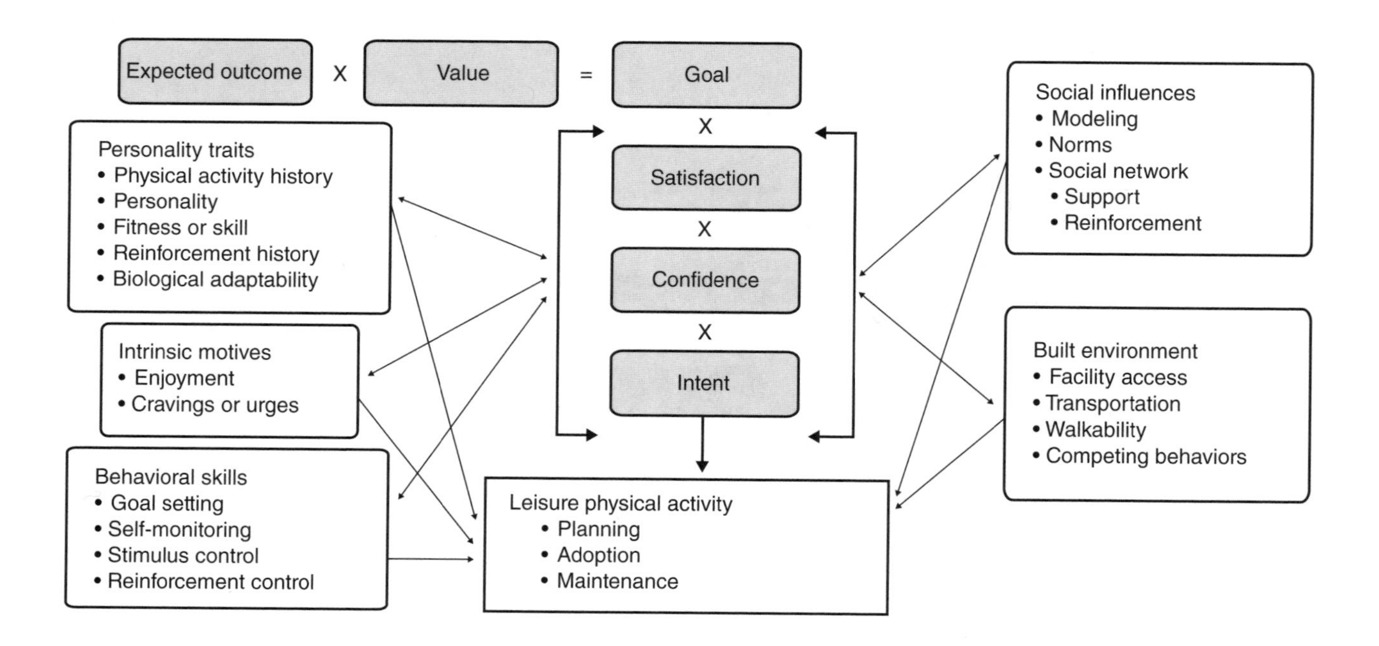

Figure 21.1 Model of interactive influences of personal, social environmental, and physical environmental influences on personal choice to be physically active.

Data from R.K. Dishman, 2008, "Gene-physical activity interactions in the etiology of obesity: Behavioral considerations," *Obesity 16(Suppl 3): S60-5.*

the perceived likelihood that performing a behavior will result in a specific outcome. Although people are more likely to form an intention to behave when they value an expected outcome of the behavior (e.g., they have a positive attitude), that likelihood is increased when a goal is set. People who set goals about being more active and who are dissatisfied with their current activity level will be more likely to adopt physical activity, especially if they have high self-efficacy about their ability to be physically active (4). Like perceived behavioral control, self-efficacy affects behavior both directly and indirectly by influencing intentions. Efficacy beliefs can also affect physical activity directly by fostering self-management (e.g., goal setting, self-monitoring, and self-reward) and indirectly by influencing perceptions about sociocultural environments that provide assistance for physical activity, which in turn directly influence physical activity. Once formed, self-efficacy, especially beliefs about overcoming barriers to physical activity, can also moderate the relation between perceptions of social facilitators and physical activity change. Thus, beliefs in personal ability to overcome barriers to physical activity can sustain physical activity in the face of increasing barriers or declining opportunities to be active.

Measurement of Social-Cognitive Moderators and Mediators

Measurement properties of the instruments used to assess moderators and mediators of physical activity usually have not been verified according to high standards using available methods, such as confirmatory factor analysis and item response modeling, to establish the factorial validity and measurement equivalence or invariance of the measures between different types of people and across time (4). Factorial validity is the degree to which the structure of a measure conforms to the theoretical definition of its construct. Multigroup factor invariance is the degree to which the configural (i.e., factor structure) and measurement (e.g., factor loadings of items and their errors) properties of a scale conform to the theoretical definition of its construct similarly between different groups of people. Longitudinal factor invariance is the degree to which those configural and measurement properties are similar across points in time. Without such evidence, nonequivalent measures confound the interpretation of research findings based on self-report measures. If people of different ages or ethnic backgrounds interpret questions differently, or if maturation or an intervention influences their interpretation of the questions, then observed differences in questionnaire scores between groups or between two or more administrations may not reflect true differences in the actual constructs underlying the questions.

Measurement of Physical Activity

Most studies have used self-reports of physical activity. Fewer, more recent, studies have used standardized observational systems or objective monitoring by accelerometry. Although some self-report measures have been validated through demonstration of correlations with objective measures, those correlations are typically modest, accounting for less than 25% shared variance. Self-report measures of physical activity used in observational, population-based studies generally were designed to categorize or rank people according to frequency of physical activity or total energy expenditure estimated crudely by metabolic equivalents (i.e., MET-hours) rather than to assess specific features of physical activity such as type, intensity, and timing that are important for understanding the direction, effort, and persistence of human motivation. Studies of physical activity are needed that compare results across time using both subjective and objective measures in order to enhance convergence of methods and exploit their unique assessment features.

Environments and Choice

Cross-sectional studies have shown that self-reports and objective measures from geographic information systems of the social and built environments (e.g., neighborhood safety and facility accessibility) are weak correlates of physical activity and overweight among population-based samples (5, 6). It has not been determined whether perceived access and actual proximity to physical activity settings have direct relations with physical activity, or indirect relations moderated or mediated by social-cognitive factors such as social support (7) and efficacy beliefs (8) about overcoming barriers to physical activity, or both direct and indirect relations.

Biology and Choice

Neurobiological mechanisms that explain how the brain regulates the choice to be physically active are poorly understood and have been studied mainly to increase understanding of central fatigue during exhaustive exercise and prolonged, strenuous exercise under extreme conditions that impair performance. In contrast, the neurobiological regulation of voluntary, nonstrenuous physical activity has received little study (9).

Reduced dopamine (DA) release or loss of DA receptors in brain appears to be related to the age-associated decline in physical activity observed among many species. The mesolimbic (i.e., ventral tegmentum–nucleus accumbens) DA system is a critical component of the forebrain circuitry that regulates appetitive aspects of motivation, including reallocation of instrumental behavior away from food-reinforced tasks that require high effort and toward the selection of less-effortful types of food seeking.

Activation of neuronal activity in hypothalamic reward regions during spontaneous locomotion by rats was established 40 years ago, and electrical self-stimulation of the ventral tegmental area has been used to artificially motivate treadmill running and weightlifting by rats. However, little is known about the role of the mesolimbic DA system in the motivation of voluntary physical activity. Genes that might help explain motivated running or spontaneous physical activity have yet to be identified. Nonetheless, early gene expression has been implicated in voluntary running. A recent study showed that c-fos and delta fosB in the nucleus accumbens were activated during wheel running in rats, and mice that overexpress delta fosB selectively in striatal dynorphin-containing neurons increased their daily running compared with control littermates (10). Delta fosB could plausibly facilitate wheel running by inhibiting the release of dynorphin by GABA (gamma-aminobutyric acid) neurons, which otherwise binds with kappa opioid receptors to inhibit DA release in the ventral tegmental area or nucleus accumbens. The hypothalamic neuropeptide orexin A stimulates both feeding and spontaneous physical activity in rats when it is injected into the lateral hypothalamus (11). Neurons that contain orexin A project from caudal hypothalamic areas throughout the neuroaxis and appear to enhance feeding and spontaneous physical activity, in part by inhibiting hypothalamic GABA release or by activating the opioid-dependent mesolimbic DA pathway between the ventral tegmental area and the nucleus accumbens (12).

Multilevel Models of Physical Activity Change

Conceptual models of physical activity should include variables measured at the level of the person, including the family and home environment, but also at the level of the community (e.g., neighborhoods, churches, workplaces, schools). See figure 21.2. Complex models are needed to describe the independent and interactive contributions of

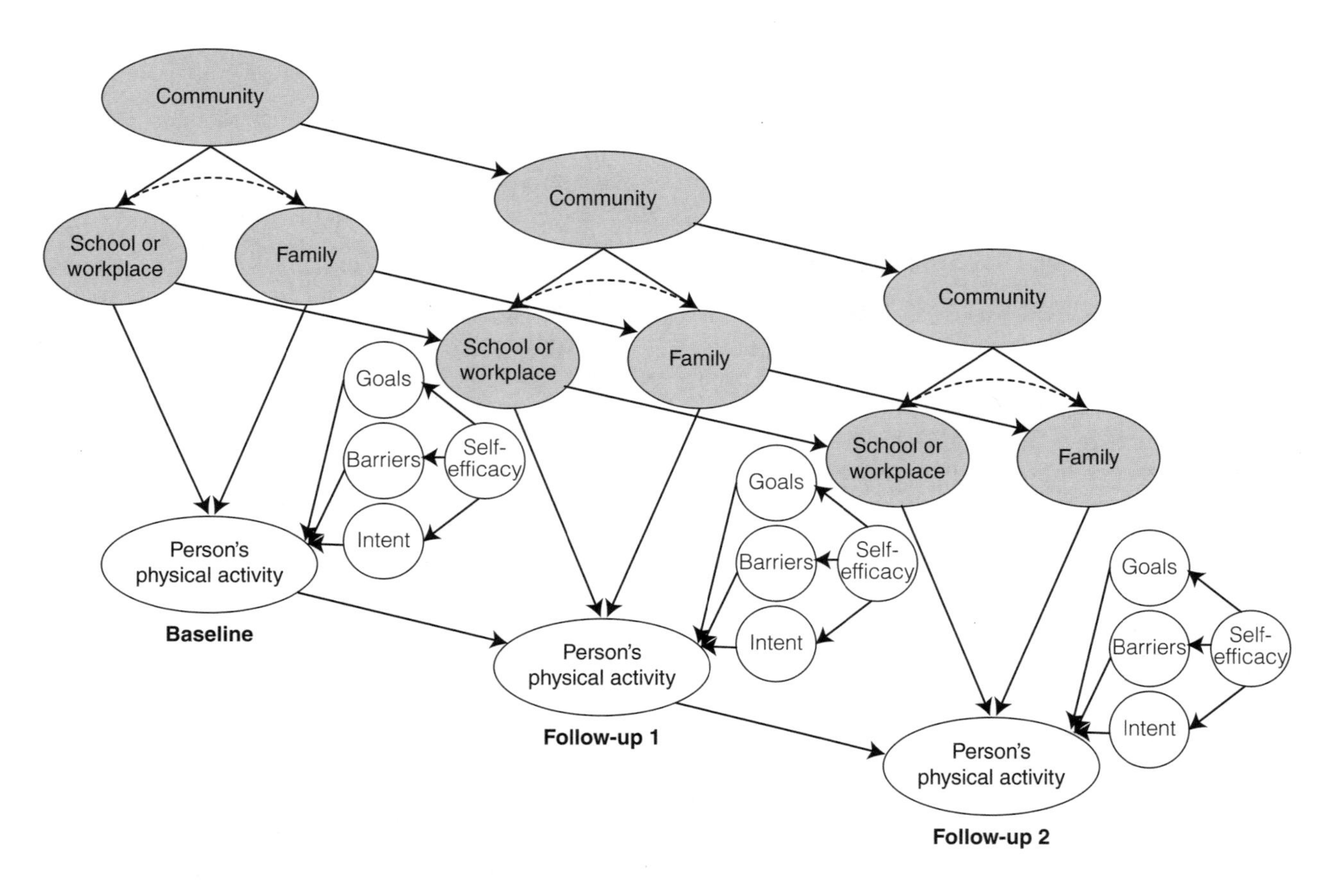

Figure 21.2 Multilevel modeling of personal- and group-level influences on social-cognitive mediation of change in physical activity.

key variables at each level to change in physical activity across multiple points in time. Such models require analytical methods that address inherently complex features, such as (a) the multilevel nature of the data; (b) the analysis of change across time; (c) hypothesis tests of independent (i.e., direct), mediated (i.e., indirect), and moderated (i.e., interactive) relations; (d) the use of different data forms including self-ratings, as well as objective measures of the physical and social environment; and (e) the need to demonstrate equivalence of the measurement properties of person-level variables between groups that describe different types of people (e.g., age, race, ethnicity, and gender) and within each of those groups across time. Common statistical techniques, such as analysis of variance and ordinary regression, cannot address all these complexities. Advanced techniques such as structural equation modeling (4) and latent growth modeling (7) provide precision for multilevel, theoretically derived analysis of mediated change in physical activity.

Summary

A contemporary challenge facing public health is the understanding of how personal motivation interacts with the physical and social environment to influence leisure-time physical activity. Meeting this challenge will require (a) inclusion of properly validated measures of putative psychological mediators (e.g., cultural values, efficacy and control beliefs, goals, intentions, enjoyment, and self-management skills) and moderators (e.g., age or maturation, personality, race-ethnicity, fitness, fatness, skill, and competing behaviors) of physical activity; (b) valid assessments of social-cognitive moderators and mediators of physical activity and specific features of physical activity exposure (i.e., type, intensity, timing, and context); (c) a search for candidate genes involved with motivational systems of energy expenditure; (d) manipulation of physical activity or prospective observation of change in physical activity at multiple times; and (e) use of statistical procedures that permit multilevel modeling (i.e., personal- and group-level variables) of direct, indirect (i.e., mediated), and moderated (i.e., interactions of mediators with external factors) relations with physical activity within complex theoretical networks that include social-cognitive and gene–environment networks.

22

Effects of the Built Environment on Physical Activity Level

James F. Sallis, PhD

Physical activity is enjoyable and welcomed in parks and health clubs, but not on the shoulders of busy roads or in classrooms. People do physical activity in specific places, and it is clear that some places are well suited for activity and some are not. Features of both the natural and built environments are likely to influence physical activity, but there is much more interest in built environments because these can be modified by policy. The natural environment includes weather and topography, and the built environment encompasses all buildings, spaces, and products that are created or modified by people. Communities, parks, roads, schools, health clubs, trails, shopping malls, playgrounds, and home exercise equipment are all part of the built environment.

Ecological models of behavior guide research and action on built environments. In contrast to models of behavior that specify only psychological and social influences on physical activity, ecological models take a broader view that built environments and policies also need to be considered (1). Educating and motivating individuals to be physically active is not likely to be effective when environments and policies create barriers to change. Ecological models can guide the use of multilevel interventions that educate individuals, change social norms, create suitable built environments, and enact policies that support physical activity. The goal is to provide people with "activity-friendly" environments that make it easy to choose to be physically active in daily life. Built-environment interventions may be needed to make a big impact on physical activity because they tend to affect all people in the environment (not just those who volunteer for a program) and to be permanent.

Built-environment attributes are expected to have highly specific effects on behavior. For example, living in a neighborhood with sidewalks may make it easier to be active but be irrelevant to healthy eating or alcohol abuse. Each of the four domains of physical activity—recreation, transportation, household, occupation—is likely to be influenced by different built-environment attributes (see figure 22.1). Improving understanding of the relation of built environments and policies to physical activity is an active field of research.

Need for Transdisciplinary Collaboration

Multiple research traditions have contributed to current built-environment research, but since none of the disciplines has the full range of needed concepts and measures, collaborative research teams are required (1). City planners and transportation planners have studied how the design of communities and transportation systems affects walking and cycling for transportation. They introduced the idea of "walkable" neighborhoods where people can walk to meet daily needs, such as shopping, because many destinations are near homes and the road network provides direct routes from place to place. By contrast, most suburbs built since the 1940s were based on zoning laws that required separation of uses, so shops and services were usually too far from homes to allow walking. Public health and behavioral science investigators studied the

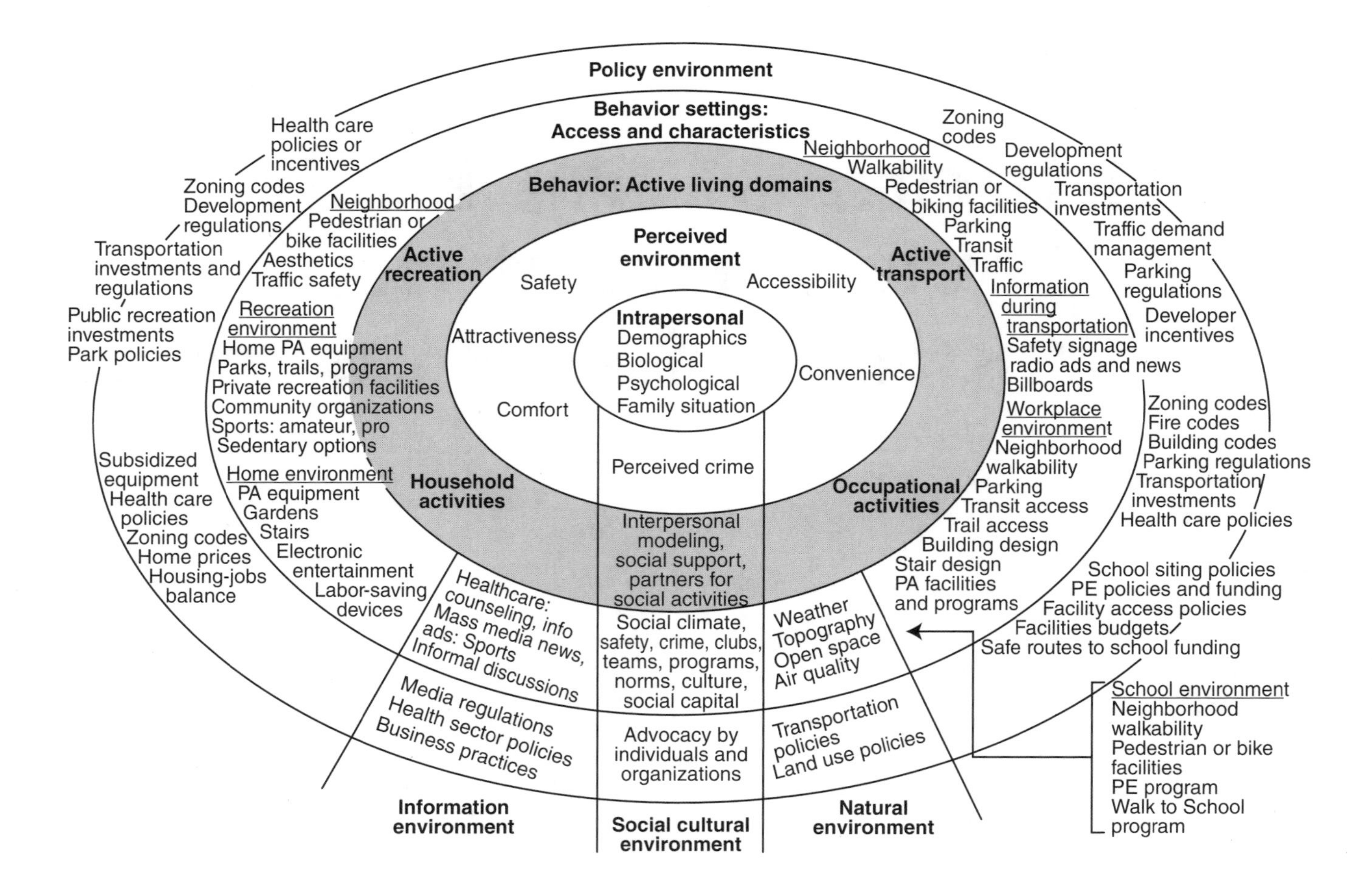

Figure 22.1 Ecological model of four domains of active living.

domain of recreational physical activity, trying to understand how access to public and private recreation facilities may encourage people to be active. Measures of the built environment in these studies were rudimentary. Parks and leisure researchers studied how people used parks and other resources. Geographers developed geographic information system software that allows detailed mapping and the analysis of spatial data.

These various research traditions merged in the early 2000s when investigators became aware of relevant work in multiple fields. Now teams of investigators are conducting true "transdisciplinary" research, developing ideas, methods, and findings that would not have arisen from any single discipline. Transdisciplinary teams are examining how a wide range of built-environment factors are related to all domains of physical activity, as well as other health outcomes. The first clear evidence of transdisciplinary collaboration was the development and evaluation of observational measures of community environments, parks, and trails that were needed for the research to advance (2).

Built Environments and Physical Activity in Adults

There is consistent evidence that adults who live in walkable communities walk and cycle more for transportation, and have higher levels of total physical activity, than those who live in low-walkable suburban areas (3). Differences between neighborhood types of 30 to 60 min per week of physical activity have been documented with both self-reports and objective measures such as accelerometers (4). The evidence was sufficient for the Centers for Disease Control and Prevention's Guide to Community Preventive Services to classify community-scale land-use changes as effective physical activity interventions (4). Cervero and Kockelman (5) introduced the "3 D" principles of designing communities that support active transportation: residential density, land-use diversity, and pedestrian-oriented designs.

Adults who live near parks and other recreation facilities generally do more recreational and total physical activity. This association has been documented in numerous studies, but due to the wide variety of methods used, it is not possible to quantify the effects. The aesthetics of recreation facilities and communities in general have been linked with physical activity (3, 4, 6). See table 22.1 for a summary of environmental attribute associations with active transportation and recreation. Other built-environment characteristics have not been studied enough or have yielded inconsistent results, including sidewalk presence and quality, other road characteristics, availability of parking, and hills.

Just a few studies of built environments have been reported for adults aged 65 years or greater. The results generally confirm that the physical activity associations with walkable community designs and proximity of recreation facilities generalize to older adults (6).

Built Environments and Physical Activity in Youth

Connections between built environments and physical activity have been documented for children and adolescents. Living in proximity to parks and other recreation facilities, as well as high aesthetic qualities, has been associated with higher physical activity among youth (6). This can be interpreted as suggesting that children need suitable places to play near their homes. Adolescents living in walkable neighborhoods are usually found to be more active than their suburban counterparts (6). More young people walk or cycle to school when they live in walkable neighborhoods (7). However, children may use cul-de-sacs as play areas because traffic is low. Thus, suburban road networks with low connectivity may simultaneously reduce walking to school and facilitate active recreation among youth (6).

For youth, there is better evidence that sidewalks are associated with high physical activity levels (6). School grounds designed for a variety of activity opportunities may stimulate more activity during free time, such as after lunch (8). Young people generally seem to benefit from the same built-environment factors as adults, but there are additional settings (e.g., schools) and activity behaviors (e.g., walking and cycling to school) that must be considered for youth.

Causal Role of Built Environments

Most of the studies on built environments and physical activity are cross-sectional, and they have been criticized because of the possibility that people who like physical activity move to walkable neighborhoods with parks. Several studies show there is some selection into activity-friendly communities, and there may not be enough such communities to meet the demand (9). However, it does not appear that self-selection can explain all the findings because, consistent with ecological models, both psychological and environmental factors seem to operate. Auto enthusiasts who lived in walkable

Table 22.1

Built-Environment Associations With Physical Activity for Transportation and Recreation Purposes, for Adults

Built-environment attribute	Active transport	Active recreation or total physical activity
Walkability: mixed land use, street connectivity, residential density	++	0
Sidewalks	?	+
Proximity of recreation facilities (parks, trails, private facilities)	0	++
Aesthetics of recreation facilities	xx	++

++: multiple findings of positive association; +: a few findings of positive association; 0: a few findings of no association; ?: inconsistent findings; xx: insufficient studies to summarize.

Reprinted from J.F. Sallis and J.Kerr, 2006, "Physical activity and the built environment," *President's Council on Physical Fitness and Sports Research Digest* Series 7(4): 1-8.

neighborhoods did not walk more, but living in suburban neighborhoods seemed to suppress physical activity among walking enthusiasts (9).

A major reason for the reliance on cross-sectional studies is that it is not possible to conduct a randomized trial in which people are assigned to live in different neighborhoods. However, there are several quasi-experimental evaluations of built-environment changes, though most are on a small scale. Building new trails usually increased trail use, especially when the trails were located in densely populated areas; introducing protected cycling lanes promoted cycling in Europe; changing road designs to slow traffic led to increased walking and cycling (10); and painting school playgrounds to stimulate active games led to long-term physical activity increases (11). These studies support a tentative conclusion that built environments can be causal influences on physical activity. However, these few studies are not sufficient to quantify the effects of built-environment changes.

Strengths and Weaknesses

Several aspects of this literature generate confidence in the results. The key findings about walkable neighborhoods, proximity of recreation facilities, and aesthetics have been replicated in numerous studies that differ in study designs, locations, and population characteristics. Many studies have used objective measures of built environments, physical activity, or both, so the literature is not dependent on only self-report measures. Transdisciplinary collaborations are producing innovations in models, measures, and analyses.

It is not well understood how built environment–physical activity associations may generalize to groups at high risk for physical inactivity, obesity, or chronic diseases. In some studies, low-income or racial-ethnic minority subgroups have not demonstrated the same associations as affluent or majority samples; but in other studies, racial-ethnic minority groups seemed to derive more benefit from activity-friendly environments (10).

Studies of built environments can be translated directly into policy recommendations that can be implemented by planning, transportation, recreation, and education agencies of government, along with recommendations for changed practices by architects, builders, landscape designers, and other private sector groups. However, the research has not been as influential as it could be. Studying broad variables like walkability, proximity to parks, and sidewalks does not provide sufficient guidance to tell designers and policy makers how to create optimal activity-friendly environments.

Future Directions

Since the turn of the 21st century, research on the built environment has become a recognized field, and consensus is rapidly developing about what kinds of built-environment changes are needed to improve physical activity in whole populations. Current research priorities are to (a) understand the built-environment characteristics that are most important for diverse population subgroups; (b) identify more specific built-environment supports for physical activity that can inform the design of optimal activity-friendly environments; (c) conduct prospective and quasi-experimental studies to improve the rigor of evidence; (d) improve understanding of how social environments (e.g., culture, social norms, media, crime) and built environments interact to influence physical activity; and (e) conduct economic and health impact assessment studies to enhance the relevance of built-environment studies for policy makers.

In a short time, the field of built-environment and physical activity research has expanded the search for solutions beyond educating the individual to a focus on environment and policy change. Researchers from many fields are using their professional relationships with practitioners to change practices, and investigators are communicating findings to policy makers in many sectors. If these efforts are successful, the groundwork will be laid for long-lasting improvements in physical activity.

part IV

Physical Activity and Risk of Obesity

A significant body of data suggests that energy expenditure associated with physical activity of all forms is an important determinant of long-term energy balance. The main objective of this section is to define the contributions of a sedentary lifestyle and low levels of leisure-time physical activity to the risk of obesity. Chapters 23 and 24 focus on the role of sedentary time in the risk of obesity in adults and children, respectively. The role of physical activity level in the risk of obesity is addressed in chapter 25 (adults) and chapter 26 (children). Subsequent chapters describe the contribution of childhood physical activity (chapter 27) and childhood obesity (chapter 28) to the risk of adulthood obesity. Chapter 29 is devoted to physical activity and the risk of obesity in older people. Next, chapter 30 addresses the important issue of the relation between physical fitness and the risk of obesity. Finally, chapter 31 is devoted to the interactions between diet and exercise on the risk of human obesity.

Sedentary Time and the Risk of Obesity in Adults

Ross Andersen, PhD

Global increases in the prevalence of obesity represent a serious health threat in both affluent and developing countries around the world. The etiology of this global epidemic remains complex and multifactorial. However, we do know that population-level increases in the prevalence of obesity occur when overall energy consumed is greater than energy expended. This global imbalance has persisted despite global public health efforts to encourage physical activity through increases in leisure-time physical activity. Bauman (1) has noted that physical activity levels have remained relatively stable in most countries over the past 30 years. During the same time period, the need for manual labor has declined, fewer people commute actively, and both adults and children are watching more television and spending more time on the computer. Recently, a large study of postmenopausal women who were followed for seven years showed that sedentary behavior was more closely correlated with a subsequent 10 lb (4.5 kg) weight gain than was leisure-time physical activity (2).

A sedentary lifestyle has been associated with an increased prevalence of obesity, a lower resting metabolic rate, higher rates of weight gain, and a greater likelihood of the metabolic syndrome. Sedentarism can be defined as the net daily activities that result in not more than 1.5 times the resting metabolic rate (3). The 2003-2004 National Health and Nutrition Examination Survey (NHANES) used Actigraph accelerometers to quantify the amount and intensity of physical activity performed by U.S. adults (4). This report showed that American adults spend an average of 7.7 h per day in sedentary activities. In addition, older adolescents and elderly adults are the two most sedentary groups. Alarmingly, these two age categories have shown the largest age-specific increases in the prevalence of obesity in many countries.

Recently, the association between sedentary behaviors and obesity was examined in a large sample of Canadian adults (aged 20-64 years) (5). Behaviors of interest included time spent on the computer, watching television, and reading. Television watching was the most popular sedentary activity for Canadian adults, with over 25% watching more than 15 h of television per week. The correlation between television watching and other sedentary behaviors was low ($r < 0.07$). Hours of television watched was linked to obesity. Only 14% of men who watched less than 5 h of television per week were obese, compared to 25% of those who reported watching 21 h or more. Similar associations between television watching and obesity emerged for Canadian women. Interestingly, the time spent reading was not significantly associated with obesity among Canadian adults. However, frequent computer users (>11 h per week) were more likely to be obese than those who used computers the least (<5 h per week).

Obesity and Sedentary Versus Active Choices

The U.S. surgeon general recommends that all Americans accumulate at least 30 min of moderate physical activity on most days of the week. The notion of accumulating activity can allow individuals the flexibility to incorporate opportunities to expend energy into the waking hours. Opportunities for physical activity include active commuting, relying less on labor-saving devices, and walking up stairs instead of riding escalators or elevators. We found that only 4.3% of obese commuters chose to walk up stairs versus riding up an adjacent escalator in an urban subway station (6). In contrast, fully 18.7% of nonobese persons walked up the stairs. In this investigation, we created a culturally sensitive sign featuring a fit African American woman climbing the stairs, which was placed between the stairs and the adjacent escalator. We reported that obese African American and Caucasian commuters were more than twice as likely to walk up the stairs when prompted with a sign. Furthermore, the obese commuters' stair use remained elevated the week after the sign was removed. These data suggest that when reminded, overweight individuals are more likely to be physically active. This study also highlights the importance of health care professionals working closely with communities in order to create opportunities for physical activity. These data are presented figure 23.1.

The U.S. Census Bureau (7) has reported that 87.7% of Americans drive their cars to work and that 77% drive alone. Public transportation was used to get to work by 4.7% of the population, while only 2.5% walked to work and 0.4% rode their bicycles. All individuals should be encouraged to explore new options for active commuting such as walking or cycling. In turn, using public transportation is often associated with opportunities for increased walking and stair use, which can help reduce sedentarism and increase activity levels.

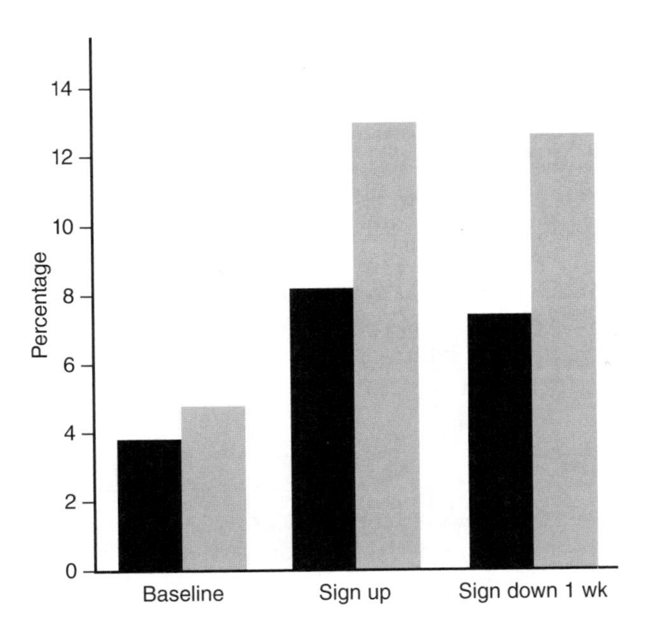

Figure 23.1 Race-specific patterns of stair use among obese Caucasian (gray bars) and African American (black bars) commuters before, during, and after the placement of culturally sensitive sign to promote the use of stairs in an urban subway station. Values are sample percentages.

Data from R.E. Andersen et al., 2006, "Effects of a culturally sensitive sign on the use of stairs in African American commuters," *Sozial-und Präventivmedizin* 51: 373-380.

Associations of Sedentary and Leisure-Time Activities With Obesity

The most common leisure-time sedentary activities are television watching, computer use, playing video games, and reading (2). It is important for health care professionals to understand that both time spent in vigorous activity and time spent in sedentary activities are *independently* and strongly associated with the risk of cardiovascular disease and markers of the metabolic syndrome (1, 8). Jakes and colleagues (8) stratified over 14,000 men and women by levels of vigorous activity and hours of reported television watching. They found that men and women who reported participating in no vigorous activity and watched the most television had the highest body mass indexes, highest blood pressures, and the worst lipid profiles. Conversely, the group who were the most active and watched television the least were also the leanest and had the lowest blood pressures. Moreover, the protective effects of an active lifestyle are diminished as hours of television watching increase (8).

Sedentary Activity and Weight Regain

Overweight and obese persons who lose as little as 10% of their initial body weight may achieve clinically meaningful improvements in many weight-related health problems. Unfortunately, these health advances are quickly reversed if weight is regained. Thus one of the greatest challenges facing obese patients who have lost weight is to ultimately manage their weight and prevent weight regain. We reported a dose–response relationship between weight regain over a one-year period and levels of activity in obese women who followed a comprehensive 16-week weight management program (9). During the one-year follow-up period, the most active third of patients met or exceeded the surgeon general's recommendations for activity in 79% of the weeks. The least active group did so only in 19% of the weeks over the follow-up period. Interestingly, the most active group lost an additional 1.9 kg (4 lb) during follow-up, and the most sedentary third regained 4.9 kg (11 lb) (9).

With use of data from the 1999–2002 NHANES (10), the prevalence of weight regain and predictors of weight regain among U.S. adults who experienced a meaningful loss of weight were examined. The study showed that 33.5% of the sample regained (more than 5%) weight in the year after losing it, and fully 59% maintained their weight. Adults who averaged 4 or more hours of screen time per day were twice as likely to regain lost weight compared to those who averaged 0 to 1 h per day. Moreover, adults who were classified as sedentary had significantly higher odds (1.8) of weight regain than those who met public health recommendations for physical activity.

The National Weight Control Registry (NWCR) comprises a nationally representative cohort of adults who have maintained a minimum loss of 13.6 kg (30 lb) for at least one year. Television watching habits were recently reported from a baseline assessment and a one-year follow-up (11). This study showed that a high proportion (62.3%) of the NWCR participants reported watching 10 or fewer hours of TV per week and 36% watched fewer than 5 h per week. Only 12.4% of these successful weight maintainers watched more than 21 h of TV per week, which is still much less than the TV viewing time of the typical American (12). In addition, increases in television viewing during the one-year follow-up were associated with significant weight regain, as were high baseline levels of TV watching.

Reducing Sedentary Time to Treat Obesity

Clearly, a strong association exists between sedentary time and obesity. A 2008 Nielsen report (12) on screen usage showed that Americans watch on average 127 h of television each month. An additional 26 h per month are spent surfing the Internet, while 2 h are spent watching videos on the Internet and 3 h are spent watching videos on mobile phones. Thus, clinicians working with overweight patients should be encouraging them to reduce the time spent in sedentary activities such as watching television, playing video games, and using the computer. To date, interventions aimed at reducing sedentary time have not been conducted in adults. Figure 23.2 shows how a model could ultimately help overweight adults manage their weight. From a metabolic perspective, doing less sedentary activity results in less time spent in a low-energy state. A perceived lack of time is the greatest barrier that adults report in connection with not participating in regular leisure-time physical activity. Therefore, reducing the time spent doing sedentary activities may result in more time available for participating in regular physical activity.

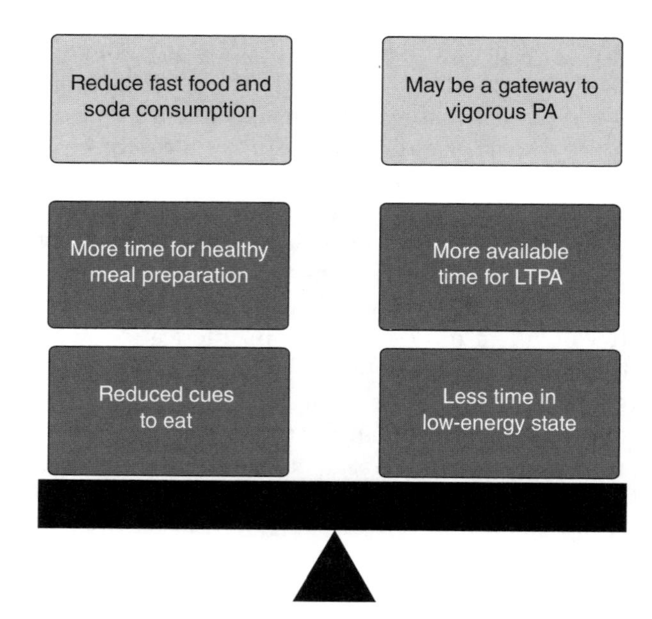

Figure 23.2 Hypothesized model of how reducing sedentary time could help obese adults manage their weight.

A few studies have examined the associations among television viewing, physical activity, and metabolic syndrome in cross-sectional samples of adults. These indicate that the time spent in sedentary activities is associated with an increased prevalence of the metabolic syndrome, independent of levels of physical activity. Persons with the metabolic syndrome should be encouraged to increase their levels of physical activity in addition to reducing the time spent watching television and in other sedentary behaviors.

Summary

Long-term weight management remains an elusive goal for many overweight adults. Health care professionals need to remind these individuals that in addition to engaging in a comprehensive physical activity program and sensible meal planning, it is important for them to limit the time spent in sedentary activities such as television watching and passive commuting in order to ultimately be successful with weight maintenance.

24

Sedentary Time and the Risk of Obesity in Children

Steven L. Gortmaker, PhD

Children and youth in the United States currently spend more than half of their waking hours in sedentary behaviors. Nationally representative accelerometer data from the United States in 2003-2004 (1) indicated that children ages 6 to 11 spent on average 6.1 h per day at a sedentary level (<1.5 metabolic equivalents); youth ages 12 to 15 and 16 to 19 spent on average 7.5 and 8 h per day at this sedentary level (figure 24.1).

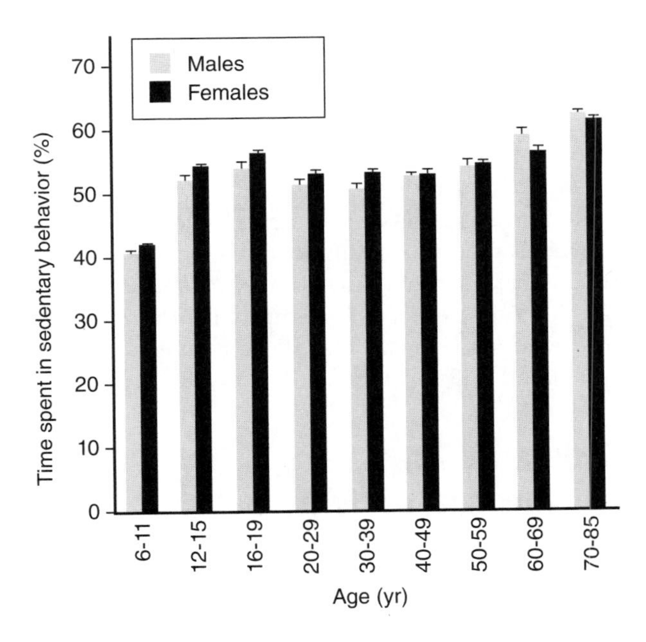

Figure 24.1 Percentage of time spent in sedentary behaviors, by age and gender, United States, 2003-2004 (1).

Charles E. Matthews, et al., "Amount of time spent in sedentary behaviors in the United States," *American Journal of Epidemiology*, 2000, by permission of Oxford University Press.

This chapter summarizes evidence for how this sedentary time affects childhood obesity. Our review of the available data indicates that a major relationship of sedentary time to obesity risk appears to operate via television time (2). Furthermore, recent evidence indicates that the major causal mechanism linking excess television viewing and obesity appears to be increased dietary intake (3, 4), a direct result of eating both while watching and at other times, and also as a result of effective marketing via the media (5). Less strong relationships are observed between other types of sedentary screen time and obesity risk, presumably because of a more limited influence of marketing. While the strongest links between sedentary time and obesity are observed for television viewing, interesting associations between sedentary time and overall physical activity levels have been reported for adults (6), still suggesting a role for the reduction of overall sedentary time as a potential pathway to decreasing obesity risk.

Television Viewing and Obesity Risk

While many aspects of diet and physical activity have been a focus of obesity research, some of the strongest evidence for a behavioral risk for obesity in children has been found in studies examining the impact of television viewing, a sedentary behavior. More than 20 years have passed since the first study linking television viewing to childhood obesity was published (7). Since that time, an abundance of epidemiologic and experimental evidence has emerged that supports the role of reducing television viewing as a primary preventive intervention for overweight in children and youth (2). Multiple observational studies document the impact of television viewing on relative weight and obesity, and these studies have been corroborated with randomized controlled trials designed to reduce levels of television viewing and reduce overweight (3, 8). Proposed mechanisms linking television viewing with obesity risk include those that affect both energy intake (food marketing leading to greater energy intake) and energy expenditure (lower resting energy expenditure and displacement of physical activity) (2).

Effects of TV Viewing on Energy Expenditure and Dietary Intake

Multiple observational studies show weak negative associations between television viewing time and levels of physical activity, and experimental studies do not show substantive evidence that decreasing television time increases overall physical activity levels. In addition, evidence for the potential impact of television viewing on lower resting energy expenditure is inconclusive (2). Thus, it appears that the effects of television viewing on overweight are in general not operating via these causal pathways. Rather, evidence indicates that the predominant causal path linking television viewing and obesity is via alterations in dietary intake. A recent randomized trial measured the effects of decreasing television time on both energy intake (measured via food records) and energy expenditure (measured via accelerometer) (3). Children (aged 4 to 7 years) who were randomized to the intervention arm showed substantial reductions in television viewing and computer use, reductions in total energy intake, and reductions in relative weight compared to the control group. The change in television viewing was related to the change in energy intake, but not to significant changes in physical activity, although the sample size was not large.

These results linking television viewing to energy intake are not surprising, as abundant evidence indicates that food advertising and marketing lead to greater energy intake (2). Most advertising geared toward children promotes high-calorie, low-nutrient foods, beverages, and meals, which, as concluded in a recent Institute of Medicine committee report, influences children to request and choose these products (5). Advertising appears to influence food eaten both while children are watching TV and while they are not (2). A prospective observational study indicated that increases in television viewing were associated with increased energy intake, and these changes were mediated by the consumption of foods commonly advertised on television (4). Every 1 h increase in television viewing was associated with more than an additional 100 kcal/day (4).

Other Sedentary Behaviors and Obesity Risk

Studies indicate that particular screen time activities like TV and computer use occupy at most half of sedentary time (1). Television was the predominant medium used by children at the time childhood obesity rates began to increase, and TV viewing is by far the most well-studied sedentary behavior in the obesity literature. As the media landscape continues to expand, however, the need to examine the impact of all types of media on physical activity, dietary intake, and obesity risk is evident. The ever-expanding media landscape includes broadcast, cable, and satellite TV; VCRs, DVRs, and DVDs; print media (books, magazines, newspapers); various audio media (broadcast, satellite, and cable radio; tapes; CDs; digital recordings); personal computers; online activities (World Wide Web, e-mail, instant messaging, gaming, music and video streaming); video games (both TV based and handheld); and portable telephones that connect to the Internet (9). The literature examining the relationship of these different forms of sedentary behavior to obesity risk, however, is relatively sparse; and no clear consensus about effects of these sedentary behaviors has emerged. Clearly a key issue to study is the potential impact of these new forms of media on advertising and marketing of food and beverages linked to obesity, as well as their relation to levels of physical activity. The impact that other media-free sedentary behaviors (such as reading and studying) have on obesity is less well studied, although early observations indicate that these activities are not associated with obesity (7).

Sedentary Time and Physical Activity Levels

Although a strong association between sedentary behaviors other than TV viewing and obesity has not been observed to date, among adults the amount of time spent in sedentary behaviors has been independently inversely associated with lower overall levels of physical activity energy expenditure (6). If true for children, these findings indicate that reducing overall sedentary behavior may still prove to be an effective approach to reducing obesity risk via increased physical activity. Although an increase in physical activity of vigorous intensity is a common obesity risk reduction recommendation, Westerterp found that the proportion of time distributed between activities of low and moderate intensity has the largest influence on the total energy expenditure (6). Matthews and colleagues calculated the potential impact of shifting sedentary time to light-intensity behaviors and concluded that even modest reductions in sedentary time could have the potential to increase expenditure and alter energy balance—assuming, of course, no effect of the increased activity on energy intake (1).

Summary

Available data indicate that the major causal pathway linking sedentary time to childhood obesity risk appears to operate via television time. Furthermore, the major mechanism linking television viewing and obesity appears to be increased dietary intake, likely a reflection of effective marketing via the media. In contrast, studies clearly indicate weak negative associations between television viewing time and levels of physical activity. Less consistent relationships are observed between other types of sedentary screen time and obesity risk, presumably because of more limited influence of marketing. While the strongest links between sedentary time and obesity are observed for television viewing, interesting inverse associations between sedentary time and overall physical activity levels have been reported for adults, still suggesting a role for the reduction of overall sedentary time as a potential pathway to decreasing obesity risk.

25

Physical Activity Level and the Risk of Obesity in Adults

John M. Jakicic, PhD

Excess body weight has been associated with an increase in health-related diseases such as heart disease, diabetes, certain forms of cancer, musculoskeletal disorders, and other related problems. Thus, it is important to consider physical activity opportunities to assist in the reduction in body weight, the prevention of weight gain, and the prevention of weight regain following weight loss. Moreover, physical activity should be considered for regulation of body weight, and physical activity may provide additional health-related benefits independent of body weight. The purpose of this chapter is to review the evidence for a link between physical activity and risk of obesity. Evidence from both cross-sectional and prospective studies is presented.

Cross-Sectional Association Between Body Weight and Physical Activity

There is a substantial literature to support the idea that physical activity is associated with lower body weight in adults (body mass index, BMI). For example, Brien and Katzmarzyk (1) provide cross-sectional evidence that physical activity may be beneficial for weight control. "Physically active" was defined as participating in at least 30 min of physical activity at least once per week. With use of this criterion to define physically active versus physically inactive, 11.8% of males who met the physically active criterion had a high BMI (≥29.2 kg/m^2) versus 20% of males who were defined as physically inactive. For females, 12.9% who met the physically active criterion had a high BMI (≥28.4 kg/m^2) versus 22% who were defined as physically inactive. Van Pelt and colleagues (2) reported that postmenopausal endurance-trained women had a lower BMI (20.6 ± 0.3 kg/m^2) compared to postmenopausal sedentary women (26.8 ± 0.8 kg/m^2). Moreover, waist circumference was significantly lower in endurance-trained (69.8 ± 0.7 cm [about 27.5 in.]) than in sedentary women (88.9 ± 2.0 cm [about 35 in.]). A similar pattern of results in BMI and waist circumference was shown for premenopausal women. However, a limitation of this study is that the endurance-trained women were highly trained athletes, which limits the understanding of how physical activity at lower thresholds may affect body weight.

Additional cross-sectional evidence supports the association of higher levels of physical activity with lower body weight. Ball and colleagues (3) reported that males engaging in high levels of leisure-time physical activity had significantly higher odds (1.76) of being in the normal BMI range (20 to 25 kg/m^2) when compared to those classified as sedentary. For females, the odds of being in a normal BMI range significantly improved in those with moderate (odds ratio = 2.31) or high levels of physical activity (odds ratio = 2.59) compared to those classified as sedentary. Moreover, occupational activity did not significantly alter the odds of being in a normal BMI range for either males or females. Lee and Paffenbarger (4) demonstrated from cross-sectional data that individuals who report levels of physical activity consistent with the consensus public health guidelines have a lower body weight than those not reporting this level of physical activity.

Additional studies consistently show similar patterns, with more physical activity being inversely associated with body weight or BMI. These studies tend to illustrate a dose–response relationship between physical activity and body weight or BMI. The extensive literature review of cross-sectional studies conducted by the Physical Activity Guidelines Advisory Committee for the recently released physical activity guidelines in the United States suggests the need for 150 to 300 min per week to maintain or significantly reduce body weight in adults (www.health.gov/PAGuidelines/Report/pdf/CommitteeReport.pdf).

Physical Activity and Weight Change

Evidence from prospective cohort studies, in which there is at least a baseline and one follow-up measure of both body weight and physical activity, appears to consistently demonstrate an association between weight change over the observation period and either change in physical activity or the single measure of physical activity at follow-up assessment. In a study conducted to examine predictors of weight gain in the Healthy Worker Project, French and colleagues (5) reported that over a two-year period, the average weight gain was 0.6 kg (1.3 lb) in women and 0.4 kg (0.9 lb) in men, with one walking session higher per week being predictive of lower body weight of 0.8 kg (1.8 lb) and 0.4 kg in women and men, respectively.

DiPietro and colleagues (6) analyzed data from 2501 healthy men (age range = 20 to 55 years) to examine the effect of physical activity dose on the trajectory of weight change; the average length of follow-up was five years. Physical activity was represented as low (<1.45 METs [metabolic equivalents] per 24 h), moderate (1.45 to 1.60 METs per 24 h), or high (>1.60 METs per 24 h) PAL (physical activity level) units. Physical activity level units are multiples of resting metabolic rate averaged over 24 h. Initial body weight was lower in those classified as high PAL at baseline compared to those classified as low or moderate PAL at baseline. Moreover, change in PAL over the follow-up period influenced change in body weight over this same period. For example, those initially classified as low PAL who increased to either moderate or high PAL showed a weight loss over the follow-up period. Moving from moderate to high PAL or maintaining high PAL resulted in the attenuation of weight gain over the follow-up period. These data suggest that maintaining relatively high levels of physical activity may attenuate weight gain in men, and that moving from a low level of physical activity to a higher level of physical activity may result in modest weight loss.

Cardiorespiratory Fitness and Weight Change

There is evidence that an improvement in cardiorespiratory fitness, which is a physiological parameter that improves with sufficient doses of physical activity, is associated with attenuation of weight gain. DiPietro and colleagues (7) reported an inverse association between change in fitness and change in body weight over an average follow-up period of 7.5 years in the Aerobics Center Longitudinal Study. Examination of data from 5353 adults (4599 men, 724 women) showed that the odds of gaining ≥5 kg (11 lb) of body weight were reduced by 14% (odds ratio = 0.86; 95% confidence interval = 0.83 to 0.89) in men and by 9% in women (odds ratio = 0.91; 95% confidence interval = 0.83 to 1.00) for every 1 min better treadmill time, which served as the measure of cardiorespiratory fitness. The corresponding odds for weight gain of ≥10 kg (22 lb) were 21% for men (odds ratio = 0.79; 95% confidence interval = 0.75 to 0.84) and 21% for women (odds ratio = 0.79; 95% confidence interval = 0.67 to 0.93). These data may reflect that it is necessary to engage in higher levels of physical activity to significantly improve cardiorespiratory fitness, and this dose of physical activity may be sufficient to attenuate weight gain or result in weight loss.

Sedentary Behavior and Weight Change

A review of the literature has demonstrated a consistent association between measures of sedentary behavior and greater obesity risk (8), with sedentary behavior represented as level of television viewing. For example, Ball and colleagues (9) followed a cohort of 8726 young women (age = 18 to 23 years) for a period of four years; results demonstrated that 41% of these young women gained weight over this period of time, which was defined as gaining ≥5% of their initial body weight. Women who reported spending ≥33 h per week sitting were 17% to 20% more likely to gain weight compared to those reporting sitting <33 h per week. When sedentary behavior is represented as the amount of television view time, a consistent relationship with an increase in body weight has been observed. Hu and associates (10) reported that baseline television viewing time was associated with an increased risk of becoming obese over a six-year follow-up period; there was a 23% increased risk of becoming obese with each additional 2 h per day of television viewing.

There may be a number of explanations for why sedentary behavior is predictive of weight gain or obesity (8). One potential explanation is that sedentary behavior occurs in place of more active behavior and therefore reduces energy expenditure. In addition, certain forms of sedentary behavior such as television viewing may be associated with increased energy intake, which produces a positive energy balance leading to weight gain. Thus, interventions may need to focus on strategies for reducing sedentary behavior, as this may result in a concurrent increase in physical activity that can significantly alter body weight and may prevent weight gain in adults.

Summary

There is clear evidence that physical activity is significantly associated with lower body weight and BMI and that physical activity may result in less weight gain over time. Moreover, there is some evidence from cross-sectional and prospective studies of a dose effect of physical activity on body weight. However, evidence from these studies suggests that the dose of activity that may be necessary to significantly alter body weight may be greater than the minimal public health recommendation of at least 30 min per day of moderate intensity physical activity (6). Thus, future research is needed to better understand the mechanisms by which physical activity alters body weight, the specific dose of physical activity necessary for the prevention of weight gain and obesity, and strategies to improve the adoption and maintenance of sufficient doses of physical activity. Moreover, the findings from cross-sectional and prospective observational studies may need replication using randomized designs to confirm the dose of physical activity that will significantly alter body weight long-term and to provide better understanding of the mechanisms affected by physical activity that alter body weight regulation.

26

Physical Activity Level and the Risk of Obesity in Children

Margarita S. Treuth, PhD

It is well documented that there has been a marked increase in obesity in children and adolescents over the last several decades. Given that energy balance is achieved when energy intake of food matches energy expenditure of the body, the concern arises when intake exceeds expenditure, a positive energy balance occurs, and weight gain results. Because activity energy expenditure or physical activity is the most variable component of daily total energy expenditure (TEE), decreased activity energy expenditure has been implicated as contributing to the rise in obesity prevalence.

Adequate levels of physical activity have therefore been promoted as a necessary component of lifelong health, recommended to begin in early childhood. There are several recommendations for children and adolescents at present. For example, the National Association for Sport and Physical Education (NASPE) and the Council on Physical Education for Children recommended at least 60 min and up to several hours per day on most days of the week (1). The activity should be age appropriate, and long periods of inactivity are discouraged (1). Another recommendation from the Institute of Medicine (IOM) (2) is to accumulate a minimum of 60 min of moderate and vigorous physical activity each day. Clearly, these recommendations point to the necessary promotion of an active lifestyle for children and adolescents.

Levels of Physical Activity in Children and Adolescents

Currently, physical activity in children and adolescents can be described in terms of the frequency, duration, and intensity of the activity with use of such techniques as self-report or accelerometers. Levels of energy expenditure and physical activity can be assessed by the doubly labeled water technique and indirect calorimetry. Recall that physical activity level (PAL) is calculated from the TEE divided by the resting metabolic rate (PAL = TEE / RMR). Physical activity levels in children and adolescents have been well documented. In a review by Torun and colleagues (3), the PAL ranged from 1.38 to 1.51 in studies including children aged 2.5 to 5.5 years. In the same review, the range of PALs from several studies for boys 6 to 13 years of age was 1.71 to 1.86, with a mean of 1.79 ± 0.06. Correspondingly for girls 6 to 13 years of age, the range was 1.69 to 1.90, with a mean of 1.80 ± 0.12. For adolescents over the age of 14 years, the mean for boys was 1.84 ± 0.05 (range 1.79 to 1.88) and for girls was 1.69 ± 0.03 (range 1.67 to 1.69).

The IOM report (2) provided a summary of the data from numerous doubly labeled water studies. The majority of the children and adolescents fell into the PAL range of 1.5 to 1.75. This would correspond to the low-active to active range. Given that these PAL values are known, it is of interest to examine how various levels of physical activity can affect energy requirements and ultimately body weight and obesity.

Impact of PAL on Energy Requirements

The report from the IOM contains the Dietary Reference Intakes for energy (2). This is important in the context of childhood obesity in that these estimates are influenced by PALs. The estimated energy requirement is defined as the dietary energy intake that is predicted to maintain energy balance in a healthy adult of a defined age, gender, weight, height, and level of physical activity consistent with good health (2). Because of growth and developmental changes in children and adolescents, requirements differ from those for adults. Therefore, the committee developed prediction equations specific for infants, children, and adolescents based on a comprehensive review of the scientific data. The report has equations for boys and girls 0 to 2 years of age (infants), 3 to 8 years of age (children), and 9 to 18 years of age (adolescents). The equations to estimate the energy requirements incorporate age, weight, height, the extra requirement for energy deposition associated with growth, and physical activity.

Four levels of activity include sedentary, low active, active, and very active based on PALs derived from previous doubly labeled water studies. These four PALs were given different physical activity coefficients in the equations. Thus to estimate the energy requirements of a child, one must choose the activity coefficient by estimating the activity level. Gender-specific equations were developed because of the differences between boys and girls in growth and fat deposition. Equations can be found in the report for each age group (infants, children, and adolescents). These equations are useful for estimating the energy needs of individuals and are fairly straightforward, as only age, weight, and height are needed. One challenge and limitation of these equations is the need to choose the appropriate activity level (and therefore the activity coefficient). If one wishes to accurately predict a child's energy requirements, it is clear that the PAL must be chosen correctly. Otherwise, a mismatch between intake and expenditure could occur, leading to obesity.

Another illustration of the influence of PAL on weight gain can be seen in the report by Butte and colleagues (4). They developed prediction equations for Hispanic children and adolescents who were followed for one year. A marked weight gain of 6.1 kg (13.4 lb) per year (range 2.4 to 11.4 kg [5 to 25 lb] per year) was observed in the Hispanic children. If the weight gain was due to a decline in physical activity with energy intake held constant, PAL would have to have decreased by 0.22 units. This would translate into about 60 min a day (range 18 to 105 min per day) of walking at 2.5 mph (4 km/h). This clearly demonstrates how a drop in PAL can influence weight gain.

Based on the recommendation from the IOM, a PAL of 1.6 is necessary to prevent excess weight gain (2). This means that in order to move from a very sedentary (PAL of 1.4) to an active lifestyle (PAL >1.6), children and adolescents must participate in moderately intense activity for a total of 60 min per day. The point is that the level of physical activity clearly affects the energy requirements, and thus it is obvious that imbalance in energy expenditure due to variations in PAL is a likely variable to explore as a potential predictor of obesity in children and adolescents.

Activity Energy Expenditure and PAL as Predictors of Obesity

Measures of physical activity, other than doubly labeled water, have been used to examine whether physical activity predicts changes in adiposity over time. Longitudinal data from the Framingham Children's Study reflect eight-year changes in body fat and physical activity (measured by accelerometry) in a large sample of boys and girls (5). The study showed smaller gains in body mass index (BMI) and skinfolds in those children in the highest tertile of average daily activity (5). A review by Must and Tybor (6) of prospective observational studies on the development of overweight in relation to physical activity, as well as sedentary behavior, presented mixed findings. However, the general conclusion was that increased physical activity was protective against relative weight and fat gains during childhood and adolescence. The authors did recommend that improved measurement methods be utilized in future studies (6).

Several well-designed longitudinal studies (7-12) on children at risk of becoming obese by virtue of parental obesity or reduced energy expenditures (measured by doubly labeled water) have examined whether energy expenditure plays a role in the development of obesity. The studies that have been conducted in children and adolescents are reviewed here, with the exclusion of studies in infants. Table 26.1 summarizes the main findings of these studies.

Childhood

Several longitudinal studies conducted in children to evaluate the role of energy expenditure (EE) in the regulation of body weight have yielded consistent results (7-12). In one study (7), body composition as well as resting and total EE by calorimetry and doubly labeled water was measured in Caucasian preadolescent boys and girls over a four-year period. Using hierarchical linear modeling and analysis of variance, the influence of sex, EE, initial fat mass (FM), and parental FM on the rate of change in FM was analyzed. The major determinants of the change in FM adjusted for fat-free mass were sex, initial fatness, and parental fatness. A reduced EE (either resting EE, total EE, or activity EE) was not a predictor of the change in FM. Johnson and

Table 26.1

Energy Expenditure and the Relation to Weight Gain in Children and Adolescents

Source	N	Characteristics of study participants	Length of follow-up	Component of energy expenditure measured	Outcome measure	Results
Goran et al. 1998 (7)	75	4 to 7.2 years; M and F Caucasian	4 years	TEE, PAEE by DLW	Rate of change in FM adjusted for FFM	Strongest predictors of change in FM were fatness levels in fathers and mothers. After adjustment for parental fatness, EE was not related to change in FM.
Johnson et al. 2000 (8)	115	4.6 to 11 years; M and F Caucasian, African American	3 to 5 years	TEE, PAEE by DLW	Rate of increasing FM relative to increasing FFM	Significant predictor of FM gain was aerobic fitness (negative relationship between aerobic fitness and FM).
Figueroa-Colon et al. 2000 (9)	47	4.8 to 8.9 years; F Race/ethnicity N/A	2.7 years	TEE, activity EE, sleep EE by calorimetry	Change in FM and % fat	No relationship between EE and change in body fat at 2 years.
Salbe et al. 2002 (10)	138	5 years; M and F Native American	5 years	RMR, TEE, PAEE by DLW	% fat	No significant predictors of the change in % body fat.
Treuth et al. 2003 (11)	101	8 to 9 years; F Caucasian, African American, Hispanic	2 years	TEE, PAEE by DLW; TEE, BMR, activity EE, sleep EE by calorimetry	FM, % fat; change in FM and % fat	Predictors of change in % fat were free-living total EE (negative) and muscle oxidative capacity (positive).
Bandini et al. 2004 (12)	196	8 to 12 years; F Caucasian, African American, Hispanic, Asian	7.2 ± 2.6 years (mean ± SD)	RMR, TEE by DLW, activity EE	Change in BMI z-score; change in % fat by bioimpedance	No relationship between RMR, AEE, TEE, and % fat; AEE was positively associated with change in BMI z-score.

F = females; M = males; TEE = total energy expenditure; AEE = activity energy expenditure; PAEE = physical activity energy expenditure; EE = energy expenditure; DLW = doubly labeled water; BMR = basal metabolic rate; FM = fat mass; FFM = fat-free mass.

colleagues (8) studied 115 boys and girls over a three- to five-year period. Measures of total EE and activity EE were examined as predictors of the rate of increasing FM relative to the increase in fat-free mass. Neither of these variables was found to be predictive of the change in FM. However, aerobic fitness was found to be a significant predictor of the change in FM, with a negative relationship observed between fitness and FM (8).

Three other longitudinal studies on this issue have been conducted in children. Figueroa-Colon and colleagues (9) studied 47 prepubertal girls using 24 h room respiration calorimetry. Two follow-up measures were taken, at 1.6 years and 2.7 years after the baseline measures. At the 1.6-year follow-up, sleeping EE was positively related to change in body fat, and activity EE was inversely related to change in body fat. However, after 2.7 years of follow-up, neither sleeping EE nor activity EE was related to change in body fat. Salbe and associates (10) studied obese Native American children at age 5 and again at age 10. Weight more than doubled over the five years, and activity EE increased over that time period. Treuth and colleagues (11) studied 101 prepubertal, multiethnic, normal-weight girls over a two-year period. Measures of fitness, baseline EE by 24 h calorimetry, total EE by doubly labeled water, and muscle oxidative capacity by nuclear magnetic resonance imaging were used to examine the predictors of weight or fat gain in these girls. Free-living total EE was negatively predictive of the change in percent fat. Muscle oxidative capacity, an indicator of fitness, was positively predictive of the change in percent fat.

In general, the published studies have yielded somewhat similar results. The differences in findings as to the predictors of weight or fat gain from these studies could be due to differences in samples (boys or girls, age, normal or overweight at baseline) and methods of analysis of the data (adjusting or not adjusting for fat-free mass).

Adolescence

In the MIT Growth and Development Study (12), EE was measured in the premenarcheal period; and body composition, dietary intake, and physical activity were measured annually until four years after menarche with a mean follow-up period of 7.1 ± 2.6 years. A linear mixed effects model was used to evaluate the longitudinal relationship between BMI z-score and percent body fat and measures of baseline resting metabolic rate, activity EE, and total EE after adjustment for potential covariates including physical activity, age, age at menarche, and diet composition. Resting metabolic rate was not related to change in percent body fat or BMI z-score. A small positive relationship of activity EE and total EE with BMI z-score but not percent body fat was observed. When the results were stratified by parental overweight, the findings were unchanged for resting metabolic rate. Total EE and activity EE were positively related to BMI z-score in girls of overweight parents. These data do not support the hypothesis that low EE is a risk factor for weight gain in girls during adolescence (12).

Future Directions

The relationship of EE and of physical activity EE to weight gain has been well studied in children and adolescents, and the general consensus for studies using doubly labeled water is that a lower TEE is not predictive of weight gain. For activity EE or PAL, the evidence is not conclusive. Many study design issues contribute to the different results. For instance, parental weight status, race-ethnicity, pubertal status, and duration of the study can influence the findings. Longitudinal studies designed to examine the relationship of EE with growth and development during childhood and adolescence must consider these factors. Perhaps the duration of these studies needs to be longer and the research continued into young adulthood. Given that EE studies are costly, this may be impractical. However, other methods of assessing physical activity (and not PAL per se as measured by doubly labeled water) can be utilized.

Summary

Research studies on the predictors of weight gain (specifically those that focus on EE measured by doubly labeled water) in children and adolescents have yielded semiconclusive results. Most studies suggest that lower TEE does not influence weight gain. Studies using self-report and accelerometry support the role of physical activity as influencing adiposity (5, 6). The use of various measures of EE and physical activity in these investigations makes drawing firm conclusions more difficult. In children and adolescents, in whom the effects of normal growth and maturation must be taken into account, the issues tend to be more complicated. However, it is certain that the increasing prevalence of obesity in the United States warrants attention to the role of EE and physical activity in the regulation of body weight.

27

Childhood and Adolescent Physical Activity and Risk of Adult Obesity

Robert M. Malina, PhD

Regular physical activity may prevent the appearance of risk factors for cardiovascular and metabolic diseases, or modify their expression, or do both. Among these risk factors, obesity is perhaps primary during childhood and adolescence. Physical activity during childhood and adolescence may also influence activity habits in adulthood. This chapter deals with physical activity during childhood and adolescence in the context of adiposity and overweight or obesity in adulthood. Physical activity during youth may reduce risk in adulthood by reducing the accumulation of adiposity and unhealthy weight gain, and also may reduce risk in adulthood if these habits track into adulthood and function to regulate weight gain and prevent adult-onset obesity.

Since the body mass index (BMI) is the most commonly used indicator of weight-for-height across the life span and is the criterion for overweight and obesity, it is imperative to understand the growth pattern of the BMI and the influence of individual differences in the timing and tempo of maturation. It is likewise important to understand age- and maturity-associated variation in energy expenditure and physical activity.

Body Mass Index as Influenced by Growth and Maturation

Height and weight increase linearly with age across childhood and adolescence. The BMI, on the other hand, declines from infancy through early childhood, reaches a nadir at about 5 to 7 years of age, and then increases linearly with age through adolescence into adulthood. The rise in the BMI after it reaches a nadir at 5 to 7 years of age has been labeled the "adiposity rebound." Some evidence suggests that an early rebound is associated with increased risk of being overweight or obese in late adolescence and young adulthood.

Change in the BMI after the rebound and through adolescence into adulthood is an additional factor that influences risk. A key factor is differential timing of sexual maturation and the growth spurt that influences the accumulation of body fat, variation in subcutaneous and visceral adiposity, and risk for overweight or obesity. Among girls more so than boys, earlier onset of puberty is associated with higher levels of fatness and higher risk for overweight or obesity. Early-maturing girls have an elevated BMI compared to late-maturing girls; this trend is already apparent in childhood. Although males ordinarily decline in percentage fatness during puberty and the growth spurt due to rapid growth of fat-free mass, fat mass increases from childhood through adolescence. Early-maturing boys tend to have, on average, more weight-for-height and, in turn, a higher BMI during childhood, and this persists through adolescence. The elevated BMI is associated with both increased fat-free mass and increased fat mass. Sexual maturation in males is also associated with a proportionally greater central adiposity, and boys advanced in maturity status have more central adiposity not only during adolescence but also in young adulthood. Maximum rate of growth of the BMI and BMI at maximum velocity during the adolescent spurt are associated with overweight (BMI ≥25.0 kg/m^2) at 35 to 45 years of age in both sexes. The rate of increase in the BMI is more strongly related to later overweight in males than in females, but the BMI at maximum velocity is a better predictor of adult overweight in females than in males (1).

Tracking Body Mass Index and Risk of Overweight or Obesity

Correlations between the BMI in childhood and BMI in adulthood are moderate but are higher during adolescence. At higher percentiles of the BMI, however, the BMI in childhood and adolescence tracks at higher levels so that the probability of many overweight and obese youth being overweight or obese in adulthood is significantly increased (1).

Energy Expenditure and Physical Activity in Childhood and Adolescence

Estimated 24 h energy expenditure (EE, kilocalories per kilogram) based on doubly labeled water declines with age, beginning as early as 4 to 5 years. The decline is especially apparent during the second decade of life. Energy expended in physical activity is the most variable component of EE. A variety of measures of physical activity indicate rather stable levels or a slight increase during childhood and a subsequent decline during the second decade. Activity levels appear to peak at about 12 to 14 years and then decline, but the timing and magnitude of the decline vary among studies and with instrument and contexts of activity (1).

Tracking Physical Activity

Physical activity is an individual characteristic that tracks at moderately stable levels across childhood into adolescence and into young adulthood; however, as intervals between observations increase, stability of physical activity levels declines (1). The relatively low-to-moderate interage correlations reflect to some extent limitations of methods of estimating physical activity and also meanings attached to physical activity at different ages by children, adolescents, and adults.

Physical Activity and Adiposity During Childhood and Adolescence

Physical activity may potentially lower adult risk of overweight or obesity by reducing adiposity during childhood and adolescence. Nonobese youth who are relatively high in physical activity tend to have less adiposity (BMI, skinfolds, estimated fatness), but enhanced activity programs in normal-weight youth appear to have a minimal effect on adiposity (2). Normal-weight youth may require a greater activity volume to bring about changes in adiposity. Experimental (intervention) studies with overweight or obese youth indicate reductions in overall and visceral adiposity with programs of moderate- to vigorous-intensity activity (2). Unfortunately, the studies do not address the issue of the amount of activity needed to maintain the beneficial effects of activity programs on adiposity.

A topic that periodically surfaces is the prevention of "unhealthy weight gain" during childhood and adolescence in order to reduce the risk of overweight or obesity during youth and in adulthood. Given the individuality of growth rate and the timing and tempo of growth and maturation during the adolescent growth spurt, it may be difficult to specify "unhealthy weight gain." However, if physical activity is associated with smaller gains in the BMI in youth (2), maintenance of smaller gains in the BMI through physical activity over time may prevent unhealthy weight gain and in turn reduce risk of overweight or obesity in adults. Two longitudinal studies suggest an important role of physical activity in the prevention of overweight or obesity in different phases of growth: More active children between 4 and 11 years have less fatness in early adolescence and may also have a later adiposity rebound (3), and an increase in physical activity during adolescence may limit the accrual of fat mass in males but not females (4).

Childhood and Adolescent Activity and Adiposity in Adulthood

Several longitudinal studies have related physical activity and inactivity during childhood and adolescence to weight status in adulthood. Results are suggestive but mixed. Adolescent physical activity (13 to 16 years) is not related to subcutaneous fatness at 27 to 29 years in both sexes but is positively related to the waist–hip ratio in females (5). Higher levels of activity during adolescence are predictive of lower skinfolds in young adulthood but not of a lower BMI (6), while higher participation in sport at 16 years is associated with a lower BMI at 30 years (7). On the other hand, higher levels of physical activity during childhood and adolescence do not reduce the risk of overweight or obesity in males at 39 to 41 years (8), and sport participation at 13 to 18 years is not associated with the BMI, percentage fat, and waist circumference at 40 years in males (9).

In contrast, indicators of physical inactivity during youth are related to risk of adult overweight or obesity. Time viewing television (>2 h per day) between 5 to 15 years is associated with an elevated BMI at 26 years of age (10), while time in sedentary behaviors (≥4 h per day) at 16 years is associated with an increase in the BMI at 30 years (7).

Transition From Adolescence Into Adulthood

There is a gap in data for the transitional years from adolescence into adulthood. This period is highlighted by the move from high school to college and career or to the workforce. Changes in lifestyle associated with the transition include, among others, an increase in sedentary behaviors and a decline in physical activities, which are risk factors for overweight or obesity. Activity during adolescence (13 to 16 years) and in young adulthood (21 years) is negatively related to subcutaneous adiposity in both sexes and positively related to the waist–hip ratio in females but not in males (5). Males active as adolescents at 14 years but inactive as adults at 31 years are at elevated risk of both overweight and obesity, while females showing the shift from adolescent activity to adult inactivity have elevated risk of obesity and specifically abdominal obesity reflected in waist circumference (11). On the other hand, males inactive at both 14 and 31 years are at increased risk of adult abdominal obesity, whereas no associations with obesity are evident in females who are inactive at both ages (11).

Conclusion

Risk factors for cardiovascular and metabolic complications cluster with obesity. Thus the prevention of overweight or obesity during youth has potential for long-term health benefits. The available data need to be evaluated in the context of age- and maturity-associated changes in indicators of overweight or obesity and in indicators of physical activity and inactivity. An additional consideration is variation in measures of physical activity and inactivity. Evidence linking physical activity during childhood and adolescence, as well physical inactivity during adolescence, to risk of adult overweight or obesity is suggestive. Increasing levels of habitual physical activity and reducing sedentary behaviors during childhood and adolescence and into adulthood can play a significant role in prevention of adult overweight or obesity.

Childhood Obesity and the Risk of Adult Obesity

François Trudeau, PhD

The purpose of the present review is to examine the literature on obesity persistence from childhood into adulthood. Another way to frame the issue is to ask, What is the risk of becoming an obese adult if the individual is obese in childhood or adolescence? This question is an important one, as it may help to identify children and adolescents at risk of becoming obese adults. It has been suggested that in the early teens we can obtain a pretty good account of the adiposity of future adults (1).

Investigators have tracked different measures of body composition—more often body mass index (BMI), sometimes skinfolds, and, more recently, waist circumference (WC)—over various periods spanning childhood to adulthood, to examine how well adult obesity can be predicted from measurements at a younger age. Tracking can be defined as "maintenance of the relative position in a group with respect to time" (2, p. 538) or "the stability or the conservation of a certain risk factor over time, or the predictability of a measurement of a certain factor early in life for values of the same risk factor later in life" (3, p. 888). Tracking of obesity can be expressed as (1) correlation coefficients of obesity measurements between two points in time, (2) relative risks and odds ratios (ORs) of remaining obese, or (3) percentages of obese or overweight children remaining in the same category as adults. Bloom argued that if a biological characteristic has an interage correlation of at least $r = 0.50$ over an interval of one year or longer, it should be considered a stable trait (4). From this criterion, analysis of the literature indicates that whether expressed as BMI or skinfold, body fatness at a young age in obese or normal-weight subjects is moderately correlated with adult values (3). However, tracking generally decreases as the time elapsed between the two measurements of body fatness increases (5, 6). For the purpose of the present chapter, we are more interested in tracking to the extreme (i.e., to verify to what extent obese children become obese adults). Accordingly, we mostly present results of obesity tracking according to the methodology used to estimate obesity. (See table 28.1 for a comparison of methodologies.)

Body Mass Index

The great majority of interage obesity tracking studies have been performed with BMI, which is the most common index of obesity in epidemiological settings but is less appropriate for the clinical evaluation of individuals, since variations in fat mass can be confounded with changes in fat-free mass (FFM). However, excess BMI in averaged data for a large sample usually reflects obesity. Another problem with BMI is that many studies have examined self-declared weight and height to calculate BMI, and there are differences between these and others monitoring real measurements (7).

With BMI, childhood to adulthood interage correlation coefficients and relative risks are lower than those calculated from adolescence (postpubertal) to adulthood. As an example, Whitaker and colleagues (8) investigated BMI and observed a progression of the odds ratios from 1.3 to 20.3 of remaining obese or very obese as an adult if the individual was obese at age 1-2 to 15-17 years. When only the

Table 28.1

Comparison of Methods Used to Follow Up Obesity From Childhood and Adolescence to Adulthood

	Body mass index	Skinfolds	Waist circumference or abdominal fat content
Methodological challenges or caveats for follow-ups	• A surrogate measure of body fat • Use of self-declared weight and height • Confounding effect of growth, race, and gender on evolution of fat-free mass/fat mass ratio	• A partial picture of fat content (subcutaneous adipose tissue only) • Possibility of racial and gender differences according to the skinfolds considered • Accuracy dependent on the operator • Lack of standards and cutoffs at the present time for all age groups	• A partial picture of fat content (abdominal subcutaneous and intra-abdominal fat) • Need to compare tracking of anthropometric (waist circumference) with dual-energy X-ray absorptiometry data • Lack of standards and cutoffs at the present time for all age groups
Advantages	• High availability of data and standards	• Possibility of predicting percentage of fat mass with an appropriate prediction equation • Relative simplicity	• May be a better predictor of future adulthood morbidity (e.g., metabolic syndrome) than the two other methods

very obese were considered, the OR of becoming an obese adult was 2 at 1 to 2 years, but up to 44.3 at 10 to 14 years old! The relative risk of remaining obese in adulthood is thus stronger for the more obese categories during childhood. Whitaker and associates (8) concluded that after 10 years of age, parental influences are more limited than previously, and that childhood characteristics become more significant in predicting adult obesity. Even in studies with fewer intermediate points, such a progression in the likelihood of remaining obese is also observed as the initial measurement is closer to the adult measurement.

Tracking of obesity from postpuberty to adulthood is somewhat greater than between prepuberty and adulthood for two reasons: the shorter time period and the adolescence growth spurt. The first reason is quite obvious: An increased time span is more likely to be accompanied by body composition changes from many causes (diseases, altered socioeconomic conditions, etc.). However, the second reason is associated with changes in body composition linked to puberty (i.e., changes in the proportion of fat vs. FFM). Tracking of obesity measurements between childhood and adulthood may be more difficult to interpret when growth spurt periods are included in follow-up, particularly the one in puberty. Indeed, the increase in lean body mass may potentially affect the validity of the ubiquitous BMI as a predictor of adult obesity. However, as the proportion of fat is greater in obese children, the effect of FFM in decreasing BMI tracking may be attenuated, explaining the better tracking in overweight and obese categories.

Effect of Gender

It has been suggested in some studies that BMI tracks higher in females than in males. Part of the reason for the gender difference in tracking coefficients for BMI could be associated with gains of lean mass, which are larger in boys than in girls during puberty. As an example, the estimated increase of FFM in males from the Trois-Rivières sample was 30.3 ± 0.9 kg (around 67 lb) versus 14.5 ± 0.8 kg (around 32 lb) for females, while the gain in the sum of four skinfolds was similar in men and women, confirming that larger FFM was responsible for the greater increase of BMI in men (9). However, gender differences in gains of BMI and skinfold thickness have not been observed consistently in obese persons. The number of studies showing better tracking of BMI in obese females is approximately the same as the number showing similar tracking for obese males and females (10).

Skinfolds

Tracking of adiposity, as indicated by various skinfold correlation coefficients *(r)*, ranges from $r = 0.26$ to 0.70 (9, 11-13). Tracking of skinfold measurements is similar over time to that of BMI: The longer the follow-up period, the lower the predictive value of the initial skinfold measurement (11, 12).

Waist Circumference

Studies tracking WC or the waist-to-hip ratio from childhood to adulthood are still rare. Since abdominal obesity as shown by WC is a more recent preoccupation for clinicians, we will have to wait for additional longitudinal data to document the tracking of abdominal obesity. However, a recent longitudinal investigation (the Fels Longitudinal Study in southwestern Ohio) demonstrated not only significant tracking of WC (and BMI) from 6 to 8 years in boys and from 9 to 13 years in girls to an average of age 51 years, but also its capacity to predict adult metabolic syndrome (14). It is worth mentioning that 5-year-old boys and 18-year-old males exceeding their age-specific WC had ORs of 8.6 and 24.1 for developing a WC over the adult male threshold of 102 cm (40 in.) at age 30 years (14). However, in females, the ORs were much smaller (1.3 and 1.5, respectively). Therefore, a gender difference in tracking abdominal obesity probably exists.

Challenges for Research

To study childhood obesity and the risk of adult obesity, data from longitudinal research are necessary. Although such data have obvious advantages, longitudinal data on previous generations have the disadvantage of not being generalizable to all generations. Abdominal obesity is becoming more prevalent during childhood and adolescence; Magarey and colleagues hypothesized that the wider prevalence of obesity in children born in the 1990s could result in higher tracking of obesity in the general population, since obese and overweight subjects will represent a greater proportion of the population (15). This may eventually lead to stronger tracking of BMI when all body composition categories are analyzed together. Another challenge for future research is to study the tracking of abdominal obesity by indirect methods like WC in conjunction with direct procedures like dual-emission X-ray absorptiometry. Finally, racial differences have been shown in tracking BMI (12) and merit further investigations. It is also important to understand the potential benefits of youth physical activity interventions to stem the perpetuation of obesity from childhood to adulthood. This question is covered in chapter 64, "Physical Activity and Body Composition in Children" by Dr. Bob Gutin.

Conclusion

From the available data, it is possible to conclude that a population with a high prevalence of obesity during childhood is at increased risk of maintaining a high prevalence of obesity during adulthood. Despite the variety of body composition measurements and of the age spans studied between childhood and adulthood, a consensus may be reached that the probability of obesity persistence from childhood to adulthood is moderate and significant. In fact, it appears that obesity tracking is significant and stable whether we consider only obese youths or all body composition categories. However, higher levels of obesity show stronger tracking for the same age ranges followed up.

29

Physical Activity and Risk of Obesity in Older Adults

Wendy M. Kohrt, PhD

This brief review will (1) provide an overview of the rates of obesity and physical activity in older adults; (2) discuss whether physical activity is effective in preventing or attenuating fat gain with aging; and (3) consider whether physical activity counteracts, to some extent, the detrimental effects of obesity on physical functional abilities of older adults.

Prevalence of Obesity in Older Adults

The prevalence of overweight and obesity in the United States and many countries around the world has been rising steadily over recent decades. Based on data collected in the National Health and Nutrition Examination Survey (NHANES), the prevalence of overweight or obesity among U.S. adults aged 60 years or older in 2003-2004 was estimated to be 71% for non-Hispanic whites, 78.8% for non-Hispanic blacks, and 78.1% for Mexican Americans (1). The prevalence of obesity, specifically, was 29.7%, 44.9%, and 36.9%, respectively, in these three racial-ethnic groups. Although much of the information in this chapter is based on studies of older adults in the United States, the problem of obesity in older adults is global. Current estimates (i.e., since 2000) of the prevalence of overweight or obesity in older adults (www.who.int/bmi/index.jsp) are high for many, but not all, technologically advanced countries (e.g., Australia 41%, Canada 52%, Chile 71%, Germany 61%, Japan 26%, Russian Federation 72%, Spain 82%).

Although the rates of overweight and obesity in older adults are high, they are similar to or slightly lower than those in middle-aged adults (1). It is logical to postulate that a plateauing or decrease in the prevalence of overweight and obesity after the age of 60 is due to poor survival among overweight and obese older adults. However, it is also likely that body mass index (BMI) becomes more variable and is a less accurate index of adiposity in older adults because of age-related factors other than adiposity that influence BMI. Based on two large cross-sectional studies of adult women and men in the United States (2) and Switzerland (3), fat-free mass appears to be well maintained until about 60 years of age, after which it declines. Thus, increases in fat mass after the age of 60 may be countered by the loss of fat-free mass. Another factor that influences BMI values in older adults is the loss of height that commonly occurs, particularly in women, as a result of osteoporosis. Although BMI remains a good indicator of the degree of adiposity among older adults (4), the fact that it is influenced by the loss of height and lean mass means that it should be used cautiously as an index of adiposity when young and older individuals are being compared.

Regardless of whether obesity is defined by BMI or by the measurement of fat mass, the prevalence is high in older adults and is projected to increase. It has been predicted that the incidence of obesity among white and black adults in the United States will have increased by about 9.3 million from 2000 to 2010, and that 8.3 million of these individuals will be 50 years of age or older (5). The forecast is that 65-year-olds of average height will weigh 2.5 to 3.4 kg (4.4 to 7.5 lb) more in 2010 than did 65-year-olds in 2000. Such projections suggest that there will also be an increase in the prevalence of age-related diseases and conditions for which obesity is a risk factor, such as diabetes, hypertension, heart disease, cancer, and osteoarthritis.

Prevalence of Physical Activity in Older Adults

The high prevalence of overweight and obesity in older adults is likely attributable, in part, to an age-related decline in physical activity level. In recent surveys, the percentage of women and men in the United States who reported getting at least 30 min per day of moderate-intensity activity on five or more days per week, or at least 20 min per day of vigorous-intensity activity on three or more days per week, was highest in young adults and became progressively smaller in older age groups (figure 29.1*a*) (6). The performance of vigorous physical activity, in particular, is markedly lower in older adults than in young adults. In the 2006 National Health Interview Survey, 80% to 90% of women and men aged 65 years or older reported that they *never* perform vigorous physical activity, defined as at least 10 min of activity that causes heavy sweating and a large increase in breathing, heart rate, or both (figure 29.1*b*) (7). Although there may be a slight upward trend in the percentage of adults who engage in regular physical activity, particularly older women and men, many still fall short of the Healthy People 2010 goal of increasing the proportion of adults who engage in regular physical activity to at least 50% (figure 29.1*a*).

Does Physical Activity Prevent Fat Accumulation With Aging?

An increase in adiposity is a usual consequence of aging in the overwhelming majority of individuals. Even those who maintain a very high level of physical activity over many years accrue fat mass. In a prospective study of master athletes, 74 male and female runners, aged 39 to 75 years at the baseline assessment, were evaluated after an average of 5.8 years of follow-up (8). The average rate of fat accrual was approximately 0.26 kg (0.57 lb) per year and was relatively similar across all ages. Running mileage decreased over the period of study from about 35 to 28 miles (56 to 45 km) per week. This decrease in training volume would have been more than sufficient to cause the fat gain that occurred if other factors that influence energy balance (e.g., energy intake and physical activity other than running) remained constant.

It is not known whether fat accrual with aging can be prevented in competitive athletes, such as runners, who maintain their training volume. However, in this context, it is important to understand

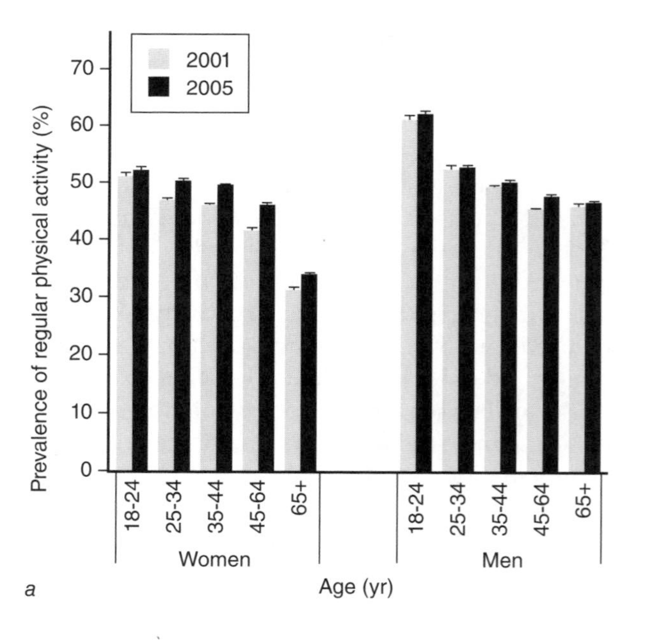

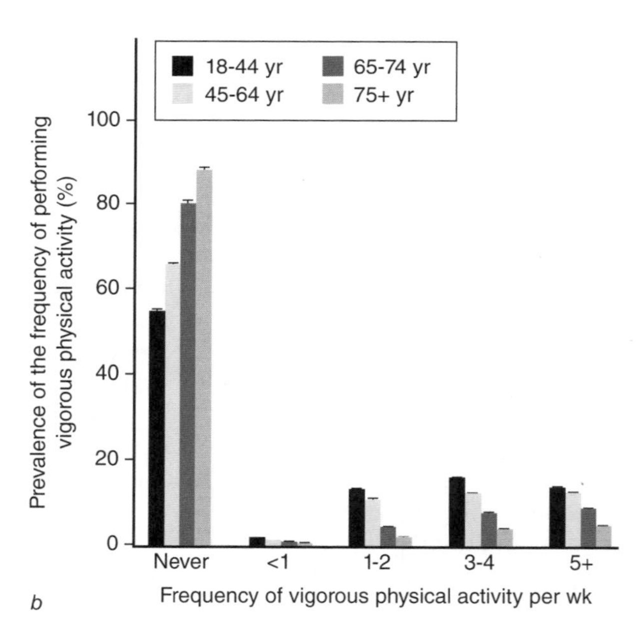

Figure 29.1 Estimated prevalence of regular physical activity (*a*) and frequency of performing vigorous physical activity (*b*) in women and men in the United States grouped by age.

Adapted from data reported by the Centers for Disease Control and Prevention 2007; and Pleis and Lethbridge-Cejku 2007.

that there are inevitable changes with aging in factors that make it challenging to maintain the same training volume over many years. For example, maximal heart rate decreases as a function of age (i.e., maximal heart rate = 220 – age) regardless of physical fitness level, and this contributes to the well-described decrease in maximal aerobic power with aging (8). A reduction in maximal aerobic power translates into a decrease in the rate at which energy can be produced (i.e., decrease in running speed). This means that the runner who maintains the same training mileage with advancing age will have a gradual increase in the time needed to complete the workout. For the individual who maintains a lifelong habit of exercising an hour a day at a vigorous intensity, there will be a progressive decrease with advancing age in the absolute amount of exercise performed and the amount of energy expended. In the absence of a compensatory decrease in energy intake, the prevention of fat gain with aging would seemingly require a progressive *increase* in time spent exercising. Importantly, even though maintaining a high level of physical activity is unlikely to *prevent* an increase in adiposity, physical activity can attenuate the rate of fat accrual and be effective in maintaining body weight in the normal range.

Does Physical Activity Reduce Obesity Risk in Older Adults?

The findings that regular, vigorous activity does not prevent fat gain with aging in competitive athletes raises the question of how effectively regular physical activity prevents fat gain in nonathletes. A large cross-sectional study of physically active and sedentary women and men across a wide age range showed that being physically active has little impact on fat-free mass, but supported the role of physical activity in diminishing the magnitude of fat accumulation; fat and fat-free mass were normalized to height to adjust for secular changes (figure 29.2) (3). The study, which was conducted in Switzerland, enrolled 3549 men and 3184 women, aged 18 to 98 years, and measured body fat by bioelectrical impedance. Physically active participants were those who reported more than 3 h per week of moderate- to vigorous-intensity (i.e., ≥4 metabolic equivalents, or METs) leisure-time activities for longer than two months. There was no BMI criterion for inclusion, but the average BMI values for women and men categorized by decade of age were all in the normal or overweight range for both the sedentary and physically active groups. For this reason, the

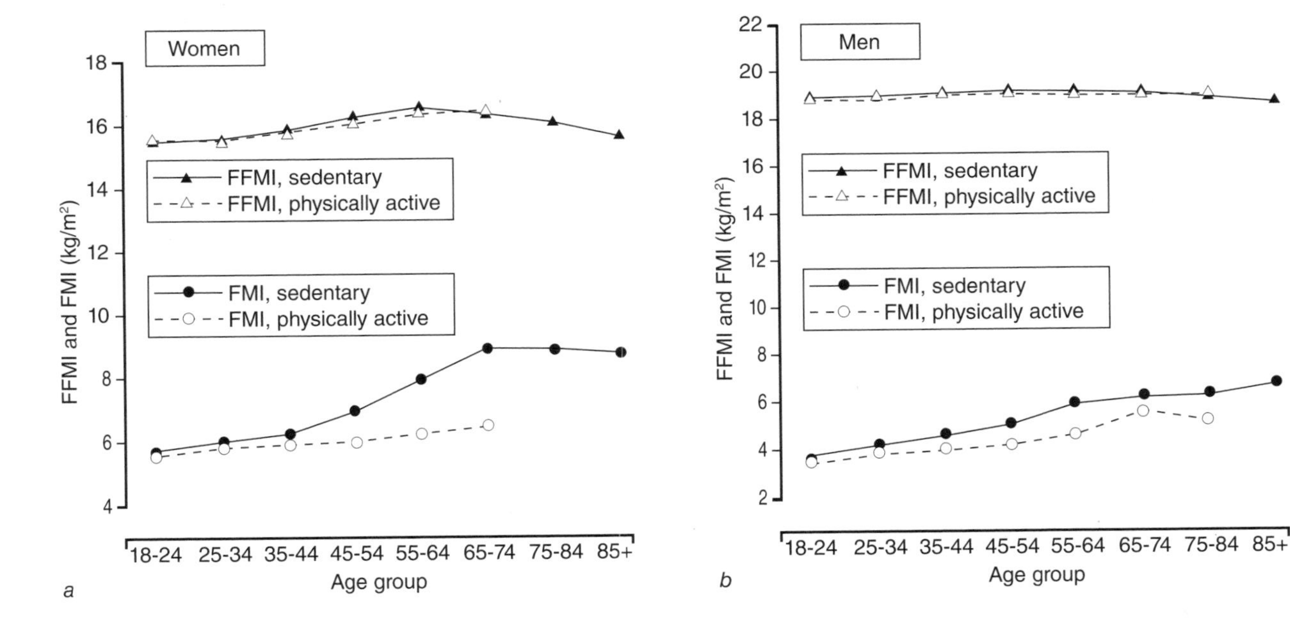

Figure 29.2 Age-related (i.e., based on cross-sectional data categorized by age group) changes in fat-free mass index (FFMI; fat-free mass in kilograms per height in meters squared) and fat mass index (FMI; fat mass in kilograms per height in meters squared) in *(a)* women and *(b)* men.

Adapted from U.G. Kyle et al., 2004, "Aging, physical activity and height-normalized body composition parameters," *Clinical Nutrition* 23: 79-88.

findings may not be generalizable to populations with a higher prevalence of obesity. Nevertheless, among the 55- to 84-year-olds in the study, BMI was about 2.5 and about 1.5 kg/m^2 lower in the physically active women and men, respectively, than in their sedentary counterparts.

The lack of an apparent effect of physical activity on fat-free mass, as a surrogate for muscle mass, is not surprising, because >90% of the physical activity reported involved endurance activities. Such activities (e.g., walking) would not be expected to build, or perhaps even preserve, muscle mass with aging; only resistance types of activities (e.g., weightlifting) are effective in increasing muscle mass. However, even if sedentary older adults have absolute levels of fat-free mass similar to those of physically active older adults, the higher levels of fat mass compromise their physical functional abilities. Traditionally, the clinical phenotype of frailty has included such characteristics as low body weight and muscle weakness. It is becoming increasingly apparent that another phenotype of frailty is that of the obese elderly adult, which has been termed "sarcopenic obesity" (9). Although a single working definition of sarcopenic obesity has not yet been adopted, the concept is that absolute levels of fat-free mass and muscle strength may be normal or even high, but be inadequate relative to the level of adiposity and thus increase the risk of physical functional disability.

Independent Effects of Physical Activity and Obesity on Physical Function

There are clearly detrimental effects of obesity on physical function in older adults, but it is less clear whether physical activity status exerts independent beneficial effects on function. Support for the benefits of physical activity on functional status of obese older adults has emerged recently from large prospective cohort studies, including the U.S. Health and Retirement Study (HRS; 8704 women and men, aged 50+ years at baseline; median follow-up 72 months), the English Longitudinal Study of Ageing (ELSA; 1507 women and men, aged 50 to 69 years at baseline; median follow-up 73 months), and the Health, Aging, and Body Composition Study (Health ABC; 2982 women and men, aged 70 to 79 years at baseline; follow-up 78 months) (4, 10).

The prevalence of overweight or obesity among the cohorts in these studies ranged from 56% to 78%. Physical activity was categorized in both the HRS and ELSA by participation in vigorous activities three or more times per week (i.e., yes/no) (10). The incidence rates of self-reported mobility impairment in the normal-weight, overweight, and obese groups in HRS were 6.2%, 8.8%, and 17.0%, respectively, in those who reported being physically active, compared with 16.8%, 18.8%, and 33.8% in the sedentary group. Similarly, in ELSA, the incidence rates of impaired measured physical function in normal-weight, overweight, and obese groups were 12.4%, 11.8%, and 15.4%, respectively, in physically active subjects versus 21.1%, 22.7%, and 31.6% in sedentary subjects. The Health ABC study divided participants into quartiles of physical activity and categorized them as high (highest quartile), medium (two middle quartiles), and low (lowest quartile). The incidence of mobility limitation was higher in obese than in normal-weight women and men, and increased in the low physical activity group when compared with the high group. Across all groups, low physical activity level was independently associated with a 70% higher risk of mobility limitation.

Summary

The decline in physical activity with aging undoubtedly contributes to the high prevalence of overweight and obesity in older women and men. Maintaining a vigorous level of physical activity over many years appears to attenuate, but not prevent, an increase in adiposity, suggesting that a small increase in fat mass with aging may be unavoidable. Available data support an important role for physical activity in maintaining BMI in the normal range with advancing age. However, it is not possible to prevent the decline with aging in the rate at which energy can be expended. This suggests that the prevention of excess weight gain with aging may require an increase in the time spent participating in moderate- to vigorous-intensity physical activities to compensate for a lower rate of energy expenditure. An important benefit of physical activity for older adults, which appears to be independent of the level of adiposity, is a reduction in the incidence of physical functional limitations.

30

Physical Fitness and Risk of Obesity

Steven N. Blair, PED

Thermodynamics dictates that the worldwide obesity epidemic is caused by a persistent positive caloric balance, in which calories absorbed exceed calories expended, in too many people over too many days. This positive caloric balance must be stored. In the human body it is stored as fat, and approximately 7700 calories are required to produce 1 kg (2.2 lb) of fat. Minor caloric imbalances lead to weight gain over time that ultimately may progress to obesity. For example, a persistent positive caloric balance of 50 kcal/day will result in a weight gain of >2 kg/year (4.4 lb/year). That is, if you consume 50 more kilocalories every day than you burn in all metabolic processes, you will consume more than 18,000 calories over the course of the year in excess of your metabolic needs. This excess energy must be stored, and this leads to the >2 kg gain. Of course ~2 kg/year does not sound impressive, but a persistent gain of this magnitude results in >10 kg (22 lb) over five years, an amount that most would consider significant.

We do not know which side of the caloric balance equation—energy intake or energy expenditure—is the primary cause of the obesity epidemic. Perhaps the most widespread hypothesis is that average daily caloric intake has increased over the past few decades, leading to a persistent positive caloric balance and thus to the obesity epidemic. It seems obvious that numerous dietary factors or habits could have led to the typical person's eating more now than in the past. Portion sizes have increased, at least in the United States; an ever-increasing proportion of meals are eaten away from home, often in fast food restaurants; vending machines with high-caloric-density snacks have become more common; and food availability is ubiquitous in modern societies. But despite these trends, there is little compelling evidence that average daily caloric intake has increased over the duration of the obesity epidemic.

There are also are many reasons why average daily caloric energy expenditure might have declined. Many changes have occurred in the workplace, from the factory floor to administrative offices, that result in less energy expenditure at work. Household and garden tasks are increasingly accomplished with less human energy than was required decades ago. Self-propelled vacuum cleaners and lawn mowers, leaf blowers, and energy-saving devices in the kitchen are prominent examples. It takes substantially less human energy to prepare a meal by putting a prepackaged dinner into a microwave oven than it took your parents or grandparents to slice, chop, stir, and cook food. Of course, sedentary leisure-time pursuits also have become widespread. Televisions have reached saturation in the United States, with nearly every household having at least one set and many having multiple sets; often there are more sets than people. Even the sedentary activity of watching TV has involved a reduction in caloric expenditure with the universal adoption of remote channel-changing devices. These are but a few examples, and the entire chapter could be used to list ways in which our

Acknowledgements
I am grateful for the assistance of Duck-Chul Lee, PhD, in preparing this chapter. He helped conceptualize the approach, performed the literature review, completed the table and figures, and commented on earlier drafts of the manuscript. I thank Gaye Christmus for editorial review and Maude Holt for assistance with manuscript preparation.

lives now require less energy expenditure than just a couple of decades ago. However, no population-based data demonstrate clearly that average daily energy expenditure has decreased in recent decades.

In summary, specific causes of the obesity epidemic are elusive. We do not know the extent to which the epidemic is due to increasing energy intake, declining energy expenditure, or a combination of the two. If the reason is the latter, a combination of changes in the two variables, we must consider how the relative proportions of energy expenditure and energy intake contribute to a positive energy balance. Is it 50-50, 30-70, 70-30, or some other combination? Answering this question is crucial to dealing with the epidemic. We cannot develop and implement rational strategies to address the problem if we do not know where to put most of our emphasis and resources. The reason for our lack of understanding of the specific causes of the obesity epidemic relates largely to our crude and imprecise assessments of energy intake and energy expenditure. Self-reports of dietary intake and physical activity are not accurate enough to demonstrate the small changes in energy balance that we need to measure. One approach to this dilemma is to seek more objective data on energy balance. Thus, the purpose of this chapter is to examine data on physical fitness and weight gain or obesity. The value of using fitness is that it is measured more accurately than is physical activity, and perhaps studies using fitness can shed additional light on the complex problem of obesity.

Methods

I reviewed the reference lists of known articles on physical fitness and body weight or obesity and also searched PubMed for 1998 through 2008 using the search terms "physical fitness" and "obesity." The search identified 798 reports, and I reviewed them to determine if they provided an objective measurement of fitness as an exposure and some measure of weight, adiposity, or fat distribution as outcomes. The great majority of studies identified in the initial search did not have data on these variables, but 26 studies were selected for further review. I wanted to report on studies both in children and adolescents and in adults. I also evaluated the studies based on such criteria as sample size, quality of the data, and type of population studied. Due to space limitations, I decided to include only 10 articles here, and the ones I chose are summarized in the table 30.1. In the following sections I briefly discuss these 10 articles in more detail.

Fitness and Obesity in Children and Adolescents

I include four studies on fitness and obesity in children and adolescents (see table 30.1). Two were cross-sectional and two were prospective designs.

Cross-Sectional Study

Chen and colleagues (1) reported on the prevalence of obesity in representative population samples in 1999 and 2001. The surveys included 13,935 Taiwanese children and adolescents (6904 girls, 7031 boys) aged 6 to 18 years in 1999 and 24,586 (12,219 girls, 12,367 boys) in 2001. The investigators measured weight, height, and physical fitness (bent-leg curl-ups for abdominal muscle strength and endurance, sit and reach for flexibility, and step test for cardiorespiratory fitness) in both surveys. Normal-weight participants performed better ($p <$ 0.05) on all fitness tests in both surveys, except for the sit and reach in the 2001 survey. These results were seen for both girls and boys, and in all age groups. Blood pressure was measured in the 2001 survey, and those who were overweight or obese and unfit were approximately three times more likely to be classified as hypertensive (blood pressure ≥140 mmHg systolic or ≥90 mmHg diastolic) when compared with normal-weight or fit individuals. The authors concluded that there is an inverse association between fitness and overweight or obesity. In a second cross-sectional study of children aged 8 to 17 years, physical fitness was associated with lower visceral and abdominal subcutaneous adipose tissue (2).

Prospective Studies

Investigators at the University of Southern California conducted two important prospective studies on fitness and other predictors of increases in adiposity in children from white, black, and Hispanic groups (3). In one study, 95 black and white girls and boys participated in extensive laboratory assessments and were followed for up to five years. The laboratory measurements included assessment of energy expenditure by doubly labeled water techniques, cardiorespiratory fitness by $\dot{V}O_2max$ via treadmill exercise testing, and body composition by dual X-ray absorptiometry. Baseline fat mass was the main predictor of increasing adiposity, but cardiorespiratory fitness also was inversely associated with adiposity increases. For each 0.1 L/min higher $\dot{V}O_2$, the investigators observed a 0.081 kg lower gain in fat mass per kilogram of lean mass.

Table 30.1

Summary Information From 10 Epidemiological Studies on Fitness and Weight

Authors	Study design	Population	Main results	Comments
CHILDREN AND ADOLESCENTS				
Chen et al. 2006[1]	Cross-sectional	19,398 boys, 19,123 girls (Taiwanese), ages 6-18 years	The normal-weight group performed better than the overweight and obese group in cardiovascular and muscular fitness tests.	Overweight and obesity are associated with poor muscular and cardiovascular fitness.
Lee and Arslanian 2007[2]	Cross-sectional	58 boys, 55 girls (U.S.), ages 8-17 years	Cardiorespiratory fitness was inversely related to total adiposity, waist circumference, and visceral and abdominal subcutaneous adipose tissue.	Fitness is associated with lower visceral and abdominal subcutaneous adipose tissue.
Johnson et al. 2000[3]	Prospective (3.7-year follow-up)	36 boys, 79 girls (U.S.), ages 4.6-11 years	There was a negative relationship between aerobic fitness and the rate of increasing adiposity.	Aerobic fitness is a significant independent predictor of increasing adiposity.
Byrd-Williams et al. 2008[4]	Prospective (4-year follow-up)	84 boys, 76 girls (U.S.), ages 8-13 years	In overweight boys, 15% higher $\dot{V}O_2max$ resulted in a 1.38 kg lower fat mass gain, but relationship was not significant in girls.	In overweight boys, greater cardiorespiratory fitness is protective against increasing adiposity.
ADULTS				
Ross and Katzmarzyk 2003[5]	Cross-sectional	3719 men, 3854 women (Canadian), ages 20-59 years	For a given body mass index (BMI), men and women in the high cardiorespiratory fitness group had lower waist circumference and sum of skinfolds than those with low fitness.	Fitness may attenuate the health risks attributed to obesity.
Wong et al. 2004[6]	Cross-sectional	397 men (U.S.), ages 30-76 years	For a given BMI, men in the high cardiorespiratory fitness group had lower abdominal fat than men in the low fitness group.	Fitness may attenuate the health risk of high BMI by reducing abdominal fat.
DiPietro et al. 1998[7]	Prospective (7.5-year follow-up)	4599 men, 724 women (U.S.), ages 20-82 years	Each 1 min improvement in treadmill time attenuated weight gain in both men and women.	Improvements in fitness attenuate weight gain.
Sidney et al. 1998[8]	Prospective (7-year follow-up)	899 men, 1063 women (U.S.), ages 18-30 years	Weight gain >20 lb was associated with a large decrease in fitness.	Fitness changes are related to changes in weight.
Mason et al. 2007[9]	Prospective (20-year follow-up)	291 men, 315 women (Canadian), ages 20-69 years	Low musculoskeletal fitness was associated with at least 10 kg weight gain.	Musculoskeletal fitness is a significant predictor of weight gain.
Brien et al. 2007[10]	Prospective (20-year follow-up)	223 men, 236 women (Canadian), at least 18 years	Higher $\dot{V}O_2max$ was associated with lower odds of obesity.	Cardiorespiratory fitness is an important predictor of future obesity.

Energy expenditure at baseline was not associated with increases in adiposity. Byrd-Williams and colleagues (4) observed an inverse association ($p = 0.03$) between baseline $\dot{V}O_2$max and increases in adiposity in Hispanic boys over up to four years of follow-up. There was no association between fitness and adiposity in Hispanic girls. These two prospective studies indicate the potential of using fitness testing to identify children at risk for increasing adiposity in the future.

Fitness and Obesity in Adults

There have been surprisingly few published reports on fitness and obesity in adults. We found two cross-sectional and four prospective studies. To our knowledge, there is only one report on muscular strength and weight gain.

Cross-Sectional Studies

Ross and Katzmarzyk (5) reported on the association of cardiorespiratory fitness and obesity in 3854 women and 3719 men aged 20 to 59 years who participated in the Canada Fitness Survey in 1981. Cardiorespiratory fitness was assessed by a submaximal exercise step test, and obesity and fat distribution by body mass index (BMI), skinfolds, and waist circumference. Within BMI strata, both women and men with higher fitness had more favorable data on waist circumference and skinfold measurements. The authors concluded that higher levels of fitness may contribute to lower health risk within various BMI strata, and that this is due to the effect of exercise in reducing abdominal obesity.

Findings from this study were supported by a report on 293 men in the Aerobics Center Longitudinal Study (6). Cardiorespiratory fitness was determined by a maximal exercise test on a treadmill, and abdominal fatness by electron beam computed tomography scans. The authors reported that men with high fitness had less abdominal fat than those in the low-fit group.

Prospective Studies

DiPietro and associates (7) reported on follow-up of a large group of women ($n = 724$) and men ($n = 4599$) in the Aerobics Center Longitudinal Study. These individuals completed at least three examinations from 1970 through 1994, during which cardiorespiratory fitness was assessed by maximal exercise testing on a treadmill. The authors calculated change in fitness between the first and second examinations (mean interval = 1.8 years) and examined this change as a predictor of change in weight from the first to the last examination (mean interval = 7.5 years). Improvements in treadmill time from the first to the second examination were associated with significantly less weight gain over the course of the study. Figure 30.1 shows a strong and significant inverse association between change in treadmill time and weight change, based on multiple linear regression models adjusted for age, height, smoking, baseline weight and treadmill time, number of clinic visits, and follow-up time. The incidence of a ≥5 kg (11 lb) or a ≥10 kg (22 lb) gain during follow-up was 16% and 4% for men and 17% and 4% for women. Each minute of improvement in treadmill time between the first and second examinations resulted in a 14% lower odds of a ≥5 kg gain in men (OR = 0.86; 95% CI: 0.83-0.89) and a 9% lower odds in women (OR = 0.91; 95% CI: 0.83-1.0). The lower odds of a ≥10 kg gain per minute improvement in treadmill time were even greater: 21% in both men (OR = 0.79; 95% CI: 0.75-0.84) and women (OR =

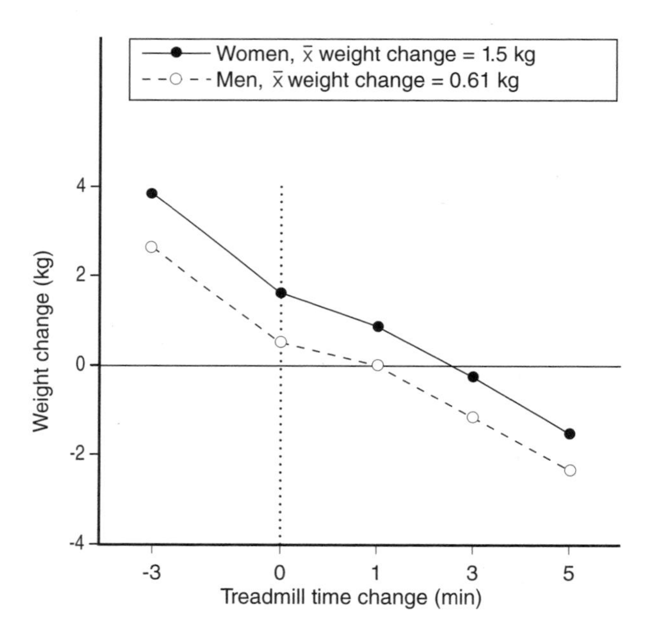

Figure 30.1 Inverse relation between change in treadmill time and weight change in the Aerobics Center Longitudinal Study. The regression estimate represents the difference in weight change (between the first and last examination) per minute change in treadmill time. Given a 1 min improvement in treadmill time, weight gain was minimized by 0.60 kg in men and women, which when added to the mean weight change for each sex resulted in a net change of 0.01 kg for men and 0.91 for women. A 3 min improvement in treadmill time resulted in a net change of –1.19 kg for men and –0.29 kg for women.

Data from L. DiPietro et al., 1998, "Improvements in cardiorespiratory fitness attenuate age-related weight gain in healthy men and women: The Aerobics Center Longitudinal Study," *International Journal of Obesity Related Metabolic Disorders* 1998; 22(1): 55-62.

0.79; 95% CI: 0.67-0.93). The findings from this study support the hypothesis that improving fitness even by a small amount results in a dramatically lower risk of substantial weight gain.

Sidney and colleagues (8) reported on seven-year changes in exercise test performance and weight gain in the CARDIA study. This project included a biracial sample (blacks and whites) of 1962 young adults (8). Results from this report are consistent with those from the Aerobics Center Longitudinal Study already described.

Katzmarzyk and colleagues have published two papers on fitness and weight gain in the Canada Fitness Survey (9, 10). One report is on musculoskeletal fitness (9) and the other on cardiorespiratory fitness (10). A strength of these reports is that they are based on a representative population sample, and the follow-up was long-term, approximately 20 years. The results from both papers are consistent with the findings of DiPietro and colleagues (7) and Sidney and colleagues (8) that fitness is inversely associated with weight gain.

Conclusion

Relatively little research has examined physical fitness and weight gain or the development of obesity. However, the findings of the available studies are quite consistent. Both cardiorespiratory fitness and musculoskeletal fitness, in children, adolescents, and adults, appear to reduce weight gain and the development of obesity. Although fitness has a genetic component, it is primarily determined by exercise habits. The advantage of using fitness as an exposure is that it is measured objectively, and probably provides better data on the role of activity and weight gain than can be obtained from the usual crude and imprecise self-reports of physical activity. Additional research is certainly needed, especially prospective studies with multiple measures of both fitness and weight over multiple years. For the moment, it is reasonable to assume that developing and maintaining both cardiorespiratory and musculoskeletal fitness, by following public health recommendations for physical activity, will have a beneficial effect on management of body weight and prevention of obesity.

31

The Interaction of Diet and Physical Activity on Obesity

Tom Baranowski, PhD

The recent worldwide increases in obesity in adults and children have spawned a profusion of obesity prevention interventions, most of which have not been effective. The conceptual mediating-variable model that underlies many of these programs (see figure 31.1) proposes that dietary and physical activity behaviors that are both causally related to and highly predictive of adiposity or obesity must be targeted for change. However, it is not clear what dietary or physical activity behaviors have caused the obesity epidemic.

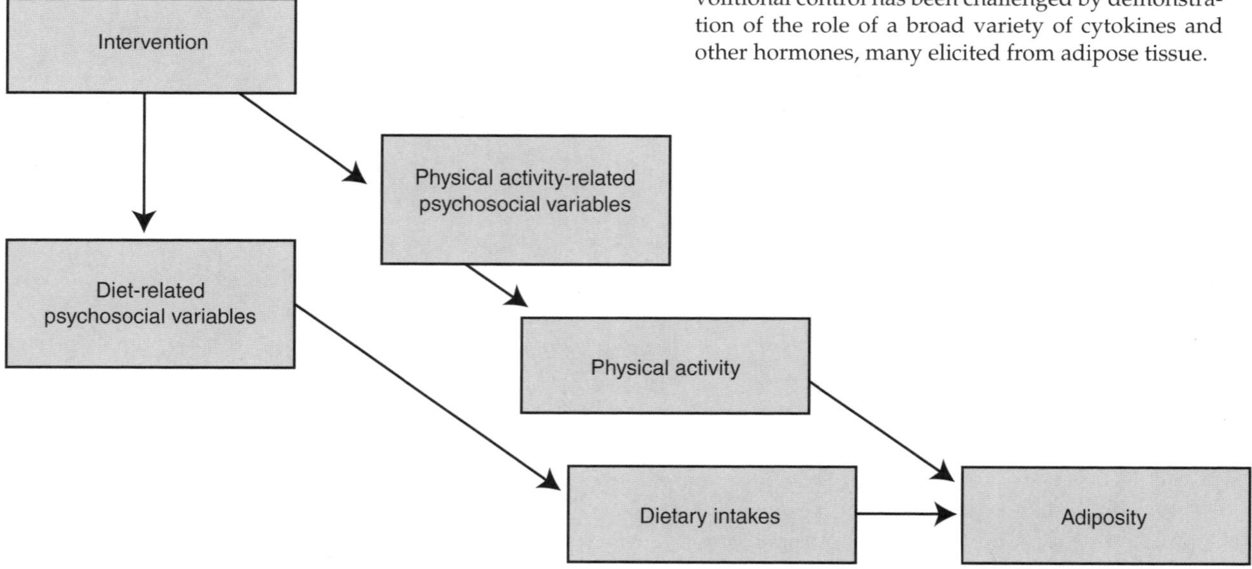

Figure 31.1 A mediating-variable model for an obesity prevention program targeting both diet and physical activity.

The Energy Balance Concept

The accepted or conventional wisdom is that obesity is the result of an imbalance in total energy intake and total energy expenditure:

> The energy balance concept is based on physics principles of conservation of energy, and usually assumes that kcal intake and kcal expenditure are linearly dependent (more $kcal_{expenditure}$ requires more $kcal_{intake}$), but otherwise orthogonal, and intake and expenditure are primarily under volitional control. A substantial number of studies have shown relationships of dietary intake and energy expenditure to adiposity, but a substantial number have not, which have often been attributed to limitations of the measures employed for both physical activity and diet. The idea of substantial volitional control has been challenged by demonstration of the role of a broad variety of cytokines and other hormones, many elicited from adipose tissue.

Energy intake and expenditure can be divided into its food group and energy expenditure components:

$$\frac{kcal_{FV} + kcal_{grains} + kcal_{dairy} + kcal_{meats} + kcal_{other}}{kcal_{REE} + kcal_{PI} + kcal_{LA} + kcal_{MA} + kcal_{VA}}$$

(where FV = fruit and vegetables; REE = resting energy expenditure; PI = physical inactivity; LA = light activity; MA = moderate activity; and VA = vigorous activity).

This ratio is intriguing because some of these food groups have lower caloric density (more water and dietary fiber content), which tends to enhance satiety at a lower caloric intake.

Emerging Questions About Diet and Physical Activity

A number of studies have challenged the assumption of orthogonality of dietary intake and expenditure. Ideas of limited volitional control have been introduced, suggesting that people who elect to be more physically active for health reasons may also be more likely to eat more healthfully. This may mean consuming lower-energy-dense and otherwise more health-promoting foods (e.g., fruit, vegetables, whole grains). This idea has opened a line of research on the interrelationships of diet and physical activity, which is important for three reasons. It is possible that there are undetected confounding effects of diet or physical activity in relation to adiposity and other chronic disease risk factors when diet or physical activity alone is studied (1); there may be synergistic effects of diet and physical activity on changes in adiposity (2); and there may be psychosocial variables (e.g., health protection motivation) that are motivators for behavior change in some subgroups of the overall population.

One of the earlier studies in this area used a cross-sectional design, with a large sample (n = 2053) of adult respondents in a randomly selected survey of households in two northeastern U.S. communities, a food frequency questionnaire, and self-reported response to a single item on physical activity. The moderately active and very active were more likely to report limiting their fat and salt intakes and eating more fruit and vegetables even after total calorie intake was controlled for (1). A study with a somewhat smaller but still large sample (n = 919) of adults with elevated low-density lipoprotein cholesterol in several communities, also in the northeastern United States, used both a seven-day and a 24 h diet recall, with a single-time "exercise" question as part of the seven-day diet recall. The level of calories consumed was not different in the more active, but these respondents consumed less high-fat foods (especially saturated fats) and more fruit and vegetables (3).

A third study took two substantial methodological leaps, from self-reported physical activity to a measure of cardiovascular fitness (from a maximum stress test) and from recalled diet to three-day diet records in a large (n = 10,741) but self-selected sample of adults attending a health and fitness center. Similar patterns were detected (4), except that the moderate-fitness group consumed the fewest calories. Among a large sample (n = 1335) of young adults (19 to 37 years old) in one southern U.S. community using a one-item assessment of physical activity and a food frequency questionnaire, more active individuals tended to consume more fruit and juices and dairy products and less percentage energy from fat (5). Again, the middle activity group consumed the fewest calories. Similar patterns have been detected among other samples of adults, with the important additional finding that poor diet and low activity were more common among those who were less educated, not married, and from ethnic minorities (6).

One of the studies in this vein among children involved a sample of overweight 8- to 10-year-old African American girls (n = 210) from across selected sites in the United States. An accelerometer was used to measure physical activity, and two 24 h dietary recalls were used to assess diet. More active girls had no higher caloric intake, but were less likely to consume dietary fat and more likely to consume carbohydrates (7).

A key issue in this line of research is whether an intervention in diet or physical activity would have positive effects on the other behaviors. In a review of the literature of mostly small-sample studies on the impact of physical activity interventions on dietary intake, no consistent pattern was detected in macronutrient intake change (8), but food groups were not reported. Given the general lack of difference in caloric intake by activity group in the previously cited studies, this finding in the review of exercise interventions is not surprising, but the review did not include food groups, which were related to physical activity. However, in a sample of 8- to 10-year-old African American girls (n = 127) experiencing an obesity prevention intervention that targeted both diet and physical activity, there was no covariability in the changes in physical activity with change in any component of diet, including total calories (9). More research with larger samples needs to address whether increased activity results in differences in food group intake or vice versa.

Very little research has looked at the interactive effect of components of diet and physical activity on adiposity or obesity. In a large nationally representative sample ($n = 748$) of 3- to 11-year-old French children using parent-reported child activity, inactivity and diet, and measured child BMI, principal components analysis revealed three lifestyle factors among the 7- to 11-year old children: "varied food and physically active"; "big eaters at main meals"; and "snacking and sedentary" (10). Children in the higher two tertiles of the varied food and physical activity factor had a significantly lower risk of overweight, even after adjustment for the individual lifestyle components (10). This provides some support for the interactive effects of diet and physical activity on adiposity. Children in the higher two tertiles of the big eaters at main meals factor had a significantly higher risk of overweight, even after adjustment for the individual lifestyle components (10). Tertiles of the third factor (snacking and sedentary) were not related to overweight.

A brief summary of this literature reveals that in large-sample cross-sectional studies of adults and children, consistent relationships were detected whereby more physical activity was correlated with healthier dietary intakes (e.g., less fat, more fruit and vegetables) but not generally with total kilocalorie intake. Large samples appear to be required to allow detection of these effects, suggesting either relatively weak effects or substantial measurement error. Since higher levels of error in reporting dietary intake were associated with higher levels of objectively measured energy expenditure (by doubly labeled water), it is possible that the relationships obtained with the use of self-reported diet and physical activity were due to some correlated errors in the self-reporting of health behaviors. Research with more objective methods (e.g., accelerometers for physical activity, dietary biomarkers) will need to resolve this possible alternative explanation. The lack of consistent relationships between dietary and physical activity changes in intervention studies in both adults (8) and children (9) warrants similar attention to methods.

Future Directions

Although diet and physical activity practices may be related, this does not necessarily mean that their joint effect (an interaction term) is greater than either effect separately (an additive effect). The single obtained relationship of a factor score combining the effects of diet and physical activity on adiposity is promising (10) and needs to be investigated in other studies. Since caloric intake has not varied across activity levels but types of food have, it is possible that some metabolic factor or bulk- or water-induced satiety is accounting for the effect, but not energy balance. Future research in this area may provide promising approaches for weight loss interventions.

It appears likely that lower levels of activity are associated with greater media use (e.g., television, cinema, video games). Media use has been related to adiposity through dietary variables, for example snacking during media attentiveness, greater exposure to commercials for higher-energy-dense foods, and media turned on during meals (which may distract from attention to internal cues of satiety). These are behavioral pathways that could account for some of the relationships among low physical activity, diet, and adiposity, and deserve further research attention.

If this developing line of research reveals that dietary and physical activity practices are interrelated and the interrelationships are associated with adiposity, research will be needed on the psychosocial factors that differentiate those with healthier joint diet and physical activity practices from those with less healthy joint practices. This should result in delineating psychosocial targets for behavior change programs. Such interventions should achieve a synergy in changing both behaviors, thereby more effectively and efficiently changing adiposity.

Conclusion

The obesity epidemic is likely on the verge of causing major increases in morbidity and perhaps mortality in national populations. Better understanding of the interrelationships among diet, physical activity, and adiposity offers the possibility of identifying key targets for change, which could substantially alleviate the problem. Remedying this problem deserves our best research efforts.

part V

Physical Activity and Biological Determinants of Obesity

Physical activity has the potential to modulate the effects of a number of tissues and pathways that are often defined as biological determinants of obesity. This section is composed of 14 chapters. Four chapters are devoted to the role of physical activity in key components of metabolic rates in humans: resting metabolic rate (chapter 32), thermic effect of food (chapter 33), substrate oxidation rates (chapter 34), and sympathetic nervous system activity (chapter 35). One chapter focuses on adipose tissue biology (chapter 36) and another on leptin biology (chapter 37). The role of physical activity regarding a number of hormones and pathways is then reviewed: hypothalamic peptides (chapter 38), gut peptides (chapter 39), thyroid hormones (chapter 40), and the hypothalamic-pituitary-adrenal axis (chapter 41). A chapter is devoted to physical activity and skeletal muscle biology (chapter 42) and another one to postexercise energy expenditure (chapter 43). Finally, two chapters deal with the genetic (chapter 44) and epigenetic (chapter 45) determinants of obesity.

Physical Activity Level and Resting Metabolic Rate

Angelo Tremblay, PhD

Resting metabolic rate (RMR) is a physiological variable that provides an estimate of the minimal energy cost of living. In a healthy population, RMR corresponds to about 1 kcal/min and necessarily represents the main component of daily energy expenditure in a context in which sedentariness has imprinted the physical activity habits of people. Its variations are primarily attributable to the amount of body lean tissue, which frequently explains a much greater fraction of the variance in RMR than any other factor. Accordingly, obese individuals are expected to display an increased absolute level of RMR, which has been interpreted by some obesity specialists as an absence of thermogenic defect in these persons. In this regard, our research experience has revealed that beyond its positive relationship with body weight or components of body weight, RMR is significantly influenced by environmental factors such as physical activity. As described in this chapter, the impact of physical activity on RMR can be substantial, but this effect is influenced considerably by other factors.

Resting Metabolic Rate in Exercise-Trained Individuals

The study of the profile of the regular exerciser is often used as a first step to document the potential effect of physical activity on a phenotype. With respect to RMR, cross-sectional comparison of trained and untrained subjects revealed a significant difference in the level of the body weight–RMR regression line between the two groups (1). This difference predicted a level of RMR relative to body weight that exceeded by 100 to 200 kcal/day the level of the sedentary controls. Further investigations showed that this difference was particularly evident when sedentary subjects were compared to athletes performing a large amount of exercise (2).

From a mechanistic standpoint, it seems that the increased RMR characterizing exercise-trained individuals is mediated by beta-adrenoreceptors. For instance, we have shown that acute administration of the beta-blockade propranolol completely abolished the difference in RMR between trained and untrained subjects (3), which raises the possibility of a tissue-specific increase in beta-adrenergic stimulation induced by exercise training. This hypothesis is indeed supported by animal research demonstrating that exercise training promotes an increase in beta-adrenoreceptors in skeletal muscle (4) but favors a decrease of these receptors in the heart (5). This agrees with the increased RMR in trained individuals who are also characterized by a reduced resting heart rate and blood pressure.

Acute Effect of Exercise on RMR

The increase in RMR observed in the trained person may be a repeated acute effect or may represent the outcome of long-term exposure to the exercise stimulus. In humans, one can measure RMR variations either by imposing short detraining on trained individuals or by subjecting inactive persons to an exercise bout. Our research experience shows that these two strategies lead to the same conclusion. Indeed, short-term interruptions of exercise in trained subjects resulted in a 6% decrease in RMR that corresponded to about 115 kcal when extrapolated over 24 h (6).

In untrained inactive subjects, acute exposure to an exercise session of prolonged vigorous exercise favored an increase in RMR that also corresponded to about 100 kcal/day (7). More recently, we performed this type of investigation to examine the acute effect of exercise intensity on postexercise RMR and lipid oxidation. We found that the stimulating effect of high-intensity exercise was significantly greater than that promoted by an exercise of similar energy cost but of lower intensity (8). However, the administration of propranolol eliminated the difference in effect between the two intensity types (8).

Taken together, these observations demonstrate that aerobic exercise of sufficient duration and intensity can acutely increase postexercise RMR. This effect, which is mediated by beta-adrenoreceptors, suggests that the increased RMR of trained individuals at least partly represents a repeated acute effect of the exercise stimulus.

A Genotype–Exercise Interaction Effect on RMR

It is a truism in biology that there are individual differences in the response of a phenotype to a stimulus such as physical exercise. Thus, our laboratory has measured the effect of exercise training on RMR in monozygotic twins. This experimental approach allowed us to determine the within-twin pair variation, that is, the variance in RMR response in individuals sharing the same genetic background. This within-twin pair in response was compared to the between-twin pair variance which is the variation in change in RMR between subjects who differ genetically. This provided a ratio of variance that suggested a role of heredity in the response of RMR to exercise training in our studies, whether in a protocol of 22 (9) or 100 (10) days. Thus, our twin studies tend to show the existence of a genotype–exercise training interaction effect on RMR, suggesting that some individuals are more responsive with respect to RMR than others because of their genetic background.

The next step in this field of research requires a more specific characterization of the genetic contribution to variations in RMR to identify genotypes or epigenetic effects that predispose to a greater response in RMR to training.

Obesity Treatment and RMR

It is well established that a weight-reducing program generally results in a substantial decrease in RMR that may exceed the decrease predicted by morphological changes (11, 12). Since a lower than predicted metabolic rate is related to a more pronounced weight gain over time, it could be suggested that weight loss prepares weight regain, which raises the question of whether exercise can prevent this excess fall in RMR in the context of a weight-reducing program. We recently considered this issue in a diet and exercise program in which RMR was measured after each 5 kg (11 lb) weight loss and at weight loss plateaus. The results showed that the decrease in RMR beyond the value predicted by body weight was substantial and highly statistically significant despite regular physical activity (Tremblay and Chaput, unpublished data).

The observation that a substantial weight loss promotes thermogenic changes that favor weight regain must be considered together with the impact of some chemical pollutants on body energy expenditure. This is particularly relevant for organochlorines (OCs), which are lipid-soluble compounds present in the body substance of all humans despite the fact that they were banned in many countries several decades ago. Organochlorines are hormone disruptors that negatively affect thyroid hormone levels and mitochondrial functionality (13). As expected, body weight loss accentuates the plasma and tissue concentrations of OCs, with potential consequences on variations in RMR. Accordingly, a recent study revealed that change in OC plasma concentrations was the main variable explaining differences in RMR in response to a weight loss of about 10 kg (22 lb) (13). In addition, a comparable study measured the change in sleeping metabolic rate resulting from changes in circulating OCs during a weight-reducing program, comparing actual effects against those predicted by morphological factors (14). This investigation showed that 50% of the variance in the greater than predicted decrease in sleeping metabolic rate was attributable to altered levels of circulating OCs. This may be a significant adaptation counteracting the expected effect of exercise in the weight-reduced obese person.

Conclusion

Physical activity stimulates postexercise energy metabolism to a level that may reach 100 to 200 kcal/day. There are, however, individual differences in the ability to increase RMR in response to exercise, and this phenomenon is at least partly explained by heredity. The enhancing effect of exercise on RMR is considered a potential benefit of including regular physical activity in a weight-reducing program. This benefit is at best limited and does not have the potential to postpone the occurrence of resistance to weight loss up to normalization of body weight in reduced-obese individuals.

Physical Activity and Thermic Effect of Food

Yves Schutz, PhD, MPH

In a very simplistic, convenient, and pedagogic way, total daily energy expenditure (EE) can be partitioned into three main components: basal metabolic rate, diet-induced thermogenesis (comprising an integration of the thermogenic effect of each meal), and the energy cost of physical activity. In healthy individuals, the diet-induced thermogenesis (DIT) component is the smallest component of the three. It is highly variable and can amount to 5% to 15% of the metabolizable energy intake (1). Thus our bodies are relatively efficient at utilizing exogenous macronutrients for storage.

The thermic (thermogenic) effect of food (TEF), also called postprandial thermogenesis, is the net increase in resting heat production and body core temperature after eating. It is essentially due to the metabolic energy cost of digesting and absorbing macronutrients (active transport of exogenous nutrients from the gut as well as gut motility) and to the energy cost of (re)synthetizing macromolecules stored in tissues in the postprandial phase, such as protein glycogen and fat.

Whereas TEF refers to the immediate effect of a meal given acutely on resting EE measured over several hours (typically 3 to 6 h), DIT includes the overall effect of the diet (usually measured over 24 h) at a different plane of nutrition (e.g., in overfeeding conditions).

The TEF is measured by indirect calorimetry ($\dot{V}O_2$ and $\dot{V}CO_2$ measurements) and is calculated as the integration of postprandial EE above the postabsorptive baseline resting value. This excess EE is divided either by the energy content of the meal fed (TEF) or by total energy intake (DIT). Its magnitude is determined by a number of factors, but mainly by (1) the nutritional characteristics of the meal given (relative and absolute macronutrient composition, energy level, associated micronutrients such as vitamins) and the nature of the food eaten, (2) endogenous host factors (genetic, body composition, endocrine milieu), and (3) the experimental conditions under which the thermogenic measurement takes place (environmental temperature).

Factors Influencing TEF in Obesity

Most studies on TEF in obesity were published 10 years ago or more. Heavily studied, particularly by our group in Lausanne (1), the relative importance of TEF for energy balance regulation was not well understood in the 20th century. Studies have yielded discrepant results regarding whether TEF is reduced or not in obesity (1). In their comprehensive review, de Jonge and Bray (2) analyzed experimental human studies on TEF that compared obese and lean subjects. They confirmed that several factors influence the magnitude of TEF: In addition to meal size, meal composition and the nature of the previous diet, the degree of insulin resistance, physical activity, and aging were identified.

In obese individuals, roughly three-quarters showed a statistically significant reduction in TEF, which was related to the degree of insulin resistance and a low level of sympathetic activity stimulation. Methodological differences also undoubtedly contribute to the discrepant results in the literature. Since TEF remains a relatively small component, a "clean" methodology is a key factor in adequately measuring postprandial thermogenesis. A more recent paper reviewed 50 studies that investigated TEF in obesity, and focused on factors related to experimental protocol and subject control that reportedly effect measurements of resting EE, postprandial EE, and the calculation of TEF (3).

Thermic Effect of Food Combined With Exercise

Physical activity and exercise elicit major perturbations in metabolism during the period of exertion itself, and this effect continues to influence energy metabolism after the exercise has been discontinued. Therefore, it is fundamental to explore the effect of meal ingestion prior to exercising or during the recovery phase.

With regard to the effect of general unstructured physical activity, the concept of "nonexercise activity thermogenesis" (NEAT) has recently received some attention. The thermic effect of food is never considered in the NEAT concept (4), but NEAT is calculated above the resting metabolic rate, which includes TEF. The hypothesis that NEAT does not potentiate TEF, however, does not take into consideration different physical activity circumstances.

Most studies investigating the interaction between meals and exercise were carried out one to two decades ago. Three main purposes were behind the exploration of a putative synergetic effect between meal ingestion and exercise: (1) to attempt to "normalize," thanks to exercise, the small defect in TEF described in certain subgroups of obese individuals; (2) to study the effect of an enhanced "substrate trafficking" between exogenous macronutrient storage and endogenous macronutrient mobilization induced by exercise; and (3) to investigate whether the combination of exercise plus food may stimulate TEF, compared to TEF under resting conditions, since insulin sensitivity increases during exercise and one factor explaining the blunted TEF in obesity is increased insulin resistance.

Exercise may affect TEF under acute exposure conditions or as a result of exercise training. Exercise can also be of different types (aerobic vs. resistance) and of differing intensity and duration. Furthermore, the chronology of exercise versus meal ingestion needs to be defined: An exercise can be performed prior to the meal or at a specific time after meal ingestion.

The experimental design for adequate exploration of the interactions between TEF and exercise is demanding, since it requires several control groups (positive and negative). At least four experimental conditions are required using a design similar to a quadratic Latin square: (1) resting, no meal; (2) meal ingestion alone while resting; (3) exercise alone (including postexercise), no meal; (4) meal combined with exercise, the latter being performed before meal ingestion or during the postprandial phase. In addition, a solid preintervention baseline is required in each situation.

Confounding factors generally include training status as well as body composition of the individuals. In addition, a rigid standardization of the diet plus exercise protocols in the days preceding the study is of the utmost importance.

The potential interactions between exercise and food ingestion have been the subject of several papers. The experimental studies of Segal and collaborators (5-7) more than 20 years ago have been of great interest since they included obese and lean volunteers, using different experimental designs in order to take into account the difference in body weight and body composition among the two groups. These studies showed a significant interaction between acute exercise and food ingestion–induced TEF, since the former potentiated the latter. However, the thermogenic synergism was largely reduced in obese subjects.

In the study by Broeder and colleagues (8), lean and obese men performed low-intensity versus high-intensity exercise with and without a meal in the postexercise period. Whereas the EE of both groups was the same during the exercise itself, the thermogenic response was higher in the high-intensity group (14%) than in the low-intensity group (6%) in the postexercise period. However, the increase in TEF remained marginal.

Some studies showed contrasting results in certain categories of subjects or under dynamic nutritional conditions: For example, the thermogenic response to a mixed meal was not potentiated by moderate exercise in obese subjects either before or after weight loss. Moderate exercise failed to influence the thermogenic response to a mixed meal in healthy elderly individuals. No clear explanation about the mechanisms explaining the potentiation of TEF by exercise has been provided.

A general hypothetical scheme (figure 33.1) summarizes the metabolic flux expected when a meal is ingested during performance of exercise. The meal induces postprandial macronutrient substrate storage whereas the exercise involves endogenous macronutrient mobilization. Both processes increase macronutrient utilization and turnover as well as fuel "trafficking." For example, exogenous carbohydrates (CHO) are partly used as fuel during the postprandial phase. When exercise occurs, CHO oxidation markedly increases, and a small fraction will be derived from "online" utilization of exogenous sources. An analysis of previous exercise studies shows that a single CHO load ingested during exercise will be oxidized at maximum rate of about 1 g/min (4 kcal/min), irrespective of the amount of CHO ingested. Therefore any excess of exogenous CHO will be stored as glycogen.

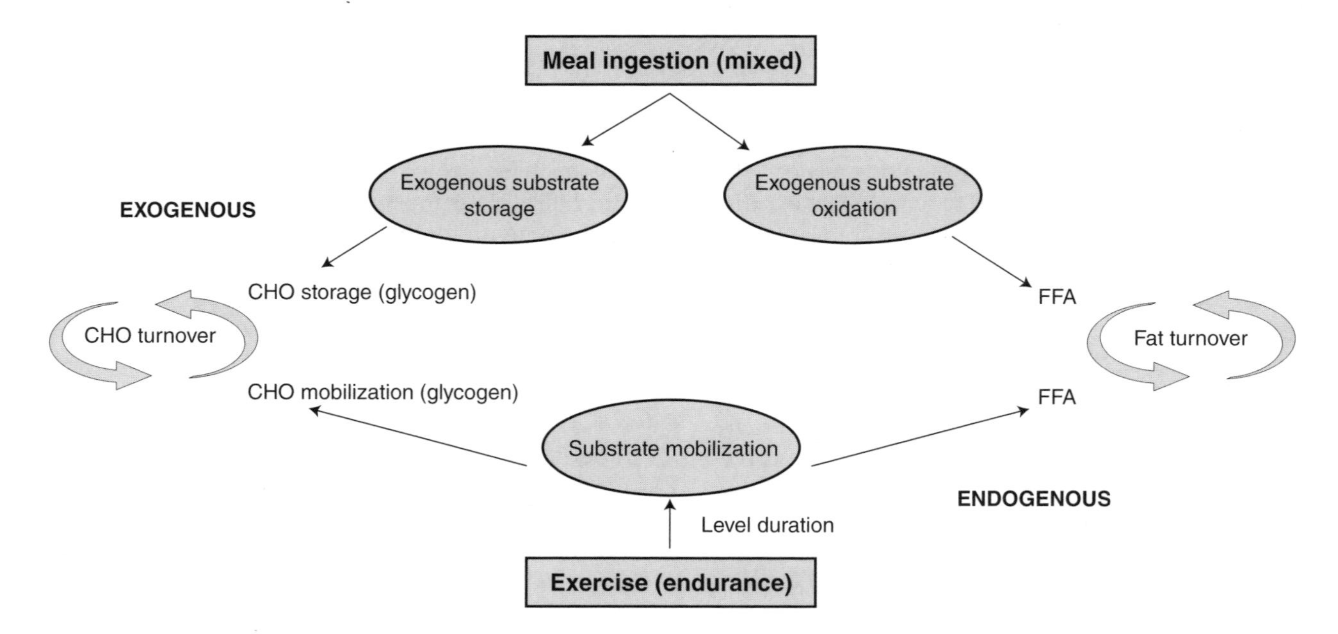

Figure 33.1 This scheme summarizes the metabolic flux expected when a mixed meal (containing carbohydrates and fat) is ingested. Feeding induces macronutrient substrate storage in the postprandial phase, whereas performing exercise simultaneously results in endogenous macronutrient mobilization. Both processes increase macronutrient utilization and substrate turnover. CHO = carbohydrates; FFA = free fatty acids.

To summarize, TEF constitutes a relatively small component of EE under resting conditions. This is even more the case during exercise: If an individual exercises at 10 metabolic equivalents (METs) and TEF represents a fraction of, say, 10% of total EE under sedentary conditions, then the contribution of TEF will be only 1% of total EE—a value that can be judged as clinically negligible.

Thermic Effect of Food and Exercise Training

In contrast to the acute response, the chronic effect of exercise on TEF has not been widely studied in relation to obesity. This situation requires evaluation of TEF longitudinally, that is, before, during, and after a training period of sufficient intensity and duration. Full adherence to such a program in obese individuals is challenging. Furthermore, changes in body composition result, so that it is difficult to separate out this effect from the effect of exercise training per se that increases cardiorespiratory fitness.

The TEF seems to be increased by aerobic training as well as by weight resistance training. Segal and colleagues (7) determined the impact of physical training (12 weeks of vigorous cycle ergometer training) on TEF at rest and after 1 h of cycling at 100 W in obese and lean men, as well as in obese, diet-controlled type 2 diabetic men. Before and after training, the TEF at rest remained lower in the obese than in the lean men. Thermogenesis was improved after short-term exercise in obese and diabetic men compared with that at rest, but TEF was not normalized.

Other studies have shown that some trained individuals have a blunted TEF associated with a reduced stimulation of catecholamine production, as well as a lower insulin response in the postprandial phase. The relationship between $\dot{V}O_2$max and DIT has not been investigated in obese subjects. Thus, short-term exercise enhances TEF in obesity but fails to normalize it completely.

A general scheme summarizing the overall effect of training on TEF is shown in figure 33.2.

Exercise, TEF, and Postprandial Substrate Utilization

An important issue is the response of substrate oxidation, as assessed by the profile of respiratory quotient, in addition to the stimulation of EE. A combination of these two factors allows estimation of the fuel mix oxidized in the postprandial phase. We may anticipate a potential shift in postprandial macronutrient oxidation with exercise. This tells us more about substrate regulation than the magnitude of the thermogenic response itself.

Studies on this issue showed that respiratory quotient remained lower if exercise preceded the meal as compared to eating a meal after no exercise. In other words, exercising before eating moderates the shift toward postprandial carbohydrate oxida-

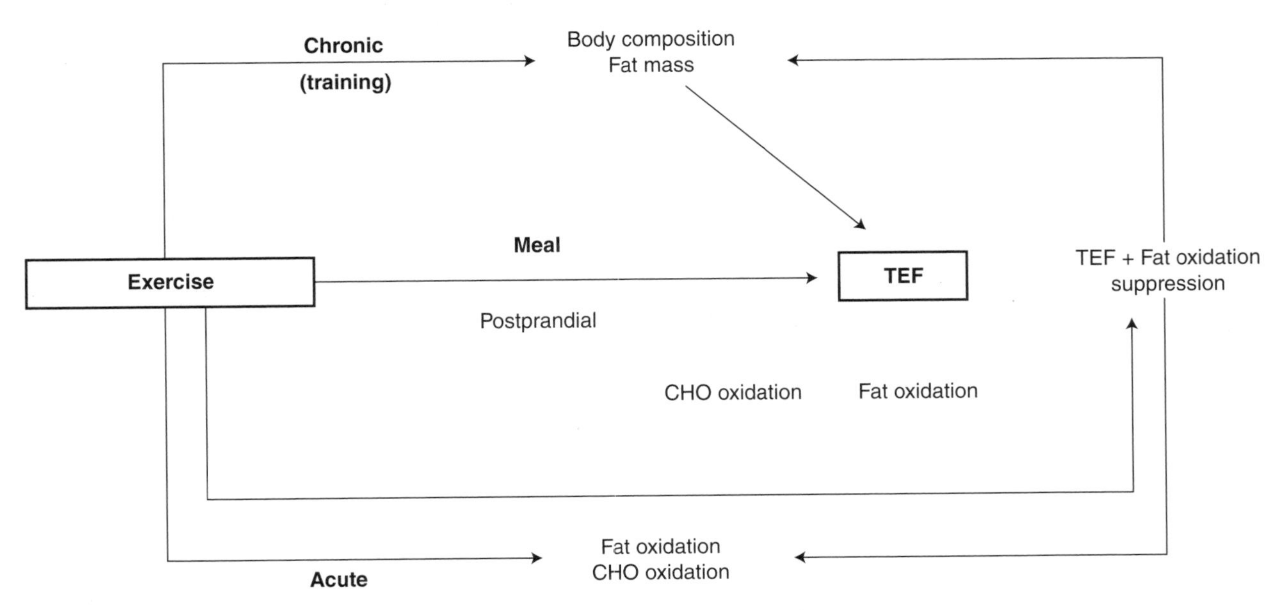

Figure 33.2 The overall concept that both acute endurance exercise (above 70% $\dot{V}O_2$max) and chronic exercise training increase the thermic effect of food (TEF) by modulating substrate utilization in the postprandial phase. CHO = carbohydrates.

tion after eating. As a result, more endogenous fat is oxidized if exercise precedes the meal than if a meal is consumed without previous exercise.

In the study by Broeder and colleagues (8) mentioned earlier, when walking preceded feeding, a significant attenuation of postprandial respiratory quotient values was observed, suggesting less suppression of fat oxidation. This is consistent with another study using high-intensity exercise on the treadmill; this investigation showed that the fuel mix oxidized after a mixed meal was characterized by a greater contribution of lipid oxidation to total EE when the meal was ingested after exercise as compared with when it was ingested without previous exercise (9). Another study using swimming as exercise stimulus demonstrated a similar effect, that is, a shift in metabolism during the postprandial phase leading to a stimulation in fat oxidation over the hours following exercise (10). These observations are consistent with the postprandial postexercise attenuation of plasma triglyceride levels generally observed. Indeed, exercise reduces the adverse effects of a high-fat (high-energy) meal on postprandial lipemia.

To conclude, exercise intensity may play a role in potentiating postexercise fat utilization independently of total EE; the higher the exercise intensity, the greater the effect. To optimize fat balance, exercising before eating is more potent than exercising during or after the meal.

Conclusions

The existence and the functional significance of impaired TEF in obesity remain controversial. Methodological factors, differences in study design, and biological intra- and interindividual variabilities largely influence TEF. The conditions during which TEF is assessed should therefore be highly standardized: postabsorptive conditions, comfortable thermoneutral ambient temperature, no previous exercise immediately before the test or on the previous day, and standardized meals the day before.

The thermic effect of food is negatively associated with relative body fat content and therefore the obese state. The heterogeneity of human obesity, regarding body composition, body fat distribution, and functional hormonal pattern (insulin resistance) can influence the magnitude of TEF. Thus, TEF may be considered a proxy for exploring the net efficiency of exogenous macronutrient utilization.

In a number of studies, it has been demonstrated that acute exercise potentiates TEF when a meal is eaten with exercise and that the intensity of exercise plays a role. The magnitude of potentiation of TEF by exercise is small. Exercise prior to a meal does not normalize completely the blunted TEF observed in the obese state. Generally speaking, both acute endurance exercise (above 70% $\dot{V}O_2$max) and chronic exercise training increase TEF.

When exercise is combined with food intake, the changes in substrate oxidation observed reflect the contrasting effects of exercise versus the meal on exogenous + endogenous substrate oxidation and storage. If fat oxidation needs to be stimulated, it seems more appropriate to perform muscular exercise before rather than after the meal. Although exercise appears to play an important role in the regulation of substrate balance, its quantitative effect on postprandial substrate utilization remains rather limited when considered over a 24 h period.

Physical Activity and Substrate Oxidation

Steven R. Smith, MD

The ability to oxidize fuel and generate adenosine triphosphate (ATP) is fundamental not only to muscular contraction but also to almost every cellular function. Each day we synthesize and consume approximately 50 kg (110 lb) of ATP. The process of oxidative phosphorylation requires not only the ATP-producing machinery (the mitochondria), but also oxygen and fuel to carry out the phosphorylation of adenosine diphosphate (ADP) to produce ATP. The primary fuels used for ATP synthesis are glucose and fatty acids, and the selection of one fuel over the other is tightly regulated. Glucose and fatty acids are channeled into parallel pathways of glycolysis and fatty acid oxidation, with key enzymes determining the relative rate of glucose versus fatty acid combustion. Substrate selection, also known as substrate switching, is governed not only by these minute-to-minute systems but also through cellular signaling to the nucleus and remodeling of the machinery needed to switch between substrate and fat and glucose oxidation. In this chapter, I will review recent advances in understanding of the relationships between physical activity and substrate selection in obesity from both perspectives.

Metabolic Inflexibility in Obesity and Type 2 Diabetes

The ability to switch from fat to glucose as a fuel is impaired in obesity and type 2 diabetes (T2DM). There is a dual defect whereby fasting fat oxidation is decreased and the switch to glucose oxidation after a meal or insulin infusion is blunted. David Kelley and Larry Mandarino termed this dual defect "metabolic inflexibility" (1). Initially described as a defect in skeletal muscle as measured by arterial–venous difference, the concept has expanded into other measures such as whole-body glucose oxidation during a euglycemic-hyperglycemic clamp. We recently found that metabolic inflexibility was present in young men with a family history of T2DM; this paralleled reduced fat oxidation during sleep (measured in a whole-room calorimeter) and was tightly linked to mitochondrial content in skeletal muscle (2). This suggests that the defects are present early in the course of the disease and are coupled with reduced mitochondrial mass and capacity for fat oxidation. We have also recently found that insulin resistance resulting in reduced glucose uptake is a major component in the reduced ability to turn on glucose oxidation in obese T2DM patients. Weight loss improved metabolic flexibility as defined by an increase in insulin-stimulated glucose oxidation but did not increase fasted fat oxidation.

Reduced Capacity for Fat Oxidation?

Given that a reduced capacity for fat oxidation, low muscle mitochondrial content, and insulin resistance go hand in hand, several perspectives can be applied to the problem of slow substrate switching observed in obesity. One view is that the control system inappropriately favors glucose oxidation and "turns down" fat oxidation. Evidence for this

Acknowledgements
The author acknowledges the contributions of the following scientists to create better understanding of substrate switching: Eric Ravussin, Jose Galgani, Cedric Moro, Sheila Costford, Sudip Bajpeyi, and George A. Bray. This work was supported by NIH DK072476, NIH AG030226-01, and U.S. Department of Agriculture 2003-34323-14010.

view comes from the "paradoxically" increased resting muscle glucose oxidation in obesity and T2DM. A second view is that the mass of mitochondria or the entry of the products of beta-oxidation into the electron transport chain may be limiting. This view is supported by the observation that mitochondrial mass is decreased in obesity and especially T2DM, as indicated by the work of Kelley and Simoneau (3). Low ATP demand may lie upstream of the observed decrease in resting ATP turnover. A more parsimonious view is that physical inactivity or *insensitivity to the signals generated during physical activity* leads to both a reduction in the mass of mitochondria (needed for fatty acid oxidation) and defects in insulin signaling (preventing glucose oxidation after a meal) (2).

Is Inflexibility "Preprogrammed" Into Muscle Cells?

Given that metabolic inflexibility is present in young individuals with a family history of T2DM and that T2DM is partly a genetic disorder, it is possible that metabolic flexibility is also inherited. One approach to exploring this concept is at the cellular level ex vivo. Many of the clinical phenotypes having impaired substrate switching in vivo are maintained in cell culture ex vivo. For example, insulin sensitivity and the preference for glucose as a fuel (4) are preserved in myoblasts cultured ex vivo under rigidly controlled conditions. Genetic and potentially epigenetic mechanisms could account for a portion of the between-subject variability in metabolic flexibility observed in vivo and should be further explored. Understanding these mechanisms might lead us to new discoveries and therefore the identification of unanticipated targets for the pharmacotherapy of obesity and T2DM.

Importance of Physical Activity in Glucose and Fat Oxidation

In the basal or resting state, fuel selection, as measured by the respiratory quotient, largely matches the composition of the diet (also called the food quotient). The time required to adjust substrate usage in response to changes in dietary macronutrients is relatively long and dependent upon both the magnitude of the change and the physical activity level. In parallel with work of Patrick Schrauwen in Maastricht, we found that three to seven days are required to increase fat oxidation when dietary fat is increased. There is a large degree of variation in the rate at which fat oxidation increases and glucose oxidation decreases. The decrease in glucose oxidation is linked to the upregulation of the pyruvate dehydrogenase kinase 4 messenger RNA (mRNA), protein, and activity. Pyruvate dehydrogenase kinase 4 phosphorylates and inhibits pyruvate dehydrogenase, a key enzyme for the entry of glucose into the tricarboxylic acid (TCA) cycle. There is a lag between the intake of dietary fat and the increase in fat oxidation and fall in glucose oxidation. This lag is greater in persons who are insulin resistant. Insulin resistance is a feature of obesity but also occurs in healthy young people with a family history of T2DM. During the lag between intake of a high-fat diet and the activation of fat oxidation, glucose oxidation continues, and by necessity glycogen stores will be decreased. J.P. Flatt's glycogenostatic model suggests that low glycogen stores will drive food intake to replenish glycogen stores, leading to a positive fat balance. Recent prospective data support this hypothesis, but it is by no means proven.

Importantly to this discussion, we found that when physical activity was increased at the same time as fat intake, the rate of "adaptation" to a high-fat diet was rapid and complete, even in those who could not "adapt" while sedentary (figure 34.1) (5). Several important conclusions can be drawn from these findings. First, it is important to know that the control systems needed to increase fat oxidation are blunted in individuals with insulin resistance as occurs in obesity; and second, modest-intensity physical activity reverses the effects of inactivity and insulin resistance on substrate oxidation. The latter message is positive and could help encourage increased physical activity in the people who need it most. There may also be a practical application of these results: Populations with high dietary fat intake may need increased levels of physical activity to prevent the cumulative effects of small daily positive energy balances that occur when fat intake exceeds fat oxidation. This suggests that physical activity or pharmacologic or nutritional treatments that increase the *capacity* for fat oxidation (as exemplified by the AICAR- and PPARs-treated mice of Ron Evans) could improve both fat and glucose oxidation and prevent weight gain in susceptible populations. Understanding the mechanisms underlying the interindividual difference in the ability to rapidly increase fat oxidation and turn off glucose oxidation may lead to novel therapies for both the prevention of weight gain and the treatment of obesity.

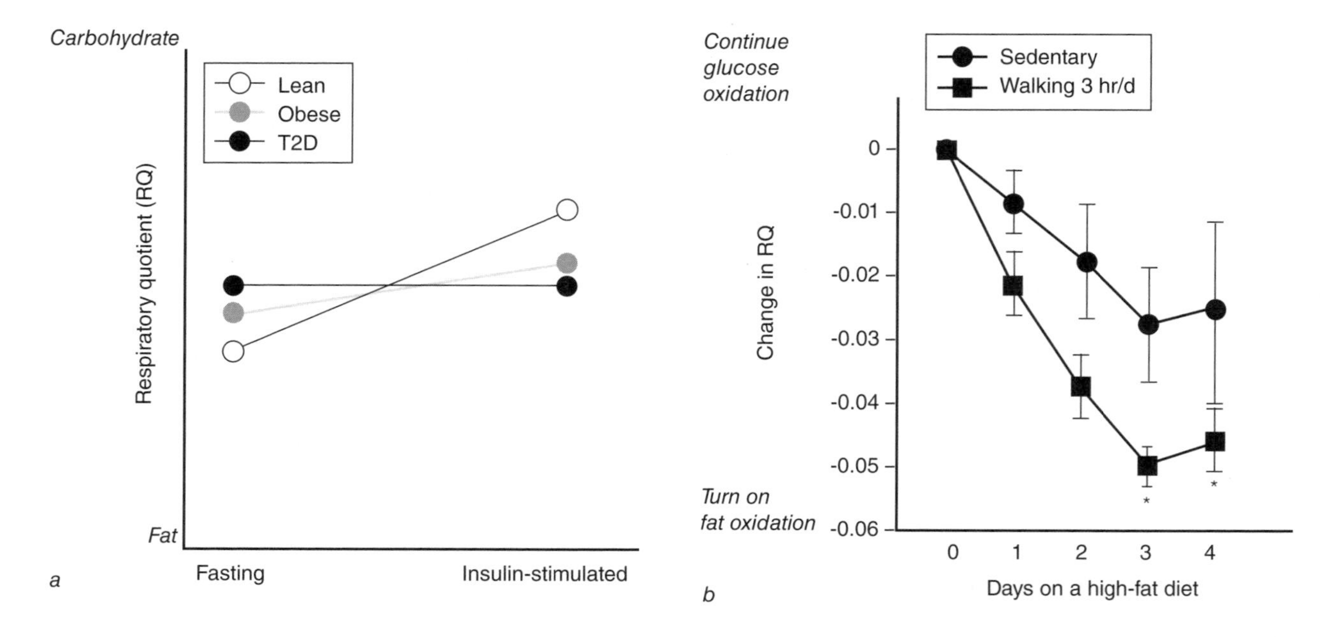

Figure 34.1 *(a)* Glucose oxidation should increase after a meal. This can be measured by indirect calorimetry as an increase in respiratory quotient. In obese subjects and even more so in patients with T2DM, insulin does not stimulate glucose oxidation. Also seen are an inability to oxidize fat and a reliance on glucose as a fuel during the fasted state. *(b)* Fat oxidation increases slowly when we consume a high-fat diet at energy balance; it takes several days to turn on fat oxidation and turn off glucose oxidation. Subjects were able to turn on fat oxidation faster while walking on a treadmill for approximately three hours per day as measured in a whole room calorimeter. Asterisk indicates a significant increase in fat oxidation as compared to that in the sedentary condition.

(a) Adapted from D.E. Kelley and L.J. Mandarino, 2000, "Fuel selection in human skeletal muscle in insulin resistance: A reexamination," *Diabetes* 49: 677-683. *(b)* Data from S.R. Smith et al., 2000, "Concurrent physical activity increases fat oxidation during the shift to a high-fat diet," *American Journal of Clinical Nutrition* 72: 131-138.

Intramyocellular Lipid Droplets as a Fuel Source During Exercise

As illustrated in figure 34.2, triglyceride is stored in intramyocellular lipid droplets, and the lipolysis of these lipid droplets is an intricate part of the control system regulating which fuel is burned during activity (carbohydrate vs. fat). These small lipid droplets lie adjacent to mitochondria in the intramyofibrillar space, and recent work suggests that the translocation of hormone-sensitive lipase to the lipid droplet regulates lipolysis during contraction (6). Exercise training in sedentary obese subjects increases the frequency of the connections between the lipid droplet and mitochondria—presumably to facilitate the direct delivery of free fatty acids to the mitochondria. This is consistent with the observation that low-intensity exercise training increases fat oxidation during physical activity, more specifically nonplasma fat oxidation (intramyocellular lipid oxidation) (7). How the connections between the lipid droplet and the mitochondria are regulated by physical activity in obesity is an area ripe for exploration. The same is true for understanding what proteins and signaling pathways are responsible for the complete "coupling" of lipolysis to mitochondrial fat oxidation.

Substrate utilization is regulated at several key points as shown in figure 34.2. In addition to the regulation of classic control points (acetyl-CoA carboxylase and pyruvate dehydrogenase), recent progress in the area of intramyocellular lipolysis and lipid synthesis points to the lipid droplet as an important buffer for the excess lipid delivery from adipose tissue to skeletal muscle in obesity. The lipid droplets are "coated" with members of the PAT family of proteins. Along with the triglyceride, diglyceride, and monoglyceride lipases, the balance between lipid storage and lipid synthesis is critical to prevent lipotoxicity, that is, the buildup of ceramides and diacylglycerides. These lipid intermediates, along with short chain acylcarnitines produced by a mismatch of beta-oxidation to complete electron transport chain oxidation, have been hypothesized to produce insulin resistance and turn off insulin-stimulated glucose oxidation. In concert with the observed decrease in insulin-stimulated glucose oxidation, fat oxidation is impaired, contributing to the reliance on glucose as a fuel. Together these defects constitute "metabolic inflexibility" or impaired substrate switching between glucose and fatty acids. Seen in both obesity and type 2 diabetes, these defects run in step with insulin resistance.

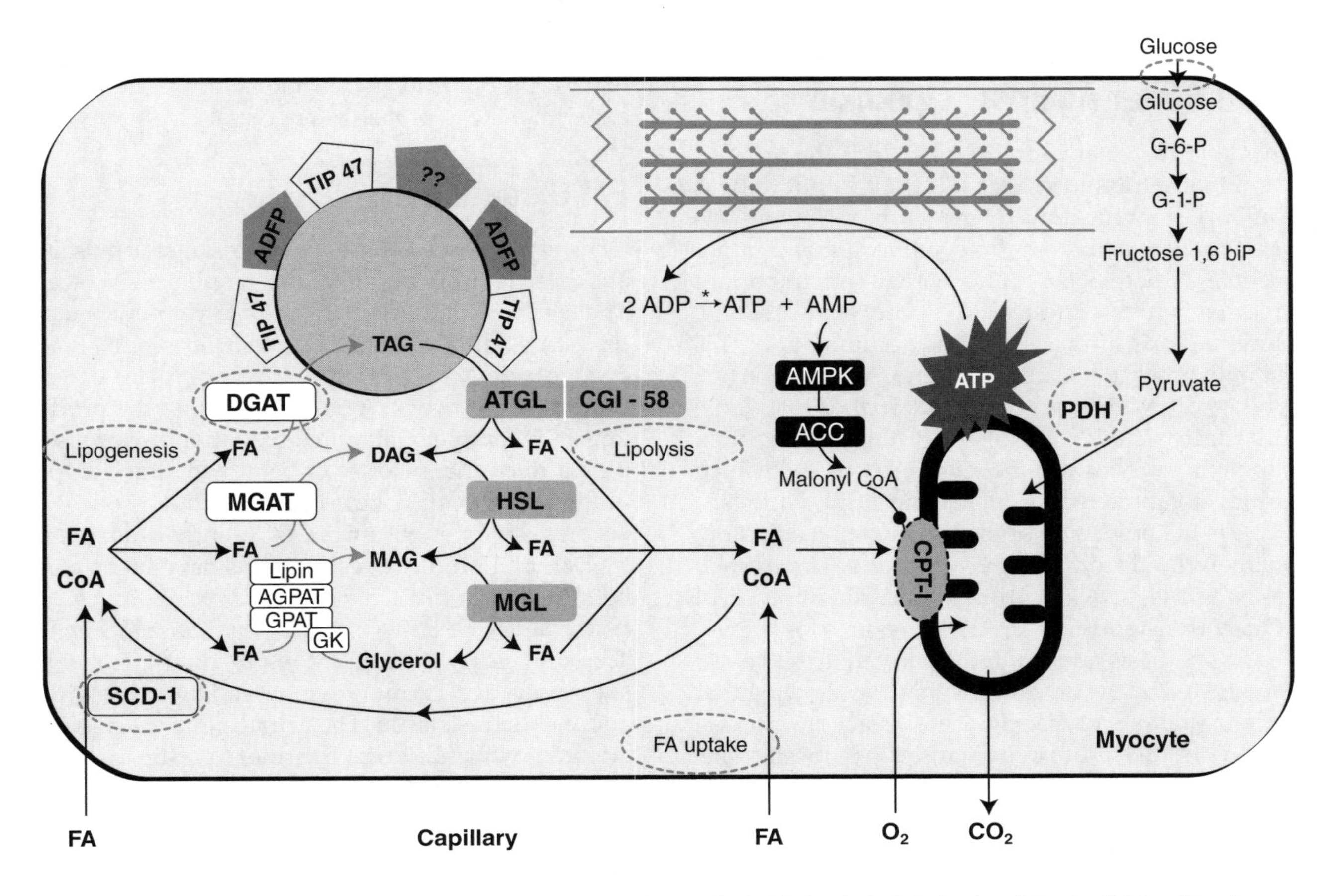

Figure 34.2 Substrate utilization is regulated at several key points marked with the dashed circles (see following list for abbreviations). Asterisk indicates the enzyme adenylate cyclase.

ACC: acetyl-CoA carboxylase

ADFP: adipose differentiation-related protein

AGPAT: 1-acylglycerol-3-phosphate O-acyltransferase

AMPK: protein kinase, adenosine monophosphate (AMP) activated

ATGL: adipose triglyceride lipase, also known as PNPLA2 patatin-like phospholipase domain containing 2; catalyzes the first step in triglyceride lipolysis

CGi-58: a.k.a. ABHD5 or abhydrolase domain containing 5, a potent coactivator of ATGL

CPT-1: carnitine palmitoyltransferase

DAG: diacylglyceride

DGAT: diacylglycerol O-acyltransferase, key enzyme in triglyceride synthesis that is postulated to prevent lipotoxicity

FA: fatty acid

FA-CoA: fatty acyl-CoA

GK: glycerol kinase

GPAT: glycerol-3-phosphate acyltransferase

HSL: hormone-sensitive lipase; catalyzes the conversion of DAG to MAG and FA and also has some triglyceride lipase activity

lipin: a.k.a. PAH-1, a $Mg2^{+}$-dependent PA phosphatase, the penultimate enzyme in the pathway to synthesize monoacylglycerol

MAG: monoacylglyceride

MGAT: monoacylglyceride acyltransferase

PDH: pyruvate dehydrogenase, key rate-limiting step in the entry of pyruvate (glucose) into the TCA cycle

SCD-1: stearoyl-CoA desaturase (delta-9-desaturase)

TAG: triacylglycerol

TIP-47: a.k.a. M6PRBP1, mannose-6-phosphate receptor binding protein 1; a member of the PAT family of lipid droplet-associated proteins

Intramyocellular Lipid Droplets as a Buffer Against "Lipotoxicity"?

Another recent advance in the field is the recognition that intramyocellular lipid serves not only as an important fuel for exercising muscle but also as a buffer between excess lipid supply and intramyocellular "lipotoxicity." Lipid storage is an important mechanism to sequester "free" fatty acids, thereby limiting toxic intracellular lipid accumulation. This should occur in the adipose tissue. As noted earlier, if the adipose tissue is dysfunctional or is so large that fatty acids "spill over" to the muscle, then lipotoxicity follows. If the muscle receives too much lipid, it can adapt by increasing lipid synthesis. This is the other side of the lipid droplet equation, namely, lipid storage. Key synthetic enzymes such as stearoyl-CoA desaturase and diacylglycerol O-acyltransferase are key control points for triglyceride synthesis and can reduce lipotoxicity. This is supported by the preclinical literature showing that overexpression of diacylglycerol O-acyltransferase prevents the lipotoxicity caused by obesity and protects from insulin resistance via a decrease in lipotoxic intermediates, namely ceramides and diacylglycerides. Insulin sensitivity is required for the activation of meal- or insulin-stimulated glucose uptake and oxidation, and therefore for substrate switching and metabolic flexibility. Clinical studies to test the hypothesis of lipid storage mediated by diacylglycerol O-acyltransferase are lacking and should be considered as a major void in the literature. This hypothesis is, however, consistent with findings in endurance-trained athletes in whom lipid storage is high and lipid droplets are increased in size and number while insulin sensitivity is also high (e.g., the "athlete's paradox").

Intramyocellular lipid droplets are "coated" with specialized proteins called PAT proteins. First discovered in adipocytes by Londos and colleagues, the perilipin, ADRP, TIP-47 (PAT) family of proteins serve the important function of regulating lipase access to the lipid droplet in adipocytes and presumably in skeletal muscle. The best-studied protein in adipose tissue, perilipin, is not present in skeletal muscle, whereas ADRP, TIP-47, and possibly OXPAT are present in skeletal muscle. The ways in which exercise and exercise training regulate the skeletal muscle triglyceride droplet at key steps (storage into the droplet, droplet localization, protein packaging, and regulation of lipolysis toward oxidation or re-esterification) need more attention at the fundamental and especially at the clinical level. The major challenge for clinical investigations lies in the development of tools to accurately measure each of these processes in vivo; stable isotope pulse-chase experiments are probably a necessary first step.

Fat Oxidation

Obesity increases lipid supply to skeletal muscle and has the potential to cause insulin resistance. The mechanism or mechanisms are just now being discerned. The activation of protein kinase C by diacylglycerides and the increase in ceramides are two candidate pathways. A new idea is that the products of incomplete fatty acid oxidation might also play a role. The majority of beta-oxidation occurs in the mitochondria, but some oxidation probably occurs in the peroxisomes. As recently illustrated by Deborah Muoio, obese animals have increased beta-oxidation but *without an increase in complete oxidation to* CO_2 *through the electron transport chain.* This suggests that the lipid intermediates, namely, short chain acyl carnitines, may lead to lipotoxicity and insulin resistance. The actual signal or signals are unknown and deserve further investigation.

Intensity and Duration of Exercise Needed to Improve Metabolic Flexibility

How much activity or exercise is needed to reverse the defects in substrate switching in obese patients? We don't know the answer to this important question. Recent data from the STRRIDE study suggest that less intense, longer-duration exercise might actually be better for increasing insulin sensitivity (8). This is consistent with the work of van Baak showing that low-intensity exercise training improves intramyocellular fat oxidation and with the work of Venables and Jeukendrup (9), which also favored low-intensity, high-volume training. Importantly, exercise training with weight loss, but not weight loss alone, improves mitochondrial function and fat oxidation in obese patients (10), whereas both interventions improve insulin action. Together, these findings support the notion that low-intensity exercise training improves mitochondrial function, fat oxidation, and insulin action in obese patients.

Will the "dose" that optimizes fat oxidation during exercise for obese patients also be "best" for mitochondrial biogenesis, insulin sensitivity, or both? The answers to these questions will require integrated investigation on the effective exercise intensity that will not only maximize short-term fat oxidation, but also increase mitochondrial mass and improve insulin action. It is possible that these will track together. How each of these individual domains, plus the function of the machinery and activity of the intramyocellular lipases that surround the lipid droplet, might change in obese subjects under different levels of exercise intensity should be explored.

Summary

In summary, recent advances in the molecular and cellular biology of substrate switching, combined with clinical data on the impact of obesity on metabolic flexibility, provide us with a dual view of substrate oxidation. Minute-to-minute signals are clearly important in substrate switching. In addition, integration of short-term signals generated by physical activity and the neuroendocrine milieu *reprogram* muscle metabolism toward insulin sensitivity and increased capacity for fat oxidation. Combined, these two interdependent pathways lead to improved metabolic flexibility and have the potential to prevent weight gain and improve health.

Physical Activity and Sympathetic Nervous System Activity

Ian A. Macdonald, PhD

The sympathoadrenal system comprises the sympathetic nervous system (SNS) and adrenal medulla. The latter acts as an endocrine gland and releases mainly epinephrine, but also norepinephrine, into the plasma when stimulated by the splanchnic nerve. The SNS innervates most of the organs and tissues of the body and is responsible for regulating cardiovascular and metabolic function, as well as influencing respiratory, gastrointestinal, visual, salivary, and thermoregulatory functions (1). In relation to exercise and physical activity, the major roles of the SNS and adrenal medullary hormones concern the regulation of the cardiovascular response to exercise, including the initiation of skin vasodilation and sweating to control body temperature and the regulation of fat mobilization from adipose tissue (figure 35.1). The cardiovascular response to exercise varies depending on the type and duration of the exercise; but the most relevant exercise in relation to energy balance and obesity is more prolonged aerobic exercise, such as walking, swimming, and cycling. These types of exercise can normally be sustained for prolonged periods of time and have the potential for utilizing body fat stores. Such aerobic exercise involves an increase in cardiac output combined with a redistribution

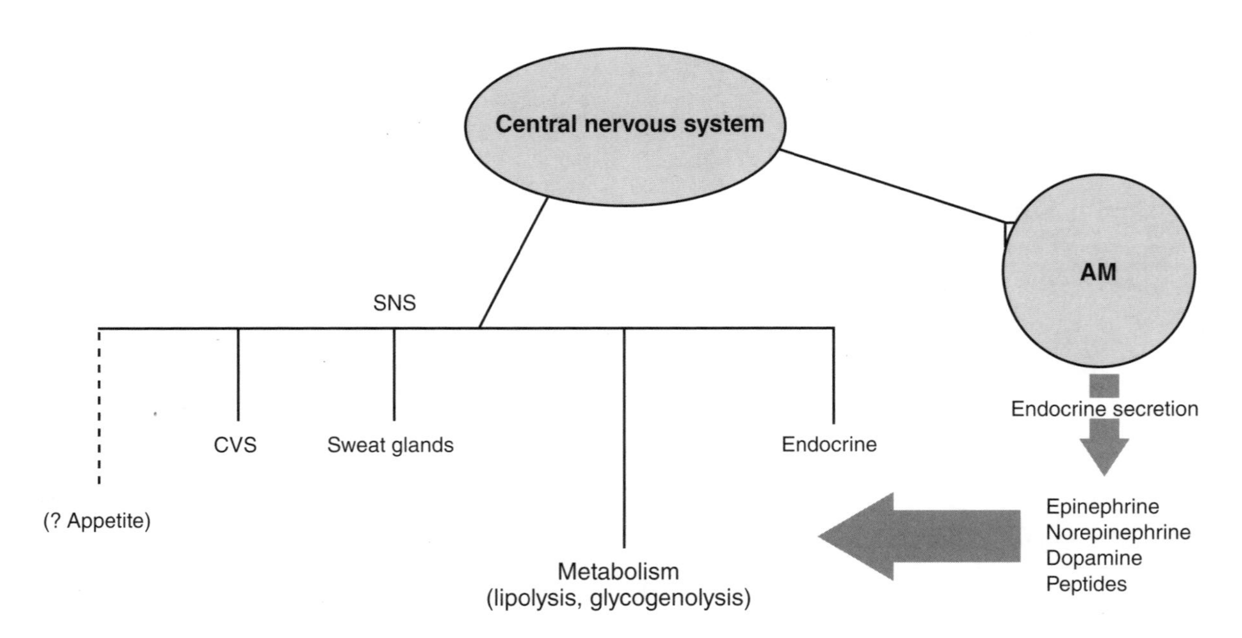

Figure 35.1 Physiological role of the sympathetic nervous system and adrenal medulla (AM) in relation to physical activity.

of this output toward the active muscles, and initially away from the skin, gastrointestinal tract, and inactive muscles. As such exercise proceeds, the resulting generation of heat leads to a rise in body temperature and an increase in skin blood flow, which together with the initiation of sweat production, produces an increase in heat loss and prevents an excessive increase in temperature. These vascular and sweating responses to exercise involve the SNS innervation of blood vessels and sweat glands with the exception of the increase in blood flow in the active muscles, which results from the local release of vasodilator peptides and metabolites.

While the primary functions of the SNS and adrenal medulla during exercise are to regulate the cardiovascular system and maintain blood pressure to sustain vital organ function, it is also clear that in this and other physiological situations the activation of the SNS is highly differentiated (2). Thus, SNS activation of skin and muscle blood vessels is separately regulated, and muscle vascular control is regulated differently in one part of the body compared to another. Muscle sympathetic nerve activity (MSNA) appears to be entirely vasoconstrictor, and recordings of MSNA have been used to assess the SNS response to a variety of situations. Using this MSNA recording technique, it is clear that leg MSNA increases in line with the increase in blood pressure seen in an acute bout of exercise, while the majority of studies involving physical training do not show any consistent effect of training on MSNA (2); the latter issue is discussed in more detail later.

Exercise and the Sympathoadrenal System

Acute exercise is associated with an increase in plasma catecholamine concentrations, which is directly proportional to the exercise intensity (3). These increases in catecholamine concentrations represent an activation of the SNS and adrenal medulla, although there are several problems in interpreting such data due to the site of blood sampling and the type of exercise performed. The principal SNS response to exercise is concerned with cardiovascular regulation, but a contribution toward adipose tissue lipolysis and increased availability of fatty acids for muscle metabolism is also important. Higher intensities of exercise promote the adrenal medullary secretion of epinephrine, which is involved in stimulating lipolysis and in promoting gluconeogenesis and glycogenolysis, thus contributing to the regulation of blood glucose. While there is some suggestion that obese persons may have reduced plasma catecholamine responses to acute exercise, it has been proposed that this does not reflect altered control of the metabolic response to exercise, as it is offset by increased catecholamine sensitivity (3). Little attention has been paid to the potential effect of altered SNS responses or sensitivity on the cardiovascular response to exercise in the obese.

Physical training is reported to lead to reduced resting and exercising plasma catecholamine levels (3), but many of these observations were made in studies that obtained the blood samples from an antecubital (forearm) vein, which is not optimal for drawing reliable conclusions about the SNS response to an intervention such as exercise. As will be seen in the subsequent section, there is a substantial amount of disagreement in the literature on the effects of training or increased physical activity on the activity of the SNS and adrenal medulla.

One aspect of undertaking regular exercise is that in order to maintain energy balance, one must have an energy intake that matches the total energy expenditure. Thus, any impact of regular exercise on the SNS or adrenal medulla could be due to a direct effect of exercise or a consequence of the high energy intake or high overall energy flux. Bell and colleagues (4) showed that a reduction in whole-body energy flux in habitually active older subjects, achieved by abstention from exercise for five days with an accompanying reduction in energy intake to maintain energy balance, was associated with a reduction in MSNA as well as a reduction in resting metabolic rate. Moreover, when a beta-receptor antagonist was administered acutely, it reduced resting metabolic rate in the control active state but not during the period of inactivity and reduced energy turnover. While it cannot be determined whether the effect of being physically active is due directly to the exercise, is secondary to the higher energy intake, or both, this study shows that in older, habitually active people, being active is accompanied by an increase in SNS activity and a higher resting metabolic rate.

A related issue is how to control for body composition when one is comparing active and sedentary individuals. Physically active, and especially endurance-trained, individuals have lower body fat content than sedentary individuals of the same weight. Alvarez and colleagues (5) reported similar levels of MSNA in young endurance-trained men and sedentary men matched for weight but not for body fat content. However, when the comparison was made between endurance-trained men and sedentary men with a matched body fat content,

MSNA was approximately 30% higher in the trained men. Previous work had also shown that the MSNA responses to a series of cardiovascular challenges were not different between trained subjects and sedentary controls with a higher body fat content. Thus, MSNA is similar between those who are endurance trained and sedentary individuals with a higher body fat content; and while one might speculate that the SNS is likely to be having different effects on metabolic function in the two states, more work is needed to identify the significance of these observations.

Physical Activity and the Sympathoadrenal System

Early studies of Pima Indians identified potential contributions of the SNS to the development of obesity and also indicated possible links with physical activity. Christin and colleagues (6) used an isotope tracer technique to measure the rate of appearance of norepinephrine into the circulation and thus assess SNS activity, finding that it was related to the level of spontaneous activity recorded while subjects were studied in a whole-body calorimeter. The activity recorded under such conditions would be more related to fidgeting than to movement of the whole body; but it is of interest that the SNS activity was not a consequence of enforced inactivity in the calorimeter, as it was assessed on a different day from the activity measurement. Such observations do not provide any mechanistic link between the SNS and physical activity, but subsequent training studies in other populations have been aimed at providing information on possible cause-and-effect links.

Cornelissen and Fagard (7) performed a meta-analysis of 72 trials that involved randomized controlled assessment of the effects of endurance training on blood pressure. A subgroup of these trials also used plasma norepinephrine concentrations as indices of SNS activation and plasma renin activity as an index of the activation of the renin-angiotensin system. Overall, there was a small but significant reduction in resting and ambulatory blood pressure in response to endurance training, with larger effects seen in studies of hypertensive subjects (approximately 6 mmHg) than in normotensives (approximately 2 mmHg). The small number of studies in which plasma norepinephrine and renin activity were assessed showed reductions in both in response to training; the decrease in norepinephrine was approximately 27%. While the use of forearm venous plasma norepinephrine is rather inaccurate as an assessment of SNS activity, it does suggest that endurance training may reduce blood pressure at least in part due to effects on the SNS and on renin release. However, the trials included in this analysis involved young to middle-aged adults (mean values 38 to 50 years), and it appears that the effects of training on the SNS may, at least in part, depend on the age of the subjects being studied.

Grassi and colleagues (8) looked at the effect of endurance training in 16-year-olds. Blood pressure and MSNA were decreased after 10 weeks of training, while the SNS response to cardiovascular disturbances was enhanced by training. Thus, in young individuals, although resting SNS activity is decreased by training, the responsiveness of the SNS is enhanced at least in relation to cardiovascular control. By contrast, Sheldahl and colleagues (9) found no effect of training on MSNA or baroreflex responses to variations in blood pressure in a group of middle-aged and older men. The training duration and the improvements in maximum oxygen uptake and reductions in resting blood pressure were similar in the two studies. It is not known whether the different effects on MSNA in young and older subjects were accompanied by differential effects of training on metabolic variables. It is interesting that neither set of results fits with the observation mentioned earlier (4) that acute cessation of exercise in habitually active older subjects is associated with a decrease in MSNA, implying that habitual activity is associated with enhanced MSNA (at least in older subjects).

Public health advice encourages people to walk at least 10,000 steps per day in order to improve health and help manage their weight. One study has assessed the impact of such walking on blood pressure and an index of cardiac SNS activity in a group of Japanese middle-aged, hypertensive men (10). Walking more than 10,000 steps per day for 12 weeks was accompanied by reductions in systolic and diastolic blood pressure of approximately 8 to 10 mmHg in hypertensive men, with no change in those with normal blood pressure. The exercise intensity with walking is likely to be much lower than that achieved in the endurance training programs used in other studies; but it was sufficient to produce a small increase in maximum oxygen uptake in the hypertensive subjects, although no information was given about possible changes in those with normal blood pressure. It is interesting to note that the initial maximum oxygen uptake values in the hypertensive subjects were very low, indicating a very poor level of aerobic fitness initially. The walking program was accompanied by a reduction in the index of cardiac SNS activity, determined from the low frequency power in the power spectral analysis of heart rate variability that was recorded in this study. Unfortunately no information was provided as to the effect of the walking program on this index of cardiac SNS activity in those with normal blood pressure.

Conclusions

A functional SNS is essential in order for someone to be able to be physically active. This enables the production of a coordinated cardiovascular and thermoregulatory response during periods of activity and also enables mobilization of fuel reserves, especially fat stores, so that prolonged activity can be undertaken. Undertaking regular physical activity is valuable in improving health status, producing a drop in blood pressure in those with hypertension and having metabolic benefits in most people. There is a general assumption that regular activity is of benefit in reducing the activity of the SNS, but this is not supported by the majority of the studies described in this chapter. In otherwise healthy middle-aged and older individuals, a physical training program is associated with either no change or even an increase in SNS activity; and an acute period of inactivity in those who are habitually active is associated with a reduction in SNS activity. It is only in studies of exercise training in hypertensive adults or in healthy teenagers that a reduction in SNS activity with training is observed. It would be of interest to investigate the effects of a physical training program on SNS activity in healthy, obese individuals who have an elevated level of SNS activity in the sedentary state. It is possible that in such individuals, increased physical activity will have an effect similar to that seen in those with hypertension.

Physical Activity Level and Adipose Tissue Biology

Isabelle de Glisezinski, PhD

White adipose tissue is considered a quasi-unlimited energy depot compared to the low amount of energy stored as glycogen. Adipocytes, the adipose tissue cells, primarily comprise a single vacuole filled with triglycerides. Under conditions of increased energy demand such as fasting or exercise, these triglycerides are hydrolyzed and released in the bloodstream as nonesterified fatty acids (NEFA). Their uptake and subsequent oxidation in other tissues, mainly skeletal muscle, generates energy to meet increased demand.

An important consideration for human studies is that lipid mobilization is heterogeneous according to adipocyte localization. In humans, adipose tissue depots are often classified as subcutaneous (superficial) and visceral (deep). The latter are generally more strongly associated with metabolic and cardiovascular diseases since NEFA release is facilitated in the visceral depots. These NEFA pour into the portal system, contributing to metabolic disturbances observed in upper body obesity. Lipolysis in subcutaneous adipose tissue (SCAT) is less facilitated. Nevertheless, SCAT represents more than 80% of total body fat mass and therefore is the main source of NEFA. Most studies on adipose tissue lipolysis assess SCAT since access and sampling are more feasible than in visceral tissue. Subcutaneous adipose tissue is generally considered a good reflection of substrate availability, particularly for physical exercise.

Adipose tissue is also considered an endocrine organ. Adipocyte secretions, known as adipokines, take part in the regulation of energetic metabolism; but this function will not be addressed here. This chapter focuses on adipose tissue lipolysis for energy provision during physical exercise, particularly in obese subjects, and the effects of physical training on lipid mobilization.

Adipocyte Lipolysis Regulation

The regulation of human adipose tissue lipolysis during exercise was previously attributed to both the increase in catecholamine levels and the simultaneous decrease in plasma insulin concentration. In fat cells, lipolysis is activated through stimulation of beta-adrenergic receptors (beta-AR) and inhibited through stimulation of alpha2-adrenergic receptors (alpha2-AR) (figure 36.1). The simultaneous activation of these two receptors modulates intracellular cyclic adenosine monophosphate (cAMP) production, which activates a cAMP-dependent protein kinase, leading to the phosphorylation and activation of the hormone-sensitive lipase (HSL). Insulin exerts its antilipolytic action through reduction of cAMP amounts as a result of phosphodiesterase-3B activation and phosphorylation of beta-AR. In addition, it has been shown that hyperinsulinemia reduces the lipolytic response to adrenaline in SCAT of normal-weight subjects. Thus, exercise-induced insulin suppression may contribute to accelerated lipid mobilization during exercise by two mechanisms: (1) decreased antilipolytic action of insulin and (2) sensitization to the lipolytic effect of catecholamines. In addition to catecholamines and insulin, other hormonal signals (atrial natriuretic peptide [ANP], growth hormone, etc.) may also contribute to exercise-induced lipolysis.

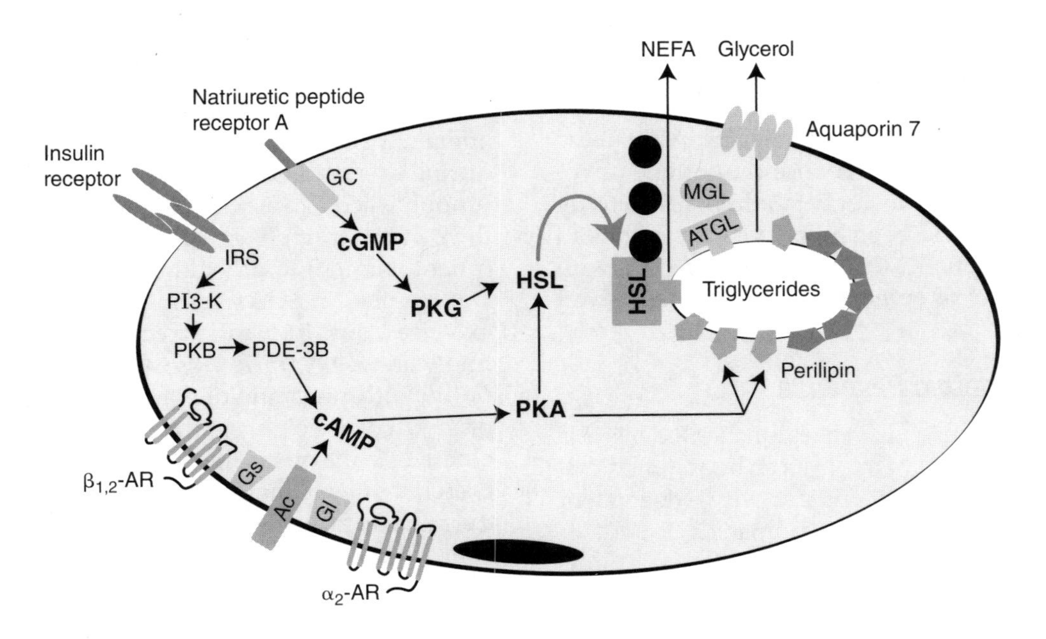

Figure 36.1 Adipocyte lipolysis regulation. AC: adenylyl cyclase; GC: guanylyl cyclase; HSL: hormone-sensitive lipase; ATGL: adipose triglyceride lipase; MGL: monoglyceride lipase; PKA, PKB, PKG: protein kinase A, B, G; PI3-K: phosphatidylinositol 3-kinase; IRS: insulin receptor substrate; NEFA: nonesterified fatty acids.

It has been shown that natriuretic peptides are potent activators of lipolysis in human fat cells. The physiological relevance of the ANP-dependent lipolytic pathway was demonstrated in young men performing physical exercise. In these subjects, circulating ANP concentrations rose two- to threefold during short exercise bouts. Atrial natriuretic peptide acts through stimulation of fat cell plasma membrane receptors (NPR-A subtype), which bear an intrinsic guanylyl cyclase activity and increase intracellular levels of cGMP, thereby activating a cGMP-dependent protein kinase. cGMP-dependent protein kinase-dependent phosphorylation of HSL stimulates lipolysis. Finally, within a concentration range observed in humans, ANP stimulates adipose tissue lipid mobilization and increases circulating NEFA levels.

Lipolysis defects in the obese state relate to hormonally stimulated, rather than basal, lipolysis. Resistance to catecholamine-induced lipolysis in SCAT has been demonstrated in obese subjects and is attributed in part to decreased expression and enzymatic activity of HSL (1). Blunted lipolysis and decreased HSL expression were shown in preadipocytes of obese subjects, suggesting that altered lipolysis could be a primary defect in the etiology of obesity. Recently, a novel triglyceride lipase, adipose triglyceride lipase (ATGL), has been identified; this enzyme plays an important role together with HSL in the control of lipolysis. Coregulation of HSL and ATGL gene expression has been observed, suggesting that the expression of ATGL may also be altered in obesity. Decreased expression of ATGL has been reported in white adipose tissue of genetically obese mice and obese human subjects. Nevertheless, in humans, HSL is thought to be the major lipase catalyzing the rate-limiting step in stimulated lipolysis (as in physical exercise), whereas ATGL would participate mainly in basal lipolysis (1).

Lipolysis and Exercise in Obesity

Lipolysis can be considered in terms of adrenergic (alpha and beta), non-adrenergic, and insulin regulation. The effects may vary by gender, magnitude, metabolic state, and assessment methods.

Beta-Adrenergic Response

Basal lipolysis of obese subjects is unimpaired; they have a normal lipolytic rate that yields higher plasma NEFA levels than in lean people, given their larger adipose tissue mass.

However, lipolysis during exercise is blunted in obese compared to lean subjects; this attenuated response to exercise in obese subjects has been attributed to reduced sympathetic nervous system activity and blunted catecholamine response. Indeed, the SCAT lipolytic response to beta-adrenergic stimulation (in situ isoprenaline perfusion) is comparable in lean and obese subjects (2). Horowitz and Klein (3) have shown that lipolytic sensitivity to epinephrine is blunted in abdominal SCAT of obese women; however, this response cannot unequivocally be attributed to lower beta-adrenergic activity since epinephrine acts on adipocytes through both beta-AR and alpha2-AR. The next section deals with this dual effect of epinephrine on both beta- and alpha2-AR.

Alpha-Adrenergic Response

It has been shown in vitro that alpha2-AR density is cell size dependent; therefore it can be assumed that alpha2-adrenergic antilipolytic activity is dependent on the degree of adiposity in humans.

Microdialysis is a method particularly suitable to study of the in vivo lipolytic response of SCAT to pharmacological stimulation during rest or exercise. Perfusion of an alpha2-AR antagonist (phentolamine) in a microdialysis probe inserted into abdominal SCAT allows assessment of the alpha2-AR contribution in situ. The blockage of alpha2-AR in obese men during a 60 min moderate exercise period compared with that in lean men demonstrates that exercise-induced lipolysis in SCAT is impaired in obese subjects and that the physiological stimulation of adipocyte alpha2-AR during exercise contributes to this impairment (4). Indeed, the involvement of alpha2-AR in the suppression of lipolysis was demonstrated by the enhancement of exercise-induced lipolysis produced by local perfusion of phentolamine. This greater activation of alpha2-AR in obese subjects could account for the impairment of adrenergic response-induced lipolysis.

To lead to a better understanding of these findings, however, the results must be interpreted according to metabolic state (fed or fasted, rest or exercise), adipose tissue depot (the alpha2-adrenergic activity is higher in SCAT, essentially femoral SCAT, than in visceral fat), and gender. It has been shown, for example, that the specific antilipolytic effect of adrenaline reported in men is absent in overweight women. We have also reported that there are clear gender differences in lipolysis-regulating mechanisms and whole-body lipid mobilization during exercise in overweight subjects (5).

Nonadrenergic Response During Exercise

Exercise promotes both sympathetic nervous system activation and cardiac release of ANP. The physiological relevance of the ANP-dependent lipolytic pathway was demonstrated in young lean men performing physical exercise. In a recent study, we compared the relative contribution of catecholamines, insulin, and ANP in the control of lipolysis during exercise between lean and overweight men (unpublished). To assess the specific role of ANP in the control of lipolysis, combined alpha-blockade (phentolamine) and beta-blockade (propranolol) was applied in SCAT during a 45 min moderate exercise bout. In lean subjects, exercise-induced lipolysis was significantly reduced (31%) in the probe with propranolol compared to the control probe. In overweight subjects, the addition of beta-blockade in the probe did not reduce lipolysis during exercise, suggesting that the increase in lipolysis during exercise is independent of beta-adrenergic stimulation promoted by catecholamines. In addition, it was shown that this nonadrenergic lipolytic component could reflect ANP-mediated lipolysis. In conclusion, beta-adrenergic lipolysis is impaired in obese subjects during exercise, and the residual catecholamine-independent lipolytic response is attributed to ANP.

Hormonal release and adipose tissue lipolysis control are dependent on exercise intensity. Indeed, the physiological relevance of nonadrenergic pathways (involving probably ANP) in the control of lipolysis is effective during low-to-moderate exercise in overweight humans; but during high-intensity exercise (70% of $\dot{V}O_2max$), only a minor involvement of this pathway is commonly observed (5).

Insulin Activity During Exercise

During exercise, adipose tissue lipolysis is attributed to increase in both catecholamine and ANP levels, but also to decreases in concentration of plasma insulin, which acts as an antilipolytic agent. Moreover, it is well known that obese subjects exhibit reduced insulin sensitivity. However, it is still unclear whether insulin resistance occurs within adipose tissue and is related to lipid metabolism disorders. We attempted to address this question by assessing lipolysis during exercise in lean and obese men under a moderate euglycemic-euinsulinemic clamp (unpublished). In normal-weight men, insulin infusion during exercise promoted a drastic reduction of exercise-induced lipolysis, suggesting that insulin exerts an antilipolytic effect even during exercise. On the other hand, the effect of a constant insulin plasma level on exercise-induced lipolysis was not apparent in obese subjects; suggesting that adipose tissue in obese subjects could be resistant to the antilipolytic action of insulin during exercise.

Lipolysis and Exercise Training in Obese Persons

We have seen that excess adipose tissue impairs exercise-induced lipolysis. Therefore, it is reasonable to ask whether this effect is reversible by exercise training.

Beta-Adrenergic Response

Three months of aerobic exercise training in obese men and four months of aerobic exercise training in overweight young men caused improvements in the control of lipolysis through an enhancement of the beta-adrenergic response. These results were observed under resting conditions: in vitro (6) and in vivo with infusion of isoproterenol (a beta-adrenergic agonist) directly in the SCAT (7).

On the other hand, endurance training does not change lipolytic sensitivity to epinephrine infusion in normal-weight men (8) or in obese women (9). As mentioned earlier, epinephrine acts on adipocytes through both beta-AR and alpha2-AR; therefore, it is also necessary to examine alpha-adrenergic antilipolytic activity to understand the effects of endurance training.

Alpha-Adrenergic Response

As mentioned previously, four months of training in overweight men improved lipid mobilization during a moderate exercise bout through decreased antilipolytic alpha-AR effect in SCAT (10). Likewise, in obese women, training reduced the antilipolytic action of alpha-AR during exercise; nevertheless, the authors concluded that this was due to a reduction of exercise-induced catecholamine increase after training (9). Indeed, even though exercise was performed at the same relative intensity before and after training, epinephrine (but not norepinephrine) levels were lower during an exercise bout after training in obese women. However, in overweight men, identical plasma catecholamine concentrations were obtained in both conditions (10). These findings suggest that epinephrine levels cannot explain the lower alpha-adrenergic response seen during exercise after endurance training (figure 36.2).

Nonadrenergic Response

We have seen that lipolysis in overweight subjects depends mainly on nonadrenergic activity (i.e., ANP activity). It was also shown that endurance training improves ANP receptor–mediated lipid mobilization at rest in overweight men, as demonstrated by infusion of ANP through a microdialysis probe in SCAT (7). It would be interesting to assess this nonadrenergic activity during exercise before and after aerobic exercise training in obese subjects, with alpha-blockade and beta-blockade perfused in a microdialysis probe in SCAT during exercise.

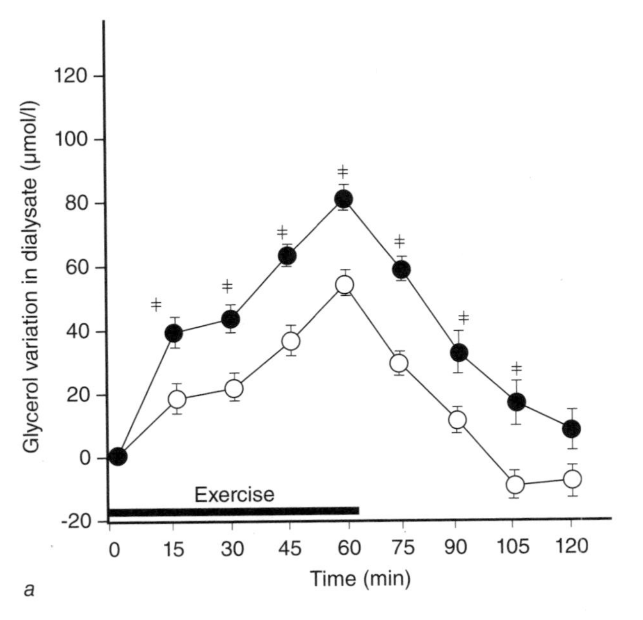

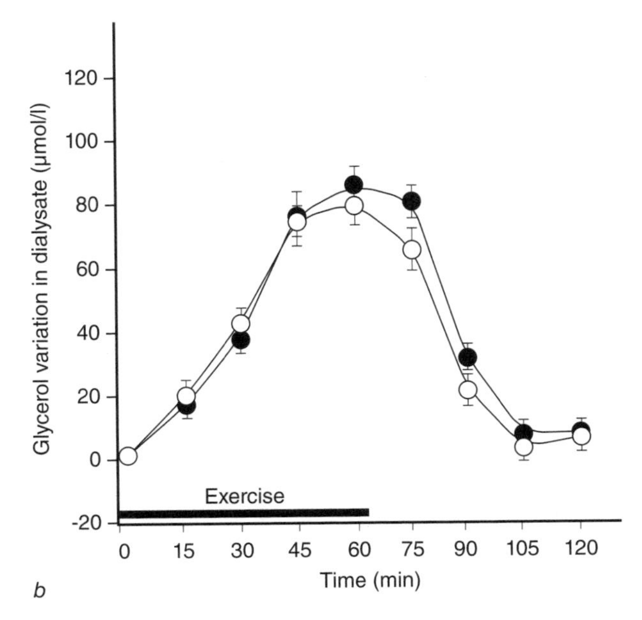

Figure 36.2 Changes in dialysate glycerol concentrations during a 60 min exercise period and recovery in overweight men *(a)* before and *(b)* after training, in control probe (❍) and in probe infused with phentolamine (●). Data are expressed as means ± SEM. *$p < 0.01$ when compared to control probe.

Strength Training Effect on Lipolysis

Most exercise training studies in obese subjects have focused on aerobic training. One study investigated the effects of strength training on SCAT lipolysis in obese men (11). Before and after three months of dynamic strength training, adrenergic lipolytic activity was assessed with infusions of isoproterenol or epinephrine alone or with phentolamine in SCAT through microdialysis probes during a euglycemic-hyperinsulinemic clamp. This experiment demonstrated an increased insulin sensitivity in SCAT and increased beta-adrenergic response, together with an increased antilipolytic action of catecholamines mediated by alpha-AR after strength training.

Conclusion

By virtue of their large adipose tissue mass, obese subjects have elevated plasma NEFA levels. However, adipocytes from obese subjects exhibit impaired lipolytic activity attributable to a strong alpha-adrenergic response, which minimizes NEFA release in the bloodstream.

Exercise, mainly endurance training, leads to enhanced NEFA oxidation and therefore decreased plasma NEFA levels. Consequently, exercise training in obese subjects leads to improved adipose tissue lipolysis not only through enhanced beta-adrenergic and nonadrenergic responses, but also through decreased alpha-adrenergic antilipolytic activity. Moreover, aerobic and strength training both improve insulin sensitivity in SCAT of obese individuals. However, the insulin sensitivity increase leads to a decrease in plasma insulin level. So, these low insulin concentrations could prevent strengthening of the insulin antilipolytic effect without impairing lipid mobilization from adipose tissue.

37

Physical Activity and Leptin Biology

David J. Dyck, PhD

Leptin, the product of the *ob* gene, is a 16 kDa cytokine-like peptide produced primarily by adipose tissue that regulates a multitude of physiological processes relevant to the maintenance of energy homeostasis, including food intake, energy expenditure and physical activity, and reproductive function. Leptin has been dubbed the "antiobesity" hormone and is considered to play a pivotal role in the mechanism defending against the excessive accumulation of body fat. Accordingly, an increase in energy consumption and the accumulation of body fat stores would lead to increased plasma concentrations of leptin, with a subsequent decrease in food intake and increase in sympathetic tone and energy expenditure. Conversely, in response to starvation and loss of body fat, a decrease in leptin would result in an increase in food intake and a decrease in energy expenditure, as well as cessation of reproductive function and somatic growth, for the purposes of conserving energy.

Various rodent models that either fail to secrete functional leptin (*ob/ob* mice) or do not express the leptin receptor (*db/db* mice and *fa/fa* rats) are characterized by extreme obesity, physical inactivity, infertility, and insulin resistance. Interestingly, rodent and human cases of lipoatrophy or lipodystrophy that are characterized by an absence or very low levels of circulating leptin also demonstrate physical inactivity, reproductive dysfunction, and insulin resistance. In general, however, documented cases of human obesity due to extremely low levels of circulating leptin are relatively rare. By far, most cases of human obesity are characterized by high plasma leptin concentrations; the positive relationship between body fat and circulating leptin is a well-accepted phenomenon. Therefore, the general assumption is that during obesity there is a resistance to the physiological and metabolic effects of leptin. Certainly, the ability to induce leptin resistance via the consumption of high-fat diets, both centrally (hypothalamus) and in peripheral tissues such as liver and muscle, has been demonstrated (see figure 37.1).

Much of the initial research examining the biological effects of leptin focused on its central effects on appetite control and energy expenditure. However, leptin also interacts with numerous peripheral tissues that express receptors, including the placenta, blood vessel endothelial cells, heart, liver, skeletal muscle, and others. Essentially, there are two major isoforms of the leptin receptor: the long isoform, which is required for full stimulation of the janus kinase-signal transducers and activators of transcription (JAK-STAT) pathway, and the short isoforms, which result in the activation of JAK2 but not STAT. Both major isoforms exist in skeletal muscle, a key tissue in the regulation of glucose homeostasis, although the majority are the short isoform. Leptin signaling in skeletal muscle leads to the activation of various kinases, including AMP-activated protein kinase (AMPK), phosphoinositide 3-kinase (PI 3-kinase), Akt or protein kinase B, protein kinase C, mitogen-activated protein kinase (MAP kinase), Jun kinase (JNK), and extracellular signal–regulated kinase. The peripheral effects of leptin are numerous and beyond the scope of this chapter. It has been postulated that one of the major roles of leptin is to prevent lipotoxicity (lipid accumulation) in peripheral tissues. A decrease in lipid stores as a result of chronic leptin administration has been demonstrated in numerous tissues including pancreas, adipose, heart, liver, and skeletal muscle; conversely, the lipid content of peripheral tissues becomes markedly increased in cases of low leptin levels (e.g., lipodystrophy or lipoatrophy).

Importance of Skeletal Muscle and the Potential Role of Leptin

For many years, abnormal lipid metabolism has been believed to be a significant factor leading to decreased insulin sensitivity in skeletal muscle, the most important tissue by mass responsible for clearing a glucose load. In skeletal muscle, impaired fatty acid (FA) oxidation leading to increased lipid stores in muscle has been considered a significant factor in the development of insulin resistance. Conversely, an increased capacity to oxidize FA, a hallmark feature of exercise training, is strongly correlated to improved insulin sensitivity. Over the past decade, the discovery of various adipokines and their role as modulators of insulin sensitivity has improved our understanding of the link between obesity and insulin resistance.

Acute Effects of Leptin Regulation of Muscle Fatty Acid Metabolism and Insulin Sensitivity

Leptin stimulates FA oxidation and triacylglycerol (TAG) breakdown in skeletal muscle while simultaneously decreasing incorporation into lipid pools. Leptin has also been shown to acutely induce the translocation of the FA transporter, fatty acid translocase (FAT/CD36), to the sarcolemma. Thus, one of the acute roles of leptin would appear to be to facilitate the clearance of FAs from the blood and their subsequent oxidation. This is consistent with the observation of fasting-induced increases in plasma leptin, which coincides with the release of FAs into the circulation. The acute stimulatory effect of leptin on FA uptake and oxidation in skeletal muscle appears to be due, at least in part, to a rapid and transient stimulation of AMPK; interestingly, acute exposure to FAs also activates AMPK in skeletal muscle. However, it

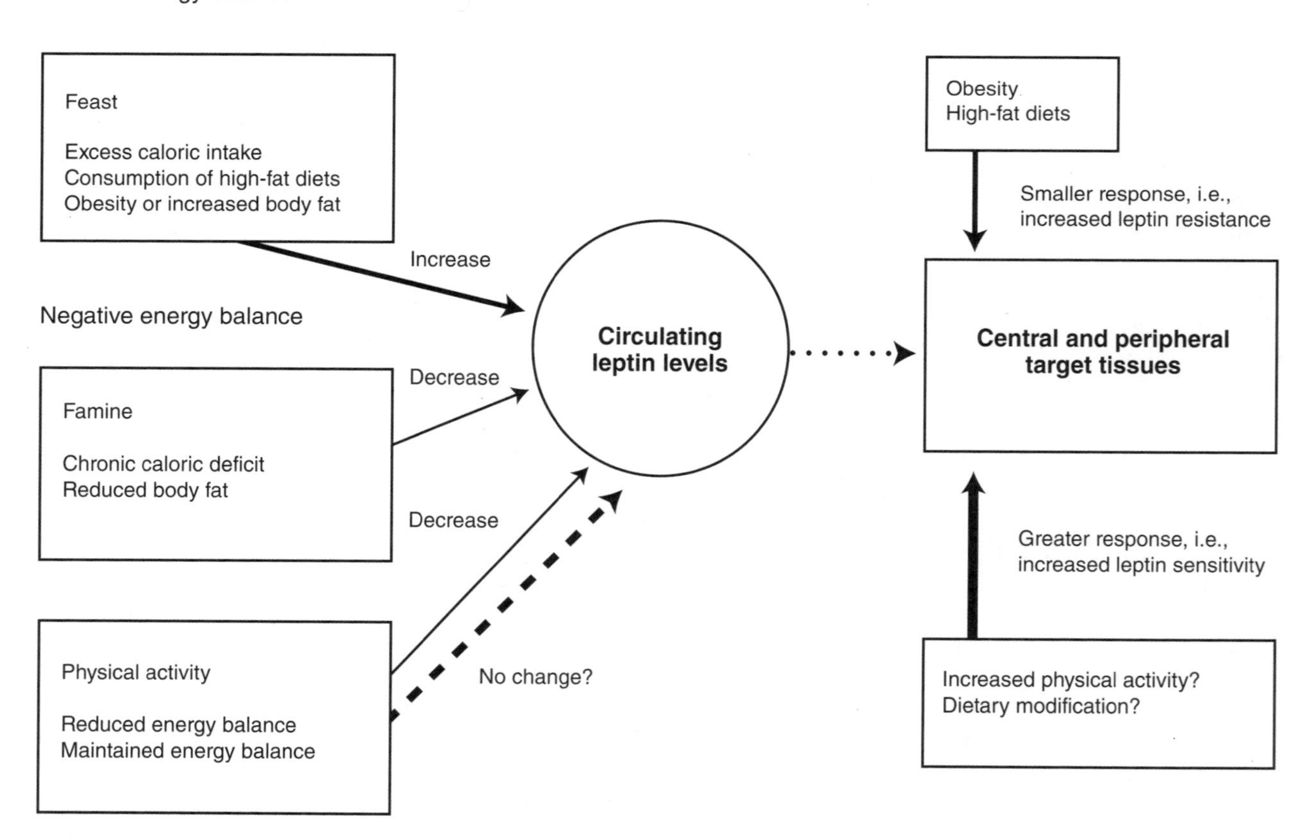

Figure 37.1 Schematic of relationship between energy balance, physical activity, and leptin levels.

is unlikely that the acute effects of leptin on FA metabolism can be fully accounted for by the activation of AMPK. For example, leptin-induced stimulation of FA oxidation can be blocked in the presence of wortmannin, a PI 3-kinase inhibitor. Conversely, leptin acutely stimulates FA oxidation in cardiac muscle but only in the absence of any stimulation of AMPK. Finally, the activation of AMPK appears to be antilipolytic rather than prolipolytic. Thus, it is likely that factors other than AMPK, such as extracellular signal–regulated kinase, are involved in the acute effects of leptin on lipid metabolism.

Chronic Effects of Leptin Regulation of Muscle Fatty Acid Metabolism and Insulin Sensitivity

The chronic metabolic effects of leptin in rodents have been examined via injection or subcutaneous implant. Chronically, leptin stimulates muscle FA oxidation and lipolysis while simultaneously blunting FA storage as TAG. However, in stark contrast to acute exposure, the chronic administration of leptin decreases the abundance of the FA transporters, FAT/CD36, and the plasma membrane-bound fatty acid binding protein (FABPpm) in the plasma membrane. Thus, the combined effect of reduced FA uptake, partitioning of FA toward oxidation and increased rates of lipolysis, results in a net decrease in intramuscular TAG following chronic leptin exposure. The mechanisms underlying leptin's chronic effects on FA metabolism have not been fully elucidated but also appear to involve AMPK. Importantly, the chronic administration of leptin to both normal and high-fat fed rodents results in significant improvements in insulin-stimulated glucose uptake and insulin signaling (1).

Regarding obesity and leptin resistance, high levels of circulating leptin characterize most cases of obesity, suggesting the presence of central or peripheral leptin resistance or both. The feeding of high-fat diets to rodents results in elevated circulating leptin levels, impaired leptin transport across the blood–brain barrier, and diminished metabolic response to peripheral leptin injections. Peripheral tissues, such as liver and skeletal muscle, also appear to develop resistance to the effects of leptin as a consequence of a high-fat diet. For example, the ability of leptin to blunt hepatic glucose production is diminished in rats fed a high-fat diet. In addition, results from our own laboratory have indicated that the acute stimulatory effect of leptin on lipolysis and FA oxidation is lost in skeletal muscle in rats fed a high-fat diet (2). Recent evidence has also been provided for the existence of leptin resistance in human skeletal muscle (i.e., leptin can stimulate FA oxidation in isolated muscle strips from lean but not from obese humans) (3).

Physical Activity and Leptin

An increased capacity of skeletal muscle to oxidize FAs and improved insulin sensitivity are hallmark features of chronic exercise training. Given the potential role of leptin as an important regulator of both these parameters, one should consider the possibility that exercise causes improved lipid metabolism and insulin sensitivity by modulating plasma leptin concentration or the sensitivity of various tissues to leptin (or both). For excellent reviews on the effects of exercise on leptin, the reader is directed to those of Kraemer and colleagues (4) and Berggren and colleagues (5).

Acute Exercise

Numerous studies over the past 10 years have examined the relationship between physical activity and plasma leptin concentrations. The effect of acute exercise (i.e., a single bout) on plasma leptin concentrations is somewhat equivocal, although studies generally show either a decrease immediately following exercise that is sustained for several hours or no effect. A potential source of variability in the outcome is the duration and intensity of the exercise bout. In general, it seems that acute exercise of a greater intensity (e.g., above lactate threshold) or duration (e.g., greater than 60 min) elicits a reduction in leptin while more moderate, less exhaustive exercise does not. It is not known precisely what signal causes the acute exercise–induced reduction in leptin levels, although candidates include stimulation of adipocyte adrenergic β-3 receptors by epinephrine and norepinephrine, increased cortisol and growth hormone, and decreased insulin. Very long bouts of exercise (i.e., several hours or more) are the most likely to induce a decrease in plasma leptin. This phenomenon appears to be closely related to the magnitude of the energy deficit created by the exercise (6). Indeed, the suppression of leptin induced by long bouts of exercise can be largely countered by the provision of sufficient calories to meet the energy needs of the exercise.

Chronic Exercise

Generally, studies employing less than 12 weeks of training in humans show no change in circulating leptin concentrations unless there is a significant loss of body fat. The findings of studies utilizing training protocols lasting longer than 12 weeks are equivocal; some studies fail to show any change in circulating leptin, while others show a decrease. However, of those that do show a decrease, a loss of body fat is often considered to be the cause. In various studies in which the loss of weight or fat (or both) that is often associated with training was prevented, a decrease in leptin was not detected (7). Thus, as with acute exercise, a decrease in overall energy balance appears to be a primary signal for the reduction in leptin concentration.

Aerobic Training to Improve Leptin Sensitivity

Despite the physiological significance of leptin resistance and its relationship to obesity and insulin resistance, there has been little research to examine lifestyle factors, such as aerobic training, as a means to prevent or reverse its development. Endurance training is known to improve skeletal muscle insulin sensitivity; recently our laboratory investigated whether aerobic training could improve skeletal muscle leptin sensitivity in rats placed on a high-fat diet. The main finding of this study was that skeletal muscle leptin resistance induced by a high-fat diet was partially reversed with endurance training as demonstrated by the restoration of leptin's ability to partition FAs toward oxidation and away from TAG storage (8). The inducement of leptin resistance by high-fat feeding also coincided with an increased expression of suppressor of cytokine signaling-3 (*SOCS-3)* mRNA, a potent inhibitor of leptin signaling. Surprisingly, restoration of leptin sensitivity with endurance training was accompanied only by a trend toward a decrease in *SOCS-3* mRNA, indicating the possible involvement of other factors in the regulation of leptin signaling or suggesting that the changes in *SOCS-3* messages are not paralleled by changes in its protein.

Summary

Leptin is an important hormone in the regulation of energy homeostasis. In addition to its well-recognized central effects, leptin interacts with numerous peripheral tissues expressing its receptor, thereby affecting many physiological and metabolic parameters. Skeletal muscle, which is the main tissue involved in the clearance of a glucose load, is also affected by leptin. Leptin stimulates FA oxidation and decreases lipid storage in muscle, which may in part be responsible for its insulin-sensitizing effects. However, the sensitivity of muscle (as well as other tissues) to the effects of leptin appear to be diminished (e.g., leptin resistance) as a consequence of consuming high-fat diets, and in cases of obesity. Whether leptin resistance actually precedes or causes skeletal muscle insulin resistance is not known. It is worthwhile noting that the majority of studies of the acute effects of leptin on muscle metabolism have utilized pharmacological concentrations in the absence of other cytokines and hormones. Thus, as always, extrapolation from isolated ex vivo studies to the in vivo situation is difficult. Physical activity may also play a significant role in the biology of leptin (see figure 37.1). Physical activity, both acute and chronic, may result in a decline in circulating leptin concentrations. This appears to be largely due to the degree of energy imbalance or deficit created by the exercise or training, although this idea is not without some controversy. Limited evidence suggests that endurance training can protect rodents from the leptin-desensitizing effects of a high-fat diet, although more research examining the effect of exercise on leptin sensitivity is clearly needed.

38

Physical Activity Level and Hypothalamic Peptides

Christopher D. Morrison, PhD

Hypothalamic neurons regulate an array of physiological processes, and the hypothalamus is particularly implicated in the regulation of energy balance. Although significant progress has been made in defining hypothalamic circuits regulating food intake (1), it is clear that the hypothalamus also regulates both energy expenditure and physical activity. Deletion or overexpression of key neuropeptides, their receptors, or the intracellular signaling systems regulating their production can markedly alter body weight homeostasis and obesity predisposition, and these effects are driven by independent yet complementary changes in both food intake and energy expenditure.

Although an effect on energy expenditure is well established, whether and how individual components of energy expenditure are altered is less clear. Many studies assess cumulative energy expenditure (oxygen consumption) without specifically assessing physical activity, yet an increasing number of studies focus specifically on physical activity. This chapter focuses on two facets of the interaction between the hypothalamus and physical activity. The first is a direct effect of hypothalamic peptides and their receptors on physical activity, and the second is the impact of physical activity on the hypothalamus.

Hypothalamic Regulation of Energy Balance

Many of the neuropeptides associated with physical activity were initially described based on their effects on food intake. Therefore a brief introduction to the basic structural and functional features of the hypothalamic regulation of feeding will be used to set the stage for a subsequent focus on physical activity. Much of our understanding of how the hypothalamus influences energy homeostasis stems from work on the adiposity hormone leptin and its interaction with neural populations within the hypothalamic arcuate nucleus (ARC). These neurons include those expressing the orexigenic neuropeptide Y (NPY) and agouti-related protein (AgRP), as well as those expressing pro-opiomelanocortin (POMC) and the anorexigenic alpha-melanocyte stimulating hormone. Other brain areas also undeniably contribute to the regulation of energy balance, but the mechanisms underlying the regulation of ARC neurons and their action on downstream sites are the most well established.

Hypothalamic neurons respond to a variety of nutritional signals. Immediate availability of fuels is signaled by glucose, fatty acids, and amino acids, for which specific sensing mechanisms have been described in ARC and other neurons of the hypothalamus. Availability of fuels in the near future is signaled from the gut by gastrointestinal hormones such as ghrelin, GLP-1, and peptide YY. Lastly, information about the level of energy stored in fat is signaled by circulating adiposity signals, particularly leptin and insulin. Arcuate nucleus neurons then transmit metabolic information to second-order targets both within and outside the hypothalamus. The major recipients of arcuate input are other hypothalamic areas, in particular the paraventricular nucleus (PVN) and lateral-perifornical hypothalamic areas. Lateral-perifornical hypothalamic area neurons contain neuropeptides associated with food intake, energy expenditure, activity, and arousal. These neurons are also highly connected with other brain areas, receiving "metabolic" information from NPY/AgRP and POMC neurons, as well as input from brain areas associated with reward, motivation, learning and memory, sensory input, circadian time, vagal and visceral sensory input, sensory motor coordination, and arousal. The PVN also represents a downstream target of ARC neurons, with relevance to food intake, energy expenditure, and physical activity. The PVN is classically associated with neuroendocrine function via the hypothalamic-pituitary axis, as well as the autonomic nervous system.

Hypothalamic Regulation of Physical Activity: General Considerations

The hypothalamic regulation of food intake has direct implications for the regulation of physical activity. Although laboratory experiments can dissociate food intake from physical activity, in natural settings the procurement of food is often a highly active process requiring extensive travel or bursts of activity (hunting) or both. Physical activity and food intake are therefore often coupled. A good example of this interaction is the food anticipatory behavior exhibited by rodents, who tend to acutely increase activity in anticipation of a scheduled meal. When discussing individual neuropeptides it becomes apparent that many that alter physical activity also affect food intake, and this overlap should be no surprise considering that food intake is a behavior that often involves physical activity.

The hypothalamus also plays an intrinsic role in circadian rhythm (2). The core machinery of the circadian clock is housed within the hypothalamus (i.e., the suprachiasmatic nucleus), and many hypothalamic sites receive direct projections from these core circadian neurons. Just as food intake and activity are often coupled with each other, they are also coupled to circadian time, with arousal, physical activity, and feeding all occurring in distinctly circadian patterns. Thus a role for hypothalamic neurons in coordinating changes in behavior and activity levels during circadian time would be expected, and hypothalamic neuropeptides, particularly orexin, are associated with changes in sleep and arousal, as well as with food intake and activity.

Lastly, the hypothalamus also contributes to neuroendocrine and autonomic output. Neurons within the hypothalamus are primary regulators of the hypothalamic-pituitary-adrenal axis (HPA). In addition, hypothalamic neurons directly impinge on the sympathetic and parasympathetic nervous systems. Physical activity involves the coordinated regulation of autonomic and neuroendocrine outflow and also interacts with stress and anxiety. These broad physiological effects of the hypothalamus provide ample room for theoretical models in which alterations in activity reflect changes in energy homeostasis, circadian rhythmicity, or stress and anxiety; and in some cases the same neuropeptide might be regulating different endpoints via signaling in different brain areas.

Regulation of Physical Activity by Hypothalamic Neuropeptides

Of the hypothalamic peptides associated with physical activity (3), orexins (hypocretins) are perhaps the most significant (4). Orexins (A and B) are produced in the lateral hypothalamus and were initially found to significantly stimulate food intake (hence the name orexin). However, it was soon discovered that orexins also increase arousal and activity. Orexin neurons project to brain areas both within and outside the hypothalamus, and physical activity is increased following direct orexin injection into a broad number of sites. Mutations in the orexin 2 receptor are associated with narcolepsy in both dogs and humans, while deletion of orexin or orexin receptors in mice leads to obesity, reduced activity, and alterations in arousal and sleep. Orexin neurons interact with "metabolic" areas, such as the ARC, and orexin expression is sensitive to nutritional status. Lastly, orexin neurons also receive projections from hypothalamic areas associated with the circadian clock. These observations suggest that orexin neurons may coordinate physical activity, feeding behavior, and circadian time. As already noted, this interaction should not be surprising, as the coordination of feeding and physical activity with circadian time is critical for animals in the wild.

Adjacent to lateral hypothalamic orexin neurons are a population of neurons that contain melanin-concentrating hormone (MCH). Expression of MCH is regulated by nutritional status and inhibited by leptin, and local injection of MCH increases food intake, body weight, and body adiposity while decreasing energy expenditure. Mice deficient for either MCH or its receptor MCHR1 are lean and show significant increases in locomotor activity, and MCH neurons project to brain areas influencing motivated behavior and motor control.

The hypothalamic neuropeptide corticotropin-releasing hormone (CRH) is produced within the PVN, and direct brain administration of CRH increases spontaneous activity while suppressing food intake. As with orexin neurons, CRH neurons are targets for circadian output from the suprachiasmatic nucleus and may contribute to circadian changes in arousal, activity, and glucocorticoid secretion. While low doses of CRH promote physical activity, higher doses suppress activity. One explanation for this discrepancy is a secondary effect on stress and anxiety, because CRH neurons are generally activated in response to stress.

Neuromedin U (NMU) is another hypothalamic peptide associated with both food intake and physical activity. It is produced throughout the brain as well as in peripheral tissues, such as the pituitary and gastrointestinal tract, and signals through at least two G-protein-coupled receptors that are also ubiquitously distributed. Though NMU has diverse biological functions, injections of NMU into the brain suppress food intake and increase physical activity, as do local injections into the PVN or the ARC. Injections into the PVN also stimulate the HPA axis and may be related to stress responses. Lastly, genetic NMU deletion leads to hyperphagia and reduced activity, while overexpression reduces food intake without altering activity.

Of the classic feeding neuropeptides (NPY, AgRP, and POMC), only AgRP is clearly associated with activity, with *Agrp*-null mice having increased activity and intracerebroventricular administration of AgRP decreasing activity. There are links between activity and the receptors for these classic neuropeptides, as *Y1*-null mice display reduced activity, and genetic variations in *MC4R* are associated with activity phenotypes in humans. In addition, recent data indicate that the deletion of the transcription factor BSX reduces *NPY* and *Agrp* gene expression and reduces locomotor activity. While the dysregulation of NPY and AgRP may contribute to the activity phenotype, it is also possible that the loss of BSX in other brain areas, such as the dorsomedial nucleus of the hypothalamus (DMH) and lateral hypothalamus, may underlie the reduced activity.

Effect of Physical Activity on the Hypothalamus

Although there is strong evidence for an effect of hypothalamic neuropeptides on physical activity and energy expenditure, there are considerably fewer data focusing on the opposite effect, that is, the influence of exercise and physical activity on the hypothalamus. Regular physical activity is clearly beneficial for the brain. Considering the propensity of many hypothalamic neuropeptides to alter food intake, body weight, and physical activity, it is almost expected that physical activity would provide feedback to the hypothalamus. However, currently very few data directly address this possibility or mechanisms that would mediate such an effect (5). Particularly problematic for this discussion are the variable approaches used to induce physical activity in laboratory models, as there are likely significant differences in the response to voluntary activity (wheel running) versus forced activity (treadmill or swimming).

Regarding direct effects of physical activity on the hypothalamus, perhaps the strongest evidence exists for CRH. Produced in the PVN and DMH, CRH influences feeding behavior, stress, and anxiety and has also been shown to alter physical activity. Because exercise affects anxiety, stress, and the HPA axis, it would be predicted that exercise might also alter CRH. Studies of forced activity demonstrate an activation of CRH neurons and an increase in CRH expression; however, voluntary activity does not appear to reproducibly increase CRH expression. Therefore, the activation of CRH neurons may not be dependent on the exercise per se, but due instead to the stress of forced exercise. However, CRH is also produced in other brain areas, including the DMH. A study of voluntary wheel running in obese OLETF (CCK A-receptor deficient) rats demonstrated increased CRH expression in the DMH but not PVN, suggesting that while CRH neurons in the PVN are activated by forced (stressful) exercise, CRH neurons in the DMH may be sensitive to voluntary physical activity (6).

Beyond changes in CRH, there are relatively few data describing changes in hypothalamic neuropeptides following physical activity. Some data support changes in POMC, oxytocin, and NPY, but the number of studies are few and the approaches used are variable. It is therefore difficult to come to any consensus regarding the effects of physical activity on hypothalamic neuropeptide levels, and the mechanisms that would underlie such an effect are also unclear. Possible candidates include changes in energy balance or nutrient flux, production of signaling molecules from the periphery, stimulation of neurotrophic factors, alterations in cerebral blood flow, or neural or somatosensory feedback to the brain. Each of these is a general possibility, and some or all of them could also affect the hypothalamus of active individuals.

Conclusions

Physical activity is intimately linked to feeding behavior, body weight regulation, circadian rhythmicity, and stress and anxiety; and the hypothalamus plays a regulatory role in each of these processes. An increasing number of hypothalamic neuropeptides have been shown to alter physical activity, although the mechanisms underlying these changes and their role in whole-animal physiology are less clear. Similarly, it is likely that physical activity might provide feedback to affect the hypothalamus, but to date relatively few experiments have addressed this possibility.

39

Physical Activity Level and Gut Peptides

Stephen C. Woods, PhD

This chapter provides a brief summary of the literature on the interaction of physical activity and peptides secreted from the gastrointestinal (GI) tract, that is, gut peptides. The area is complex for several reasons. The best-documented action of the gut peptides is to facilitate the digestion and utilization of nutrients during meals. This is enabled in part by shunting blood flow away from skeletal muscle and into the GI tract. Many gut peptides (e.g., gastrin, cholecystokinin, secretin, glucagon-like peptide-1 or GLP-1) have relatively low basal levels in the plasma much of the time, increasing mainly during meals—a time when physical activity tends to be low (1). Conversely, acute physical activity often suppresses or delays eating, secondarily influencing gut peptide activity (1). Such correlations with prandial activity might lead to the prediction that physical activity is inversely correlated with gut peptide activity, or that increased physical activity per se would cause an acute reduction in the secretion of these compounds (perhaps as a secondary consequence of reduced food intake at that time), or both. As discussed later, this scenario is not in fact the case, as the relationship between activity and various gut peptides is more complex.

An important counterexample to what might be expected pertains to the hormone ghrelin, a peptide secreted mainly by the stomach. Ghrelin levels increase with fasting with an additional increment prior to anticipated meals, and its levels decrease once a meal commences (2). Further, in contrast to many other gut peptides, exogenous ghrelin increases food intake when administered to animals or humans. Ghrelin might therefore be anticipated to have an opposite relationship with physical activity relative to that of other gut peptides.

Another consideration pertains to body fat. There is a presumed association between the increased incidence of obesity (presumably related at least in part to an increase of food intake and digestion-related activity) and a decrease in average physical activity in Western societies (3). Based on this, the first-order prediction might be that individuals who eat more food and consequently have increased activity of gut hormones also have reduced physical activity. Chronic exercise in fact often does reduce body fat, as well as changing other parameters that could secondarily influence gut hormones. On the other hand, animal models indicate that when meals are anticipated, exercise ("activity-based anorexia") increases considerably in the interval just prior to food availability and can in fact interfere with consumption (4); and meal anticipation is associated with cephalically increased secretion of many gut peptides as well (2), such that there is not a straightforward relationship.

Another caveat is that commonly used assays often do not distinguish active from inactive forms of gut peptides. Therefore, because these hormones have very short half-lives, assessed total levels may reflect secretion without accurately reflecting functional activity. A final consideration is that the relationship between physical activity and gut peptides is a two-way street. While altered exercise may elicit changes of gut peptide secretion, the administration of certain gut peptides might in turn influence activity level or related energy expenditure (e.g., nonexercise activity [5, 6]).

As a final point, physical activity or exercise and related variables are not constant from study to study. Exercise varies in kind (e.g., aerobic vs. nonaerobic), duration, and intensity; subjects vary in their degree of fitness (and hence as to whether exercise is a stressor or a stress reliever); and exercise in animal models may not be voluntary. Such factors render comparisons among experiments difficult, especially when the database and literature are sparse. The overall point is that one needs to take numerous factors into consideration when reviewing physical activity and GI hormones.

Gastrointestinal Peptides

Gastrointestinal peptides are secreted from enteroendocrine cells lining the lumen of the GI tract; they facilitate digestion, enhancing its efficiency by customizing the secretion of enzymes and cofactors added to the chyme to best reflect the blend of nutrients that has been eaten. At the same time they control the physicochemical properties (pH, osmolarity), mechanical mixing, and flow of chyme along the tract to ensure the maximal absorption of energy-rich compounds. Gastrointestinal peptides act hormonally, influencing distant tissues, or in a paracrine manner within the GI tract itself, or in both ways. Many gut peptides are macronutrient selective in that their secretion is more sensitive to one or another macronutrient. At least 16 different kinds of gut hormone-secreting enteroendocrine cells have been described, but the secretions of only a few have been considered with regard to interactions with physical activity.

Ghrelin

Ghrelin, an endogenous ligand of the growth hormone secretagogue receptor, is unique among the GI peptides in that its administration increases food intake. Secreted mainly by the stomach, and unlike most other GI peptides, ghrelin crosses the blood–brain barrier and might be anticipated to directly affect central circuits influencing physical activity. In this regard, it is known that ghrelin gains access to regulatory control areas of the ventral hypothalamus. Ghrelin secretion is inversely correlated with body fat, and its levels increase with food deprivation and meal anticipation. Ghrelin is unique in that it is acylated as it is secreted, and only the acylated form is biologically active. Most ghrelin assays measure total ghrelin (acylated plus nonacylated), making interpretation of the results questionable.

Reports on the interaction of physical activity and ghrelin are mixed. Several reports indicate that when humans undergo prolonged training or fitness programs, plasma ghrelin (like plasma leptin) changes, but the changes are what would be expected given the concomitant changes of body weight. The conclusion in these reports was that a program of prolonged physical activity has little or no lasting impact on ghrelin per se. In contrast, Kim and colleagues found that a chronic increase of exercise resulted in increased total as well as unacylated (desacyl) ghrelin with no change in acylated ghrelin (7). Weight loss also occurred, and they concluded that the increase of total ghrelin in some reports is probably misleading because active hormone may not actually change. Several reports indicate that acute aerobic exercise has little effect on plasma ghrelin; there is, however, evidence that acute resistance exercise in humans causes a decline in plasma ghrelin (8).

In animal models, several early reports indicated that administering ghrelin or a specific ghrelin antagonist has little impact on locomotor activity or on metabolic rate. Later reports using pharmacological and genetic approaches showed a subtle effect whereby ghrelin, acting within the brain, causes an acute increase of general activity followed by a several-hour reduction of endogenous activity (see review in [5]). Since ghrelin is also synthesized in the brain, it is not known whether these effects are related to GI-originating ghrelin. One report indicated that forced exercise in rats lowered plasma ghrelin in parallel to decreased body weight.

Cholecystokinin

Cholecystokinin (CCK) is secreted from duodenal cells in response to ingested lipids and proteins, and its hormonal actions include stimulating bile flow and exocrine pancreatic secretion into the duodenum. The parent CCK molecule can be cleaved to several shorter peptides that have varying degrees of potency at CCK receptors. Administration of CCK prior to meals in humans and animals elicits premature satiation and reduces intake, and administration of selective CCK1 receptor antagonists increases meal size; however, it is not clear whether chronic treatment can successfully cause weight loss.

Few experiments have addressed CCK and physical activity, and the conclusions are mixed. Highly trained athletes are lean, secrete less CCK in response to a standardized meal, and experience less satiation than nonathletes. Acute normoxic exercise reportedly increases plasma CCK (whereas hypoxic exercise decreases CCK); exercise to exhaustion elicits increased CCK, and this is exaggerated after a meal. Rats and mice with compromised CCK receptor activity have lower spontaneous activity, although mice lacking CCK have normal physical activity and metabolism.

Glucagon-Like Peptide-1$_{3\text{-}36}$

Glucagon-like peptide-1$_{3\text{-}36}$ (GLP-1) is secreted from cells in the distal intestine in response to ingested carbohydrates. Because GLP-1 increases in the plasma before significant food reaches that section of the gut, it is thought that a neural reflex elicits the increment. Plasma GLP-1 levels become increased with acute strenuous exercise (3) and remain elevated for at least 1 additional hour, a span when ratings of hunger are reduced. Several other experiments indicate that either acute bouts of exercise or regimens of increased aerobic exercise over several days result in an increase of basal GLP-1 as well as an increased increment in response to an acute meal in both normal and overweight subjects. One report indicates that high-intensity running has no impact on basal plasma GLP-1, but the subjects in that experiment were highly trained athletes and may not represent the general population. Therefore, although the literature is small, the findings on the effect of activity on GLP-1 are generally consistent and indicate that GLP-1 is increased. Whether the increment is secondary to an increase of plasma glucose seems unlikely (see [3]), and it is not clear whether GLP-1, which is known to induce a feeling of satiation, contributes to any reduced feeling of hunger during or soon after exercise or both.

There are few data on the obverse, that is, on whether a certain amount of GLP-1 activity influences general activity. What is known is that mice lacking GLP-1 receptors have increased metabolic rate and increased general activity (9). The implication is that GLP-1 and activity are normally positively associated.

Gastric Inhibitory Polypeptide

Glucose-dependent insulinotropic polypeptide, or gastric inhibitory polypeptide (GIP), is secreted by intestinal cells during meals in response to increases of glucose, amino acids, or fats in the intestinal lumen. There is one report that GIP does not change after a bout of exercise in humans (10), although other GI peptides did change in that experiment (e.g., GLP-1). Mice lacking GIP-secreting cells have increased energy expenditure, but general activity has not been reported; and mice lacking GIP receptors have increased metabolic rate as well as increased general activity.

Peptide YY

Peptide YY (PYY) is cosecreted from the same intestinal cells that secrete GLP-1; and like GLP-1, PYY has been reported to reduce food intake. Peptide YY levels increase in response to acute exercise in humans (3).

Other Gastrointestinal Peptides

Oxyntomodulin, like GLP-1, is a product of the preproglucagon gene. Its administration causes an increase of general activity and energy expenditure in humans. Basal levels of the stomach peptide gastrin are normal in highly trained athletes, and both basal and meal-stimulated gastrin levels increase in the blood following strenuous exercise. Levels of secretin have been reported not to change in response to exercise.

Conclusions

The literature on the interaction of physical activity and peptides secreted from the GI tract suggests that there is no simple relationship between activity and the secretion or levels of gut peptides. Any conclusions in this regard must be tentative until experiments use more sensitive assays to differentiate the lengths of some peptides (e.g., CCK), as well as distinguishing between the active and the inactive forms (e.g., ghrelin, GLP-1, GIP). Conclusions based on manipulating activity in humans and assessing hormone levels often differ from conclusions based on pharmacological or genetic interventions in animals. With regard to particular gut peptides, the bulk of evidence, although controversial, suggests that any effect of altered activity on functional levels of plasma ghrelin may be secondary to changes in body weight. Basal and meal-stimulated CCK and GLP-1 increase after exercise, but levels of most GI peptides are relatively uninfluenced by physical activity.

Physical Activity Level and Thyroid Hormones

Anthony C. Hackney, PhD, DSc

The thyroid gland is considered one of the most critical endocrine tissues of the human body. The glandular cells of the thyroid secrete three hormones into the circulatory system: thyroxine (T4 [3,5,3',5'-tetraiodothyronine]), triiodothyronine (T3 [3,3',5-triiodothyronine]), and calcitonin (produced by the C cells of the thyroid). The hormones T4 and T3 are essential for normal bodily function across a broad context due to their ability to modulate the actions of metabolic rate, as well as their effects upon overall carbohydrate, lipid, and protein metabolism. The third thyroid hormone, calcitonin, is a key regulator in the circulating humoral levels of calcium and is particularly effective via its function on osteoblastic cellular activity. This chapter focuses only on T4 and T3 due to their extensive, dynamic impact on metabolism and consequently on the energy storage status of the body and an individual's ultimate risk for developing obesity.

Regulation and Physiologic Function

The release of the thyroid hormones (T4 and T3) is under the control of thyroid-stimulating hormone (TSH, also known as thyrotropin), a glycoprotein-based hormone released from the anterior pituitary gland. The anterior pituitary releases TSH in response to thyroid-releasing hormone (TRH) from the hypothalamus within the brain. Consequently, the control and release of these hormones involve a negative feedback loop that directly includes TRH, TSH, and T4 and T3. This regulatory loop is referred to as the hypothalamic-pituitary-thyroidal axis (1). Thyroid-stimulating hormone uses a cyclic adenosine monophosphate (cAMP) means of secondary messenger signaling to the thyroid gland cells to initiate the production and release of T4 and T3. Once these latter hormones are released into the circulatory system, they exist in both a carrier-bound form and a free, unbound form. The carrier proteins for the bound forms of the hormones are thyroid-binding globulin (accounting for 70%), thyroxine-binding prealbumin (10-15%), and albumin (15-20%). Amounts of the unbound, free forms of the hormones are relatively small in relation to the total hormonal amounts (free T4 ~0.03%; free T3 ~0.3%); however, these are the effective and most biologically active varieties of the hormones.

Both T4 and T3 act directly at their target tissue cells to stimulate and affect the function of these tissues (e.g., metabolic activities, such as those related to and essential for muscular contraction) (1, 2). All tissues of the human body except the brain, anterior pituitary gland, spleen, testis, uterus, and the thyroid gland itself appear to be strongly influenced metabolically by the thyroid hormones. For example, thyroid hormones increase oxidative phosphorylation metabolism at the mitochondria. These hormones also increase tissue responsiveness to the catecholamines, which can influence metabolic rate and have a cardiogenic effect and accordingly increase heart rate and contractility of the myocardium. The thyroid hormones, moreover, increase all aspects of lipid metabolism in the body, especially within skeletal muscle. Elevated blood levels of thyroid hormones can also enhance hepatic glycogenolysis, as well as influence protein degradation. Conversely, low circulating thyroid hormone levels can promote protein synthesis and the conversion of glucose to glycogen. Based upon this information, it should be apparent that thyroid hormone changes can profoundly influence tissue growth and development across the age span of humans (1-3). See figure 40.1 for a summary of major effects.

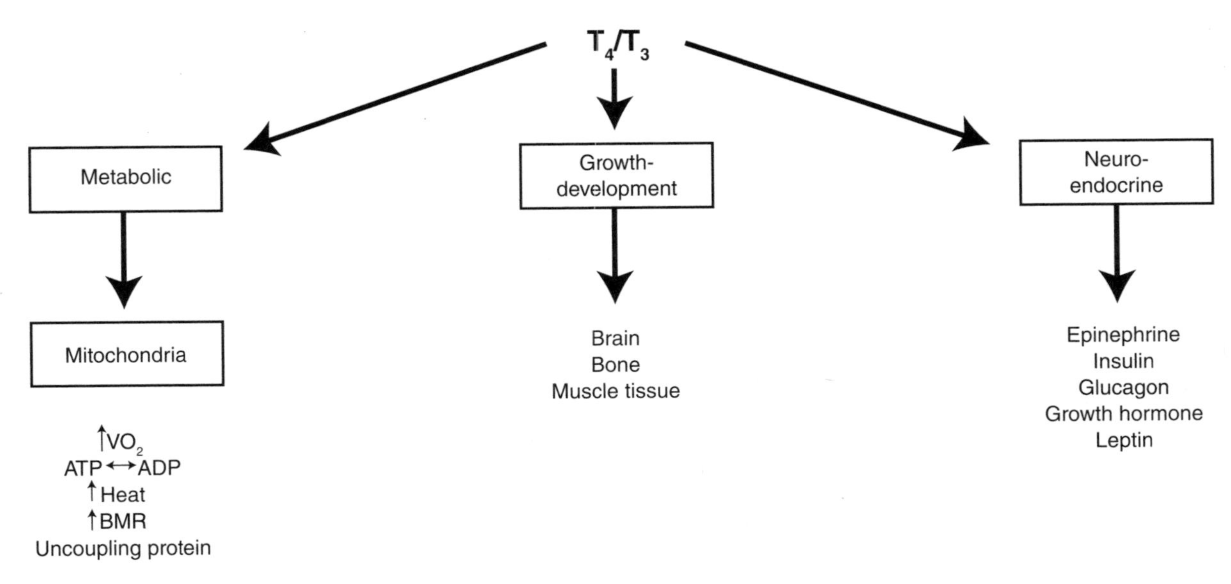

Figure 40.1 A summary of the principal physiological roles of the thyroid hormones in humans.

In the blood of healthy individuals, circulating concentration of T4 exceeds that of T3, but T3 is the more biologically viable of the two metabolic thyroid hormones. The turnover rate of the circulating thyroid hormones is very low relative to their existing large extracellular hormonal pool. Consequently it can be complex and challenging for scientists to detect changes, even relatively large ones, in thyroid gland activity following a perturbation of some type (1).

Physical Activity

Technically some experts view "physical activity" and "exercise" as not necessarily mutually exclusive terms (e.g., someone running may be doing both physical activity and exercise; someone gardening may be viewed as doing physical activity but not necessarily exercise). However, for the purposes of this brief chapter, they are considered essentially one and the same, and the following discussion is applicable to physical activity and exercise.

During short-term exercise (<20 min), blood TSH concentrations rise progressively with increasing workloads (i.e., during incremental exercise); a critical intensity (oxygen uptake, VO_2) threshold of approximately 50% of $\dot{V}O_2max$ is necessary to induce significant changes in the blood concentrations (3, 4). Even though TSH appears to rise with incremental exercise, most of the research literature on short-term exercise indicates that the concentrations of total and free T4 and T3, as well as rT3 (reverse 3,3',5'-triiodothronine, produced by the conversion of T4 in peripheral tissues), and the clearances of these hormones are essentially unaffected by such exercise (3). Some research has indicated significant increases for total levels of the hormones during the recovery from such exercise; however, these findings in part appear to be brought about by the exercise-induced hemoconcentration of the blood.

An increase in TSH levels might be expected to stimulate the thyroid gland, but physiologically there is an inherent delay in the stimulus–secretion response of the thyroid gland (1). Thus, if blood sampling is for too short a period of time, it would not be unanticipated for research studies to show the blood concentrations of T4, T3, and rT3 as essentially unchanged by a single bout of exercise (2, 5, 6). However, Galbo (4) has reported small but significant transient increases in T4 and T3 to maximal graded exercise testing ($\dot{V}O_2max$ test). These research findings are regrettably confounded by the fact that in many situations, the influences of environmental factors, dietary practices, and diurnal hormonal secretion patterns were not controlled extensively and thus could not be entirely separated from the influence of exercise alone. Furthermore, when free hormonal levels are examined, it is unclear whether observed exercise-induced changes in thyroid concentration in the blood might be due to changes in the concentrations of binding proteins. Although work on hormonal binding proteins in general suggests that for short- to moderate-duration (<30-45 min) exercise, free hormonal changes exceed binding protein effects alone (2).

The influence of prolonged exercise (>60 min) on thyroid gland function, despite extensive research, is still controversial. Several investigations, which used different exercise protocols, showed no effect on the concentration of TSH in the blood (2, 4, 6). In contrast, other studies demonstrated that TSH concentrations increased progressively with higher-intensity workloads and reached an elevated steady-state level after approximately 40 min into a prolonged exercise session (2, 4). As previously mentioned, an increase in TSH might be expected to cause an increase in T4 or T3, but physiologically the issue is complicated by the delayed response in the thyroid hormones to a TSH stimulus (1).

Berchtold and colleagues (6) reported that during very prolonged submaximal exercise (~ 3 h), total T4 remained constant but then declined significantly during recovery. In the same study, the level of T3 was found to actually decline continuously during exercise. Other researchers have reported total T3 to remain unchanged but total T4 to be increased by 60 min of a prolonged submaximal exercise session (5). Galbo (4) suggested that strenuous, prolonged exercise will result in an increase only in the concentration of free T4. These divergent findings are difficult to interpret due to the highly varying exercise (i.e., duration and intensity) and blood sampling protocols employed in the various studies. Obviously more work is necessary in this area to clarify these contradictory results and to determine the physiologic interpretation and consequences of the responses.

Evidence suggests that acute bouts of anaerobic exercise may influence the thyroid hormones. Hackney and Gulledge (5) reported that high-intensity (~110% $\dot{V}O_2$max) intermittent exercise increased total T4 levels for several hours into the recovery from such exercise. These changes were not due to hemoconcentration alone; but the degree of the changes due to increased T4 secretion, a suppressed metabolic clearance rate (MCR) for the hormone, or both of these factors, was unclear.

Research on the effects of resistance exercise on the thyroid hormones is sparse. The existing studies are limited by some of the factors noted earlier and have yielded contradictory findings. McMurray and associates (7) performed a well-controlled study that produced interesting results. The investigators examined total T4 and T3 responses immediately after an intensive resistance training session and for 12 h into recovery (during the nighttime). Transient but significant elevations in T4 and T3 were seen immediately after exercise, apparently primarily related to hemoconcentration. However, later significant nocturnal elevations in T3 levels were observed as compared to control night measurements taken after a day on which no resistance exercise had occurred. Physiologically, the elevations of T3 levels suggest that in the recovery from resistance exercise there is an increase in metabolism, most likely associated with tissue repairs and an increase in protein turnover (1).

On the basis of most of the research, the effects of a chronic exercise training program on TSH and thyroid gland hormones are unclear (2). Some investigators report that TSH is not influenced by training, but others have shown that the rate of T4 secretion is higher in exercise-trained individuals than in untrained individuals (2, 4). Conversely, Galbo (4) reported that short-term, intensive exercise training periods resulted in significant reductions in thyroid hormones. Other studies have not confirmed these results and have shown that the rate of T4 secretion is similar in trained and in sedentary control subjects (1, 2, 8). Basal and TRH-stimulated blood TSH concentrations were also reported to be similar in trained and weight-matched animals and humans (2). The discrepancies in these findings may relate to the fact that the studies failed to completely account for nutrient balance influence on thyroid turnover rate. That is, a negative energy balance has been shown to substantially reduce thyroid hormone levels (1). However, Baylor and Hackney (9) followed elite athletes during a 20-week training-competitive cycle and found that resting free T3 and TSH were reduced during intensive competitive periods even though no significant changes in body mass occurred. The turnover rate of thyroid hormones appears increased in training athletes (2, 8), as is perhaps the tissue sensitivity to T3 (1, 8), which could account for the finding noted earlier regarding free T3. However, more research is warranted to clarify and fully explain the relationship between exercise training and thyroid function.

Lipid Metabolism

As noted earlier, relative to lipid metabolism, the thyroid hormones (in particular T3) can stimulate metabolic rate and affect adipose tissue lipolysis. This latter effect seems to be mediated via interactions of the thyroid hormone with the catecholamines (epinephrine and norepinephrine). The combined effect of the thyroid hormones and catecholamines enhances mobilization of triglycerides from adipose tissue and increases fatty acid lipogenesis in the liver (1, 3). Furthermore, research suggests that increased T3 exposure causes augmentation in mitochondrial density, which can elevate resting metabolic rate and enhance the capacity for Beta-oxidation and lipolysis. Consequently, elevated T3 (under isocaloric conditions) can result in a weight loss, while decreased T3 levels can result in weight gain. In a similar fashion, weight loss results in a suppression of circulating T3 levels. Even the anterior pituitary hormone TSH can increase lipolysis by activating hormone-sensitive lipase, but the effect appears relatively weak (1, 3). Select key metabolic hormones interact with the thyroid hormones to affect lipid metabolism. For example, growth hormone (GH) administration stimulates the conversion of T4 to T3 and elevates energy expenditure. Studies also show an association between leptin and the thyroid hormones. Decreased leptin causes an increase in neuropeptide Y (NPY), which results in decreased thyroid hormone production (1, 3).

Obesity and Exercise

In the resting state, obesity is associated with adaptive changes in many hormonal secretions. Although hypothyroidism is reported with some body weight gain, surveys of obese individuals show that less than 10% are hypothyroid. In untreated obesity, total and free T4, total and free T3, TSH levels, and the TSH response to TRH are essentially normal, although some discrepancies have been reported in the literature (3, 10). Treatment of obesity with hypocaloric diets causes changes in thyroid function that resemble "sick euthyroid syndrome" (i.e., changes consist of a decrease in total T4 and total and free T3 with concurrent increases in rT3) (10). Untreated obesity is also associated with low GH levels, but levels of insulin-like growth factor 1 (IGF-1) appear normal. The GH-binding protein levels are increased, and the GH response to GH-releasing hormone is decreased. Cortisol levels can be abnormal in people with abdominal obesity, who exhibit an increase in urinary free cortisol but normal or decreased blood cortisol (perhaps due an increase in cortisol MCR) and normal adrenocorticotropin concentration (2, 10). There is also an increased response to corticotropin release hormone in obesity. The increase in cortisol secretion seen in people with abdominal obesity may contribute to the risk of development of metabolic syndrome (i.e., insulin resistance, glucose intolerance, dyslipidemia, and hypertension) (10).

Relative to exercise, obese individuals appear to have a capacity to oxidize lipid similar to that of normal-weight people; however, those who are obese do show different hormonal responses to exercise than normal-weight people. The most substantial changes are a blunted sympathetic nervous system and catecholamine response, meaning that there is less of a stimulus to the overall lipolysis process (3). Obesity also appears to blunt the exercise GH response (moderate-intensity activity) somewhat, which further suppresses the potential lipolytic response to exercise. The thyroid hormone response in obese persons who exercise does not appear to be drastically different from that in normal-weight individuals (3). In light of the thyroid hormone changes seen in obese persons at rest, it might be expected that more profound exercise variance would be reported for the thyroid hormones. However, exercise research studies on this topic are limited in number. Furthermore, many research protocols have not involved examination of the thyroid hormones in an extended postexercise recovery period, during which significant changes have been reported for normal-weight individuals. Nevertheless, in theory, the blunted sympathetic nervous system, catecholamine, and GH exercise responses of persons who are obese most certainly could have an impact on the effectiveness of the thyroid hormones in influencing overall lipid metabolism.

Conclusion

In conclusion, thyroid hormones are critical to normal physical function in humans, especially relative to energy expenditure and lipid metabolism. For this reason, they play a vital role in regulating aspects of exercise responses. Obesity causes significant disturbances within the resting thyroid hormone profile, but exercise thyroid responses of obese individuals do not seem to vary drastically from those of normal-weight persons. Extensive research, however, is needed in order to substantiate or refute this assertion.

Physical Activity Level and the Hypothalamic-Pituitary-Adrenal Axis

Denis Richard, PhD

The hypothalamic-pituitary-adrenal (HPA) axis is the (neuro)endocrine entity responsible for the secretion of the adrenal glucocorticoids cortisol (main active glucocorticoid in human and nonhuman primates) and corticosterone (main active glucocorticoid in rats and mice) through the coordinated and successive secretions of corticotropin-releasing factor (CRF) and adrenocorticotropic hormone (ACTH). Corticotropin-releasing factor and ACTH are two peptides released from the hypothalamus and the pituitary, respectively, in response to physical or psychological stimuli that disrupt the body's homeostasis (1). Hypothalamic CRF neurons project to the external layer of the median eminence, where they discharge CRF into the portal blood in the vicinity of the pituitary corticotrophs, which release ACTH into the general circulation. Adrenocorticotropic hormone reaches the adrenal cortex and triggers the secretion of glucocorticoids, which are capable of genomic and nongenomic actions via two types of intracellular receptors, namely the glucocorticoid and mineralocorticoid receptors. The glucocorticoid and mineralocorticoid receptors mediate the initiation and repression of gene transcription, as well as the nongenomic control of HPA axis activity through negative feedback exerted at the level of both the pituitary corticotrophs and CRF neurons via direct or hippocampus-mediated action (2). Glucocorticoid levels follow a circadian rhythm; in humans, cortisol peaks in the morning, whereas in laboratory rats and mice, corticosterone reaches its highest daily value in the afternoon, that is, prior to the nighttime when rodents enter their daily active period. Hypothalamic-pituitary-adrenal axis activity is generally increased to cope with enhanced neuropsychological and metabolic needs.

The HPA Axis and Stress

The HPA axis is stimulated by challenges disrupting body homeostasis. These challenges are known as "stressors" and are as varied as hemorrhage, infections, exhaustive physical activity, sleep deprivation, and emotional concerns. Activation of the HPA axis (probably more appropriately than activation of the sympathetic nervous system) characterizes the response to stress, which can be seen as the ultimate reaction to the disruption of the body's homeostasis caused by physically or psychologically demanding stimuli.

Activation of the HPA axis is triggered by the stimulation of CRF neurons located within the parvocellular division of the paraventricular hypothalamic nucleus (PVH) (3). Incidentally, activation of the hypophysiotropic CRF neurons constitutes a sound criterion by which to recognize the presence of stress. In stressed rats and mice, expression of the proto-oncogene c-*fos* (a reliable neuronal activity marker) can indisputably be detected within the hypophysiotropic CRF neurons (4). Stress also induces expression of the CRF type 1 receptor (CRF_1 receptor) in these PVH neurons (4). The CRF_1 receptor, which is not constitutively expressed in the PVH, is abundant within the pituitary gland corticotrophs, where it mediates ACTH secretion (3). The CRF system also includes the CRF_2 receptor, the CRF-binding protein, and the CRF-receptor ligands urocortin (UCN), UCN II, and UCN III (3). It is noteworthy that the CRF-containing neurons are not limited to the corticotropic PVH neurons, as they are found in several areas of the brain (3). Neurons containing CRF and UCNs form a network of cells that participate in the integrated response to stressful events, which is characterized by activation of the sympathetic nervous system, suppression of the reproductive function, and anxiety-like behaviors (3). The stimulation of the pituitary-adrenal axis nonetheless remains the most typical effect of CRF.

The release of glucocorticoids in response to stressful events constitutes the ultimate response of the HPA axis to cope with stress. Glucocorticoids, which are produced by cells of the zona fasciculata of the adrenal cortex, lead to protein catabolism, promote glucose mobilization mainly through gluconeogenesis, inhibit the inflammatory response, and exert a negative feedback on the activity of the HPA axis. The release of glucocorticoids appears to be beneficial acutely as it is likely required to recover homeostasis following stressful conditions. However, persistent hyperglucocorticoidism can be deleterious (1). Indeed, HPA axis hyperfunction or dysfunction ultimately leads to metabolic, immune, and reproductive abnormalities. Persistent HPA axis activity stimulates fat intake; suppresses thermogenesis; provokes muscle catabolism; promotes visceral or ectopic fat distribution (and its associated complications including insulin resistance); leads to bone demineralization and inhibits the inflammatory response; and causes mood disorders while promoting anxiety, bipolar disorder, posttraumatic stress disorder, depression, burnout, and chronic fatigue syndrome (see figure 41.1).

The HPA Axis, Exercise, and Exercise Training

Exercise, when it is sufficiently intense or prolonged, represents a physiological stimulus upon the HPA axis (5, 6). Exercise stimulates the PVH CRF neurons as ACTH levels rise during strenuous exercise. Treadmill running studies performed in rats, although the results should be interpreted cautiously (as treadmill running likely causes exercise-unrelated stress), have emphasized the ability of exercise to stimulate the hypophysiotropic CRF

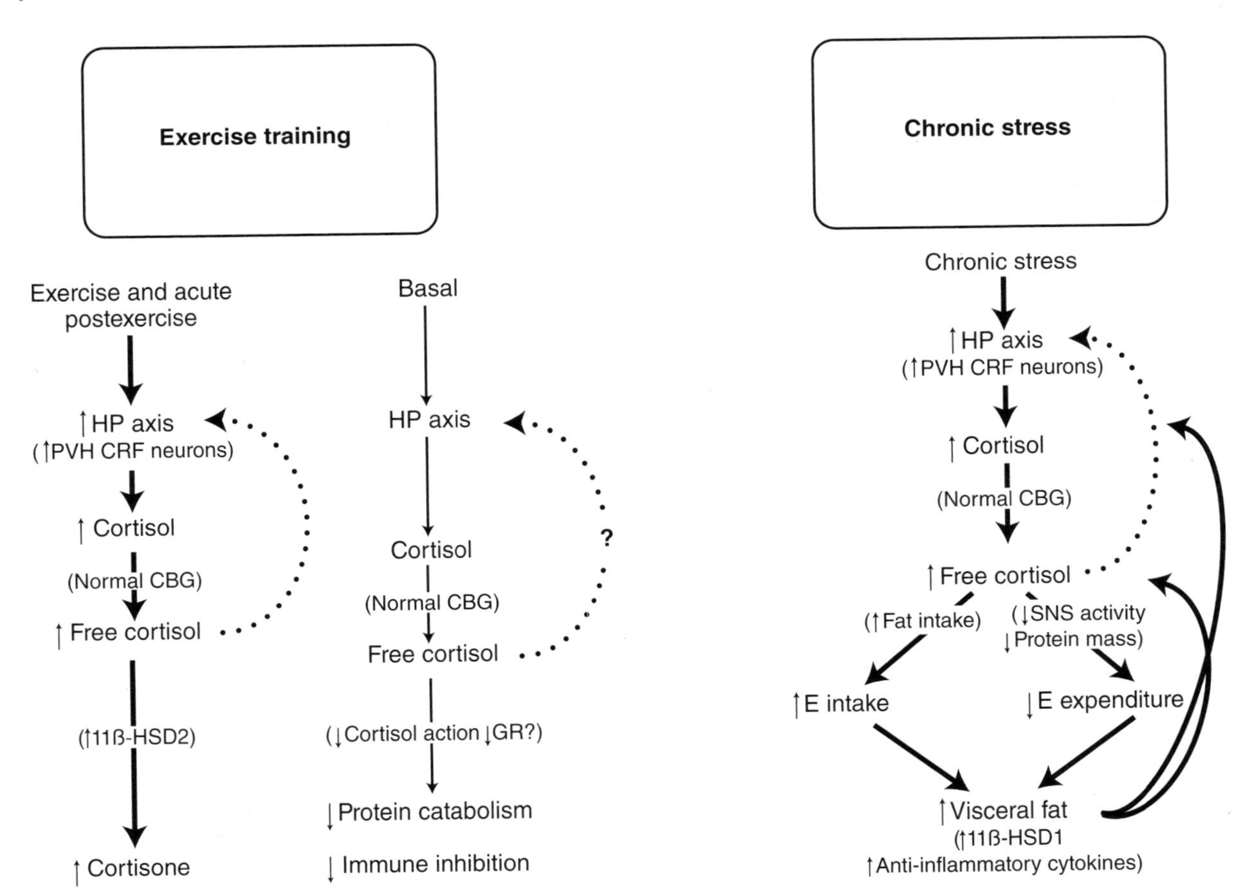

Figure 41.1 Hypothalamic-pituitary-adrenal axis and energy metabolism in relation to exercise training and obesity.

neurons (4). Treadmill running in rats leads to c-*fos* expression in the PVH as well as the expression of the CRF_1 receptor in the PVH.

Intense aerobic exercise leads to cortisol elevations that persist for a few hours after the cessation of exercise (5). Activation of the HPA axis during exhaustive exercise constitutes a coherent reaction to the energetic, metabolic, anti-inflammatory, and vascular demands of exercise (5). Exercise-induced activation of the HPA axis has also a strong psychological component; completion of an exhaustive exercise is psychologically demanding and creates a combination of perceived mental effort, exertion, and discomfort, which likely is a major component of any stress. The anti-inflammatory and psychostimulant properties of the glucocorticoids are likely to contribute to the frequent use of these steroids by athletes (5).

Exercise training (repetitive, regular practice of exercise) reduces the HPA axis activity for a given exercise workload. Exercise training does not, however, prevent activation of the HPA axis when the intensity of exercise is high relative to a given aerobic capacity (5). In highly trained athletes such as marathon runners, triathletes, or elite cyclists, significant elevations of cortisol have been reported to be present 2 h after the end of prolonged exercise sessions (5). Exercise training does not, however, lead to chronic hypercorticosolism; under nonexercising conditions, exercise-trained individuals exhibit 24 h urinary cortisol levels that are similar to those of age-matched sedentary subjects (5). Urinary cortisol closely reflects the production rate of cortisol over a given period since there is a constant fraction of free cortisol, which is filtered by the kidney. Exercise-trained subjects exhibit normal nycthemeral and circannual HPA axis activity rhythms. Furthermore, urinary cortisol levels of exercise-trained individuals are also normal overnight, that is, when the general hormonal profile is anabolic and therefore mostly favorable for exercise recovery and protein synthesis (5). It is noteworthy that urinary cortisol, similar to salivary cortisol, is free cortisol and represents the extracellular bioavailable cortisol. A large fraction of cortisol is bound to corticosteroid-binding globulin (CBG). There is no evidence that exercise training modulates free cortisol availability through the production of CBG.

Exercise training may alter the sensitivity to glucocorticoids. Compared to sedentary men, exercise-trained individuals exhibit a decreased sensitivity to glucocorticoids 8 and 24 h after exercise (5). Considering the effects of glucocorticoids in enhancing protein catabolism and in suppressing the immune response, it would seem reasonable to speculate, as suggested by Duclos and colleagues (5), that the decreased sensitivity occurs as a beneficial adaptation to protect muscle and other tissues (including the brain) against glucocorticoids at a time when the catabolic or anti-inflammatory actions of the steroids are no longer needed. Interestingly, the sensitivity to glucocorticoids returns to normal during exercise and in the acute recovery period, that is, when the circulating levels of corticosteroids are high and likely required to mobilize energy and to limit the inflammatory response associated with exercise-induced muscle damage and muscle inflammatory reactions (5). It is noteworthy that the restricted inflammatory response occurring immediately following exercise has, however, a detrimental side, leading to an increased susceptibility to viral and bacterial infections.

In most conditions, exercise training does not affect the cortisol/cortisone ratio, which strongly depends on the activity of the tissue-specific enzymes 11β-hydroxydehydrogenase 1 (11β-HSD1) and 11β-HSD2 [5]. 11β-HSD1, which is found in numerous tissues including white adipose tissue, converts inactive cortisone to active cortisol; in contrast, 11β-HSD2, which is mainly found in kidney, inactivates cortisol to cortisone. Increased production of cortisol associated with exhaustive exercise sessions in highly trained athletes is readily balanced by a compensatory enhanced inactivation of cortisol into cortisone. The activity of 11β-HSDs (in particular that of 11β-HSD2) therefore appears to swiftly adapt to prevent cortisol from deviating markedly from normal values.

Whereas exercise-trained individuals do not exhibit cortisol/cortisone ratios that markedly differ from those of sedentary individuals, overtrained athletes do exhibit abnormal ratios (5). Indeed, overtrained subjects demonstrate abnormally high cortisol/cortisone ratios in the precompetition season, that is, when the training load reaches a zenith. During the competition season, overtrained athletes manifest extreme fatigue and blunted performance. A high cortisol/cortisone ratio in overtrained athletes suggests a lack of increase in 11β-HSD2 activity to compensate for exercise-induced cortisol production. Overtrained athletes also show abnormal HPA axis sensitivities to exercise when bouts of exercise are performed consecutively. Indeed, in a protocol consisting of two consecutive bouts of exercise separated by 4 h, aimed at assessing the pituitary adrenal capacity, it was found that overtrained subjects exhibited an exaggerated hormonal response on the first session and a suppressed response on the second bout of exercise (5).

The HPA Axis, Energy Homeostasis, and Obesity

The involvement of the HPA axis in the regulation of body weight has been emphasized in the obese Zucker rat and *ob/ob* mouse. In these mutants, which show markedly attenuated or no actions of the adipose-derived hormone leptin, the HPA axis is hyperactive, and adrenalectomy-induced corticosterone removal prevents or alleviates obesity. Corticosterone is capable of anabolic effects as it blocks thermogenesis while stimulating food intake. It opposes the catabolic effects of leptin, which can in turn participate in the control of the HPA axis activity (7). Leptin has the ability to reduce the HPA axis activity through actions that can be exerted at every level of the HPA axis. Leptin can attenuate the response to stressful events, including food restriction, which causes a psychogenic stress in obese Zucker rats and *ob/ob* mice (8). Leptin signaling on the HPA axis can be exerted at the PVH level on hypophysiotropic CRF neurons, whose adrenalectomy-induced overexpression is markedly blunted by leptin (9).

One of the most significant deleterious metabolic consequences of HPA axis hyperactivity is the increase in visceral fat deposition due to excess circulating levels of corticosteroids. Visceral fat distribution is indeed one of the most visible characteristics of individuals exhibiting high levels of corticosteroids, such as patients with Cushing's syndrome. In these patients, resolution of excessive cortisol production normalizes fat distribution and most of the deleterious effects of visceral obesity. There is also strong evidence that chronic stress generates visceral obesity in humans (10). In fact, stress induces elevation of cortisol, which not only enhances visceral fat per se but also creates a milieu highly favorable to the deposition of intra-abdominal fat by reducing production of the growth hormone and the activity of the hypothalamic-pituitary-gonadal and thyrotropin-triiodothyronine axes (11). It is noteworthy that visceral fat could also stimulate the activity of the HPA axis (12). Such an effect could be modulated through certain cytokines emerging from the abundant macrophages disseminated within the visceral fat depots.

Conclusion

The HPA axis plays an important role in energy metabolism in relation to exercise training and obesity (figure 41.1). In humans, strenuous, long-lasting exercise represents a stressful event that stimulates the HPA axis from the hypophysiotropic PVH CRF neurons. The levels of free cortisol remain high for a certain period of time following exercise. The levels of CBG-bound cortisol and free cortisol parallel each other, suggesting that the levels of CBG are not affected by exercise. Similarly, the cortisol/cortisone ratio is not influenced by exercise, suggesting a normal accelerated catabolism of cortisol through an increase in 11β-HSD2 activity. Exercise training (repetitive or regular practice of exercise) does not markedly affect the activity of the HPA axis and in consequence the levels of cortisol and free cortisol. It appears, however, to decrease sensitivity to glucocorticoids, which over the long term could be beneficial by preventing both protein catabolism and inhibition of the immune system. Whether this decreased sensitivity to glucocorticoids can blunt the strength of the negative feedback on the HPA axis is uncertain. Hyperglucocorticoidism subsequent to sustained stimulation of the HPA axis favors visceral fat deposition. This is inherent to the action of glucocorticoids, which stimulate fat intake while blunting energy expenditure through a reduced lean mass and a decrease in sympathetic nervous system activity. It is noteworthy that visceral fat can increase the production of cortisol through its 11β-HSD1 activity or the production of cytokines capable of stimulating the HPA axis.

Physical Activity Level and Skeletal Muscle Biology

David A. Hood, PhD

Skeletal muscle is one of the most adaptable tissues in the body. In response to regular physical activity, muscle alters its phenotype to better match the energy demands and force requirements of the exercise. This adaptational change does not occur after one bout of exercise but is a consequence of persistent, regularly performed exercise over the course of several weeks. Nonetheless, it should be appreciated that muscle adaptation begins during the recovery phase following the first period of exercise. In contrast, during prolonged periods of physical inactivity, muscle cells can atrophy, and they revert to a less metabolically efficient tissue.

The alterations in phenotype brought about by changes in physical activity level can potentially occur in three fundamental areas of skeletal muscle biology: (1) myofibrillar isoform shifts, (2) calcium-handling kinetics, and (3) energy provision (figure 42.1). These changes are a result of exercise-induced alterations in gene expression. That is, exercise increases the expression of some genes, perhaps one encoding an isoform within a family of proteins, while downregulating the expression of another gene product, perhaps encoding an alternative isoform. This form of regulation can occur at many steps in the "pathway" of phenotype expression, from the gene to the mRNA to the protein. In addition, not only are synthetic (or anabolic) pathways affected, but so are those of degradation. For example, a twofold increase in the synthesis of a specific protein need not be the only way to increase the level of that protein within the muscle cell. A decrease in the degradation rate of the protein by twofold could potentially provide an equal effect, in the face of unchanged synthesis rates. A similar argument could be made at the mRNA level of expression. Thus, the adaptive response of muscle phenotype requires consideration at all levels of phenotype expression for a full understanding of the processes involved. Indeed, with regard to response to an increase in regularly performed physical activity, the focus of exercise physiology research is most often on synthetic (i.e., anabolic) rather than degradation (i.e., catabolic) pathways. However, when muscle mass is reduced during periods of extended inactivity (e.g., immobilization, denervation, microgravity) or disease (e.g., cancer cachexia), research focus most often turns to degradation pathways as the major issue affecting muscle phenotype. In reality, both pathways of protein turnover should be considered during the adaptation of muscle to an alternative phenotype resulting from changes in physical activity level.

In response to exercise signals, nuclear DNA is transcribed into mRNA, and the mRNA is translated into protein. This is the pathway of gene expression that is accelerated or decelerated in response to exercise, leading to adaptation within muscle. The proteins can be targeted to organelles such as the sarcoplasmic reticulum (SR) or mitochondria, or

Acknowledgements
Funding for research in the author's laboratory is provided by NSERC Canada and by CIHR. The author is grateful to Giulia Uguccioni for her assistance with the manuscript. David A. Hood holds a Canada Research Chair in Cell Physiology.

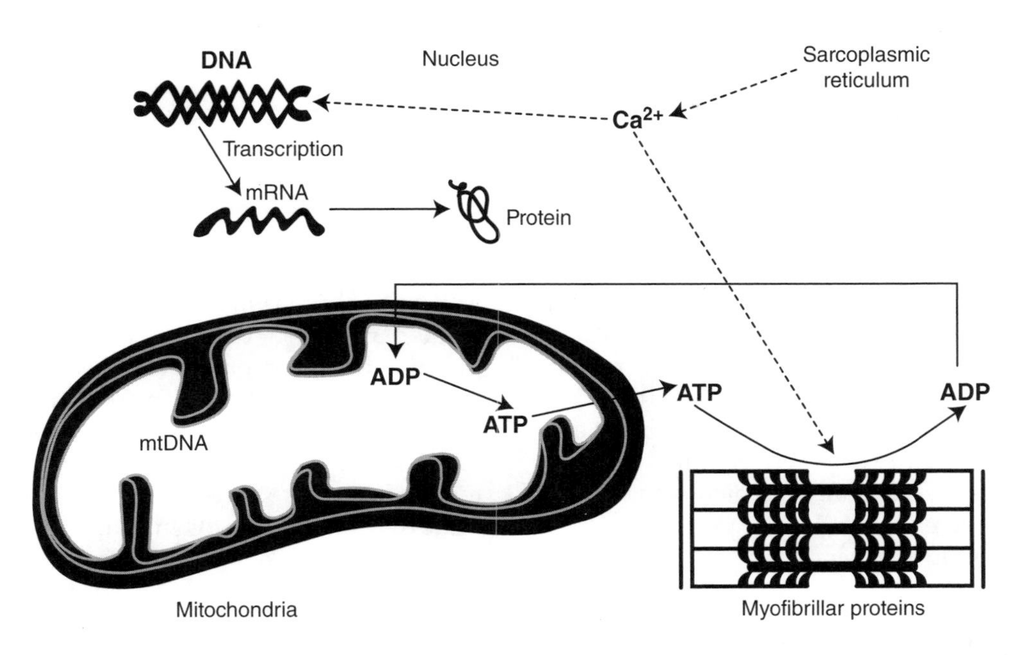

Figure 42.1 Cellular organelles and structures involved in muscle adaptations to exercise and disease.

can remain cytoplasmic, for example myofibrillar proteins like actin and myosin isoforms. The SR (responsible for calcium-handling kinetics) and the myosin heavy chain isoforms (type I, IIa, or IIx) are the main contributors to muscle contractile speed (i.e., fast- vs. slow-twitch muscle). Mitochondrial adaptations to chronic exercise are well established. It is now also well known that mitochondrial dysfunction accompanies aging and several metabolic diseases, such as type 2 diabetes.

In this brief review, unless otherwise indicated, increase in physical activity level refers to the adoption of an exercise program that is fully attainable for previously sedentary individuals, involving a frequency of four times per week for 20 to 40 min per day, at an intensity corresponding to 50% to 75% of $\dot{V}O_2$max, for a duration of at least two to three months. In contrast, inactivity refers to extended periods (months to years) of rarely performed bouts of physical activity, ultimately leading to an energy balance disequilibrium, weight gain, and possible carbohydrate and lipid metabolic disturbances.

Myofibrillar Isoforms

A large number of proteins are responsible for contractile performance and the generation of muscle force. These include the myosin heavy chain (MHC), actin, and troponin and tropomyosin isoforms. By far the greatest amount of research has been devoted to myosin and its isoforms. Much less is known and much more research is required concerning the regulatory contractile proteins and their adaptive responses to physical activity in human muscle. In the adult, MHCs consist of the type I (slow) and type IIa and type IIx (both fast) isoforms of the protein, which are about 200 kDa in size. The three isoforms have differing speeds of cross-bridge cycling. The most dramatic difference is between the type I and the type IIa isoforms; these differ by about four- or fivefold. Despite being classified as fast isoforms, the type IIa and type IIx MHCs differ in their speeds of actin and myosin interaction, by approximately twofold. This can likely be translated into differences in contraction economy (i.e., fuel utilized to maintain a given amount of tension). It is known that slow myosin isoforms are 1.5 to 3 times more economical than fast isoforms. This represents an alternative energy-sparing mechanism, unrelated to energy production by metabolic pathways. More economical actin–myosin cross-bridge interactions represent an equally effective means of energy conservation.

Regardless of the type of exercise training (i.e., resistance or endurance), the adaptive response of MHC expression is similar. Type IIx MHC isoforms are replaced by type IIa isoforms, thus increasing the proportion of fibers that contain a greater majority of IIa MHC. These then become classified as type IIa fibers using traditional histochemical methods. This is a result of a reduction of gene expression in MHC type IIx, while that of type IIa is increased within the same muscle fibers. With chronic muscle disuse, the opposite trend appears to occur (1).

There is evidence that obese individuals possess a greater proportion of type II compared to type I fibers (2). Epidemiological studies indicate that a higher proportion of type II fibers is correlated with insulin resistance, higher blood pressure, and a lower $\dot{V}O_2$max (3). The insulin insensitivity associated with a high proportion of type II fibers is likely related to a reduced number of insulin receptors on the cell surface, as well as a decrease in the amount of the insulin-responsive GLUT 4 isoform in type II fibers compared to type I fibers (4). While moderate programs of physical activity have little effect on this proportion of fiber types, type II fibers can nonetheless adapt to exercise by (1) increasing their oxidative capacities to enhance lipid oxidation, (2) inducing an increase in GLUT 4 protein, and (3) increasing their insulin sensitivity.

Calcium-Handling Kinetics

When a motor unit is recruited and an action potential propagates along the sarcolemma and down the T-tubule of individual fibers, calcium release is promoted from the SR. In fast-twitch fibers, the triggering of this release results from a direct interaction of the voltage-sensitive dihydropyridine receptor (DHPR) protein with the calcium release channel, the ryanodine receptor (RyR). A series of conformational changes initiated by the DHPR promotes the opening of the RyR and the subsequent liberation of calcium from its bound storage site to calsequestrin within the SR. The kinetics of calcium release in fast-twitch muscle are rapid because of the one-to-one interaction of the DHPR with the RyR in this fiber type. In contrast, in slow-twitch muscle, the ratio is less than 1, approaching 0.5 (5). This diminished interaction attenuates the release of calcium from the SR of slow-twitch muscle fibers and contributes to the reduced twitch tension kinetics in this fiber type. Once released, calcium diffuses to bind to troponin C, thereby allowing tropomyosin to undergo a conformational change resulting in the uncovering of the actin binding site. This permits actin and myosin interactions to occur. In the absence of subsequent action potentials, the released calcium can diffuse back to the SR or can undergo facilitated diffusion by binding to parvalbumin, a protein expressed at high levels in fast-twitch muscle fibers. The relative absence of parvalbumin in slow-twitch fibers contributes to the slower rate of relaxation in slow-twitch fibers. The reuptake of calcium by the SR is dependent on the action of the SR calcium ATPase protein (SERCA).

Fast-twitch muscle contains the SERCA1 isoform (6), which has more rapid uptake kinetics than the SERCA2 isoform predominant in slow-twitch fibers. Estimates of the energy cost of calcium cycling indicate that up to approximately one-third of the adenosine triphosphate (ATP) cost of contractions can be used for calcium mobilization within the muscle fiber (7). In response to chronic contractile activity in animal models, muscle fibers express greater amounts of the slow isoform of the SERCA pump and downregulate the faster SERCA1 isoform. Although this is not an adaptation to moderate levels of physical activity, it is known that the activity of the SERCA pump can diminish with muscle fatigue produced by high-intensity contractile activity in humans (8). In addition, with age, the ratio of DHPR to RyR is reduced. This contributes to the reduced force capability of muscle in older individuals, as well as to a slower muscle phenotype. There is no evidence in the literature that these proteins are altered within skeletal muscle in obesity or type 2 diabetes; however, more research in this area is needed.

Energy Provision

Skeletal muscle fibers possess a range of mitochondrial contents and glycolytic capacities. Type I muscle fibers tend to have the highest mitochondrial content (5% to 7% by cell volume) and the lowest glycolytic enzyme capacity, while type IIx fibers have the lowest mitochondrial volume (about 2% to 3%) but are well adapted for glycogen breakdown and glycolytic ATP provision. One of the most dramatic adaptive processes within muscle is the ability of this tissue to respond to increased energy demands through the synthesis of more mitochondria (9). This process is known as physical activity–induced mitochondrial biogenesis, and it represents an expansion of the mitochondrial membrane network within muscle cells, both in the region of the sarcolemma (i.e., subsarcolemmal or SS mitochondria) and between the myofibrils (intermyofibrillar or IMF mitochondria).

If the intensity of the exercise program is moderate, mitochondrial biogenesis will proceed only in the muscle fibers recruited. Thus, only type I and some type IIa fibers will adapt. With a more intense physical activity regimen, all three muscle fiber types will adapt to the program by an increased synthesis of mitochondria. The result is an increased capacity for oxidative ATP provision per gram of muscle. In addition, the capacity of muscle to oxidize lipids increases considerably over time as the training program progresses. This occurs because of an induction in the enzymes of beta-oxidation, as well as the proteins involved in free fatty acid uptake and transport within the muscle cells. The higher mitochondrial content brought about by training results in a smaller disturbance of metabolism when exercise is performed on an acute basis. The consequence is a reduced reliance on glycolytic flux, less formation of adenosine monophosphate (AMP), attenuated AMP kinase activation, reduced production of lactic acid, and a lower rate of glycogen utilization. These are among the most important factors that lead to improved endurance performance with less fatigability. In older individuals or those who have metabolic diseases such as diabetes or obesity, the consequence is an enhanced quality of life.

The adaptive response of muscle to regular physical activity enhances the uptake of plasma lipids. In concert with this, the higher muscle mitochondrial content per gram of tissue enhances the oxidation of plasma membrane-transported as well as intracellular lipids in the form of intramuscular triglycerides. This increased rate of lipid oxidation reduces the accumulation of lipid-derived metabolites within the muscle cell that can impair insulin signaling (10). Indeed, a single bout of exercise can reduce the cellular level of ceramide and diacylglycerol, two metabolites known to activate protein kinase activity on serine residues of the insulin receptor (11). This inhibits insulin receptor tyrosine phosphorylation and the subsequent activation of the insulin signaling pathway toward GLUT 4 translocation. Inhibition of this translocation of GLUT 4 from an endosomal location toward the sarcolemma, a step required for insulin-stimulated glucose transport into muscle cells, would result in reduced rates of glucose uptake. In contrast, the removal of this inhibition by exercise likely contributes to an enhanced postexercise insulin sensitivity. Indeed, the effect of a single bout of exercise is a short-lived (about 24 h) effect on postexercise insulin sensitivity, which is beneficial for glucose homeostasis in individuals with preexisting insulin resistance. The insulin signaling pathway activity is enhanced, and glucose uptake into muscle is increased as result of the rise of insulin levels in the blood back to resting levels, as well as the greater p38 MAP kinase–induced GLUT 4 translocation to the plasma membrane (12).

In response to repeated bouts of physical activity, the benefit of individual exercise periods on glucose metabolism is one of the most important health benefits of regular physical activity. This form of physical activity produces an even more robust and long-lasting insulin sensitivity, in part because of an induction of GLUT 4 protein, as well as important enzyme components of the insulin signaling pathway, via changes in gene expression. Because muscle is such a large consumer of glucose within the body, this adaptation has a profound effect in improving whole-body glucose homeostasis. This beneficial effect can be observed in normal individuals, as well as those with preexisting insulin resistance.

Summary

Skeletal muscle is a large tissue, occupying up to 40% of body mass. Therefore, it is an important contributor to whole-body metabolic rate. In response to acute physical activity, changes in gene expression are initiated that accumulate with repeated exercise bouts. The extent of these changes ultimately leads to phenotype alterations. These adaptations are dependent on the intensity, frequency, and duration of exercise and the period of time over which the training takes place. Adherence to a regular exercise program is essential to maintaining the adaptations observed, since exercise-induced benefits are lost with a reversion to prolonged inactivity. The benefits of the adaptations include reduced fatigability, improved lipid metabolism, and increased insulin sensitivity. These changes can be observed even with relatively mild exercise regimens in previously sedentary individuals. More vigorous and longer-term exercise programs have the potential to produce further adaptations involving calcium handling and MHC expression, with the possibility of improving the economy of muscle contractions.

Postexercise Energy Expenditure

Elisabet Børsheim, PhD

The continued elevation in O_2 uptake after exercise is referred to as "excess postexercise O_2 consumption" or EPOC (1). The phenomenon was first observed a century ago when Benedict and Carpenter (2) had two subjects sleep in a respiration calorimeter 7 to 13 h after performing severe exercise and observed an 11% increase in resting metabolic rate. It was initially thought that postexercise elevation in energy expenditure was an important factor in daily energy expenditure, but this picture was later modified by better-controlled studies.

The EPOC consists of rapid and prolonged components (1, 3, 4). The rapid component can be defined as the sum of components that decay within 1 h after exercise, whereas the prolonged component can be defined as the sum of components that decay mono-exponentially over several hours (figure 43.1). Note that EPOC is an acute effect of exercise and is different from the more persistent training effect on resting metabolic rate.

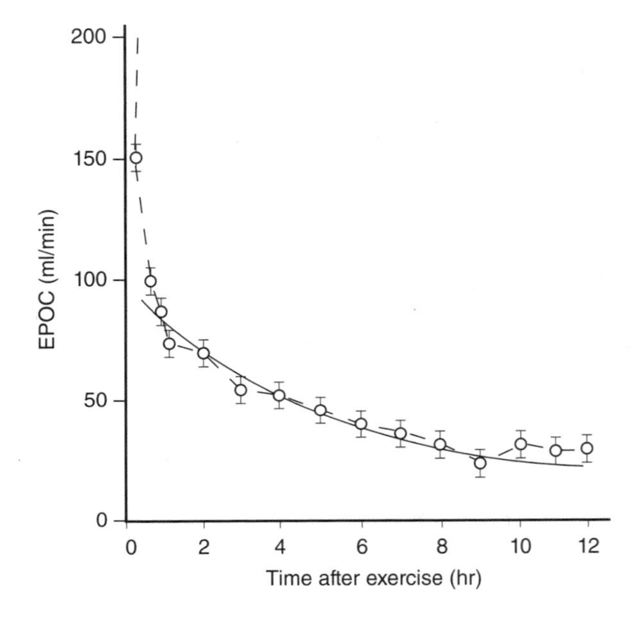

Figure 43.1 Time plot of excess postexercise O_2 consumption (EPOC) after exhaustive submaximal exercise in young men (80 min at 75% of maximal O_2 uptake, $n = 12$). Solid line shows the prolonged EPOC component.

Reprinted, by permission, from R. Bahr, 1992, "Excess postexercise oxygen consumption-magnitude, mechanisms and practical implications," *Acta Physiologica Scandinavica* Suppl 605: 1-70.

Methodological Considerations

When one is measuring EPOC, accurate control over the preexperimental conditions and excellent reproducibility in the indirect calorimetry measures are of highest importance (4). Validity and reproducibility of the calorimetric equipment are seldom reported in the EPOC literature. Further, a resting-time control study is preferred to account for diurnal variation in resting energy metabolism. However, a single preexercise value is frequently used as baseline, perhaps underestimating EPOC. The subject's resting position—seated or recumbent—may also contribute to variations in measurement error and thus the ability to detect EPOC (4). Further, recovery O_2 uptake may be measured continuously or at discrete time points, for a predetermined time or until baseline is reached (4). Techniques to determine when O_2 uptake reaches baseline also vary (5).

Even with these issues taken into consideration, some well-controlled studies showed that EPOC lasted for hours, whereas in others it was concluded that EPOC is brief and minimal (4, 5). The conflicting results can be resolved if differences in exercise intensity and duration are considered. Factors like exercise mode, subject characteristics, and diet may also influence EPOC.

Aerobic Exercise

The absence of a sustained EPOC is a consistent finding in studies with low exercise intensity and duration; only after higher-intensity exercise is a significant and sustained EPOC observed (4, 5). In comprehensive and well-controlled studies by Bahr (3) and Gore and Withers (6) using cycling and running, respectively, a curvilinear relationship between EPOC magnitude and exercise intensity was observed. Only intensities above 50% to 60% of maximal O_2 uptake ($\dot{V}O_2max$) induced an EPOC lasting several hours after exercise. Further, at or above this intensity level, a linear relationship between EPOC size and exercise duration was seen. After the most exhaustive bouts (80 min running at 70% of $\dot{V}O_2max$ [6] and 80 min biking at 75% of $\dot{V}O_2max$ [3]), EPOC lasted for 7 to 12 h and amounted to 300 to 700 kJ (~72 to 167 kcal). It was concluded that for weight loss to occur, similar workloads must be performed habitually; but untrained or overweight subjects are unlikely to tolerate this type of regimen.

The fact that exercise must be of a certain intensity in order to yield the linear relationship between exercise duration and EPOC magnitude points toward a synergistic rather than additive interaction between intensity and duration. This is illustrated in figure 43.2, which shows mean EPOC values from a range of cycling studies (4). The figure illustrates that intensity must be of a certain size to achieve a significant EPOC and that the highest EPOC are found when exercise duration is also substantial.

A few studies have compared EPOC after split exercise sessions to that after a continuous bout, generally indicating that total EPOC is larger after split exercise (4). However, the larger EPOC is small in relation to the exercise energy expenditure, and exercising for a few more minutes may make up for this. Further, there will be one O_2 deficit for each session. Thus, the relative difference in recovery energy expenditure after split versus continuous sessions will be smaller than the EPOC difference, since some of the O_2 is used to pay the O_2 debt.

Supramaximal Exercise

Given the relationship between exercise intensity and EPOC, it is no surprise that supramaximal exercise, even of short duration, can produce even larger EPOC than lower-intensity exercise of greater total work (4, 5).

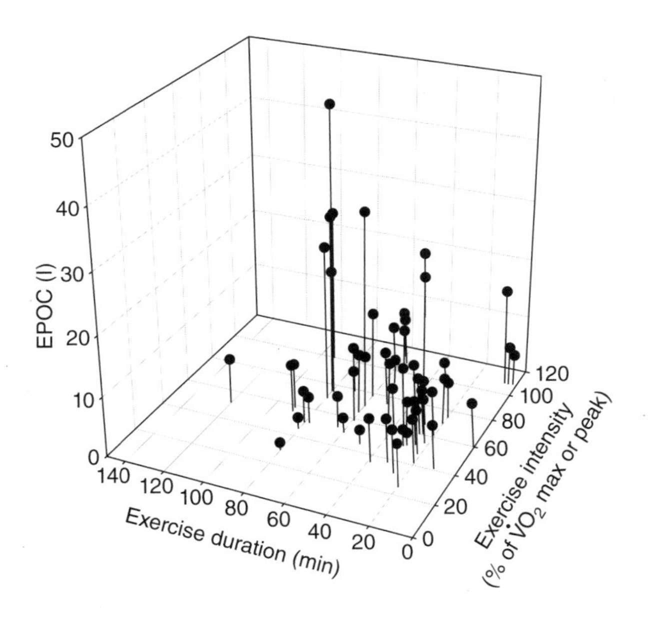

Figure 43.2 Relationship between exercise intensity, exercise duration, and excess postexercise O_2 consumption (EPOC) magnitude. Plot shows mean EPOC values from studies that have used cycling exercise and either present EPOC values or allow estimates of EPOC. Note that recovery O_2 uptake was not measured until it had returned to resting control values in all the studies included (for more details on studies, see [4]).

Reprinted, by permission, from E. Børsheim and R. Bahr, 2003, "Effect of exercise intensity, duration and mode on post-exercise oxygen consumption," *Sports Medicine* 33: 1037-1060.

Resistance Exercise

An understanding of the consequences of intensity and duration of resistance exercise on EPOC is confounded by a large diversity in protocols (e.g., circuit training vs. multiple sets, weights, sets, repetitions, length of rest periods, etc.). These factors influence the exercise energy cost but are difficult to quantify precisely. Furthermore, each session or set of resistance exercise will have its own EPOC during the period between exercises, which one must include in order to determine total energy expenditure. All in all, EPOC after resistance exercise seems to be influenced by exercise intensity, since a more prolonged and substantial EPOC has been observed following hard versus moderate resistance exercise (4, 5).

Comparison of EPOC between aerobic and resistance exercise is difficult since it is unclear how work volumes relate to each other. When either estimated exercise energy cost or O_2 uptake has been used to equate continuous aerobic and intermittent resistance exercise, the latter has produced a greater

EPOC. However, indirect calorimetry may underestimate the cost of the resistance bout because of the considerable anaerobic component.

Mechanisms Underlying EPOC

Excess postexercise O_2 consumption is the sum of many underlying mechanisms, some still unknown. Hence, the influence of individual factors on EPOC occurs through an effect on these causal mechanisms. Most of the mechanistic studies have been done with cycling protocols, and less is known about EPOC after resistance exercise.

Some metabolic processes thought to be responsible for the rapid EPOC component are well defined: resaturation of blood and muscle O_2 stores; adenosine triphosphate (ATP) and creatine phosphate (CrP) resynthesis; lactate removal; and elevated circulation, ventilation, and body temperature (1, 3). Thus, the classic O_2 debt hypothesis is one of several factors explaining the rapid component.

Mechanisms for the prolonged EPOC component are less well understood. Sustained increased circulation, ventilation, and body temperature may contribute, but the cost is low (<1 L of O_2) (3). After prolonged exhausting exercise, there is a relative shift from carbohydrate to fat as substrate source. Since the energy equivalent of O_2 is lower with fat versus carbohydrate as substrate, this shift has been estimated to account for 10% to 15% of EPOC after such exercise (3). Further, an increased rate of the energy-requiring triglyceride–fatty acid cycle has been observed after similar exercise (4), and this can also account for a considerable part of EPOC (3).

It is unclear whether glycogen resynthesis is a mechanism contributing to the prolonged EPOC. No difference in EPOC size in the fed state versus the fasted state (when glycogen resynthesis is low) was found after prolonged exhausting aerobic exercise (3). Further, controversies surround the interactions of food and exercise on energy expenditure, perhaps caused by large intra- and interindividual variations in the thermogenic effect of a meal.

Several hormones may potentially contribute to EPOC. Propranolol given to dogs before exercise to block the catecholamine effect on beta-adrenoceptors led to an attenuated EPOC (7). However, because propranolol was administered before exercise, the physiological effects of exercise were different between the control and propranolol study groups. Thus, a clearly important indirect effect of catecholamines on EPOC is via stimulation of various processes during the exercise itself, which are reversed slowly during recovery, even if there is no increased sympathoadrenal activity in this period. When beta-adrenoceptor blockers were given to active young men during rest with or without previous exercise, O_2 uptake was reduced to a similar extent (4), but this may be different in inactive subjects. Further, an extended depression in arterial insulin concentration may be important for the increase in fat mobilization, oxidation, and cycling after prolonged exercise (4).

Protein synthesis and breakdown are elevated during recovery from both aerobic and resistance exercise (4). Protein synthesis is energetically expensive and thus may explain part of the prolonged effect of resistance exercise on energy metabolism. For instance, in experienced weight trainers, resting metabolic rate was found to be elevated for 38 h after high-intensity exercise (5). Further studies of this issue are necessary. Finally, it has been speculated that a change in energy efficiency via the activity of uncoupling proteins can contribute to prolonged EPOC (4).

Exercise intensity and duration affect the EPOC components differently. High-intensity, short-duration exercise mainly affects the rapid component, whereas more prolonged exhausting exercise stimulates mechanisms present also beyond the first hour of recovery. This remains to be fully explained, but is understandable from what is presently known about the mechanisms causing EPOC.

Effect of Subject Characteristics

It is reasonable to hypothesize that subject characteristics (e.g., trained vs. untrained, men vs. women, young vs. old, obese vs. lean) may influence several of the mechanisms underlying EPOC (e.g., substrate shift from carbohydrate to fat). Despite attempts to study the influence of subject characteristics on EPOC, findings have generally not been linked to mechanisms.

Training Level

No existing study design provides a definite answer regarding whether training status affects EPOC, since comparing groups of various fitness levels at the same absolute exercise intensity results in trained subjects working at a lower relative intensity, which will influence EPOC. Further, comparing trained and untrained persons at the same relative intensity with equal total work means that the untrained subjects must work longer, and this may also influence the results. Still, it appears that O_2 uptake in trained individuals (vs. untrained individuals) returns more rapidly to resting levels during exercise at either the same relative or absolute work rate; but studies following more strenuous exercise are needed.

Gender

Few studies have compared EPOC in men and women. Thus, the gender effect on EPOC is not fully clarified. Again, the question about use of absolute or relative workloads must be addressed. Further, energy expenditure may vary with menstrual phase, although the findings in the literature on this issue are controversial. Thus, controlling for menstrual cycle appears to be important in studies with female subjects.

Age

The EPOC has not been compared between young and old and has generally not been measured in children.

Body Weight and Obesity

Studies of the effect of obesity on EPOC are needed. A few studies have compared EPOC after short (30 min) exercise in obese versus lean men and demonstrated a reduced EPOC in obese men. A recent study suggested that the difference was related to a lower growth hormone response and elevated cortisol response to exercise in the obese (8). Further, EPOC after 20 min treadmill exercise at 70% of $\dot{V}O_2$max was negatively related to fat mass in men of similar lean body mass (9). Exercise O_2 uptake relative to body weight was, however, greater for the low-fat-mass group. Similar EPOC studies in women have not shown differences between lean and obese subjects.

Effect of Diet Composition

In most EPOC studies, the subjects have been fasted, and prolonged diet was not controlled. If subjects participated in several studies, the diet was controlled only to the extent that the same food was ingested on the last day before each study. The diet composition is generally not reported. Interestingly, the studies showing an extended EPOC are mostly from Scandinavia and Australia, whereas American studies often show no prolonged EPOC. Some of this is clearly explained by differences in exercise intensities and durations, and some probably by the fitness level of participants, with only the more fit subjects managing to perform the exhaustive exercise needed to produce a prolonged EPOC. A third hypothetical explanation is that the difference is related to diet. A recent study (10) provides some preliminary support for an effect of dietary fat composition on EPOC, with a lower EPOC following a high saturated fat versus an unsaturated fat diet (28 days). Further studies of the effects of diet on EPOC are warranted.

Implications for Total Energy Expenditure and Obesity

The recovery increase in energy expenditure per se after exercise appears to be unimportant in relation to energy balance and weight loss. However, it should be pointed out that there are no data from prolonged studies of the effect of EPOC on energy balance. It has been estimated (4) that EPOC amounts to 50 to 100 kJ (~12 to 24 kcal) after moderate exercise (≥1 h, ~50% of $\dot{V}O_2$max), with a resultant energy loss of about 11,700 kJ/year (about 2800 kcal/year) if exercise is performed three times per week. This represents only about 311 g fat. More demanding exercise (≥1 h, ≥70% of $\dot{V}O_2$max) can result in about 700 kJ (167 kcal) more expended during recovery, corresponding to about 109,200 kJ/year (about 26,000 kcal/year) (exercise three times per week) and equivalent to about 2.9 kg fat. However, overweight or untrained persons are not likely to undertake such demanding exercise. Although overweight may result from a small positive energy balance over time and EPOC may contribute to the opposite when strenuous exercise is performed regularly, it must be concluded that EPOC is negligible in relation to weight loss in the overweight or obese person. For an athlete on a regimen of high training, the situation may be different. Also, it is well known that exercise per se has an important role in weight regulation. Thus, energy expended during exercise, as well as exercise programs that stimulate compliance, should be emphasized in obesity prevention and weight loss.

Genetics of Obesity

Ruth J.F. Loos, PhD

Over the past three decades, the prevalence of obesity has risen dramatically worldwide, reaching epidemic proportions in most industrialized countries. There is no doubt that a changing environment, which promotes excessive food intake and discourages physical activity, has been driving this rapid increase. However, not every person who is exposed to this obesogenic environment grows obese. Some individuals are resistant and remain lean under any circumstance; others easily gain weight when adopting an obesity-promoting lifestyle, and some will gain more than others (figure 44.1). This observation has highlighted the variation in the susceptibility to weight gain that is likely to be genetically determined. Figure 44.1 illustrates the genetic susceptibility to obesity in restrictive (food intake is limited and physical activity is high) and obesogenic (food intake is high and physical activity is limited) environments. Indeed, obesity is a typical common, multifactorial disease that arises through the joint actions of multiple genetic and environmental factors.

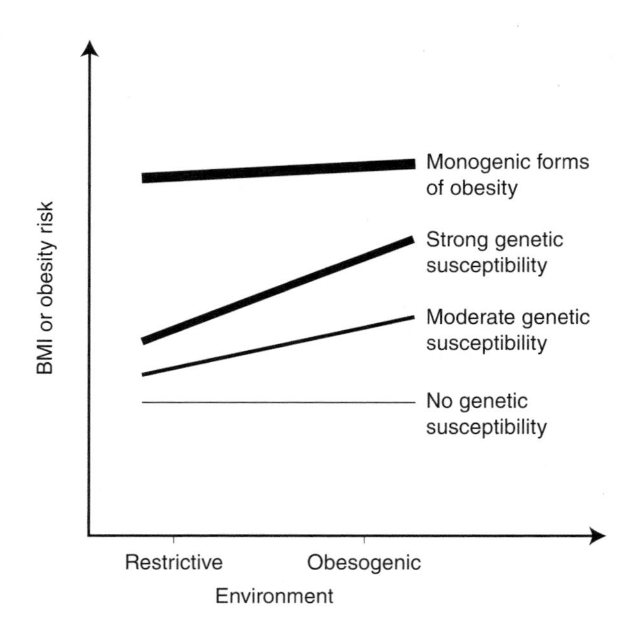

Figure 44.1 Genetic susceptibility to obesity in restrictive and obesogenic environments. In a restrictive environment (limited food intake and high physical activity), most people have normal body mass index (BMI), with little variation in BMI in the population. In an obesogenic environment (high food intake and limited physical activity), most but not all people gain weight (and thus increase BMI), and the amount of the weight gain depends on one's genetic susceptibility to obesity. Individuals with monogenic forms of obesity will be obese under any condition.

Adapted, by permission, from R.J.F, Loos and C. Bouchard, 2003, "Obesity–is it a genetic disorder?" *Journal of Internal Medicine* 254: 401-425.

Evidence for a Genetic Contribution to Obesity

The first studies that provided evidence for a genetic contribution to obesity risk were based on familial relatedness and on ethnic diversity without actually determining the genetic variation at the molecular deoxyribonucleic acid (DNA) level (figure 44.2).

Descriptive epidemiological studies have shown that certain populations have a substantially increased susceptibility to becoming obese (e.g., about 69% of Pima Indians in Arizona are obese) compared to other populations who live in the same obesogenic environment but who have a different genetic background (e.g., about 30% of white Americans are obese). Familial aggregation studies estimate the recurrence risk of obesity between family members, which is compared to the obesity risk in the general population. These studies have shown that individuals with an obese first-degree relative are two to three times more at risk of being obese than the general population. The increased familial risk suggests that genes are involved in the development of obesity; however, it may also reflect a shared familial lifestyle. Heritability studies have attempted to disentangle the contribution of genes and shared environment by comparing concor-

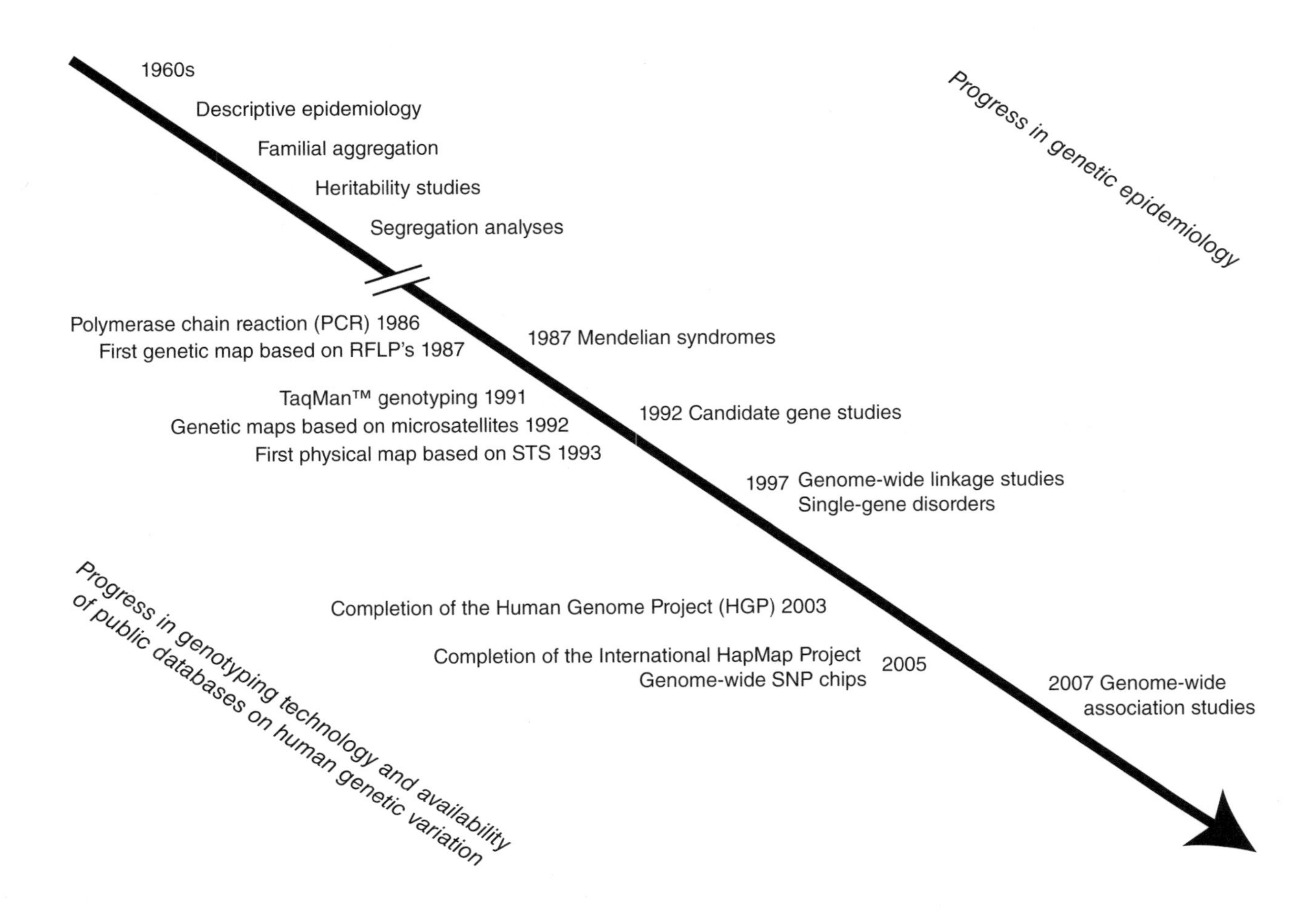

Figure 44.2 Progress in genetic epidemiology is paralleled by the increased availability of public databases that contain information on genetic heterogeneity in humans and by advances in genotyping technology. RFLP: restriction fragment length polymorphism; STS: sequence tagged site.

dances in traits or diseases between individuals with variable degrees of relatedness. Family and twin studies have estimated that genetic factors account for 40% to 70% of variation in common obesity (1).

Mendelian Disorders and Single-Gene Disorders

Until recently, most progress in obesity genetics was made through the study of monogenic forms of obesity, which are generally very severe with an early onset. To date, about 30 Mendelian disorders have been reported for which obesity is one of the clinical manifestations in addition to developmental anomalies and mental retardation (e.g., Prader-Willi and Bardet-Biedl syndromes, Albright hereditary osteodystrophy). At least 11 monogenic forms have been described for which morbid obesity is the dominant feature (e.g., mutations in *MC4R, POMC, LEPR, LEP*) (2). Although individuals with such monogenic forms represent only a small fraction of the obese population, the identification of genes that cause these disorders has helped to reveal novel regulatory pathways that control appetite and energy homeostasis. Interestingly, most of the monogenic obesity mutations seem to affect the appetite control centers in the brain, leading to hyperphagia. In particular, mutations in the pro-opiomelanocortin *(POMC)*, melanocortin 4 receptor *(MC4R)*, leptin *(LEP)*, and leptin receptor *(LEPR)* genes have been extensively described. Of those, only variants in *MC4R* have also been unambiguously associated with common obesity as well (3, 4).

Genes for Common Obesity

Common obesity is a heterogeneous condition and, unlike the situation with monogenic obesity, it is expected that many genetic variants contribute to this condition, each conferring only a modest risk. The two main approaches applied to search for genetic variants for common obesity are (1) the hypothesis-driven approach, by means of candidate gene studies, and (2) the hypothesis-generating approach, by means of genome-wide scans. The advancement in genetic epidemiological gene discovery strategies can largely be ascribed to progress in genotyping technology and the availability of public databases on genetic variation in the human genome (figure 44.2).

Despite a relatively high heritability and the serious efforts put into searching for genes for common obesity for over 15 years, progress in the field has been slow and, until recently, success rather limited. Genome-wide association, however, the latest gene-finding strategy, has sparked enthusiasm and has already contributed to the discovery of more than 400 genetic susceptibility loci robustly associated with nearly 100 common diseases; and it is poised to do the same for common obesity.

Candidate Gene Studies

The candidate gene approach is hypothesis driven and is based on current understanding of the pathophysiology that underlies obesity. Genes and pathways that have been shown to contribute to obesity-related traits through the study of animal models, cellular systems, or monogenic forms of obesity are tested for association at the population level.

Since the first candidate gene studies for obesity and related traits were reported in the early 1990s (figure 44.2), the number of proposed obesity susceptibility genes has grown steadily. The latest update of the human obesity gene map presented 127 candidate genes for obesity-related traits, but only a few of these have been unequivocally confirmed (2). The main reason for the limited success of this approach is that sample sizes are often small ($n < 1000$) and thus underpowered to identify genuine associations with small effect sizes. It is, however, encouraging to see that in recent years an increasing number of association studies have been performed in larger populations ($n > 5000$) and that the initiative to carry out meta-analyses is more often taken.

For example, common variants in *MC4R* and *PCSK1* (prohormone convertase 1/3), mutations of which lead to monogenic forms of obesity, have been unambiguously associated with obesity by meta-analyses of published data and by new large-scale association studies (3, 5). With data on almost 40,000 individuals, it was shown that carriers of the *MC4R* 103I-allele and of the 251L-allele have a 20% and 50% reduced risk of obesity, respectively (3, 4). Two potentially functional variants in the *PCSK1* gene, N221D and Q665E-S690T pair, were found to increase the odds of obesity by 34% and 22% respectively for each additional risk allele in 13,659 individuals of European descent (5). Furthermore, a meta-analysis on the extensively studied W64R variant in the beta3-adrenergic receptor *(ADRB3)* showed that the 64R-allele was consistently associated with a 0.24 kg/m^2 increased body mass index (BMI) in East Asians, but not in white Europeans, when data from 97 studies including 44,833 individuals were combined (6).

Large-scale studies are also powered to prove that an association is truly negative. Three such studies, two including >8000 and one including >5000 individuals, could not confirm that the K121Q variant in *ENPP1* (ectoenzyme nucleotide pyrophosphate phosphodiesterase), an inhibitor of insulin-induced activation of the insulin receptor, is associated with BMI or obesity, despite having more than 95% power to identify previously observed effects (7). Also the potential role of the -174G>C interleukin 6 (IL-6) variant in obesity was refuted by a meta-analysis combining data of 26,944 individuals from 25 populations (7).

Yet, even with considerably large samples, results can remain inconclusive. Two studies, each with >4000 individuals, yielded only a suggestive association between *NPY2R* promoter variants and obesity risk, such that further replication will be needed for more definitive answers (7).

Genome-Wide Surveys

Genome-wide surveys, which are hypothesis generating, include genome-wide linkage and genome-wide association studies. Through screening of the whole genome, these studies aim to identify new, unanticipated genetic variants that are associated with obesity or related traits.

Genome-Wide Linkage Studies

Genome-wide linkage studies have been conducted for more than 10 years, and they differ from genome-wide association studies mainly in study design and marker resolution. They require populations of related individuals to test whether certain chromosomal regions cosegregate with a disease or trait across generations. Typically, a genome-wide exploration requires around 400 highly polymorphic markers, genotyped across the whole genome at 10-centimorgan (cM) intervals. This allows identification of broad intervals often containing hundreds of genes. When strong evidence of linkage is found, the region is narrowed down with a denser marker set to eventually pinpoint the gene that underlies the linkage with the trait or disease.

Since the first genome-wide linkage scan for body fat percentage in 1997, more than 60 scans have been described, showing more than 250 loci linked to obesity-related traits (2). However, none of these could be narrowed down sufficiently to pinpoint causal genes or genetic variants predisposing to obesity. This suggests that genome-wide linkage might not be an effective approach to identify new, unanticipated genetic variants for common obesity. This conclusion is further supported by a recent meta-analysis that combined 37 genome-wide linkage studies with data on more than 31,000 individuals from 10,000 families of European origin (8). Despite substantial power, no locus could be unequivocally linked to BMI or obesity (8).

The two most important limitations of genome-wide linkage are that (1) it requires data on related individuals who are often hard to recruit, which in turn affects the sample size and power; and (2) it relies on meiotic recombination of genomic regions that are transmitted across generations. If pedigrees are small, recombination does not break up chromosomes sufficiently, and therefore transmitted regions will be relatively large; this impedes fine-mapping of the causal gene.

Genome-Wide Association Studies

Genome-wide association studies are carried out using multistaged designs with at least a discovery stage and a follow-up or replication stage. The discovery stage is the actual genome-wide analysis and entails high-density genotyping of hundreds of thousands of genetic variants, typically single-nucleotide polymorphisms (SNPs), across the genome; each SNP is tested for association with a trait or disease, such as BMI or obesity. In the replication stage, associations that meet the genome-wide significance threshold ($p < 5.0 \times 10^{-7}$) are taken forward for replication in additional populations to confirm the initial observation. Unlike linkage studies, genome-wide association studies can be performed in populations of unrelated individuals.

The completion of the Human Genome project and of the International HapMap, together with advances in high-throughput genotyping technology, has facilitated the design of SNP chips that interrogate the entire genome at a resolution previously unattainable (figure 44.2). The first generation of high-resolution genome-wide association studies has already resulted in an unprecedented chain of newly indentified susceptibility loci for various common diseases (9). Also in the field of obesity genetics, we have recently witnessed the discovery of the first two, robustly replicated, obesity susceptibility loci: *FTO* (fat mass and obesity associated gene) (10, 11) and a locus at chr18, a short 188 kb downstream of the *MC4R* gene (12).

FTO was identified by two independent genome-wide association studies (10, 11). The first study, by Frayling and colleagues (10), was a genome-wide association scan for type 2 diabetes. A cluster of common variants in the first intron of *FTO* showed a highly significant association with type 2 diabetes. After adjustment for BMI, this association was completely abolished, suggesting that the *FTO*–diabetes association was mediated through BMI. Follow-up analyses confirmed this observation, as the *FTO*–BMI association was consistently replicated in 13 cohorts including more than 38,000 individuals (10). The second study, by Scuteri and colleagues (11), was the first large-scale high-density genome-wide association study specifically designed for BMI. In the discovery stage, which included more than 4000 Sardinians, variants in *FTO* and *PFKP* (phosphofructokinase platelet type) showed the strongest associations. In the follow-up stage, only the *FTO* variants replicated significantly (11).

The frequency of the *FTO* risk alleles in populations of European descent is high, with 63% of the population carrying at least one risk allele and 16% carrying two. Each risk allele increases BMI by about 0.40 to 0.66 kg/m^2 units or about 1.3 to 2.1 kg (2.9 to 4.6 lb) in body weight for a person 1.80 m (about 71 in.) tall. The odds of overweight and obesity increase by 18% and 32% per risk allele, respectively. Similar associations have been reported in East Asians, in whom the prevalence of the risk allele is much lower than in populations of European descent. Carrying *FTO* risk alleles, however, does not imply that one is destined to become obese. An *FTO*–lifestyle interaction study showed that when physically active, individuals homozygous for the *FTO* risk allele did not have an increased BMI compared to noncarriers (13). Although more work needs to be done to unravel the physiological mechanisms through which *FTO* confers obesity, the first lines of evidence from human and animal studies suggest that *FTO* is abundant in the hypothalamic nuclei and regulated by feeding and fasting.

Exactly one year after the discovery of *FTO*, a second obesity susceptibility locus was identified. This locus was the most significant association signal, after *FTO*, in a meta-analysis of seven genome-wide association studies for BMI comprising 16,876 individuals (12). The locus resides in a noncoding, intergenic region at chromosome 18q21,

188 kb downstream of *MC4R*, a gene well known for its role in monogenic severe childhood-onset obesity. Subsequent replication analyses firmly established this locus by confirmation of association in more than 60,352 adults and 12,863 children (12). Each risk allele at this locus increased BMI by 0.22 kg/m^2 units or 710 g in body weight for a person 1.80 m (about 71 in.) tall. The odds of being overweight and obese increased by 8% and 12%, respectively, for each additional risk allele. The effect size at this locus is roughly half that of *FTO* variants (12), while the associations were more pronounced in childhood and already seen from age 7 years onward (12). Although the pattern of phenotypic associations of this 18p21 locus of common variants showed some similarity with that seen in individuals with rare severe coding mutations in *MC4R*, it remains unclear whether this locus is indeed functionally linked to *MC4R*.

Future Directions

With genome-wide association, we are at the beginning of a new era that is poised to fundamentally improve our pathophysiological understanding of obesity. There is no doubt that many more obesity susceptibility loci remain to be uncovered that are currently hidden amid low-frequency signals. Several new developments in human genetics are set to provide further avenues for continued gene discovery. First, genome-wide association studies will need to scale up sample sizes to be sufficiently powered to identify less common variants and variants with small effects. International collaborative initiatives such as the Genomic Investigation of Anthropometric Traits (GIANT) consortium are set to rapidly increase the discovery of more obesity susceptibility loci. Second, reducing the phenotypic heterogeneity (e.g., by measuring more accurate markers of adiposity) will further increase the power to identify new loci. Third, further progress in genotyping and sequencing technology, to capture more comprehensively rare variants and copy number variants, will allow an even more exhaustive examination of the genome. Fourth, extending analyses to populations of different ethnic backgrounds may not only bring new insights into ethnic-specific susceptibility but also be a valuable tool for fine-mapping of causal variants given the varying linkage disequilibrium patterns across ethnicities.

Translation of these new genetic insights into clinical practice continues to be an ambitious goal; and while the effect size of new loci is small, it is important when predictive testing is the ultimate endeavor. As sample sizes increase, the average effect size of newly discovered loci will progressively decrease, which will pose further challenges for the implementation of the new genetic knowledge within personalized treatment. A major challenge is the follow-up of the new loci into experimental analyses to uncover the underlying physiological and molecular mechanisms.

Conclusions

Genome-wide association promises to greatly enhance our understanding of the genetic basis of common obesity. Large-scale data integration through international collaboration will undoubtedly continue to increase the discovery of obesity susceptibility loci, which in turn will expand our insights into the pathogenesis of this condition. Translation of these findings into clinical practice, however, will become the challenge of the future.

45

Epigenetic Effects on Obesity

Peter W. Nathanielsz, MD, PhD, ScD

This chapter covers the recent origins of the concept that offspring phenotype is determined by developmental programming of function that results from environmental challenges during fetal and early postnatal life. The focus is on the ways in which offspring systems that regulate appetite and physical activity may be programmed. *Developmental programming* can be defined as the response to a specific challenge to the mammalian organism during a critical developmental time window that alters the trajectory of development, qualitatively, quantitatively, or in both ways, with resulting persistent effects on phenotype.

The long-standing debate concerning the relative influences of nature and nurture on phenotype is clearly reaching a new stage of understanding as a result of human epidemiologic findings and carefully controlled animal studies relating the conditions experienced during development to alterations in adult phenotype. The mammalian organism passes more biological milestones during development than at any other time in life. It is therefore very understandable that the precise timing and sequence of maturational events during this critical period may have important consequences and that as a result, altered development will program adult function and may predispose to adult disease such as obesity (1).

In 1961, Barraclough performed a classical study that highlights the key principles of developmental programming (2). He injected female rats with a single dose of testosterone on the first day of neonatal life and observed that when they reached puberty, they did not have the normal reproductive cycles demonstrated by control, saline-injected female rats. The reproductive pattern of these androgen-sterilized females was the acyclic pattern of reproductively competent males. If he injected the same dose of testosterone at neonatal day 20, females demonstrated normal fertility at puberty. Thus an event occurring in a critical developmental time window had a persistent effect on function and lifetime health.

Barraclough's experiment shows that consequences of a challenge to the developing organism in a restricted yet critical time window may not be immediately apparent. His important observation of pronounced yet delayed effects explains why it is necessary to develop animal cohorts that can be studied over long periods of time. The timing of onset of various consequences of the developmental challenge may be very different for different systems. It is thus necessary to follow animals postnatally to determine the full nature of consequences of challenges experienced during development. The unwanted consequences of developmental programming may not be immediately apparent for a second reason also: The mammalian organism has great powers of compensation. We have shown that exposure of the sheep fetus to levels of glucocorticoids inappropriate to the present stage of development produces fetal hypertension at the time of glucocorticoid exposure. After birth, lambs that have experienced this type of glucocorticoid exposure during fetal life show arterial endothelial dysfunction at 5 and 8 months, but they are initially normotensive because they compensate for the endothelial dysfunction and become hypertensive only after 3 years of age (3, 4).

During development, there are critical periods of vulnerability to suboptimal conditions. Vulnerable periods occur at different times for different tissues. Cells dividing rapidly are at greatest risk. There is much evidence that the developmental environment profoundly alters offspring body composition, especially fat content, although this has not been studied as extensively as the effects of developmental programming on cardiovascular and endocrine function. There is also suggestive but as yet inconclusive evidence that several factors, including altered offspring appetite, as well as the interaction and balance of hormones that predispose to obesity and physical activity, are contributory factors to the predisposition to obesity.

Altered Regulation of Appetite and Physical Activity

Obesity results from an imbalance of energy intake and energy utilization—metabolic rate and physical activity. More is known about development of the hypothalamic centers that regulate the drive to eat than about neural mechanisms that regulate physical activity. In primates, including humans, appetitive centers and the satiety drive develop in late gestation, although the development continues into the postnatal period. In rodents, development is mostly postnatal and hence easier to manipulate and observe experimentally. This ease of study should not obscure the need for investigations on every aspect of the differences in exposures and outcomes that affect the trajectory of development in altricial, polytocous species (such as the common experimental rodents) compared to precocial species that generally produce single offspring. It is because of these differences in mechanisms that comparative physiology becomes very important in providing clues and suggesting mechanisms.

When rat litter size is reduced to four pups to produce overfeeding in the maternal lactation period, offspring exhibit hyperphagia and adiposity, hyperleptinemia, hyperinsulinemia, and resistance to leptin feedback. Unfortunately there is very little information that relates the effects of such manipulations to offspring physical activity. In the rat, a nutrient restriction paradigm in which maternal diet is globally reduced by 70% throughout pregnancy compared with that of controls, followed by a normal diet during lactation, results in obese offspring that spend much less time wheel running than offspring of control mothers (5). Although this maternal nutritional challenge is very severe, it shows that development can alter offspring physical activity. In this study, however, the time relationship between the development of obesity and the altered physical activity was not explored to determine potential causality. The crucial question is, "Did the rats become fat because they performed less physical activity, or was the decreased activity a consequence of the obesity?" Feeding the offspring of the undernourished mothers a high-fat diet led them to perform even less physical activity. Another study provides evidence that lipid composition in the maternal diet may also be important, since offspring of rats fed a saturated-fat, lard-rich diet exhibit decreased physical activity (6). In contrast, offspring of pregnant mice fed a diet rich in polyunsaturated fat show increased swim test activity (7).

Poor maternal nutrition has been shown to alter fetal skeletal muscle development, providing one potential mechanism whereby overall physical growth is decreased. Weanling rats born to mothers fed a highly palatable obesogenic cafeteria diet during pregnancy and lactation had a smaller skeletal muscle mass with fewer muscle fibers compared with offspring of rats fed the normal laboratory diet. Increased intramuscular lipid content and adipocyte hypertrophy were also present (8). Since insulin sensitivity is highly dependent on an adequate skeletal muscle mass and since muscle lipid content is directly related to insulin resistance, these consequences of overnutrition during development provide a potential mechanism that would predispose offspring to insulin resistance in later life.

In keeping with this mechanistic view, maternal nutritional challenges that result in reduction in offspring muscle fiber number before birth have been shown to have persistent metabolic effects on offspring postnatal life. Utilization of glucose and fatty acids is a primary function of skeletal muscle, which thus plays a central role in regulating insulin sensitivity and development of resistance in obesity. Reduced skeletal muscle fiber number and muscle mass impair metabolism of glucose and fatty acids in response to insulin stimulation, and thus predispose offspring that experienced nutrient deficiency during development to diabetes and obesity in later life (9, 10). Human infants who are small at birth are at greater risk for type 2 diabetes and obesity. It is known that low resting energy expenditure is associated with increased incidences of obesity and diabetes, and skeletal muscle mass is positively related to the resting energy expenditure, further revealing the importance of muscle mass in determining the body's overall efficiency in utilization of fatty acids and glucose. Although reduced skeletal muscle development accompanied by increased visceral fat deposition has the potential to increase the likelihood of offspring obesity, it does have positive advantages for survival of offspring to the reproductive years and thus assists in maintaining the species despite the decreased longevity in the individual.

General Mechanisms of Epigenetic Modification of Gene Function

Recent studies in rodents and nonhuman primates have provided evidence for some of the epigenetic mechanisms responsible for persistent changes in tissue-specific gene expression that are set up during growth and development with potential consequences later in life. These changes contribute to the well-documented plasticity that occurs at the various developmental stages in fetal and neonatal life. There are three major ways in which epigenetic influences can act: (1) by producing changes in the methylation of promoter region of cytosine–guanine nucleotides in key specific genes, (2) through changes to chromatin structure by histone acetylation and methylation, and (3) altered microRNA control by posttranscriptional mechanisms through production of changes in the methylation of promoter region of cytosine–guanine nucleotides in key specific genes. These epigenetic DNA modifications are usually stable throughout the life of the offspring and often emerge later in life to modify cell activity when a further change, such as the increased hormone production at puberty, occurs.

Developmental programming has been shown to produce epigenetic changes in chromatin structure by introducing covalent modifications of histones that alter gene regulation without alteration in nucleotide sequence. In one study in the Japanese macaque, a maternal high-fat diet produced a threefold increase in fetal liver triglycerides and associated histological changes of nonalcoholic fatty liver disease that were accompanied by significant hyperacetylation at H3K14, lower histone deacetylase 1 gene expression and protein, and in vitro histone deacetylase functional activity. Alanine aminotransferase, a key enzyme in hepatic carbohydrate–protein interaction, was increased in the liver of fetuses of mothers fed a calorie-dense diet compared with controls (11). These studies clearly show that maternal diet has major epigenetic effects related to offspring obesity. To date, there are no parallel published studies either directly on fetal skeletal muscle or on the neural mechanisms regulating physical activity.

When we consider the effects of developmental programming on physical activity in offspring, it is important to bear in mind two of the general principles of developmental programming discussed in detail elsewhere (1). First, in multiple studies and with use of several models, it has been demonstrated that female and male offspring outcomes can differ in response to the same challenge. In one recent study of the effects of maternal diet–induced obesity in mice on offspring metabolic function and physical activity, mice were fed either standard chow containing 3% fat and 7% sugar or a palatable obesogenic diet of 16% fat and 33% sugar for six weeks prior to mating and during pregnancy and lactation. Offspring of control and obese dams were weaned onto standard chow. Offspring of obese mothers were hyperphagic and exhibited less locomotor activity accompanied by increased adiposity, with reduced activity present only in female offspring (12). Male offspring showed a greater increase in fat deposition than female offspring, although their activity was not reduced. Thus decreased physical activity was only one factor in the development of obesity and occurred only in females.

The developing fetus is very active in utero. Ultrasound technology shows that the fetus has well-developed behavioral patterns that include limb and breathing movements. It is now clear that fetal activity is part of the normal developmental program and that in the absence of muscle movement, muscle development is decreased. For example, lung development is retarded in fetuses with decreased breathing movements. Although firm evidence is still lacking, it is likely that fetal skeletal muscle activity is required for normal fetal muscle fiber development. It is a general principle that normal development is dependent on continuing normal activity. Each phase of development provides the required conditions for subsequent development.

Several other mechanisms that lead to functionally significant structural changes in important organs have been described in animal studies. The normal blood supply to the organ may be compromised in ways that decrease the growth rate of that organ during development and alter its ability to increase the level of its activity in response to challenges in later life. Feedback of homeostatic systems may also be impaired as the result of development of too many or too few hormone receptors, as has been shown in the resetting of offspring pituitary adrenal function. Since the level of circulating glucocorticoid is important in regulating fat deposition, these changes have important implications for the development of obesity in later life.

Conclusion

In conclusion, there is now compelling human and experimental animal evidence for the general concept of developmental programming in many physiological systems. Predisposition to obesity is clearly influenced by conditions that are present during an individual's development. Altered appetite control has been shown to be a part of the mechanism involved. In addition, development of a smaller skeletal muscle mass will impair insulin sensitivity and contribute to the predisposition to obesity. Finally, there is emerging evidence that an individual's preferred level of physical activity may be programmed during development as well.

part VI

Physical Activity, Behavioral, and Environmental Determinants of Obesity

This section of the book explores some important behavioral and environmental correlates and determinants of obesity. New evidence linking duration and quality of sleep with physical activity and obesity is presented in chapter 46. Chapters 47 and 48 describe the associations between occupational work and mode of transportation with physical activity, respectively, and the implications for obesity. Chapter 49 describes the emerging evidence for a potential role of the built environment in obesity. Considerable work has been conducted on the relationship between socioeconomic status and obesity; and chapter 50 presents a synopsis of some of the complex relationships that are emerging from different global regions. This section of the book concludes with chapter 51, which describes some important interactions between physical activity levels and dietary intake, as well as the implications of these interactions for weight gain and obesity.

Physical Activity Level, Sleep, and Obesity

Shawn D. Youngstedt, PhD

The startling rise in the worldwide prevalence of obesity over the past few decades does not seem to be completely attributable to changes in physical activity and caloric intake, and interventions that have targeted these behaviors have had limited success. Thus, there is an impetus to study other behaviors that might contribute to obesity. One of the most recently recognized potential risk factors for obesity is insufficient sleep. Parallel with the rise in obesity has been a modest decline in population sleep duration over the past few decades (1). The aims of this chapter are to discuss evidence linking sleep duration with obesity; to discuss potential mechanisms mediating these associations; and to discuss potential interactions of sleep, physical activity, and obesity.

Epidemiologic Association of Sleep Duration With Obesity

Excellent recent papers have reviewed epidemiologic associations of sleep with overweight or obesity in children (2) and adults (1). The evidence from studies of children is very consistent in showing an independent association of short sleep with overweight or obesity (1, 2). This association has been more evident in boys versus girls and in younger versus older children (2). Particularly compelling are prospective studies showing that short sleep is a significant risk factor for subsequent weight gain (3). Short sleep duration has been less strongly associated with childhood obesity than some other factors, especially parental obesity. Nonetheless, the epidemiologic risks of obesity associated with short sleep have been found to be comparable to those of other known factors, such as physical activity and caloric intake, even when these factors have been measured objectively (3).

Epidemiologic studies in adults have often shown a "U"-shape, with both short sleep (<6 h) and long sleep duration (>8 h) independently associated with greater risk of overweight or obesity than sleep of 7 to 8 h (1). Figure 46.1 presents an example of this association from the Quebec Family Study (4). The model used in this investigation was adjusted for age, sex, baseline body mass index, study phase, length of follow-up, resting metabolic rate, smoking habits, employment status, education level, total annual family income, menopause status, shift-working history, alcohol intake, coffee intake, total caloric intake, and participation in vigorous physical activity. These associations of both short sleep and long sleep duration with overweight or obesity have been observed in the majority of cross-sectional studies in adults, though less consistently than for children and also less consistently in older versus young adults. Nonetheless, prospective studies all showed that short sleep (but not long sleep) was significantly predictive of weight gain (1). Critics of these data have noted that the average effects (e.g., body weight differences) have been modest (5). In any case, it is noteworthy that the risks associated with short or long sleep have been comparable to those of other known risks of overweight or obesity measured in these studies, such as physical activity and caloric consumption.

There are numerous limitations of these epidemiologic studies. As with all epidemiologic investigations, these studies cannot prove that short or long sleep causes obesity. One of the most overlooked factors, which could partly explain the epidemiological associations, might be anxiety or "stress," which is often associated with both sleep loss and increased body weight. The association of sleep duration with obesity might be partly explained by reverse causa-

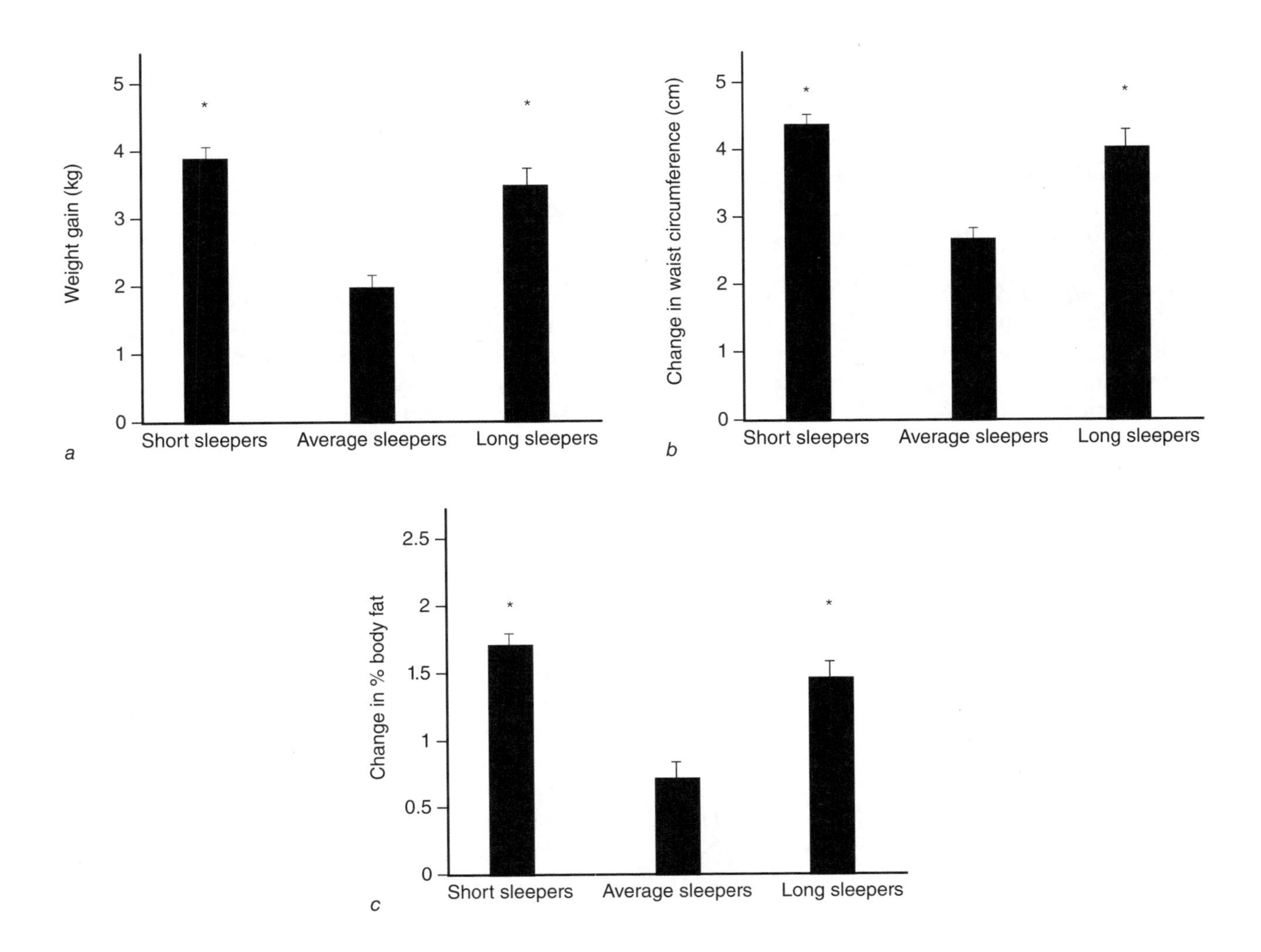

Figure 46.1 Mean *(a)* weight gain, *(b)* change in waist circumference, and *(c)* change in percentage of body fat by sleep-duration group over 6 years of follow-up in adults from the Quebec Family Study. *Significantly different from the 7 to 8 h sleeping group (average sleepers), $P < 0.05$.

Reprinted, by permission, from J.P. Chaput et al. 2008, "The association between sleep duration and weight gain in adults: A 6-year prospective study from the Quebec Family Study," *Sleep* 31: 517-523.

tion. For example, obesity is clearly associated with a more disturbed breathing, which in turn impairs sleep. Another limitation is that studies have relied almost exclusively on self-reported sleep duration, which is notoriously inaccurate in comparison with objective sleep recording. Nonetheless, the available evidence is fairly consistent and compelling. The data provide strong rationale for experimental investigation of the effects of sleep manipulation on body weight or obesity.

Potential Mechanisms

There are several potential mechanisms by which short sleep duration could cause increased body weight. First, both cross-sectional (6) and experimental evidence (7) indicates that less sleep is associated with lower plasma levels of leptin, a food satiety hormone, and higher levels of ghrelin, a hunger hormone. Consistent with this mechanism are data indicating increased subjective hunger and sugar craving following short-term sleep restriction in humans (7). However, experimental evidence of these effects is limited to studies of severe short-term sleep restriction. Moderate, chronic sleep restriction might not elicit similar changes, as suggested by studies that have shown no differences between short- and average-duration sleepers in self-reported caloric intake (1).

Second, shorter sleep duration might contribute to increased body weight simply by allowing more time to eat. Anecdotally, sleep restriction (e.g., associated with studying) is often associated with increased caloric consumption that is beyond the minimal increase in energy expenditure associated with sedentary wake activity compared with sleep.

Third, short sleep could promote weight gain indirectly by causing physical fatigue, which is one of the most significant barriers to physical activity. Epidemiologic studies have shown that the association of short sleep with obesity is independent of physical activity.

However, the role that low physical activity might play in mediating the epidemiological association of short sleep with obesity might also be underestimated by the assessment of exercise in these studies, which has typically been limited to one or two questions regarding total daily exercise. Given that only a small percentage of the population engages in regular planned exercise, it might be difficult to detect differences between normal and short sleepers with these gross metrics. More relevant physical activity metrics, which might be more sensitive to change associated with fatigue, might be those that assess nonexercise activity (e.g., housework), which comprises the vast majority of total daily energy expenditure.

Long Sleep and Obesity

Physical inactivity might also play an important role in the association of long sleep with obesity. First, "long sleep" involves ≥60 min more of completely sedentary behavior, an amount that has been associated with obesity when quantified using other markers of sedentary behavior, such as television watching. Second, epidemiologic studies have consistently shown an association of long sleep with low levels of daytime physical activity (1). These data are consistent with survey and experimental evidence that extra time spent sleeping tends to elicit feelings of lethargy (8), and that long sleep could contribute to low daytime activity simply by reducing the time available for physical activity. Finally, our study of older long sleepers (≥8.5 h) demonstrated an increased level of physical activity following eight weeks of modest restriction of time in bed (TIB) (90 min per night) (9).

Influence of Exercise on Sleep and Implications for Obesity

Epidemiologic studies have consistently demonstrated a significant association of exercise with better sleep. Experimental studies have revealed that acute and chronic exercise elicits quite modest, though statistically significant, improvements in sleep (10). For example, meta-analysis has indicated that acute exercise increases nocturnal sleep duration by approximately 10 min (10).

The modest effects of exercise on sleep observed in the extant experimental literature could be attributed partly to ceiling effects (10), as studies have generally focused primarily on good sleepers for whom substantial improvements in sleep are unlikely. There is limited evidence that exercise might elicit greater, though still modest, improvement in sleep in people with disturbed sleep. However, there is little compelling rationale for expecting that exercise could promote weight loss indirectly via these modest improvements in sleep. Indeed, in perhaps the only experimental study to examine this issue, a sample of obese women showed a small decrease in body weight following chronic moderate aerobic exercise training, but this was associated with a reduction in sleep duration (11). Nonetheless, there could be interactions of sleep, exercise, and other behaviors that could influence body weight.

There is no compelling empirical support for either of the conflicting claims that morning or evening exercise could promote better weight loss by increasing metabolism throughout the day or during a time of otherwise low metabolic rate, respectively. Apparently of far greater importance for weight control is having a regular exercise habit, which is often associated with having a consistent time of day for exercise.

However, if people must curtail sleep in order to exercise, for example in the early morning, this curtailment could attenuate weight loss benefits of exercise. Conversely, exercise could attenuate or prevent increased body weight associated with sleep loss.

Neither surveys nor experimental studies support the common assumption that evening exercise will disrupt sleep. A potential weight control advantage of evening exercise is that it might be more likely to replace sedentary behavior, which is most common in the evening.

Future Directions

Extensive epidemiologic evidence linking sleep duration with obesity provides a strong rationale for considering sleep in future epidemiologic and experimental studies of obesity. Perhaps the most important limitation that should be addressed is the lack of experimental evidence that altering sleep influences body weight. Sleep extension might be especially simple to implement compared with other putative weight loss behaviors. Research might first target overweight children, whose sleep habits might be more easily modifiable than those of adults.

It will be important to establish whether the changes in leptin and ghrelin also occur following chronic moderate sleep restriction, which is a far more common condition. There is also a need for more prospective epidemiologic studies, particularly those that track changes in sleep, which can vary considerably over time. There should be more focus on physical activity as a potential factor mediating the association of obesity with both short sleep and long sleep. In particular, further study of nonexercise energy expenditure is warranted.

47

Physical Activity Level and Occupational Work

David R. Bassett Jr., PhD

In this chapter, we consider the role of occupational physical activity (OPA) throughout history, examine recent time trends in OPA, and review the relationships between OPA and obesity. While many studies focus solely on leisure-time physical activity (LTPA), it is also important to take OPA into account. In many parts of the world, OPA and transportation are major components of the total physical activity energy expenditure. This is particularly true for developing nations and lower socioeconomic groups in developed nations.

Historical Perspective

Human cultural development is often divided into three phases: hunter-gatherers, traditional agrarian societies, and modern industrialized civilizations. Much of what we know about hunter-gatherers and labor-intensive farmers has been learned through the study of anthropology, historical records, or modern research on groups that still adhere to those lifestyles.

Historically, hunter-gatherers have had very high levels of physical activity. Studies of the Kung bushmen (Africa), Canadian Inuit, and the Ache of eastern Paraguay indicated that many hours per day were spent in physical activity. The activities included carrying water, collecting firewood, hunting, preparing meat and hides, and a nomadic lifestyle. By all accounts, obesity was rare in these societies. Food shortages, in addition to high levels of physical activity, kept people from gaining weight.

Labor-intensive agriculture was a common lifestyle for thousands of years, and it is still practiced in some parts of the world today. These traditional farmers often spend 8 to 12 h per day engaged in moderate-to-vigorous physical activity. Doubly labeled water studies on farmers in underdeveloped countries show very high levels of energy expenditure. Obesity in these traditional farming societies is rare.

With the advent of the industrial revolution, the former agriculture-based economy turned into one based on industry and manufacturing. Rapid growth of cities and increasing job specialization took place during this period. An interesting retrospective study shows the effect that these changes had on obesity. Helmchen (1) studied records of approximately 15,000 Civil War veterans; these men were Union soldiers who submitted to height and weight measurements at five-year intervals in conjunction with receiving a pension. In all age groups, the prevalence of obesity increased by about 1% of the population every five years from 1880 to 1910. Obesity rates were lowest for those in the most strenuous occupations, as follows: manual laborers (2.6%), farmers (2.9%), artisans or craftsmen (4.6%), service workers (4.6%), low-level professionals (10%), and high-level professionals (11.5%). This is logical since manual laborers would have had higher levels of energy expenditure and lower salaries (hence less food-purchasing power) than professional workers.

Recent Time Trends in Modern Industrialized Civilizations

Levels of OPA have declined over the past several decades in industrialized nations. Brownson and colleagues (2) examined U.S. Department of Labor statistics and found that the percentage of adults in highly active occupations declined by 30% between 1950 and 2000, while the percentage in less active occupations doubled over the same period (figure 47.1).

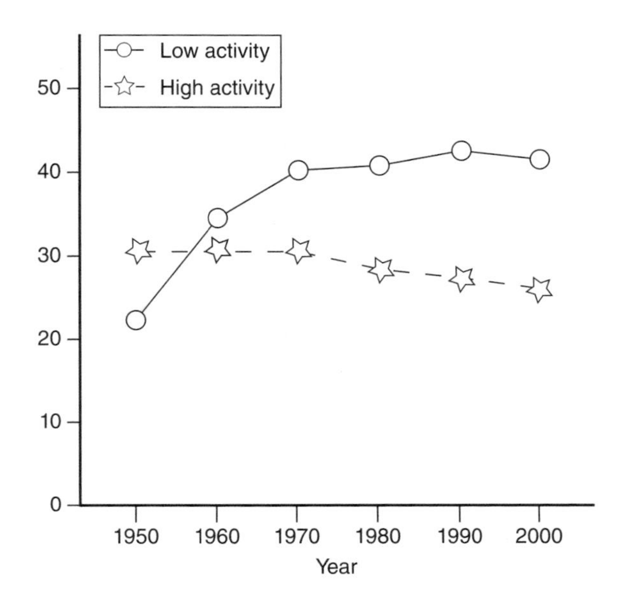

Figure 47.1 Trends in occupational physical activity in the United States, 1950-2000 (2).

Reprinted, with permission, from the Annual Review of Public Health Volume 26 © 2005 by Annual Reviews. www.annualreviews.org.

The Minnesota Heart Survey showed that OPA declined in both men and women from 1980 to 2000, whereas energy expenditure from lifestyle activities and LTPA increased (3). Only 1% of the participants performed 60 min of physical activity per day (a level of activity that would have been considered low by historical standards). Similar trends have been reported for Italy, Finland, Norway, and Great Britain over the past 30 years. In summary, there is strong evidence that OPA has declined in recent decades, although this decline has been partially offset by increases in LTPA.

Relationships Between Occupational Physical Activity and Obesity

Mayer (4) studied OPA and adiposity of mill workers in India in the 1950s. He found that sedentary workers (confined to a stall or a desk) were the heaviest. Workers who performed light, medium, or heavy OPA were lean, and there were no differences in body weight among them.

Morris and Heady (5) described the link between OPA and adiposity in London busmen. Sedentary bus drivers had larger average waists, chests, and body weights than the active conductors who walked up and down the aisles of the double-decker buses. This relationship held for all age groups; hence it was unclear whether higher levels of OPA protected the conductors against obesity or whether individuals with an inherent propensity to be heavy became drivers. Taylor (6) conducted a similar study of clerks and switchmen who worked for the railroads. The sedentary clerks had larger relative body weights than the active switchmen.

Recent studies show a more complex relationship between OPA and obesity. Oppert and colleagues (7) found that in French adults, OPA was positively associated with body mass index (BMI). In other words, women with highly active jobs were actually heavier than those with less active jobs! However, after adjustment for socioeconomic status (SES, as reflected by education level), the associations were no longer significant. Similar findings have been reported in young Dutch men and in Norwegian adults. In 2000, King and coworkers (8) examined data from the U.S. National Health and Nutrition Examination Survey, using a six-member committee to group occupations into those involving (a) high occupational activity, (b) low occupational activity, or (c) an uncertain amount of occupational activity, based on the U.S. Department of Labor occupation descriptions. Multivariate logistic regression was used to control for age, gender, race, smoking, alcohol consumption, LTPA, and income. Compared to those who engaged in no LTPA and had low levels of OPA, those who engaged in no LTPA and had high levels of OPA were 42% less likely to be obese (figure 47.2). These studies show that it is important to control for confounding variables, such as LTPA and SES, in order to see the true relationship of OPA to obesity in modern society.

The studies just discussed were cross-sectional, but the National Longitudinal Survey of Youth followed individuals from 1979 to 2000, providing further evidence that OPA is linked to obesity. In this study, job characteristics and anthropometric measurements were recorded. The results showed, for instance, that a male who spent 18 years working in the lowest OPA category would end up with a higher BMI (by 3.5 kg/m^2) than one in the most active category (9).

How Jobs Influence Caloric Expenditure

Jobs can elicit different caloric expenditures, as shown by the Compendium of Physical Activities (10). The compendium lists the energy expenditure of various activities in metabolic equivalents (METs). Light activities include seated office work (1.5 METs), typing (1.6 METs), and standing while talking (1.8 METs). Moderate activities include walking at 3 mph (3.3 METs) and operating a punch press (5 METs). Occupational tasks such as felling trees (8 METs), baling hay (8 METs), and digging ditches (8.5 METs) fall into the vigorous category. However, there are few jobs left in industrialized societies that require prolonged, vigorous effort. Automation and use of machines have reduced the need for manual labor and have contributed to increases in body weight. In addition, occupational stress, such as work on rotating shifts, can create metabolic disturbances that lead to obesity.

Two recent studies point to the hazards of sitting occupations with regard to obesity. Mummery and colleagues (11) observed that Australian workers who sit at least 6 h per day are twice as likely to be obese as those who sit less than 45 min per day. While prolonged sitting is hazardous to one's health, standing in one place is also contraindicated. Ergonomics experts warn of the hazards of prolonged standing, including neck, back, and foot pain. Thus, we now realize that "postural fixity" has adverse effects on health.

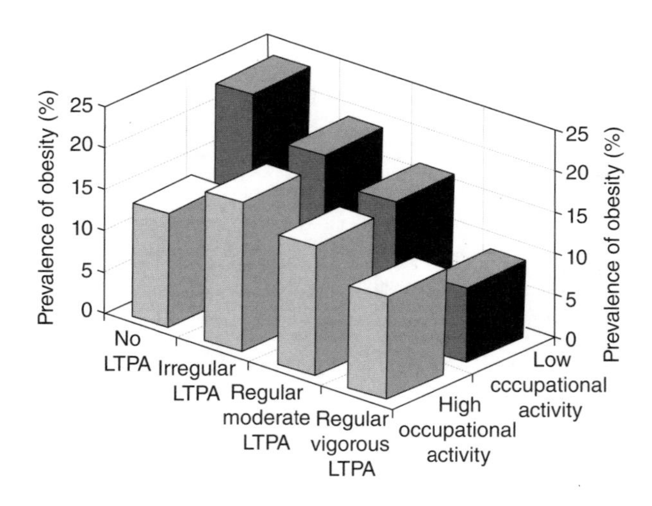

Figure 47.2 Obesity prevalence for leisure-time physical activity level and occupational physical activity level for U.S. adults. No LTPA = less than one bout per week; irregular LTPA = one to four bouts per week; regular LTPA = five or more bouts per week at a moderate (regular moderate LTPA) or vigorous (regular vigorous LTPA) level

Reprinted from S.D. Youngstedt and D.F. Kripke, 2004, "Long sleep and mortality: rationale for sleep restriction," *Sleep Medical Review* 8: 159-174.

Worksite Interventions

Since many people spend 40 h per week at work, this is a natural place to intervene for obesity prevention and treatment. Rather than designing workplaces to reduce OPA and make our lives easier, efficiency experts should design them to allow more opportunities for physical activity. Attractive, accessible stairwells with point-of-decision prompts serve to increase stair climbing. In addition, many corporations now offer employee fitness or wellness programs, and some have even added monetary incentives to exercise. Promoting active commuting by providing bicycle parking, shower facilities, and free distant parking with walking trails into the workplace are all helpful in increasing the physical activity and fitness of employees. Promoting nonexercise activity thermogenesis by providing sit-to-stand desks and even treadmill work stations for office workers might also be beneficial.

Conclusion

In summary, OPA has declined over the past three centuries due to mechanization and labor-saving devices. In past times, hard physical labor provided the working class protection against obesity, while the wealthy, leisure class was more likely to get fat. Today, however, little physical labor is required of workers, and wealthier, more educated individuals who have the discretionary money and time to exercise are now leaner. While most public health efforts to increase physical activity focus on LTPA, there are also opportunities to increase OPA in the workplace.

48

Physical Activity Level and Mode of Transportation

Catrine Tudor-Locke, PhD

The undeniable growth in the study of the built-environment effects on physical activity (chapter 22) and obesity (chapter 49) has spawned a targeted interest in the unique role of personal transportation modes. The logical mechanism (presented in figure 48.1) underlying the specific attention to transportation is tied directly to its association with physical activity (i.e., by *active transportation* modes characterized as *self-transportation* including walking and cycling, and *public transportation* requiring at least minimal walking or cycling contributions for access to destination points) or conversely, physical inactivity (i.e., by *passive transportation* modes, primarily in a motorized vehicle taken from door to door); and in turn, the role that these behaviors play in the acknowledged global obesity epidemic. Active commuting provides a feasible and natural opportunity for integrating physical activity into daily life, and therefore its role in prevention of weight gain should be intuitive. However, insert the key words "physical activity," "transportation," and "obesity" into a PubMed online search (April

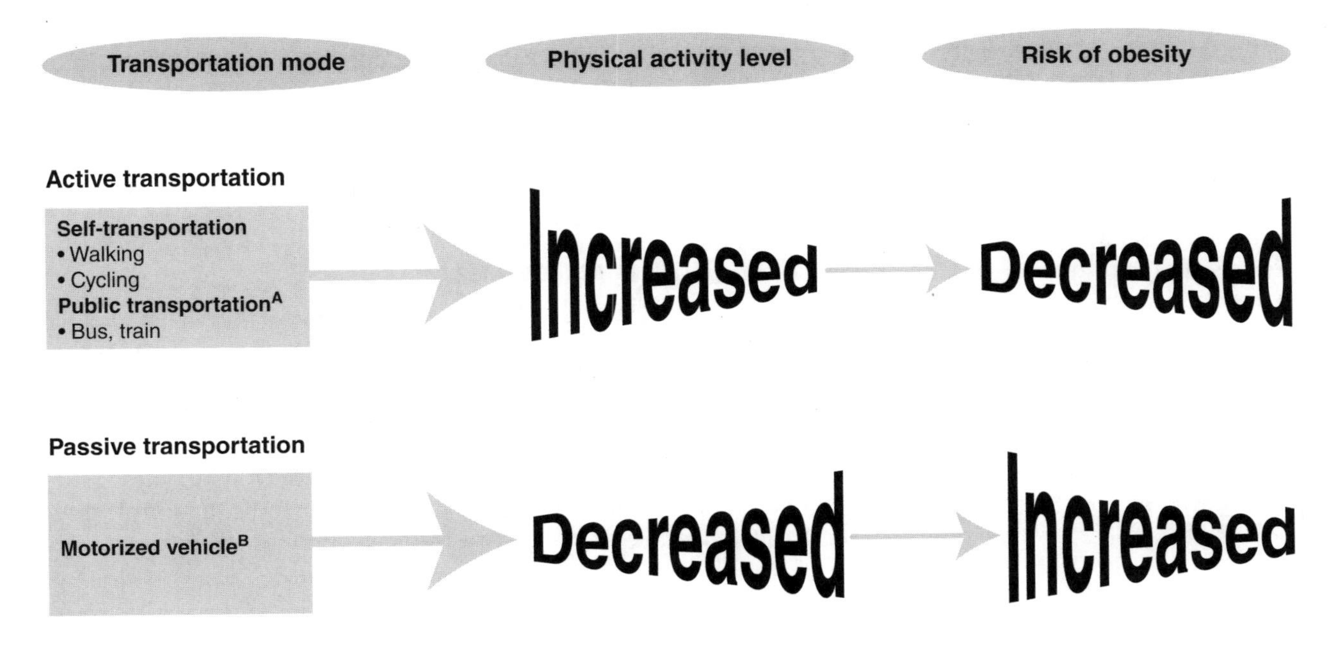

Figure 48.1 Logical mechanism [A]assumes self-transportation to destination point; [B]assumes door-to-door.

10, 2008), and only 47 articles are identified. Indeed, *Preventive Medicine* published a themed issue in early 2008 focused on self-transportation, public transportation, and health, heralding the emergence of this novel facet of obesity research. Although at least a third of the articles identified in the PubMed search just mentioned centered on children's daily commute to school, this brief chapter is limited to transportation, physical activity, and obesity highlights in adults.

Active Transportation

Active transportation, also known as active commuting, includes the use of public transportation and assumes at least some degree of necessary self-transportation (e.g., walking or cycling) in the process of arriving at a destination point.

Walking

In most national surveys, walking for exercise is consistently the most commonly reported leisure-time physical activity among adults. In contrast, walking for transportation is an incidental activity that is likely to be overlooked using these traditional survey methods.

Impact on Physical Activity

An examination of Australian Time Use data revealed that walking for transportation was indicated on a higher proportion of days compared to walking for exercise (20% vs. 9%) (1). Based on participant subsamples (i.e., "doers"; those actually performing the activity), walking for transportation occurred over 2.3 ± 1.4 bouts per day (12.5 min per bout) for a total of about 28 min per day, and walking for exercise over 1.2 ± 0.5 bouts per day (47 min per bout) for a total of about 56 min per day. Walking for transportation offers opportunities for repeated brief bouts of daily physical activity that are more frequently (i.e., number of days) performed compared with walking for exercise.

Impact on Obesity

Considering walking and cycling to work together, Lindstrom (2) showed that these active transportation behaviors were protective against overweight and obesity in both Swedish men (OR [odds ratio] = 0.62; 95% CI [confidence interval], 0.51 to 0.76) and women (OR = 0.79; 95% CI, 0.67 to 0.94) compared to a car driving reference category. Frank and colleagues (3) have reported that body mass index (BMI)-defined obesity around Atlanta was significantly associated with walking distance. Specifically, each additional kilometer walked per day was associated with a 4.8% reduction in the likelihood of obesity. The collected evidence suggests a direct preventive relationship between active transportation, specifically walking for transportation, and obesity.

Cycling

Bicycling for transportation provides an opportunity to cover longer distances (e.g., up to 5 km [3.1 miles] comfortably) than walking. Unfortunately, exposure to car traffic makes cycling for transportation a dangerous prospect in some areas. Furthermore, the relatively more intense effort, coupled with the associated equipment required for safe and comfortable cycling (e.g., bicycle, helmet, pump, weather-appropriate clothing, bicycle storage options), increases users' needs for adequate facilities (including shower and change facilities) at destination points.

Impact on Physical Activity

Cycling represents a unique opportunity for vigorous-intensity activity. Unfortunately, only 1% of transportation trips in 1995 were taken by bicycle in the United States compared with 20% in Denmark and 28% in the Netherlands (4). Not surprisingly perhaps, student bicycle commuters to an urban university attain more objectively monitored time in moderate-to-vigorous physical activity per day than those who commute by car (85.7 ± 37.0 vs. 50.3 ± 23.9 min, $p < 0.001$) (5). Cycling offers additional prospects for physical activity; however, relatively few people engage in this as a mode of transportation except in certain countries.

Impact on Obesity

Wen and Rissel (6) focused specifically on cycling to work and determined that men who cycled to work were less likely to be overweight or obese (compared to those who drove to work), with an adjusted OR of 0.49 (95% CI, 0.31 to 0.76). These relationships were not found in women. These findings are consistent with those indicating protective effects of active transportation on obesity.

Public Transportation

The term public transportation used here specifically refers to rail and bus systems serving members of the general public. It presumes a common contribution from self-transportation (i.e., walking or cycling) to reach destination points.

Impact on Physical Activity

Although only about 3% of Americans reported walking to and from public transportation services in the 2001 U.S. National Household Transportation Survey (NHTS), Besser and Dannenberg (7) estimated that these transit users accumulated a median of 19 min in total walking time and that almost a third exceeded 30 min. A study of commuters from New Jersey to New York determined that train commuters took on average 30% more steps per day than car commuters (8). Further, 40.4% of the train commuters achieved at least 10,000 steps per day (commonly accepted as active) compared to 14.8% of the car commuters (8). The self-transportation required to arrive at and depart from public transportation destination points can contribute to physical activity.

Impact on Obesity

Wen and Rissel (6) determined that Australian men (but not women) who used public transportation were less likely to be overweight or obese compared to a driving reference group (adjusted OR = 0.65; 95% CI, 0.53 to 0.81). Lindstrom (2) also showed that the protective effect of public transportation on overweight and obesity appeared to be significant in Swedish men (OR = 0.72; 95% CI, 0.61 to 0.86), but not women. At this time, the evidence supporting a connection between public transportation use and obesity is limited to men.

Passive Transportation

Motor vehicle dependency is a function of both ownership and use (e.g., annual distance traveled). Globally, the United States shows the highest dependence on motor vehicles, followed by Australian and Canadian cities, with European and Asian cities being more public transit oriented (9). The gap in motor vehicle dependence between U.S. cities and other cities internationally is more evident in terms of use rather than ownership; U.S. cities are 70% higher in car use (12,336 ± 1772 vehicle kilometers traveled) than their nearest rivals, the Australian (8034 ± 591) and Canadian (7761 ± 1257) cities (9).

Impact on Physical Activity

Wen and colleagues (10) reported that the 69% of Australians who reported driving to work in the 2003 New South Wales Health Survey were less likely to achieve recommended levels of physical activity than the self-reported non-car users (56.3% vs. 44.3%, respectively). As already noted, car commuters took approximately 2000 fewer steps per day getting from New Jersey to New York compared with train commuters (8). Use of personal motorized transportation reduces opportunities for physical activity.

Impact on Obesity

With regard to use, an Australian study showed that driving a private car to work was associated with being overweight or obese (adjusted OR = 1.13; 95% CI, 1.01 to 1.27) (10). Frank and colleagues (3) reported that each additional hour spent in a car per day in an Atlanta population was associated with a 6% increase in the likelihood of obesity (defined by BMI). Regular and incrementally excessive car use has been linked to the probability of obesity.

Summary

Increased physical activity realized through preferential selection of self- and public transportation modes as opposed to motorized vehicle options presents a largely untapped prospect for addressing population obesity. Specifically, there appears to be an abundance of missed opportunities to substitute transportation-related walking for the numerous short trips taken in the course of daily life. Although cycling presents a unique opportunity to actualize the accepted additional health benefits of vigorous physical activity, safety and convenience concerns limit its potential uptake for the time being, especially for women. Public transit use typically requires at least some duration of self-transportation while also permitting travel to longer-distance destinations. Restricting oneself to motor vehicle dependency creates a substantial physical activity deficit that would require concerted cognitive effort to replace on a daily basis through leisure-time pursuits. The evidence supports the underlying mechanism that active transportation (acting through increasing physical activity) is protective in terms of obesity whereas passive transportation (acting through decreasing physical activity) appears to be detrimental. Future research will illuminate determinants and barriers to active versus passive transportation with an eye toward shifting behaviors in favor of a more propitious energy balance.

49

Effects of the Built Environment on Obesity

Neville Owen, PhD

It is established that built-environment attributes can be associated with obesity. This chapter illustrates current evidence and highlights relevant conceptual and methodological nuances. Other chapters address other factors (endocrine and immune function, for example) that plausibly may contribute to built environment–obesity relationships.

The main emphasis of the present chapter is on understanding—in the context of built environment–obesity relationships—the role not only of lack of physical activity *(too little exercise)* but also of sedentary behaviors *(too much sitting)*. These distinct classes of behavior each can act uniquely to influence metabolic energy expenditure in ways that promote obesity. Understanding the *effects* of the built environment on obesity via these behaviors is an aspirational scientific goal. The majority of the evidence currently available is on cross-sectional relationships.

Associations of the Built Environment With Obesity

A recent review addressing the evidence on built environment–obesity relationships by Papas and colleagues (1) identified 20 research papers in which the associations of measured body weight with objective measures of the built environment were examined. Of those studies, 17 showed associations of one or more built-environment attributes with obesity; 19 showed only cross-sectional associations. Several of the studies reviewed examined associations of neighborhood *walkability* with obesity.

What is neighborhood walkability? Its key elements are argued to be proximity of destinations and street connectivity (2). *Proximity* is related to mixed land uses that create shorter distances between residences and destinations such as stores or workplaces. *Connectivity* is directness of routes to destinations, and more walking routes are available where there are grid patterns of interconnecting streets. These attributes are also related to *population density*, which tends to be higher in older, typically inner urban residential areas that have greater street connectivity; also, population density can determine the commercial viability of retail outlets that function as walking destinations. These elements are argued to be synergistic, and they influence the convenience of walking or biking distances between complementary activities. They may be assessed objectively through use of geographic information systems software and databases (2).

Papas and colleagues' (1) review shows that living in an area with the following built-environment attributes has significant associations with adiposity measures:

Either Higher Body Mass Index or Increased Odds of Overweight and Obesity

- A low-walkable neighborhood
- Greater distance to recreational, fitness, or play spaces

Acknowledgements
Thanks to Takemi Sugiyama for assistance with the preparation of chapter materials, for careful reading of drafts, and for helpful advice.

- Living on a highway or a street with no sidewalks
- High-sprawl neighborhoods
- Less mixed land use
- Lower levels of neighborhood greenery
- Living in rural and outer urban residential areas
- Access to physical activity facilities

Mixed Findings

- Greater proximity of supermarkets, grocery, and convenience stores
- Both higher and lower levels of population density
- Greater travel distances to shopping opportunities
- Higher levels of neighborhood crime

Where studies examined time spent in automobiles, there typically were significant relationships with lack of local destinations and with other attributes of urban sprawl, and positive associations with obesity (1). Sociodemographic attributes were found to have mixed and complex roles in moderating these relationships. The review, employing rigorous study selection criteria, identified several associations of different built-environment attributes with either higher body mass index (BMI) or increased odds of obesity, some of which were inconsistent (1). The recent overview and commentary of Brug and associates (3) and an earlier review (4) arrived at similar conclusions.

Research

The following are three examples of studies on built environment–obesity relationships. They are from diverse cities, illustrating some of the evidence that is currently available and highlighting some relevant research nuances.

■ *A study in Perth, Australia.* Giles-Corti and colleagues (5) surveyed a socially diverse sample of 1803 adults living in an Australian capital city. Physical environment features were either objectively determined through meticulous direct observation methods (e.g., type of street on which the respondent lived, presence of sidewalks, access to recreational facilities) or self-reported perceptions (e.g., whether there were walking or cycle paths within walking distance or a 5 min drive from the respondent's home, whether there was a store within walking distance). Body mass index was derived from self-reported height and weight. Both objectively assessed environmental attributes such as living on a busy highway or on streets with partial or no sidewalks, and the perceived distance of walking trails from home (e.g., walking trails not within a realistic walking distance) were significantly associated with being overweight. Other attributes associated with obesity included objectively assessed poor sidewalk quality and presence of garbage, perceived poor access to recreation facilities, and perceived lack of any store within walking distance. The authors identified the need for research on why directly observed and perceived factors associated with overweight and obesity can differ, and advocated the use of environmental and policy approaches in obesity prevention.

■ *A study in El Paso, Texas.* Rutt and Coleman (6) examined associations between objectively determined environmental attributes and BMI in a predominantly Hispanic adult sample (n = 452). Within the group studied, there was a statistically significant association between living in areas with *greater* land-use mix and obesity. This is in contrast to the findings of a number of other studies that have been reported. Rutt and Coleman explain this apparently anomalous finding of a positive relationship of land-use mix with obesity as a particular consequence of carrying out the study in a low- to moderate-income minority community. In such communities, it would be expected that those who are relatively more affluent would have more ready access to automobiles, more discretionary television viewing time, and also access to different food choices compared to the less affluent. The findings highlight the need to control for individual characteristics related to social disadvantage and neighborhood deprivation in studies of built environment–obesity relationships.

■ *A study in Atlanta, Georgia.* Frank and colleagues (7) examined the associations of built-environment attributes, neighborhood travel preferences, physical activity, and automobile use with obesity. Residents (2056 adults) of areas of Atlanta for which built-environment and transport infrastructure attributes could be characterized using geographic information systems software completed a travel behavior survey. There were significant relationships between objectively determined neighborhood walkability and obesity, but only after the study participants' preferences for living in high-walkable or automobile-dependent neighborhoods were taken into account. Those who did not have a preference for a walkable neighborhood environment did little walking, and there were no differences related to neighborhood walkability in obesity prevalence. However, among those who both preferred and lived in a walkable neighborhood, obesity prevalence was lower (12%) than in respondents with a preference for living in an automobile-dependent neighborhood (22%). Automobile use among those who both preferred and lived in a walkable environment was lower (26 miles or 42 km a day, on average) than among those who preferred and lived in an automobile-dependent environment (43 miles or 72 km a day), which suggests a contribution of sedentary time to the built environment–obesity relationship. Similar to the findings of an Australian study of the behavioral concomitants of neighborhood walkability (2), built environment had significant associations with obesity, but these were not independent of personal residential preferences. As others have argued (2-4), the findings of this study highlight the need to take into account personal factors (particularly preferences for different types of neighborhoods that may reflect broader behavior patterns and lifestyles) in studying built environment–obesity relationships.

Significance of Sedentary Behaviors

Generally, findings from studies of built environment–obesity relationships suggest consistently that lack of physical activity and poor food choices are important mediating influences (1, 2). However, time spent in sedentary behaviors (predominantly sitting, which is characterized by metabolic energy expenditure levels less than 2 METs) is also a distinct and important influence. An adult can walk for 30 min each day and be classified as sufficiently physically active for health benefits. However, within the balance of waking hours, there are substantial interindividual variations in metabolic energy expenditure; these result from the relative proportions of time allocated to sedentary behaviors (particularly driving an automobile or watching television) versus time allocated to other behaviors that require standing or light ambulatory activities. The study by Frank and colleagues (7) showed time spent in automobiles to be significantly associated with obesity.

Driving automobiles and watching television take up, on average, significant portions of adults' nonwork time, and the reduced energy expenditure associated with these ubiquitous sedentary behaviors is a significant contributor to obesity; for example, the Australian study described earlier showed that watching 3 or more hours of television a day was significantly associated with both overweight and obesity (5). The energy expenditure levels from these high volumes of sitting time are significantly less than those associated with light- and moderate- to vigorous-intensity activities (simply standing or walking for errands, for example) that have the potential to replace them.

Preliminary evidence indicates that built-environment attributes are related to sedentary behavior: A recent Australian study showed that women living in low-walkable communities had significantly higher odds of high daily volumes of television viewing time after the influences of area-level socioeconomic status and relevant sociodemographic attributes were statistically controlled for (8). Understanding how such environment–sedentary behavior relationships operate in the causal pathways to obesity provides ample scope for further research investigation.

A Better Understanding

Gaining a better understanding of the effects of the built environment on obesity requires investigation of *how* environmental attributes act to promote a lack of physical activity and increase the time spent in sedentary behaviors—particularly time spent sitting in automobiles, watching television, and using the computer and Internet. Models of obesogenic environments and ecological models of health behavior provide helpful, although somewhat broad, conceptual guidance. However, as has been pointed out in recent reviews and commentaries (3, 4), ecological models do not provide explanations of the *mechanisms* by which environmental factors act to influence the relevant behaviors. Future models of the determinants of obesity, populated by the relevant data, will identify the ways in which multiple, complex environmental factors are important; also, these models should be able to identify how environmental attributes interact with residential preferences (2, 7) and with social and personal factors (5, 6).

Such complexities must be accounted for in future studies. In this context of complex determination, there is a scientifically important place for parsimony. Such an approach (9) might usefully focus on the cues to action and the consequences of the behaviors that are associated with built-environment correlates of obesity. Several such candidates are identified in the cross-sectional studies described earlier (1, 4-7).

Walkable destinations and the attributes of the routes by which they may be accessed can provide several cues, such as *instrumental cues,* that relate to the feasibility and efficiency of walking or biking. Aesthetic attributes can provide *evaluative* and *affective cues* that may make active behavioral choices less attractive (potentially promoting preferences for watching television or driving an automobile rather than spending time outdoors or walking or biking to destinations). Built-environment attributes can provide *normative cues* about behavioral choices that are expected of people (e.g., large parking lots and the absence of sidewalks around shopping centers), as well as *cues about outcomes* of behavioral choices (prolonged waiting to cross busy roads, exposure to exhaust fumes). Ecological models of the determinants of particular health behaviors focused on such *behavior settings* factors, together with a behavioral choice framework (4, 9, 10), have the potential to provide relevant insights.

Conclusions

There are substantial conceptual and methodological challenges for studies of environment–obesity relationships (1-4, 9). The cross-sectional nature of much of the evidence that is available is the major limitation. Studies with prospective designs are needed. It would be particularly valuable to examine (using data from "natural experiments") how body weight and relevant behaviors change when people move from low- to high-walkable neighborhoods or from high- to low-walkable neighborhoods. Current evidence mainly has to do with how environmental attributes relate to physical activity rather than to sedentary behaviors. Studies on the "food environment" and individual and social factors related to food choices concurrently with physical activity would be valuable. Further, improved direct methods of assessing environmental attributes need to be developed; too many studies have relied solely upon self-reported perceptions of built-environment attributes as the exposure variable. Conceptual approaches for identifying the mechanisms by which the built environment might act to influence obesity require further development. The environment–obesity research field is new and interdisciplinary and shows great promise. It includes researchers from epidemiology, behavioral science, urban planning and transportation research, and several other disciplines, working collaboratively and developing integrated methodological approaches that will facilitate the future comparability of study findings (1, 9).

50

Socioeconomic Status and Obesity

Youfa Wang, MD, PhD

Numerous studies have examined the relation between socioeconomic status (SES) and obesity in various population groups across countries (1-10). Such studies not only can help in advancing understanding of the distribution of the obesity problem in populations and the causes of obesity, but also can assist in developing effective intervention programs for the prevention and management of obesity. During recent years, the growing global obesity epidemic has fueled interest in the complex relation between SES and obesity as revealed by earlier research. Overall, the current literature shows that obesity is related to SES; however, the association may vary by gender, age, and country as well as by the type of SES measures being examined. Thus far, several comprehensive reviews have summarized findings from related studies (1-6).

Global Perspectives on Adults

The first landmark comprehensive review, conducted by Sobal and Stunkard (2) in 1989, has been widely cited in the obesity literature. The authors reviewed findings from 144 studies, noting that most of the researchers reported on the relationship only in the course of investigating other factors. Sobal and Stunkard observed that the relationship between SES and obesity differed between developing and developed countries, as well as according to gender. Developed societies showed a consistently inverse association between SES and obesity for women, while the relation for men and children was inconsistent. In developing societies, a strong positive relation was observed for women, men, and children. Recently, a comprehensive study has sought to update the Sobal and Stunkard review (1).

McLauren (1) reviewed studies published during the years 1988 through 2004, aiming to expand the scope of the 1989 paper by using a three-category format (Human Development Index, HDI) of low, medium, and high national economic development, as well as more detailed SES indicators (eight SES categories in all). The review identified 333 relevant published studies representing 1914 primarily cross-sectional associations. In addition to documentation of the absolute data, correlations were made between body size and HDI, resulting in no association, an inverse association (high body size with low HDI or low size with high HDI), or a positive association (high with high, etc.). On the basis of country and sample, McLauren classified the level of development in each study as high, medium, or low using the 2003 HDI assigned by the United Nations Development Program. Analysis of these associations revealed several main findings (table 50.1).

First, patterns for women in higher- versus lower-developed countries were generally less striking than those observed in the 1989 review; that is, the association became weaker over time. This is consistent with our analyses of comparable data collected in the United States between 1971 and 2002 (8-10). It was hypothesized that this change may have been due to globalization of food markets, economic growth, and so on.

Second, the overall pattern of results was of an increasing proportion of positive associations (e.g., high weight with wealth) and a decreasing proportion of inverse associations (low weight with wealth) as one moved from countries with high HDI to low HDI (that is, in poorer countries as compared to more prosperous countries, adiposity rather than leanness was more strongly associated with high SES).

Table 50.1

Associations Between Socioeconomic Status (SES) and Body Size, According to Different Indicators*

SES indicator	Positive		Negative		Nonsignificant or curvilinear		Total	
	N	%	N	%	N	%	N	%^
Men and women							280	
HIGH HDI								
Education	1	2	31	65	16	33	48	26
Employment	0	0	1	13	7	88	8	4
Income	5	8	19	31	38	61	62	34
Occupation	1	4	16	59	10	37	27	15
Overall	10	5	86	47	87	48	183	100
MEDIUM HDI								
Education	8	32	7	28	10	40	25	29
Employment	1	50	1	50	0	0	2	2
Income	17	63	3	11	7	26	27	31
Occupation	3	38	1	13	4	50	8	9
Overall	43	49	14	16	30	34	87	100
LOW HDI								
Education	0	0	0	0	1	100	1	10
Employment	0	0	0	0	0	0	0	0
Income	5	71	0	0	2	29	7	70
Occupation	1	50	0	0	1	50	2	20
Overall	6	60	0	0	4	40	10	100
Women							939	
HIGH HDI								
Education	4	1	220	72	81	27	305	42
Employment	7	16	17	38	21	47	45	6
Income	9	6	69	49	64	45	142	19
Occupation	2	1	100	68	44	30	146	20
Overall	23	3	457	63	251	34	731	100
MEDIUM HDI								
Education	32	31	39	38	31	30	102	59
Employment	0	0	0	0	0	0	0	0
Income	24	71	1	3	9	26	34	20
Occupation	1	17	3	50	2	33	6	3
Overall	75	43	46	26	52	30	173	100

SES indicator	Positive		Negative		Nonsignificant or curvilinear		Total	
	N	%	N	%	N	%	N	%^
LOW HDI								
Education	31	100	0	0	0	0	31	89
Employment	0	0	0	0	0	0	0	0
Income	1	50	0	0	1	50	2	6
Occupation	1	100	0	0	0	0	1	3
Overall	33	94	0	0	2	6	35	100
Men							695	
HIGH HDI								
Education	14	6	126	50	114	45	254	45
Employment	3	9	2	6	28	85	33	6
Income	20	24	12	14	51	61	83	15
Occupation	7	6	49	39	70	56	126	22
Overall	53	9	209	37	302	54	564	100
MEDIUM HDI								
Education	12	24	3	6	35	70	50	39
Employment	0	0	0	0	3	100	3	2
Income	26	59	1	2	17	39	44	34
Occupation	4	27	3	20	8	53	15	12
Overall	50	39	8	6	70	55	128	100
LOW HDI								
Education	1	100	0	0	0	0	1	33
Employment	0	0	0	0	0	0	0	0
Income	1	100	0	0	0	0	1	33
Occupation	1	100	0	0	0	0	1	33
Overall	3	100	0	0	0	0	3	100

*Based on data from McLauren (1); body size includes both continuous (e.g., body mass index, BMI) and categorical (e.g., obesity defined as BMI ≥30) measures; some studies may contain multiple associations. The percentage values apply to each SES indicator and should be read across each row; based on country and sample, the level of development in each study was classified as high, medium, or low according to the 2003 Human Development Index (HDI) assigned by the United Nations Development Program (www.undp.org) to characterize and rank countries on a number of attributes, including life expectancy at birth, school enrollment and adult literacy, and standard of living based on the gross domestic product. High-HDI category included countries such as the United States; medium-HDI, countries such as Brazil and Saudi Arabia; and low-HDI, Cameroon and Zambia. For consistency with the 1989 review (2), traditional subcultures within a larger developed society were classified as being at a lower stage of development. In this case, American Indian subgroups were classified as having a medium HDI (one instance each) although the studies took place in the United States (high HDI).

^Percentages apply to the entire HDI category and should be read down the column; the percentages may not add up to exactly 100% because of rounding.

This review showed that research findings varied by SES indicators; for example, the largest associations for women in high-HDI countries were with education and occupation (inverse); in contrast, the largest associations for women in medium- and low-HDI countries were with income and material possessions (positive). Across all HDI strata, education was the SES indicator most often studied (e.g., 47% of all associations in women). A percentage value was derived from the number of positive, inverse, or no associations compared with the total for the given indicator. However, the indicators were not weighted by their proportion of the whole. For example, in high-HDI men, 40% of the associations with possessions were inverse. This was similar to the percentage with occupation (39% inverse); however, there were only 10 observations with possessions and 126 with occupation.

Both Sexes

In high-HDI countries, inverse and nonsignificant or curvilinear associations were approximately equally common (47% vs. 48%), and both were more common than positive associations (5%). Inverse associations were most often observed with education (65%; 31/48 observations), occupation (59%; 16/27), and area-level indicators of SES (52%; 17/33), whereas nonsignificant or curvilinear associations were most often observed for employment (88%; 7/8). For medium- and low-HDI countries, positive associations were more common (49% and 60%, respectively); and in medium-HDI countries, this reflected a large proportion of positive associations with income (63%; 17/27) and area-level indicators of SES (62%; 13/21). In low-HDI countries, the greatest association was with income (71% positive, 5/7).

▪ *Women:* For women in high-HDI countries, the majority of associations (63%) were inverse, and these were especially prominent for education (72%; 220/305), area-level indicators (71%; 10/14), occupation (68%; 100/146), and composite indicators (67%; 31/46). From high to medium to low HDI status, the proportion of positive associations increased from 3% (23/731) to 43% (75/173) to 94% (33/35), respectively. In medium-HDI countries, this positive association was particularly prominent for income (71%; 24/34) and material possessions (86%; 12/14). In low-HDI countries, the vast majority of associations were with education (89%; 31/35), all of which were positive (100%; 31/31).

▪ *Men:* In high- and medium-HDI countries, the predominant findings (based on associations >55%) were nonsignificant or curvilinear. This was particularly true for associations in high-HDI countries with employment (85%; 28/33), income (61%; 51/83), occupation (56%; 70/126), and area (100%; 3/3) and with education (70%; 35/50), area (60%; 3/5), material possessions (80%; 4/5), and employment (100%; 3/3) in medium-HDI countries. The association in high-HDI countries with education was inverse (50%; 126/254). Positive associations were uncommon and weak among studies in high-HDI countries but could be seen in medium-HDI countries (income, 59% [26/44]; composite indicators, 83% [5/6]). Low-HDI countries provided only three associations (education, income, and occupation), and all were positive (100%).

Another review (4), which focused on studies conducted in adult populations from developing countries that were published between 1989 and 2003, argued that obesity in the developing world tended to shift toward the groups with lower SES as the countries' gross national product (GNP) increased; this is consistent with McLauren's findings (1). Obesity could no longer be considered solely a disease of high-SES groups. The shift of obesity toward women with low SES occurred at an earlier stage of economic development than it did for men. The crossover to higher rates of obesity among women of low SES began much sooner than for men and was found at a GNP per capita of about U.S.$2500, the midpoint value for lower-middle-income economies.

Socioeconomic Status and Weight Change in Adults From Developed Countries

Although cross-sectional studies can provide useful evidence regarding the existence of a relationship between SES and obesity, they cannot prove a causal relationship: Obesity status may affect some SES indicators, as it has been reported that education, occupation, and income may be adversely affected by overweight (3). Therefore, longitudinal data are more desirable for studying causal relationships (i.e., studying the relationship of SES to weight change over time). A recent review (3) identified 34 relevant articles, with a total of 135 distinct tests of the associations in developed countries, on the relationship of various measures of SES with weight change over time in adults. Overall findings of these studies supported the hypothesis that lower SES is associated with greater likelihood or amount of weight gain, lower likelihood or amount of weight loss, or lower likelihood of weight maintenance, although there were racial differences (discussed next) between blacks and nonblacks. The methodologically strongest studies (objective adiposity data with a follow-up period of four years or more) showed relatively consistent inverse associations between occupation and weight gain for men and women among nonblack samples. Of 70 such studies, 43 (61.4%) supported the hypothesis, with 25 (35.7%) showing no association between SES and the outcomes and two (2.9%) supporting the opposite hypothesis that lower SES is associated with a lower likelihood of weight gain or obesity. When SES was assessed using education, the evidence was slightly less consistent. When income was used, the findings were inconsistent, perhaps due to the difficulties of accurate income assessment and the few studies that used income.

However, large ethnic differences existed, with almost a perfect reversal in black men. In contrast to 61.4% of the studies supporting the association and 35.7% showing no association in nonblacks, 32.2% (19 tests) in blacks supported the hypothesis, 62.7% (37 tests) showed no association, and 5.1% (three tests) supported the reverse hypothesis. In black women, 6.9% (four tests) supported the reverse hypothesis, and the remainder were equally divided: 46.9% (30 tests) were supporting and 46.9% showed no association. Finally, five studies presented results only with the two sexes combined (11 tests); four tests supported the hypothesis, one supported the reverse hypothesis, and six showed no association.

A Global Perspective of Children in Developed Countries

A growing number of studies have examined the associations between SES and obesity in children and adolescents, in particular in developed countries, while studies from developing countries are still very limited. The 1989 review examined 34 studies from developed countries published between 1941 and 1989 (2) and indicated inconsistent associations between SES and childhood adiposity: inverse association, 36%; no association, 38%; and positive association, 26%. Thus, based on these earlier studies (the proportions were similar), it is difficult to conclude that there is a clear relationship between SES and childhood obesity.

Most recently, Shrewsbury and Wardle (5) examined 45 studies from Western developed countries, published since 1989, that tested the association between SES and adiposity in childhood (5 to 18 years). A main finding of this systematic review was that, compared to results in the 1989 review, the associations between SES and adiposity in children have become predominately inverse and positive associations have disappeared. None of the studies reported positive associations in their unadjusted analyses; 19 studies (42%) showed an inverse association, 12 (27%) showed no association, and 14 (31%) showed a mixture of no associations and inverse associations across subgroups. Some of these studies used such SES indicators as parental education, parental occupation, family income, composite SES, and neighborhood SES. Findings of the investigations that used parental education showed more consistent results; 15 of these 20 studies (75%) indicated an inverse association.

Recent Findings in the United States

The United States has excellent nationally representative data collected since the 1960s, including the National Health and Nutrition Examination Survey (NHANES) series, the Behavioral Risk Factor Surveillance System (BRFSS), and the Youth Risk Behavior Surveillance System (YRBSS), among others (6), that can be used to study the association between obesity and SES in both children and adults. This can provide useful insights especially in developed countries.

Our team has carried out a number of studies on the association between SES and obesity, as well as the changes in the association over time in the United States in adults (8), children, and adolescents (7, 9, 10) based on regular statistical analysis approaches (7, 9) and innovative health economic approaches (10). These investigations are predominately based on nationally representative data from the multiple rounds of the NHANES conducted since 1971 (NHANES I, 1971-1974; NHANES II, 1976-1980; NHANES III, 1988-1994; and NHANES 1999-2002). We found that the associations are complex, varying by gender, age, and ethnic groups, and that they have changed over time. The associations weakened over the past three decades when the prevalence of obesity increased dramatically. There are considerable variations in the changes in the associations across gender and ethnic groups (7-10).

In adults, the disparity in obesity across SES based on education has decreased since the 1970s (see figure 50.1). In NHANES I (1971 to 1974), there was as much as a 50% relative difference in obesity prevalence across the low-, medium-, and high-SES groups; but by 1999-2000, this had decreased to 14%. The trend was more pronounced in women. The trends of diminishing disparities in obesity were

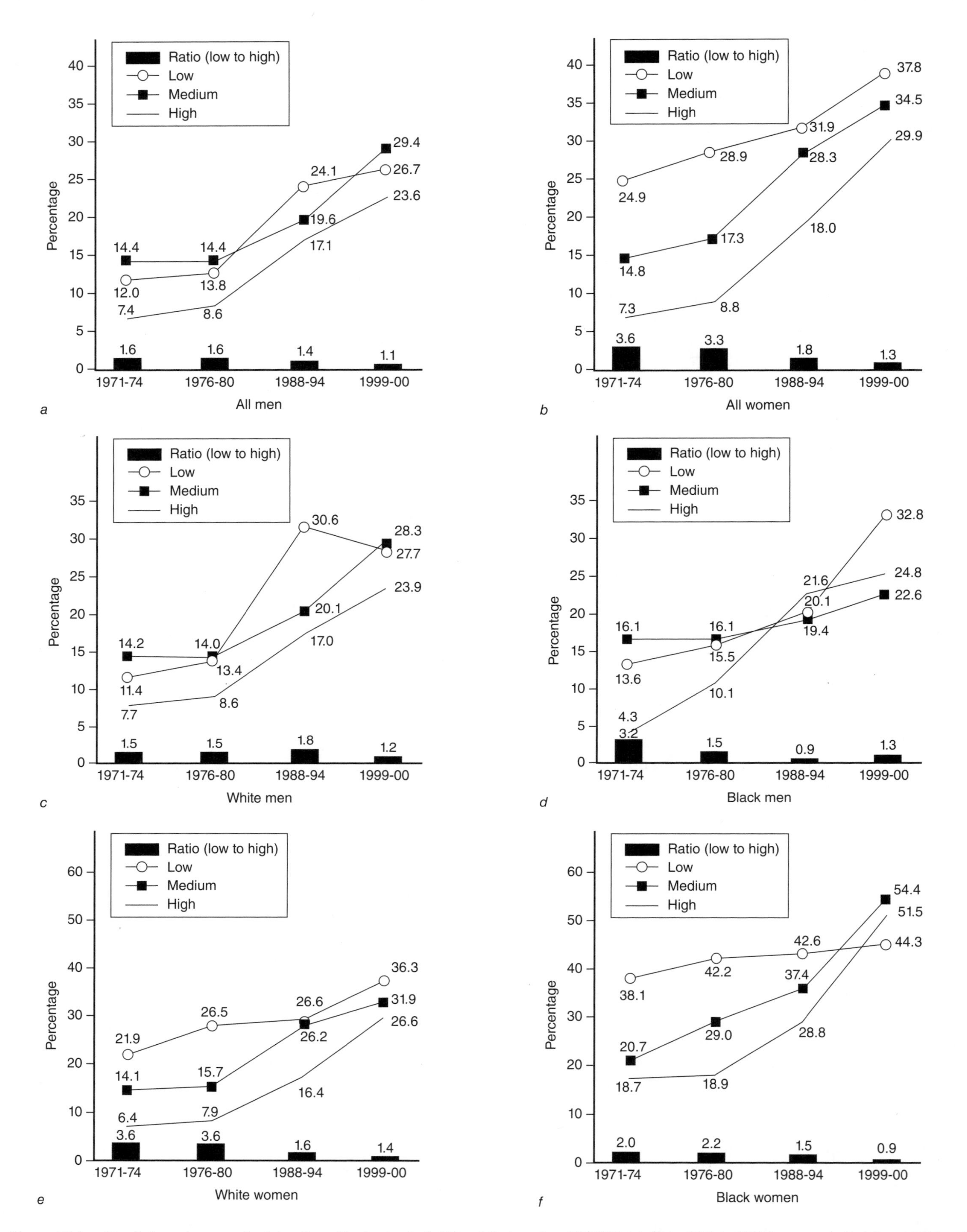

Figure 50.1 Trends in socioeconomic status disparities in obesity in U.S. adults based on NHANES data from 1971 to 2000 by sex and ethnicity: prevalence (%) obesity (BMI ≥30 kg/m^2) and low- to high-SES ratio of the prevalence. *Ratio = prevalence in the low-SES group/prevalence in high-SES group; based on ref 6 and

Youfa Wang et al., "The obesity epidemic in the United States gender, age, socioeconomic, racial/ethnic, and geographic characteristics: A systematic review and meta-regression analysis, *Epidemiology Reviews* Vol. 29, pg. 6-28, by permission of Oxford University Press.

also revealed by our logistic and linear regression analyses. The odds ratios (ORs) converged to 1 from the 1970s to 2000. In most sociodemographic groups, the relationship between body mass index (BMI) and SES (coefficients in linear regression analysis models) has weakened over time.

For U.S. children and adolescents, on the basis of NHANES data collected between 1971 and 2002, we found that the patterns of SES disparity of overweight varied across age, ethnic, and gender groups and have changed over time. Socioeconomic status disparities have decreased since the early 1990s with the rise of the obesity epidemic. An inverse association existed only in white girls; African American children with high SES were at increased risk. Socioeconomic disparities in overweight have changed over time, showing an overall trend of weakening association. Compared with figures for the medium-SES group, the adjusted ORs and 95% confidence intervals were 0.79 (0.47, 1.33), 1.08 (0.73, 1.61), 1.24 (0.73, 2.09), and 1.04 (0.82, 1.33) in NHANES I, II, and III and in the 1999-2002 NHANES for the low-SES group; they were 0.66 (0.43, 1.00), 0.60 (0.35, 1.03), 0.42 (0.23, 0.76), and 0.99 (0.68, 1.43) for the high-SES group. Between 1988-1994 and 1999-2002, the ratio in the prevalence of overweight between adolescent boys with a low or high SES decreased from 2.5 to 1.1 and between girls from 3.1 to 1.6. Similarly, for U.S. adolescents, our analysis using concentration index (which ranges between −1 and +1; the closer to 0, the smaller the inequality or the weaker the association) detected changes in the SES inequality of overweight and showed considerable gender and ethnic differences. For boys, the concentration index varied from 0.04 in NHANES I to −0.04 in NHANES 1999-2002; for girls, the concentration index varied from −0.12 in NHANES I to −0.18 in NHANES III. Among whites, SES disparity peaked in NHANES III and declined thereafter. Patterns in black and Mexican American adolescents were mixed.

Conclusions

In summary, the studies published over the past several decades show complex patterns in the relation between SES and obesity, with considerable differences between population groups and countries, as well as changes over time. More research needs to investigate the mechanisms through which SES may affect obesity. Recent findings may indicate that individual characteristics such as SES may play a smaller role than expected in the current global obesity epidemic and that broad social-environmental factors play an important role. In developed countries, strategies for obesity prevention and management should target all SES groups from a societal perspective, while in developing countries immediate efforts are needed to target groups of higher SES. In the context of the worldwide epidemic of obesity, design and implementation of consistent public policies on the physical, economic, and sociocultural environments that make healthier choices of food and physical activity feasible for all should be supported.

51

Physical Activity Level and Dietary Intake

Conrad P. Earnest, PhD

Obesity is a well-recognized consequence of energy mismanagement in humans. In its classic sense, energy balance (EB) is the sum of energy intake (EI) minus energy expenditure (EE). For decades, this relationship has been a foundation and running theme for research efforts aimed at producing and maintaining weight loss. Indeed, strategies for obesity treatment often include manipulations of diet and exercise. In a recent meta-analysis of long-term weight loss after diet and exercise, Curioni and Lourenco (1) looked at the effectiveness of dietary interventions and exercise for long-term weight loss in overweight and obese people. In their report, the authors examined overweight and obese adults randomized to clinical trials using diet and exercise interventions versus diet alone. An important aspect of this analysis was that all trials included a follow-up of one year after intervention; whereby the authors found that diet combined with exercise, versus diet alone, was associated with 20% greater initial weight loss following the intervention as well as a 20% greater sustained weight loss after one year. Nonetheless, despite the initial success of this combined therapy, almost half of the initial weight loss was regained in all groups after one year.

Despite the contribution of exercise to EB, results from various clinical trials have been somewhat disappointing with regard to weight loss when expected weight loss has been compared to the actual weight loss of participants attempting to lose weight using an exercise intervention. Specifically, the actual versus the expected weight loss in many trials shows a high degree of variance and the intervention is not always effective for all participants. This has led to the hypothesis that during periods of exercise-induced EE, compensatory mechanisms exist that offset the energy deficit imposed by the exercise stimulus. Individuals who defend their body weight by adjusting for the exercise-induced EE are termed "compensators" (2).

What Is Compensation?

The compensation construct is not new, but stems from the set-point theory proposing that body weight will return to baseline after an imposition on EB. According to the hypothesis proposed in 1980 by Epstein and Wing (3, p. 385), "Exercise may stimulate the appetite so that persons who exercise increase their eating and do not lose as much weight as expected." While a specific mechanism for compensation has not been clearly pinpointed, several potential mechanisms have been identified that help to explain this phenomenon. It is clearly possible that no one mechanism will explain the variability of all compensatory weight loss responses but that instead the answer lies in a combination of physiologic mechanisms and behavioral mechanisms.

An example of the potential for variable response is nicely demonstrated by King and colleagues (2), who identified and characterized the individual variability in compensation for exercise-induced changes in EE during a 12-week exercise intervention. In this study, overweight and obese sedentary men and women were prescribed exercise five times per week for 12 weeks under supervised conditions. Body mass and anthropometry, resting metabolic rate, total daily EI, and subjective appetite sensations were measured at weeks 0 and 12. When all subjects' data were pooled, mean body mass was significantly reduced by about 3.7 kg (8 lb) as predicted, which suggested no compensation for the increase in EE. However, examination of the results showed that participants within this study cohort demonstrated a large degree of individual variability in weight change, ranging from −14.7 to +1.7 kg (−32 to +3.7 lb).

Further analysis showed that when participants were classified as either compensators (actual weight loss less than predicted weight loss) or noncompensators (actual weight loss more than or equal to their predicted weight loss), compensators and noncompensators were characterized by different metabolic and behavioral compensatory responses, such as a trend toward decreased resting metabolic rate for the compensators and a small increase in the noncompensators. In addition, EI and average daily hunger sensations increased in the compensation group, whereas EI decreased in the noncompensating group.

Does Dieting Mean a License to Eat?

When one examines the effectiveness of physical activity and dietary intake for weight loss, it is easy to conclude that eating behavior can sabotage the most well-intended weight loss goals. However, eating behaviors are more complex than simply "making poor food choices." Muddying the waters are other phenomena that may lead to an "overcorrection" of the EE imposed by participation in exercise. Candidate mechanisms for eating behaviors include increased EI due to snacking, the ingestion of larger meals or larger portion sizes, and choice of foods that offer increased caloric density for a given food volume. This latter phenomenon leads to an interesting caveat, as increased EE may create a potential drive for high-fat, energy-dense foods in some individuals. For example, in lean subjects undertaking strenuous exercise (70% $\dot{V}O_2max$), King and Blundell (4) found that despite an initial suppression in appetite following vigorous exercise, participants consumed more calories when presented with high-fat, low-carbohydrate foods. Interestingly, there was no difference in the total food volume eaten when compared to that of a low-fat, high-carbohydrate diet. This study shows that food choice following vigorous exercise can completely reverse the negative EB imposed by an exercise session and may have important public health care implications, as it deconstructs the notion that exercise is a "magic pill" that creates a "license to eat."

Reinforcing the observations of King and Blundell, Tremblay and colleagues (5) also examined the effects of exercise on EB under dietary conditions differing by the relative lipid content of foods. In this study, nine healthy men performed 60 min of treadmill exercise followed by a 48 h observation period during which they maintained their habitual daily activities and had free access to a low-fat diet, a mixed diet, or a high-fat diet. In examining EB, defined as the sum of excess EE during exercise and a 48 h period of postexercise energy monitoring, the authors observed a very distinct difference between EI when participants had access to low-fat (−6.4 MJ; −1528 kcal), mixed-energy (−4.5 MJ; −1074 kcal), and high-fat diet conditions (0.9 MJ; 2.5 kcal).

Murgatroyd and colleagues (6) examined EE and substrate oxidation on five occasions over two days and ad libitum food intake with diets of 35% energy as fat and 60% energy as fat, both with imposed activity and without imposed activity, and a fixed overfeeding condition at 35% dietary fat with free activity in eight male volunteers. The aim of the study was to investigate the relationship between macronutrient composition and physical activity on energy metabolism and fat balance. The investigators showed that participants consumed the same energy (11.6 MJ/day, 2770 kcal) regardless of activity level on the 35% fat diet. However, participants consumed more energy on the 60% fat diet (14 MJ; 3343 kcal) than on the 35% fat diet. These results imply that exercise can induce a substantial energy deficit when accompanied by a low-fat diet, and further suggest that the increase in EI associated with a high-fat diet is sufficient to fully compensate (or overcompensate) for the energy deficit resulting from the energy cost of exercise and the increase in postexercise EE.

Does Exercise Volume Matter?

When we consider the use of exercise as an adjunct to creating an energy deficit, several subsets of questions become evident. One question concerns the potential differences between genders. A second question pertains to the intensity of the exercise stimulus undertaken by the exercise participant. In essence, it could be argued that higher exercise intensities may impose a greater physiologic stress that in turn would provoke a greater degree of compensation as the body seeks to defend strenuous perturbations in EB. For example, we recently found that postmenopausal women exercising at 150% of the physical activity dose recommended by the National Institutes of Health Consensus Development Panel showed a markedly lower degree of weight loss than predicted compared to women exercising at 50% and 100% of the recommendation (figure 51.1).

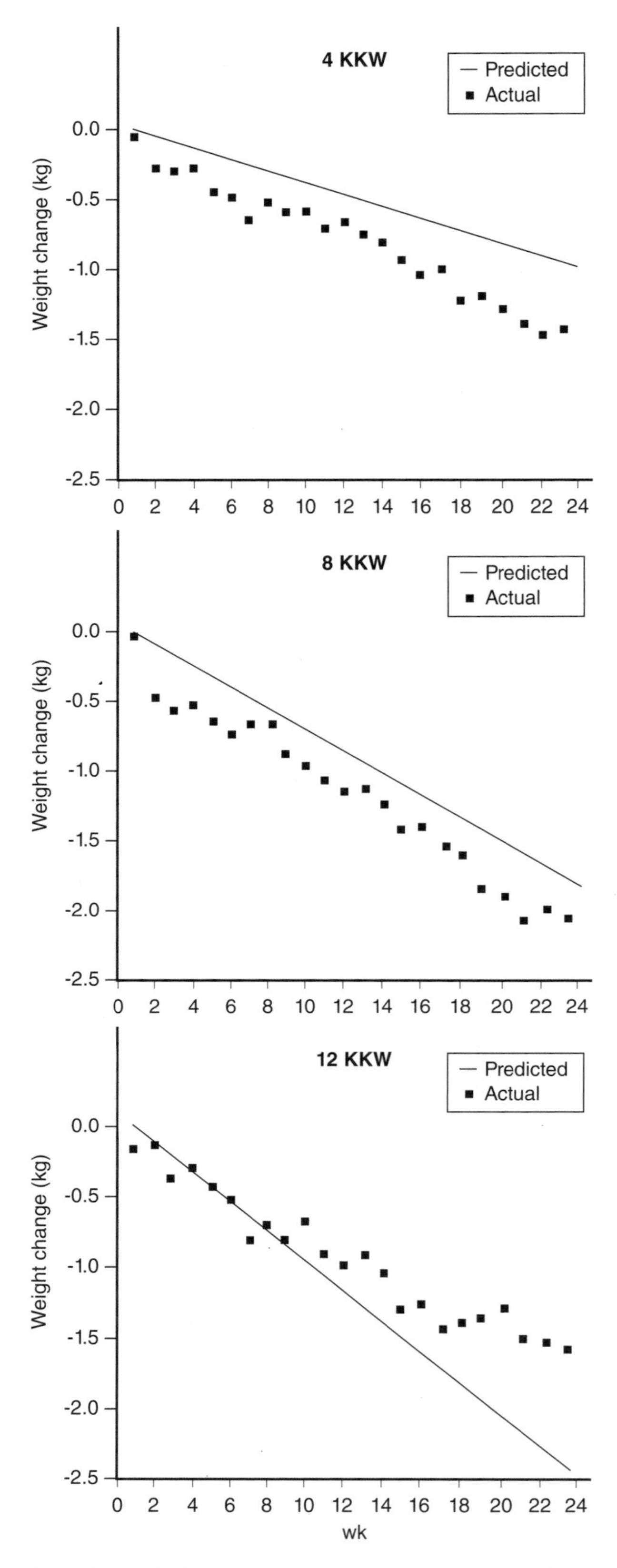

Figure 51.1 Data representing expected versus actual weight loss in the Dose-Response to Exercise in Women (DREW) trial.

Adapted from T.S. Church et al., 2007, "Effects of different doses of physical activity on cardiorespiratory fitness among sedentary, overweight or obese postmenopausal women with elevated blood pressure: A randomized controlled trial," *Journal of American Medical Association* May 16 297(19): 2081-2091.

In an effort to examine the effect of differing exercise levels on the EI of free-living women, Stubbs and colleagues compared a no-exercise control group to groups that performed a lower (1.9 MJ/day; 453 kcal/day) or a higher volume of exercise (3.4 MJ/day; 812 kcal/day) for seven days (7). During the study, participants ate ad libitum and self-weighed their food, also recording hourly sensations of hunger and appetite. Despite the greater EE of the high-exercise group, the authors observed only weak treatment effects for hunger. Nonetheless, EI values increased in a significant stepwise fashion for the nonexercise control group (8.9 MJ/day; 2125 kcal/day) to the moderate-level exercise (9.2 MJ/day; 2197 kcal/day) to the high-level exercise group (10.0 MJ/day; 2388 kcal/day). Thus, even though EE increased via exercise, it also produced a partial compensation in EI.

In a second study, the same research group used a similar approach in free-living men (8). Six men were examined three times during a nine-day protocol including prescriptions of no exercise, medium exercise level (1.6 MJ/day; 382 kcal/day), and higher exercise level (3.2 MJ/day; 764 kcal/day). Similar to findings from the previous study, EE increased throughout the day primarily due to the exercise regimen. However, contrary to their female counterparts, the men in this study did not show a compensatory increase in hunger, appetite, or body weight with EI, which was nearly identical between the control (11.2 MJ/day; 2675 kcal/day) and exercise conditions (11.6 MJ/day; 2770 kcal/day). Thus, during short-term exposure to high EE conditions, lean men appear able to tolerate a considerable negative EB through exercise without invoking compensatory increases in EI.

Refining Energy Balance

Overall, physical activity increases EE and in turn can modulate EB in favor of weight loss via the creation of an acute energy deficit. With continued research efforts, it has become increasingly apparent that exercise-induced weight loss is not as simple a matter as introducing exercise into the EB equation. Though the EB equation holds, in theory, physiologic and behavioral manifestations point toward those individuals who compensate for their exercise-imposed energy deficit through several potential mechanisms. Attempting to account for these mechanisms is clinically important in order to counteract the increasing prevalence of obesity, as the elaboration of these compensatory responses will influence the weight loss associated with programming efforts aimed at creating an energy deficit. Though some may be tempted to eschew exercise as a means of facilitating weight loss, perhaps a more prudent approach is to attempt to account for potential compensatory mechanisms that may disrupt the weight loss process. Further, it should be pointed out that despite the presence of compensatory mechanisms surrounding exercise and weight loss during short-term studies, longer studies are still missing that might attest to the weight loss once exercise becomes more habitual with a greater amount of time for adaptation. That is, longer periods of consistent exercise may allow for more consistent weight loss that is more resilient to EB perturbations.

As an example, Skender and colleagues (9) compared two-year weight loss trends using diet, exercise, and a combination of the two interventions. They examined two-year follow-up data in 127 men and women who were overweight, lived in an urban community, and had been assigned randomly to the experimental conditions. The dietary intervention for this study was a low-energy eating plan aimed at producing 1 kg (2.2 lb) per week of weight loss. The exercise component involved training in walking and a home-based program of up to five exercise periods per week. After one year, no significant differences were noted among the three groups, as the diet-only group lost 6.8 kg (15 lb), the exercise-only group lost 2.9 kg (6.4 lb), and the combination group lost 8.9 kg (19.6 lb). During the second year, the diet-only group regained weight and reached a weight that was about 1 kg above baseline. However, the combination group regained less weight, reaching a level 2.2 kg (4.9 lb) below baseline, while the exercise-only group regained to a level 2.7 kg (6 lb) below baseline. Though subjects in both exercise conditions regained weight after two years, the results suggest that dieting alone is not associated with long-term weight loss and that the addition of exercise produces a better net weight loss after two years due to a smaller amount of weight regain.

In a similar fashion, Fogelholm and colleagues (10) examined the effects of walking on weight maintenance after a period of very low EI in premenopausal obese women with the aim of maintaining weight loss following a low-calorie intervention. A total of 82 participants initiated a 12-week walking training program of moderate intensity after weight reduction via a low-EI diet. After the exercise intervention, the women followed a 40-week maintenance program: Participants were randomized to a control group consisting of no increase in habitual exercise and counseling focused on diet and relapse prevention; a walk group targeted to expend 4.2 MJ per week (about 1000 kcal per week), coupled with diet counseling; and a walking group assigned to an exercise walking program of 8.4 MJ per week (about 2000 kcal per week), coupled with diet counseling. After the intervention, the subjects were followed up for two years. The mean weight loss after weight reduction was 13.1 kg (29 lb). Compared with the end-of-study weight reduction, weight regain at the two-year follow-up was 3.5 kg (7.7 lb) less (95% CI, 0.2 to 6.8) and waist circumference regain 3.8 cm less (95% CI, 0.3 to 7.3) in the 4.2 MJ per week group versus controls. No statistical differences were noted for the 8.4 MJ group. In a study that yielded findings contrary to these, Borg and colleagues (11) observed no effect for weight maintenance with the addition of resistance training to walking when examining weight loss maintenance in obese, middle-aged men. However, the authors did note that the primary reason cited for weight regain was poor long-term adherence to prescribed exercise.

Summary

Overall, physical activity and the dietary manipulation of EE provide a stimulus sufficient to promote weight loss and weight management. It is also reasonable to conclude that when exercise and diet become habitual behaviors, the maintenance of weight loss is achievable and sustainable. However, behavioral changes, such as continuing to exercise and making better dietary choices, are difficult to maintain for many individuals over time as evidenced by the high degree of relapse or weight regain experienced by many. An important body of evidence has begun to better define those individuals who are nonresponders to changes in EE via a variety of compensatory mechanisms. While some of these compensatory mechanisms are purely behavioral, some people appear to compensate because of physiologic mechanisms that are not fully understood. An improved understanding of the underlying physiological mechanisms will help better identify and benefit those individuals who desire to lose and maintain weight using exercise.

part VII

Physical Activity in the Prevention and Treatment of Obesity

It is commonly accepted that sedentary time and low physical activity energy expenditure are correlated with weight gain and risk of obesity. This general theme is addressed in 13 chapters. These chapters highlight the role of physical activity in the prevention of weight gain (chapter 52), for weight loss (chapter 53), in comparison to diet-induced weight loss (chapter 54), in combination with pharmacotherapy (chapter 55), and as an adjunct to bariatric surgery (chapter 56). Other chapters discuss the effects of physical activity on depot-specific fat loss (chapter 57), visceral fat and ectopic fat deposition (chapter 58), maintenance of lean mass (chapter 59), and weight loss maintenance (chapter 60). Subsequently, special topics are introduced. Physical activity and weight control during pregnancy (chapter 61) and postpartum weight loss (chapter 62) are the focus of two chapters. These are followed by chapters on physical activity and birth weight (chapter 63) and body composition in children (chapter 64).

Physical Activity in the Prevention of Weight Gain

Mikael Fogelholm, ScD

According to the laws of thermodynamics, obesity is a consequence of long-term (over weeks, months, or even years) positive energy balance, that is, an intake exceeding energy expenditure. Although an overconsumption of energy from a diet with high energy density is possible in all cases, excess eating is at least intuitively more likely when total energy expenditure is low.

The daily energy expenditure can be divided into three main parts: resting energy expenditure, diet-induced thermogenesis, and physical activity. More than half of the daily energy expenditure is related to resting metabolic rate in most individuals, whereas the proportion of physical activity is normally 15% to 30% of total energy expenditure. However, physical activity is the only part of daily energy expenditure that can be increased considerably by voluntary muscle work.

Secular Trends in Physical Activity and Obesity

Physical activity can be divided into work and recreational (leisure) time, and the latter further divided into commuting (from home to work, to shops, etc.), exercise training, and lifestyle physical activity. Simultaneously with an increased trend in the prevalence of obesity, there has been a decrease in work energy expenditure and commuting physical activity (1). The former is a consequence of increased use of machines, robots, computers, and so on. Increased use of cars is likely to be the main reason for decreased commuting physical activity. In contrast to some beliefs, regular exercise training has probably even increased since the 1980s (1), but this has not counterbalanced the decreased energy expenditure in work and commuting physical activity.

Consequently, ecological studies using data on trends suggest that decreased work, commuting, and lifestyle physical activity could be related to the observed increase in the prevalence of obesity (1). Moreover, numerous cross-sectional studies (2) have shown that higher levels of both vigorous exercise and walking are independently related to lower prevalence of obesity. Unfortunately, neither ecological nor cross-sectional studies demonstrate causalities.

Prospective Studies on Physical Activity and Weight Change

Ideally, the role of physical activity in prevention of obesity should be studied using randomized trials. Unfortunately, this is difficult, since a normal process of significant weight gain takes several years. Therefore, prospective follow-up has been the main design used to study the connections between physical activity and weight change.

This chapter examines the role of physical activity in prevention of weight gain particularly in adults. The chapter is based mainly on results from a systematic review of the role of physical activity in prevention of weight gain in adults (3), completed with some more recent original studies. All prospective studies were at least two years in duration, with a mean duration of seven years.

Most of the studies cited in the review assessed physical activity at baseline. Moreover, many studies used the change in physical activity (from baseline to follow-up), and some assessed weight change during the study in relation to physical activity at the end of the follow-up.

In the few studies with physical activity assessment at the end of the follow-up, all but one showed that high physical activity was related to smaller weight gain. However, these studies do not demonstrate the role of physical activity in preventing weight gain. Rather, the results could suggest that weight gain may have an effect on physical activity, that is, that individuals with weight gain end up with low levels of physical activity.

Most of the studies relating change in weight to a simultaneous change in physical activity showed a reciprocal connection; that is, decreased physical activity was related to weight gain. About a third of the investigations did not yield any relations between changes in physical activity and weight. The interpretation of causality is difficult, since the "change-against-change" analysis is in fact similar to that in cross-sectional design. With only two assessment points (baseline and follow-up), it is not possible to tell whether a change in physical activity preceded weight gain or vice versa.

The majority of studies used physical activity at baseline as a predictor of weight change during the follow-up. The results were typically adjusted for baseline body mass index (BMI), age, smoking, and sociodemographic factors. The results were distributed over the three possible outcomes: Some studies supported the hypothesis that high physical activity predicts lower weight gain, some did not show a relationship, and some even showed that high physical activity predicted more weight gain during the follow-up.

The results were not always consistent even within a single study. For instance, Blanck and colleagues (4) reported that high baseline recreational physical activity was related to less weight gain in normal-weight but not overweight postmenopausal women. Droyvold and colleagues (5) found that both high and low physical activity levels predicted weight gain, whereas moderate physical activity seemed to have a preventive role. Finally, Parsons and associates (6) obtained opposite results for men and women.

Most of the studies cited in the review (3) assessed leisure-time physical activity, exercise training, or both. Sedentary behavior was assessed in one study, which did not find an association between TV watching and weight gain. In contrast, two more recent studies indicated that high duration of nonoccupational sedentary behavior (4) or TV watching (7) was related to higher risk for weight gain.

Reasons for Conflicting Results

Although intuitively the positive role of physical activity in prevention of weight gain seems clear, and although cross-sectional studies strongly connect high physical activity with low prevalence of obesity, the results of the prospective studies are less consistent. Why do some studies not show an association between baseline physical activity and weight gain, and why do some even suggest that higher levels of physical activity are related to more, not less, weight gain?

One interesting finding is that the associations between weight gain and physical activity are clearer when the design does not allow any conclusions on causality. This may be explained by the fact that physical activity and weight change are interrelated in two ways. Higher levels of physical activity may prevent weight gain, which makes physical activity easier. However, if weight increases were due to poor dietary habits, for example, increased weight may reduce physical fitness, which makes exercise more difficult. Hence, while physical activity may be regarded as important for the prevention of weight gain, weight maintenance is also important for the continuation of a physically active lifestyle.

The limited potential to increase energy expenditure is perhaps another factor explaining the obscure relationships between physical activity and weight change. In most studies and for most individuals, leisure exercise training is the main varying component in total physical activity. It is very difficult to change occupational energy expenditure (most jobs are sedentary) and quite difficult to change commuting behavior (due to poor urban planning, most individuals may find it almost impossible to walk or cycle from home to work or shops).

The general recommendation for health-enhancing physical activity, 30 min brisk walking daily (8), translates to an increase in energy expenditure of at most 150 kcal daily. This small increase is easily compensated for by consumption of high-energy-dense foods or by a slight decrease in light physical activities. Some expert groups have suggested that 45 to 60 min, rather than 30 min, of brisk walking daily may be needed to prevent weight gain. This corresponds to an increase in daily energy expenditure of 200 to 300 kcal. Unfortunately, none of these recommendations are based on hard scientific evidence.

Inaccurate assessment of physical activity is a methodological limitation that may explain a part of the obscure relationship between physical activity and weight change. A main assessment problem is that most studies have queried only about recreational, not total, exercise activity. The reason for this approach is evident: Variations in light activities are difficult to capture with a retrospective questionnaire, although these activities may be relevant for total daily energy expenditure.

A single measurement of physical activity at baseline does not tell us about possible changes in activity during the follow-up: Some highly active individuals will probably reduce their physical activity during the follow-up, and some who were highly active at baseline will increase physical activity. These changes may be related to weight changes, but also numerous other reasons may lead to an increase or decrease in physical activity (e.g., moving to a location with better or worse possibilities for physical activity, a new and physically active or inactive boy- or girlfriend). Changes in physical activity may even explain why some studies have connected high physical activity with weight gain.

Randomized, long-term weight gain prevention interventions have typically used a combined dietary and physical activity regimen. Although the results of these interventions are promising, they do not illustrate the independent role of physical activity. Community interventions have similar problems (for a review, see [9]). Most interventions with data on weight were not originally designed to prevent obesity but rather to reduce cardiovascular risk factors. Consequently, promotion of physical activity was just a part of a lifestyle intervention. Some of the large community studies succeeded in increasing physical activity levels, but clear effects on body weight or obesity were not reported. It is possible that the observed increase in physical activity was not large enough to have a significant effect on energy balance.

The Public Health Message

An intervention with increased leisure exercise training may not be enough to prevent weight gain. The reasons are related to a limited potential to increase energy expenditure through exercise and to poor maintenance of increased physical activity habits. However, there are still reasons to believe that physical activity may help in preventing weight gain if several components of daily physical activity can be changed and the change maintained.

A change in "many components of physical activity" means that a potentially successful intervention should address exercise habits, other leisure physical activities (such as commuting activities), and also the time spent in totally sedentary activities. This conclusion is consistent with the new physical activity recommendations (8). It may be reasonable not to set a specific recommendation for physical activity levels in preventing obesity, since the level needed is related to the dietary habits of the individual. However, one may still recommend a combination of structured exercise (usually of moderate-to-vigorous intensity) and less structured lifestyle activities (2).

In a recent longitudinal study by Brien and colleagues (10), good cardiorespiratory fitness was positively related to better weight maintenance. Vigorous exercise activities and hence maintenance of good physical fitness may therefore be important for the prevention of weight gain. Good cardiovascular fitness is also strongly related to health, regardless of the weight status.

Figure 52.1 shows the main conclusions and also recommendations from the studies presented in this short review. Total energy expenditure is a result of different components of physical activity, and these components may also have an effect on each other. Vigorous activities especially improve physical fitness, which reciprocally enables physical activity. Hence, the figure shows the complexity of prevention of weight gain and the importance of avoiding a nonproductive cycle (weight gain reducing fitness and physical activity, which increases weight gain, and so on).

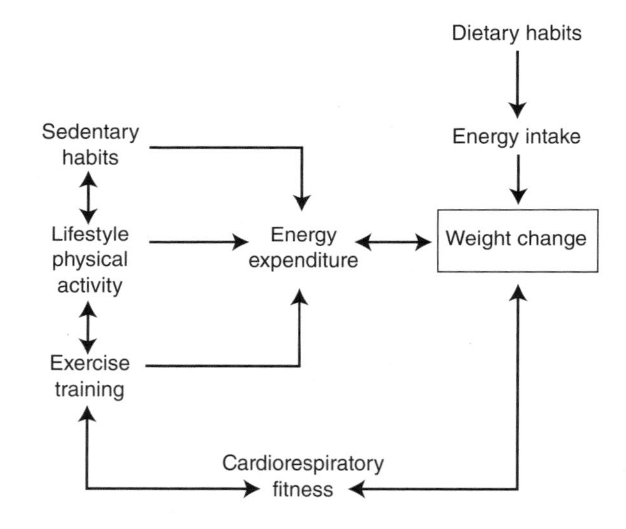

Figure 52.1 An outline of the interrelations between different components of physical activity, physical fitness, diet, and energy balance.

In addition to factors shown in the figure, individual (e.g., motivation) and environmental factors (e.g., possibilities to be physically active and to make prudent food choices) affect diet and physical activity. An effective intervention to prevent obesity should preferably include health education and counseling (individual approach) combined with multiple environmental changes. Despite some ambitious community interventions (9), the real landmark study on prevention of obesity, with all needed individual and environmental components, has yet not been done.

53

Physical Activity for Weight Loss

Robert Ross, PhD

Leading health authorities suggest that the utility of increasing physical activity as a strategy for obesity reduction depends in large measure on body weight reduction. Indeed, there is a strong consensus that a loss of 5% to 10% of body weight is required to achieve a health benefit. Although an increase in physical activity (exercise) without alteration in energy intake is associated with significant reductions in body weight in a dose–response manner, a majority of exercise-induced weight loss studies do not achieve a 5% weight loss. Accordingly, whether an exercise-induced weight loss of less than 5% is a useful strategy for reducing obesity requires consideration. Furthermore, the potential influence of gender, race, and age on the association between physical activity and weight loss needs to be addressed. These issues are the focus of this review.

Physical Activity, Weight Loss, and Obesity Reduction

It is generally accepted that a decrease in leisure or nonleisure physical activity (or both) contributes to the worldwide increase in obesity prevalence, as indicated by body weight or body mass index (1). Accordingly, it is intuitive to suggest that an increase in physical activity levels would be associated with a decrease in body weight and the accompanying adiposity. Indeed, the independent role of physical activity (exercise) as a treatment strategy for obesity has received considerable attention in recent years, and several comprehensive reviews have considered the utility of physical activity to induce weight loss (1-3). These are insightful reports, and the reader is encouraged to review them.

Early reviews of the literature suggested that the reduction in body weight associated with an increase physical activity alone (i.e., no caloric restriction) was marginal (1 to 2 kg [2.2 to 4.4 lb]), and concluded therefore that obesity reduction by physical activity in the absence of caloric restriction was not a useful strategy for the treatment of obesity. Subsequently, a careful inspection of the exercise studies revealed that, for the most part, few of the early studies prescribed an exercise program in which one would expect marked weight loss (1). Conversely, for those studies in which the exercise program prescribed did result in a meaningful negative energy balance, weight loss was substantial (1).

More recently, several reports confirm that increasing physical activity level without caloric restriction is associated with clinically significant weight loss (i.e., greater than 5%). Racette and colleagues (4) report that overweight men and women randomized to a 20% exercise-induced energy deficit for 12 months achieved an 8.4% reduction in body weight, which was associated with a 24% reduction in total adiposity. These observations are consistent with earlier laboratory trials in which increasing exercise-induced energy expenditure (supervised exercise) by an order of 500 to 700 kcal per day was associated with 8% and 6.5% reductions in body weight in men (5) and women (6), respectively. The work of Slentz and colleagues (7) extends these observations by demonstrating that physical activity is associated with significant reductions in body weight and fat mass in overweight and obese middle-aged men and women in a dose–response manner. In that study, obese middle-aged individuals who exercised for an average of 175 min per week lost significantly more body weight and total fat mass than those who exercised for an average of 114 min per week. However, it is noteworthy that the participants in this trial were instructed to maintain their baseline weight (i.e., participants consumed additional food calories in an attempt to avoid weight loss) throughout the intervention (7). The consumption of compensatory calories would lead to a significant underestimation of the effects of exercise per se on body weight and fat mass and thereby confound interpretation of these results.

Data from the Midwest Exercise Trial (8) in overweight and obese young adults provide sexually dimorphic results in response to 16 months of moderate-intensity physical activity (45 min per day, five days per week) training (8). In agreement with previous studies, men who expended an average of 3300 kcal per week during the 16-month intervention lost 5.2 kg or 11.4 lb (5.5%) of body weight, 96% of which was due to fat loss. Paradoxically, the women in this study who expended an average of 2200 kcal per week showed a resistance to weight loss. However, this amount of energy expenditure prevented the gain in body weight observed in the female control subjects. Although these observations suggest that women may be resistant to exercise-induced weight loss compared to men, they are counter to the observations of Ross and colleagues (6), who report that exercise without caloric restriction is associated with substantial reductions in body weight and adiposity in middle-aged, abdominally obese women. Furthermore, of the 18 randomized controlled trials presented in figure 53.1, four collapsed across gender after yielding no significant differences between genders. Further investigation is required to determine whether women are indeed resistant to exercise-induced weight loss in comparison to men.

Observations in older women suggest that a 12-month physical activity program is associated with modest reductions in body weight and total fat mass (9). However, among the exercising women, a dose–response relationship was noted between weekly physical activity duration and subsequent fat loss. Specifically, highly active women (>195 min per week) lost 4.2% of their total body fat, compared with losses of 2.4% in intermediate-active (136 to 195 min per week), 0.6% in low-active (≤135 min per week), and 0.4% in control subjects.

In summary, the findings from a number of recent randomized controlled trials confirm earlier reviews (1-3) and suggest that with few exceptions, physical activity without energy restriction is associated with a corresponding reduction in body weight and fat mass. From figure 53.1, it is also clear that exercise performed for as little as 200 min per week is associated with weight loss. Further, the evidence suggests that approximately 450 min per week or 45 to 60 min of daily physical activity is necessary to attain the recommended 0.5 kg (1.1 lb) per week

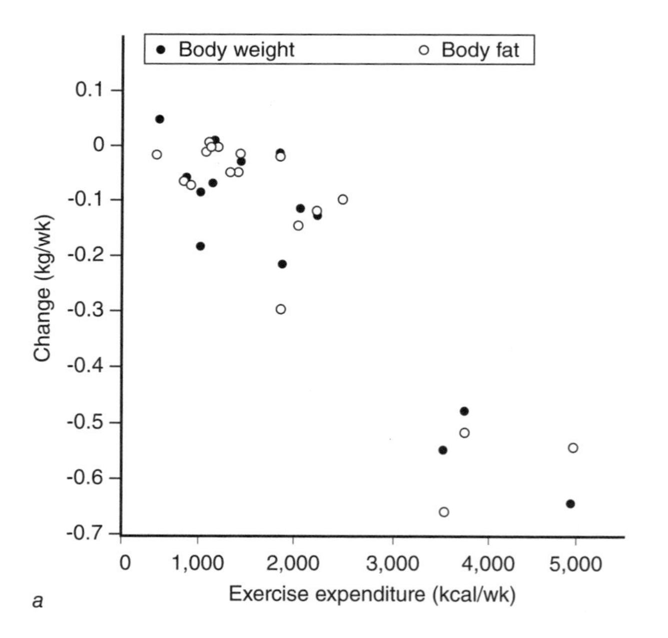

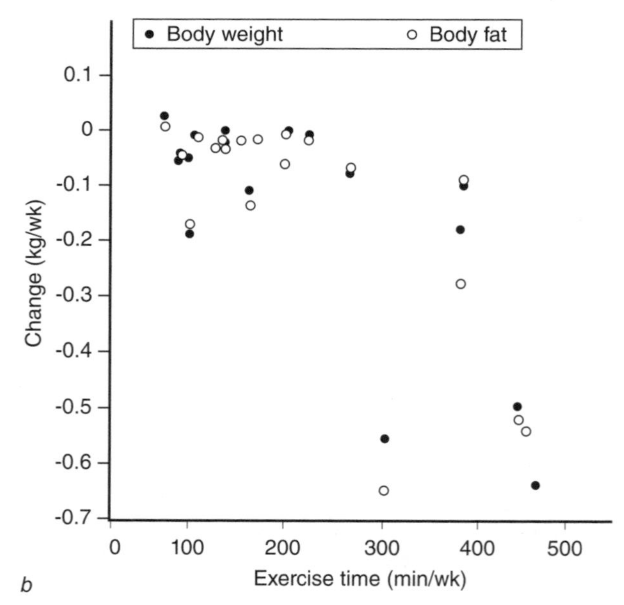

Figure 53.1 Dose–response relationship between *(a)* weekly exercise energy expenditure and *(b)* weekly minutes of exercise, and the corresponding changes in body weight (closed circles) and body fat (open circles).

Adapted from R. Ross and I. Janssen , 2007, Physical activity, fitness and obesity. In *Physical activity and health*, edited by C. Bouchard, S. Blair and W. Haskell (Champaign, IL: Human Kinetics), 173-190.

rate of weight loss. Thus, although a 30 min increase in daily physical activity provides benefit across a wide range of health outcomes, this level of activity may be below the threshold required to induce significant weight loss for a substantial proportion of adults living in today's "obesogenic" environment—at least weight loss at a rate consistent with the 0.5 kg per week recommendation. Indeed, the observations summarized here suggest that approximately 45 to 60 min of daily physical activity may be necessary to bring about significant reductions in obesity for many overweight or obese individuals. These observations are consistent with the Institute of Medicine recommendation that an increase of about 60 min of daily physical activity is required to induce weight loss for most individuals.

Although inspection of the findings summarized in figure 53.1 suggests that increasing levels (i.e., volumes) of physical activity are associated with weight loss in a dose–response manner, it is important to note that these observations are derived from interpretation of distinct reports. Absent from the literature are rigorously controlled, randomized trials that carefully examine the separate effects of dose or intensity (or both) of exercise on body weight in men and women. This represents a major gap in knowledge with implications for the development of public health messages. Indeed, unanswered questions include the following: With a given exercise dose, does higher exercise intensity result in greater weight and fat loss? With a given exercise intensity, does higher exercise dose result in greater weight and fat loss? Clear identification of the separate effects of exercise dose and intensity on obesity reduction under controlled conditions is important for development of optimal lifestyle-based strategies that can subsequently be tested in long-term effectiveness trials.

Physical Activity Without Weight Loss and Obesity Reduction

There is a strong consensus that a weight loss of 5% to 10% of body weight is required in order to obtain clinically meaningful improvement across a wide range of health outcomes; indeed, the benefits of this magnitude of weight loss are well documented. Nevertheless, it is apparent from the findings summarized here (figure 53.1) that sustained increases in physical activity in obese men and women are often associated with weight loss well below the minimal target of 5%. It is important to consider, therefore, whether increasing levels of physical activity with minimal or no change in body weight should continue to be interpreted as a failed attempt at obesity treatment. For example, of the 13 trials from figure 53.1 in which the average weight loss was <5%, 10 yielded statistically significant reductions in total body fat, abdominal obesity, or both, while nine of the trials showed statistically significant improvements in cardiometabolic risk factors. Thus, although weight loss remains the desired outcome of chronic physical activity in overweight and obese individuals, it is apparent that weight loss is not the only gauge of success in obesity treatment.

This point is reinforced by Ross and colleagues (5, 6), who investigated the effects of moderate-intensity physical activity on changes in total and regional fat distribution in individuals who maintained body weight. In order to achieve weight maintenance during the intervention, the participants consumed compensatory kilocalories equivalent to the amount expended during physical activity. The length of each intervention was roughly three months and consisted of an energy expenditure of approximately 3500 kcal per week. The primary findings suggest that in obese Caucasian men and women, as well as men with type 2 diabetes, significant reductions in total and abdominal obesity (i.e., waist circumference) occur consequent to physical activity despite minimal change in body weight. Similar findings were reported by Giannopoulou and colleagues (10), who studied the effects of 14 weeks of aerobic physical activity without weight loss in a small sample of postmenopausal women with type 2 diabetes. Despite minimal change in body weight, physical activity in this study resulted in a significant decrease in waist circumference. Given that waist circumference is a strong correlate of abdominal obesity and visceral fat and explains health risk beyond that explained by body weight or body mass index alone, the observation that total and abdominal obesity is reduced in response to weight loss that is well below the 5% target is clinically relevant.

It should be noted, however, that the reductions in total and abdominal obesity observed in exercisers who maintain body weight are substantially less than those observed in exercisers who lose weight. Thus from a clinical perspective, those at obesity-related health risk should be advised that physical activity-induced weight loss is likely associated with the greatest benefit. However, given the challenges associated with attaining substantial weight loss for many obese individuals in today's environment, it is equally important to recognize that obesity and related health risk can be markedly reduced in response to minimal weight loss (e.g., weight loss below 5% of body weight). Indeed, combined with the established observation that improvements in cardiorespiratory fitness are associated with reduced morbidity and mortality independent of body mass index and abdominal obesity (11), it is apparent that physical activity with or without weight loss will attenuate obesity and related health risk. This is good news, offering those in the allied health professions the opportunity to inform individuals who seek to reduce obesity-related health risk that irrespective of weight change, physical activity conveys a significant health benefit and thus is an excellent treatment option.

Summary

Current knowledge suggests that increasing physical activity without altering caloric intake is associated with weight loss in a dose–response manner. However, it is noteworthy that of the 18 randomized controlled trials for which data are presented in figure 53.1, only four included subjects aged over 60 years and only one included non-Caucasian subjects. Thus whether age and race influence the association between physical activity and body weight reduction is largely unknown. It is also apparent that increasing physical activity is often associated with significant reductions in obesity and related cardiometabolic risk factors despite a failure to achieve the consensus recommended 5% weight loss. These are important observations that require consideration when one is interpreting the utility of exercise to decrease body weight, obesity, and related comorbid conditions.

54

Weight Loss Induced by Physical Activity Versus Diet

Joseph E. Donnelly, EDD

Energy balance is determined by the amount of energy intake relative to energy expenditure. If energy intake and energy expenditure are equal, weight is maintained. Amounts of energy intake greater than energy expenditure result in positive energy balance and weight gain, and energy intake less than energy expenditure results in negative energy balance and weight loss. The overall model of energy balance appears straightforward; however, the contributing components of energy intake and energy expenditure are not static but instead are dynamic and interactive. Attempts to increase or decrease one component of energy intake or energy expenditure may result in a compensatory change in another component, and the intended effect may be attenuated or absent altogether (figure 54.1). For example, if an individual restricts energy intake and this results in lower levels of resting metabolism and daily physical activity, little if any weight loss may result. The potential for compensation frequently leads to the observation that weight loss by energy restriction, increased physical activity, or a combination of the two seldom achieves the level calculated by the simple equation of energy intake minus energy expenditure.

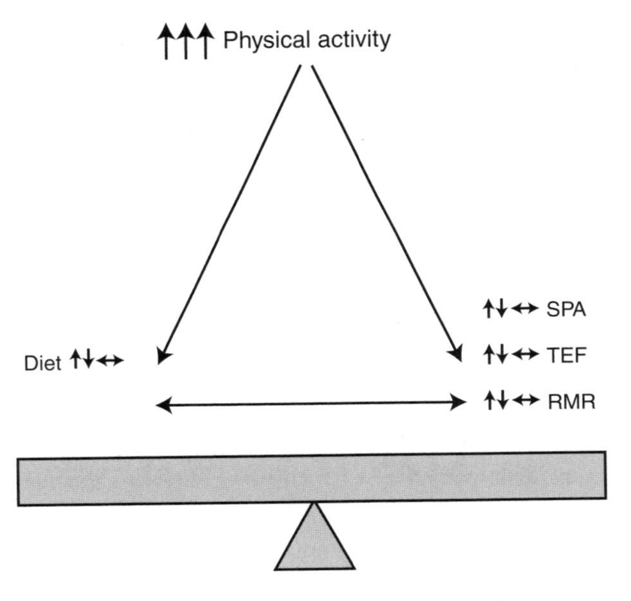

Figure 54.1 Potential compensation in energy balance in response to exercise. SPA = spontaneous physical activity; TEF = thermic effect of food; RMR = resting metabolic rate.

Donnelly, Joseph; Blair SN, Jakicic JM, Manore MM, Rankin JW, Smith BK. Appropriate physical activity intervention strategies for weight loss and prevention of weight regain for adults. *Medicine and Science in Sports and Exercise.* 2009; 41(2): 459-471.

Negative Energy Balance for Weight Loss

Obesity results from a sustained positive energy balance and is currently exhibited in about 66% of adults in the United States. In the attempt to reduce weight, a negative energy balance may be achieved by energy restriction, physical activity, or both. Attempts to achieve weight loss are frequent; at any given time, 29% of men and 44% of women are engaged in weight reduction programs. To lose weight, 90% of these individuals use energy restriction, and 66% use physical activity (1). Energy restriction decreases the amount of energy available to sustain body weight, and physical activity provides an expenditure of energy beyond that used to sustain daily living. Thus, each component by itself or in combination may result in a negative energy balance and weight loss.

Energy Gap Caused by Energy Restriction or Physical Activity

The difference between the amount of energy necessary for weight maintenance and the amount provided is termed the "energy gap," and the magnitude of the energy gap will determine the amount of weight loss. It appears that a greater energy gap may be created by energy restriction compared to energy expenditure. Diet restrictions of 500 to 1000 kcal/day are commonly employed for weight loss and can be expected to result in a 5% to 10% loss from baseline weight (2). The energy gap that may be provided by energy expenditure is generally smaller than that shown for diet restriction. An individual who walks 30 to 60 min/day may achieve 200 to 500 kcal/day of energy expenditure of exercise (including resting metabolic rate). However, many individuals are either unwilling or unable to walk at this level. Thus, people are likely achieve a greater energy gap using diet restriction compared to physical activity and therefore a greater weight loss (2).

Although many individuals cannot achieve a large energy gap through physical activity, some are capable of achieving and sustaining relatively high levels of physical activity. For example, Ross and colleagues showed that individuals who achieved a 500 to 700 kcal/day negative energy balance through physical activity lost weight (about 7%) comparable to that of individuals who achieved the same negative energy balance through energy restriction (3). Donnelly and colleagues (4) showed that men who had ad libitum diet intake and exercised five days per week, 45 min per day, at 75% of the heart rate reserve for 16 months lost 6% from baseline weight (5.2 ± 4.7 kg [11.5 ± 10.4 lb]). The corresponding negative energy gap measured by doubly labeled water was –346 kcal per day at 16 months. These levels of weight loss achieved with exercise are similar to weight loss achieved through energy restriction and are likely to confer significant additional health benefits through improvement of cardiovascular disease and diabetes risk factors. Additionally, when weight is lost due to physical activity alone, fat-free mass may be preserved to a greater extent than with weight loss using energy restriction alone. For example, in the study by Donnelly and associates (4), only 0.2 kg (0.44 lb) fat-free mass was lost as measured by dual-energy X-ray absorptiometry.

Energy Gap Caused by Energy Restriction and Physical Activity

Public health organizations recommend both energy restriction and physical acitvity for weight loss (5). The combination of energy restriction and physical activity has the potential to increase the energy gap more than energy restriction or physical activity alone. Reviews and experimental studies using both energy restriction and physical activity compared to either alone generally show somewhat greater weight loss for energy restriction and physical activity combined (2). Interestingly, physical activity has not been shown to increase weight loss when combined with severe energy restriction of less than 1000 kcal/day (6).

Energy Restriction Plus Physical Activity and Fat Mass Loss

The ideal weight loss regimen would have a high percentage of weight loss as fat mass since fat mass appears to be related to negative metabolic outcomes (e.g., insulin resistance). The most prevalent study design for investigating benefits of physical activity on fat mass loss is equivalent energy restriction between groups with physical activity added to the energy restriction for one of the groups, and the results appear equivocal. Recently, Redman and colleagues (7) reported the results from a carefully controlled trial that investigated differences in fat mass between a group receiving energy restriction and a group receiving energy restriction plus physical activity with estimated total energy expenditure equal for both. Changes in fat mass were assessed with dual-energy X-ray absorptiometry and computed tomography. No between-group differences were found for weight loss or for changes in total fat mass or abdominal fat.

Resistance Training and Weight Loss

The energy expenditure of resistance training is lower than that typically shown with endurance exercise; however, resistance training may increase energy expenditure by promoting an increase in muscle mass (figure 54.2). Muscle mass has an energy requirement of about 15 to 25 kcal/kg per day (8) and would utilize energy not just during exercise but throughout the day. Thus, although the energy expenditure of resistance training is relatively low, the accumulated energy expenditure across 24 h may be substantial enough to affect weight.

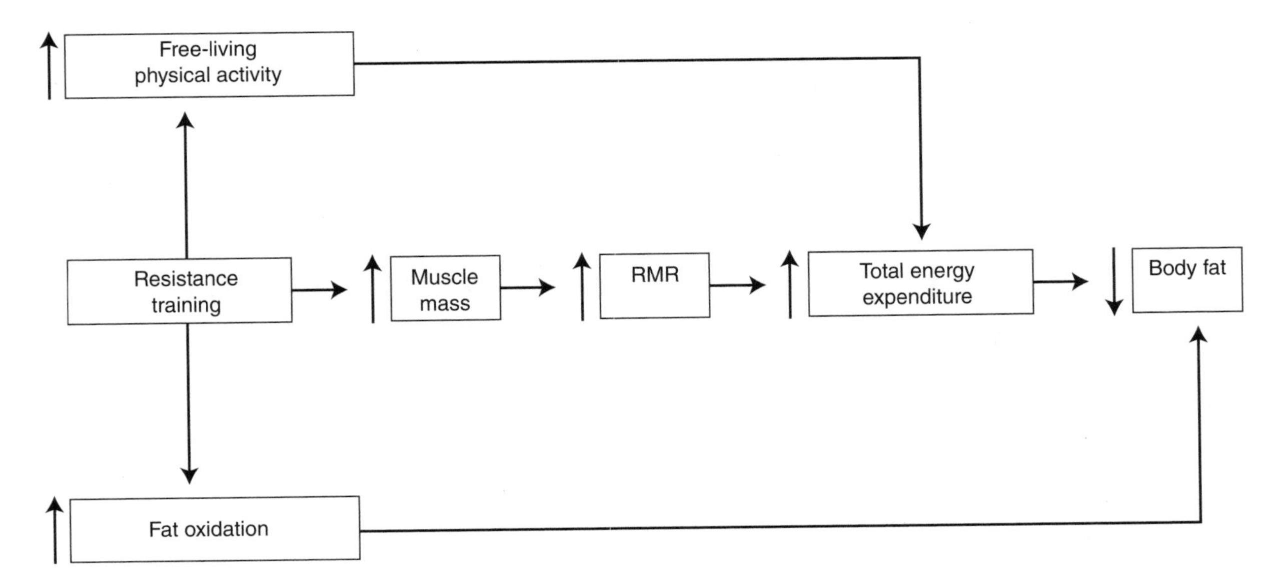

Figure 54.2 Potential influence of resistance training on energy expenditure.

Reprinted, by permission, from J.M. Jakicic et al., 2001, "Appropriate intervention strategies for weight loss and prevention of weight regain for adults," *Medicine and Science in Sports and Exercise* 33: 2145-2156.

Examination of the literature generally does not support this notion. Weight loss in resistance training studies is minimal, and resistance training contributes little to weight loss when used in combination with energy restriction (9). This implies one of the following: The increases in muscle mass are insufficient to substantially increase energy expenditure; the length of the studies is too short to show an effect; or other components of energy balance changed (compensated), such as energy intake or activities of daily living. No studies have established the magnitude of increase in muscle mass that might be necessary to provide a realistic chance to alter energy balance by a level unlikely to be attenuated by compensation in either energy intake or expenditure. Additionally, no studies have used state-of-the-art energy balance techniques to determine if and where compensation has taken place.

Resistance training used with energy restriction has been shown to preserve fat-free mass to a greater extent than energy restriction alone. However, similarly to the situation with physical activity and energy restriction, resistance training combined with severe energy restriction will not preserve fat-free mass more than energy restriction alone (6).

It should be noted that although resistance training has a minimal effect for weight loss, it is consistently associated with improvements in body composition through greater fat loss and increases in fat-free mass (10).

Summary

Individuals typically do not create an energy gap with physical activity comparable to that with energy restriction, and therefore physical activity generally provides less weight loss than energy restriction. However, when physical activity is combined with moderate energy restriction, weight loss is enhanced compared to that occurring with either physical activity or energy restriction alone. Furthermore, physical activity promotes health and may stimulate retention of fat-free mass during conditions of moderate energy restriction. Resistance training has not been shown to be effective for weight reduction. However, resistance training is associated with increased fat oxidation, increased fat-free mass, and reduction of many chronic health risk factors. All major health organizations recommend a combination of energy restriction and physical activity for weight loss.

Role of Physical Activity in Pharmacological Weight Loss

Frank L. Greenway, MD

The role of exercise in the context of pharmacologically induced weight loss is an extremely understudied subject. Most clinical trials evaluating pharmacologic agents for the treatment of obesity randomize subjects to drug and placebo groups. Most of these trials prescribe a calorie-restricted diet and recommend physical activity, usually of moderate intensity, like brisk walking for 30 to 60 min on most days of the week. The physical activity prescribed is typically similar for the drug and placebo groups in the trial. This study design makes it impossible to separate the effects of physical activity from the effect of the diet and the drug. This chapter focuses on the reasons for lack of careful investigations of the separate effect of exercise in the context of trials for obesity drugs, discusses the postulated role of exercise in pharmacologic obesity interventions based on exercise trials for the treatment of obesity, and points out fertile areas for future research into the interface of exercise and pharmacologic agents to treat obesity.

Lack of Research

Clinical trials of pharmacologic agents for the treatment of obesity usually focus on the safety and efficacy of the obesity drug. One reason for this is that the ultimate goal of many trials of obesity pharmaceuticals is to attain approval from the U.S. Food and Drug Administration (FDA) for an indication to sell the pharmaceutical as a prescription drug for the treatment of obesity. Since these trials are usually sponsored by pharmaceutical companies interested in registering and selling the drug, there is little interest or incentive to tease out the separate effects that the drug and exercise might contribute to any weight loss.

Clinical trials must be approved by institutional review boards or ethics committees. These groups require that something be offered to the placebo group, feeling that people enroll in obesity trials to lose weight and that some benefit must accrue to those who are randomized to the placebo. The drugs presently approved for the treatment of obesity confer very modest weight loss—less than 5% more weight loss than placebo (1). The FDA uses two criteria to judge the efficacy of obesity drugs. First, the obesity drug should lead to 5% more weight loss than placebo, since 5% has been demonstrated to be a medically significant weight loss (2). The second criterion is the percentage of subjects on the drug who achieve a 5% weight loss compared to the placebo group. This second criterion is the only one met by the presently approved obesity drugs because of the very modest efficacy of these drugs. The greater the ancillary weight loss in a clinical trial, the greater the overall weight loss in the trial; but as the weight loss increases in both the drug and placebo groups, the separation between drug and placebo groups is reduced. Thus, there is economic incentive for the sponsors of trials testing the efficacy of weak obesity pharmaceuticals to minimize the intensity of the ancillary weight loss intervention, including exercise, to maximize the separation between drug and placebo.

While pharmaceutical companies have little incentive to study the separate effect of exercise in the context of weight loss induced by their drug, funding agencies often feel that trials involving pharmaceuticals should be funded by industry. This policy encourages the use of scarce federal and private research funds for the study of mechanistic issues that industry has little incentive to explore. Although this reasoning may be logical, it leaves a funding gap for exploring the separate effects of exercise and diet in trials of obesity pharmaceuticals. A study was performed to compare the effects of sibutramine, behavior modification, and the combination of the two. This study clearly demonstrated that behavior modification added to the weight loss seen with sibutramine alone (3). Although the behavior modification advice did include exercise, a trial to determine the separate effect of physical activity from other aspects of the behavior program was not conducted. A trial with four arms in which the subjects are randomized—an exercise group, a group that maintain their regular activity, a group that receive a pharmacologic obesity intervention, and a group that combine the drug with physical activity—would be necessary to tease out the contribution of exercise to the overall program. A long-term trial of this nature would give important information. One of the challenges involved in doing such a trial is the notorious inaccuracy of self-reported physical activity or of self-reported dietary intake by obese individuals. Thus, to do this trial properly, the researcher would need to monitor exercise to ensure compliance. Such a trial with four arms and the statistical power to answer the question would involve considerable expense.

Potential Role of Exercise

A trial reported by Pavlou and colleagues randomized obese subjects to a diet with or without an exercise program. During eight weeks of weight loss there was no significant difference in the weight loss between the two groups; but during 18 months of weight maintenance, those subjects who maintained their exercise at the end of the weight loss period or who started exercise at the end of the weight loss period maintained their weight loss, while those who did not exercise or stopped exercise after the weight loss period regained their weight (4) (figure 55.1). The minimal effect of exercise on weight loss was borne out in a meta-analysis of exercise studies for the purpose of weight loss; 80% of the studies showed less than a kilogram of additional weight loss, and none as much as 2 kg (4.4 lb) greater weight loss than in the nonexercising groups (5). This suggests that adding exercise in the context of a clinical weight loss trial of an obesity drug

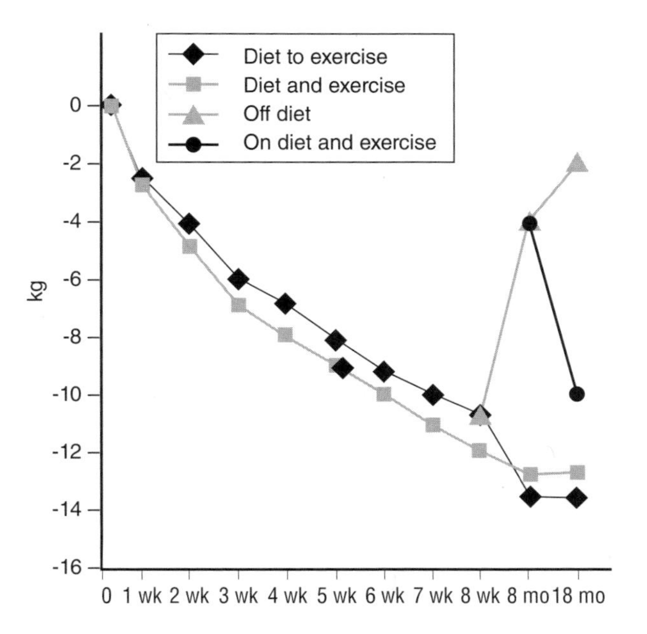

Figure 55.1 Exercise added little to weight loss over eight weeks, but the addition or continuation of exercise after weight loss maintained the loss. Stopping the diet after weight loss resulted in weight regain, but weight gain was reversed with the resumption of diet and exercise eight months following the end of the weight loss program.

Adapted from K.N. Pavlou, S. Krey, and W.P. Steffee, 1989, "Exercise as an adjunct to weight loss and maintenance in moderately obese subjects," *American Journal of Clinical Nutrition* 49: 1115-1123.

would be unlikely to enhance weight loss, but might enhance the maintenance of the weight loss. The use of exercise to enhance weight loss maintenance is supported by another trial showing a dose relationship between exercise and weight maintenance after a weight loss program (6). The natural history of obesity is a gradual weight regain over time. Long-term treatment with obesity pharmaceuticals after the weight loss plateau shows a gradual, parallel weight gain in the drug and placebo groups (7). One could postulate that an exercise program might attenuate this gradual weight gain; but few long-term trials of obesity drugs have been done, and none have been done to explore the effect of exercise on weight maintenance in the context of drug treatment.

Recent exercise trials suggest that above a certain threshold of intensity, food intake increases to compensate for the caloric reduction induced by the exercise (8). This would make one wonder if exercise should be used below the compensation threshold to accelerate weight loss induced by pharmaceuticals. It is also possible that the amount of exercise required for maintenance of a weight loss might be reduced by an obesity pharmaceutical intervention. These are issues worthy of further research.

Future Directions

New obesity combination drugs such as topiramate with phentermine and bupropion with zonisamide are being developed that lead to more than a 10% weight loss with minimal intervention by a behavior modification, diet, or exercise program (9, 10). Thus, drugs are becoming available that are strong enough to achieve a 5% greater weight loss than placebo, even with a strong ancillary weight loss program. In fact, one company has a trial in progress combining a group behavior modification program (designed to give a 7% to 8% loss of initial body weight) with one of its obesity combination medications in a clinical registration trial for an obesity treatment indication.

With combination obesity drugs and greater than 10% losses of body weight, the door is open to explore the role of exercise in the context of pharmaceutical trials for the treatment of obesity. These more effective drugs will not put the company in jeopardy of failing to make the 5% greater weight loss than placebo when ancillary treatments are superimposed. Hopefully, with the advent of these more effective obesity drugs, we will see trials addressing the effect of exercise in pharmaceutically induced weight loss, studies that address the optimal dose of exercise to maximize weight loss without caloric compensation, and studies to explore the role and dose of exercise for optimal maintenance of weight loss induced by pharmaceutical agents for the treatment of obesity.

Clearly, we know that exercise is healthy for people of various weights and that intentional weight loss is medically beneficial for the obese. Much progress is being made in the pharmacologic treatment of obesity. Now that more effective weight loss drugs are being developed, it is important to define the role and the optimal dose of exercise to be used in conjunction with obesity pharmaceuticals, especially during long-term weight maintenance.

56

Role of Physical Activity in Surgical Weight Loss

Paul E. O'Brien, MD

All forms of bariatric surgery act primarily by reducing energy availability through decreased intake, reduced absorption of macronutrients, or both. Studies of nonsurgical weight loss confirm that improved weight loss outcomes are achieved when exercise is combined with reduced energy intake compared with reduced intake alone. It would therefore be expected that an added exercise program would enhance the weight loss outcomes of bariatric surgical procedures. In this chapter, the rationale for encouraging exercise and increased physical activity in the weight loss surgery patient is examined, the data available to support or refute this rationale are reviewed, and the types of exercise programs most likely to give optimal effect are discussed.

Current Types of Weight Loss Surgery

There are three established weight loss surgical procedures that are able to generate substantial and durable weight loss.

Roux en Y gastric bypass (RYGB) is the best known and longest studied. It was introduced in 1976 by Dr. Edward Mason. There has been a surge of interest in this type of surgery since a laparoscopic approach was introduced in 1995, and it is now the most common form of bariatric surgery performed in the United States. The procedure consists of creating a small pouch of proximal stomach that is drained by a roux loop into the mid-jejunum, thereby bypassing the distal stomach and proximal small gut.

Biliopancreatic diversion (BPD) was introduced in 1979 by Dr. Nicola Scopinaro. It remains the least often performed of the major bariatric procedures because of its potentially severe metabolic consequences. The typical procedure consists of resection of approximately two-thirds of the distal stomach and division of the terminal ileum at 250 cm from the caecum, anastomosis of the distal end of this point of division to the proximal gastric remnant, and anastomosis of the proximal end of the point of division to the terminal ileum 50 cm from the caecum.

Laparoscopic adjustable gastric banding (LAGB) was introduced by Dr. Guy Bernard Cadiere in 1992. A ring of silicone is placed around the cardia of the stomach. Adjustment of the volume of saline filling an inner balloon creates a sense of satiety, reducing the wish to eat. When eating does occur, a sense of fullness is achieved with small volumes of food. This is now the most commonly performed weight loss surgical procedure worldwide.

Rationale for Exercise in Bariatric Surgery

Total daily energy expenditure is the sum of resting energy expenditure (REE), the thermic effects of food, and the energy expenditure of activity. Exercise after bariatric surgery is promoted primarily as a method of increasing the energy expenditure of activity, but it also contributes to maintaining muscle mass and thereby contributing to the REE.

Increased Energy Utilization

Weight loss and weight maintenance reflect the energy balance. The most obvious effect of exercise is to increase energy utilization. A 30 min period of moderate-intensity exercise, such as brisk walking, is estimated to use 150 to 300 kcal depending on body weight (1). This would represent an average of 20% of energy intake in a typical post-weight loss surgery patient (2).

Maintenance of Fat-Free Mass

As REE accounts for 60% to 70% total energy expenditure, the loss of muscle mass that accompanies any weight loss has a negative influence, and mitigation of this through exercise should lead to a more favorable outcome. In any weight loss program, it is important to maintain the fat-free mass (FFM), as the muscle component of the FFM is responsible for a significant fraction of the REE. This is also important for the regulation of core body temperature, the preservation of skeletal integrity, and the maintenance of function and quality of life as the body ages. An ideal bariatric surgical procedure should minimize the loss of FFM. A recent systematic review of the loss of FFM after bariatric surgery showed that BPD was associated with 26%, RYGB with 31%, and LAGB with 18% loss of FFM (3). The effect on FFM was independent of the initial body mass index and the magnitude of total weight loss.

The capacity for these values to be modified through exercise has been subject to very little study. Metcalf and colleagues (4) selected 50 "exercisers" and 50 "nonexercisers" from patients who had had a BPD procedure. Exercising at least three times per week for at least 30 min defined the separation of the two groups. Body composition was measured using bioelectrical impedance (BIA). The exercisers had no better weight loss but did show greater fat loss and, according to BIA, showed an increase in FFM. This latter finding needs to be interpreted with caution as the validity of BIA in this situation has been questioned (5), and systematic review of studies using more robust methods has shown substantial loss of FFM as a part of bariatric surgical weight loss.

Metabolic Effects

An extensive range of effects has been described. An important observation, relevant to the postsurgical patient, is the continuation of benefit after the exercise is finished. Excess postexercise oxygen consumption (EPOC) relates to an increased oxygen demand for metabolic and tissue repair, control of the increased body temperature, and elevation of cardiac and respiratory rates. The increased energy expenditure of EPOC prolongs the weight loss benefit of an exercise session (see chapter 43).

Psychological Effects of Exercise

Through improved self-esteem and mood, exercise facilitates people's confidence in adhering to the partnership and following the rules required for optimal surgical weight loss effect.

Does Exercise Make a Difference?

Do physical activity and exercise influence the outcomes after weight loss surgery? In spite of a history of weight loss surgery stretching back over 50 years, we find that this question remains relatively untested, and the answer is still uncertain.

Our group at the Centre for Obesity Research and Education (CORE) has shown that regular physical activity appears to facilitate weight loss after LAGB. Colles and colleagues (2) measured physical activity after LAGB using the Baecke Physical Activity questionnaire, the Physical Component summary score of the Medical Outcomes Trust Short Form SF-36, and a pedometer. For the Baecke, separate work, sport, and leisure index scores were calculated, plus a total score. A total of 127 patients were studied prior to and at 12 months after the LAGB procedure.

Throughout the study period, voluntary activity, assessed by the Baecke Work Index, remained constant, while Baecke Leisure and Sport Index scores increased, contributing to an improvement in the Baecke total score (table 56.1). At 12 months, the Baecke Leisure Index score was positively correlated with the percentage of excess weight lost and dietary restraint and negatively related to total energy intake, dietary disinhibition, hunger, and the number of perceived barriers to exercise. Those reporting the return of "old eating habits" and overeating due to stress also had a lower Leisure Index score. There was also a direct relationship between percentage of excess weight lost and daily pedometer step counts, suggesting that regular walking is associated with better weight loss outcomes.

Several other studies have shown a correlation between exercise and weight loss. Latner and colleagues (6) surveyed 65 female patients by telephone interview at a mean of 16 months after RYGB. They reported an increase in exercise activity from 0.7 to 2.8 sessions per week. Regression analysis showed the level of exercise to be a significant predictor of weight loss. Pontiroli and colleagues (7) performed a

Table 56.1

Changes in Physical Function and Activity Levels During the First Postoperative Year After Laparoscopic Adjustable Gastric Banding Placement

	Preoperative	At 12 months	*P*-value
SF-36 PCS score	37.2	49.2	0.001
Pedometer step count	6061	8716	0.01
Baecke Work Index score	2.49	2.46	ns
Baecke Sports Index score	1.69	2.11	0.001
Baecke Leisure Index score	2.11	2.74	0.001
Total Baecke score	6.30	7.32	0.001
Total barriers to exercise	3.44	2.27	0.001

Continuous variables presented as mean; $N = 129$ except for pedometer step count, where $N = 48$. PCS: physical component summary.

Adapted from S. Colles, J.B. Dixon, and P.E. O'Brien, 2008, "Hunger control and regular physical activity facilitate weight loss after laparoscopic adjustable gastric banding," *Obesity Surgery* 18: 833-840.

prospective study of 172 consecutive LAGB patients, seeking to relate the weight loss outcomes to a range of psychological and compliance variables. They reported that compliance with the rules regarding exercise was strongly associated with better weight loss. The actual exercise rules and the method for measuring compliance were not presented.

In France, a nationwide survey of 1236 bariatric surgical patients (8) showed that the relative risk of not achieving 50% of excess weight loss was 2.3 times greater in those who did not increase their physical activity after bariatric surgery. However, not all studies report a correlation between physical activity and weight loss. Larsen and colleagues (9) used a Baecke Sports Index questionnaire for measuring physical activity in 157 patients at 34 months after LAGB. They also used a Physical Exercise Belief questionnaire to identify the barriers to exercise. Participation in an exercise program was greater when the participants believed in the health benefits of exercise and had less fear of injury or feelings of embarrassment. However, the investigators found no correlation between the level of exercise and either weight loss or physical health.

Recommendations for Exercise Programs in Bariatric Surgical Patients

Because the trend of the data indicates that exercise is a net benefit after weight loss surgery, the recommendations that we should give to the weight loss surgery patient need to be considered. How often should the patient exercise, for how long, and at what level of intensity?

There are no data available that provide an answer specifically for the weight loss surgical patient. We therefore need to draw on the general data for exercise in the obese. The American College of Sports Medicine provided a position paper in 2001 based on an extensive review of the literature at that time. The recommendation was a minimum of 150 min of moderate-intensity exercise per week, progressing to 200 to 300 min per week. People should aim to complete 30 min each day, which can be accumulated in 10 to 15 min periods of activity.

This is the recommendation we provide for our patients after LAGB placement. Walking vigorously is the most easily available, most acceptable, and most commonly used form of exercise. Other aerobic exercises such as exercise bike or treadmill exercise and swimming or water aerobics should be considered. We encourage the patient to seek increased activity throughout the day, and the pedometer is an inexpensive and probably useful way of motivating and monitoring the level of daily activity (10, 11). We encourage resistance exercises as a supplement to endurance work. The relative merits of each for the weight loss surgery patient remain to be established. The offer of various forms of exercise provides better likelihood of compliance.

However, the obese individual has a number of barriers to exercise, and these need to be recognized in the design of a program, or compliance will not occur. Barriers include the following:

- Physical impairment, most commonly associated with degenerative joint diseases, and severe shortness of breath on exertion
- Fear of injury, and exacerbation of existing back, hip, and knee problems
- Lack of experience, confidence, and motivation, as many have never exercised before in their life
- Lack of time, being too busy, unfavorable weather, darkness, family commitments (conditions that are always available as an excuse for not doing something one doesn't want to do)
- Embarrassment about attending a gym or walking in the street
- Cost associated with gym membership, personal trainers, and home equipment

Attention to these barriers is important. The established values of exercise need to be stressed and priority in allocation of time and budget encouraged. At the same time it is important to emphasize that regular physical activity does assist in achieving weight loss as well as improving general health and the feeling of well-being and is one of the strongest predictors of long-term weight loss maintenance (12).

Future Directions

There are many gaps in the evidence required to define the role and optimal application of exercise and activity in the weight loss surgery patient. As weight loss surgery becomes a part of the mainstream of management of the disease of obesity, there is an urgency to the filling in of these gaps. Studies should include careful definition of the benefits in weight loss, health, quality of life, and survival associated with adherence to good exercise and activity. What is the optimal intensity, duration, and frequency? What are the relative roles of aerobic and resistance exercise? What changes in body composition occur with weight loss in the presence or absence of an exercise program? How can we optimize compliance for a group who is not usually exercise oriented? With the appropriate collaborations in place, these and other important questions can be answered.

57

Physical Activity and Depot-Specific Fat Loss

Victor Katch, PhD

Maintenance of negative caloric balance, at least for some time period, via increased physical activity (exercise), reduced caloric intake, or both, determines amount of body mass loss. That it is possible to gain or lose adipose or lean tissue without negative caloric balance suggests the possibility that fat, muscle, or both are regulated independently of total caloric balance and perhaps locally. There are differential and conflicting reports on changes in fat from different sites (e.g., visceral vs. subcutaneous) whether induced via diet or exercise alone or in combination. While changes in waist girth (representing subcutaneous abdominal fat [SAT]) and sagittal diameter (representing visceral adipose tissue [VAT]) appear well correlated, they seem to be dependent on the amount of body mass loss. Indeed, percentage weight loss represents the strongest predictor of preferential VAT loss, but with greater weight loss the effect is attenuated (1).

As early as 1895, Checkly (2) wrote that "fat disappears in those areas of the body where muscles are active and in proportion to their activity" (88). Indeed, individuals whose job requires standing and moving but who exercise little and do not restrict calories often exhibit relatively muscular legs with proportionally greater subcutaneous fat in the abdominal area and upper arm region.

Some studies support exercise-induced reduction in adipose tissue depots and changes in fat distribution (3-5), while other data show no exercise-induced effects (6-12).

Despite the general consensus that spot reduction does not occur, there is recent evidence that exercise-induced relative loss of fat is higher in visceral and abdominal subcutaneous adipose tissue than in femoral adipose tissue resulting from general exercise. Further, it shows that specific exercises may contribute to enhanced (specific) subcutaneous adipose tissue lipolysis, which has renewed interest in this area.

A careful review of the literature reveals controversy regarding subject selection, measurement issues, experimental design problems, exercise modality used to train subjects, and data interpretation that can account, at least in part, for the discrepancies among studies examining the concept of spot reduction.

Subject Issues

Few spot reduction studies employ strict subject inclusion and exclusion criteria. In light of gender differences in subcutaneous tissue lipolysis, exercise-induced body composition changes, rate of body mass changes, race differences in fat deposition and loss, differences in body fat patterning, and biochemical differences in adipose tissue response to acute and chronic exercise, it is difficult to ascertain the extent to which inclusion and exclusion criteria, gender, age, race, and even body composition status (body mass index, percent fat, fat patterning, or muscle mass) influence results. To date, research has not resolved these issues. The gender issue becomes particularly perplexing in light of apparent gender differences in metabolism, fat patterning, and muscle.

The number of subjects in any one study is limited (with few exceptions), and generally no estimates of the number of subjects necessary to show statistical significance are reported. This suggests that in some studies, sample sizes may not have been sufficient to allow detection of statistical significances. However, the total number of subjects in the various exercise-induced spot reduction studies appears sufficient to lead to consensus; that no clear consensus has emerged is confusing and perhaps points to the small number of definitive studies in the area.

Measurement Issues

Different studies have used different testing techniques, involving both surface measurements (mostly skinfold thickness or circumferences) and other more elaborate measurements including magnetic resonance imaging (MRI), computed tomography, dual X-ray absorptiometry, ultrasound, fat cellularity measures, and, more recently, subcutaneous blood flow (^{133}Xe washout technique) and subcutaneous tissue lipolysis (calculated from interstitial and arterial glycerol concentrations and blood flow).

The method of measurement, obviously, represents a matter of great importance (13). The issue of measurement reliability raises concerns regarding both surface and more elaborate methodologies that have been used to support or refute the efficacy of spot reduction. While more is known about the effects of unreliability of surface anatomy techniques (skinfold and girth measurements) and its impacts on the magnitude of change necessary to show significant differences, rarely are the effects discussed or estimates reported. Much less is known about errors involved in the techniques. These techniques entail multiple measurement steps, analyses, and data manipulations that lend themselves to multiplication of errors, which may mask the degree of differences necessary to show significance. Future researchers need to clarify and report on reliability data and to comment on the effects of unreliability in determining intervention effects.

Several investigators have noted differences in estimates of spot reduction depending on methodology. For example, subcutaneous fat did not change in a sample of adult men and women assessed by skinfold before and after 16 weeks of resistance training, whereas subcutaneous fat decreased when assessed by MRI, computed tomography, or dual X-ray absorptiometry (12). The authors concluded that subcutaneous fat could be detected only via MRI, computed tomography, or dual X-ray absorptiometry and not with anthropometric methods, particularly skinfold thickness measurements. This raises the interesting question of measurement validity, a topic that has rarely been the center of attention. Reported low correlations ($<r = 0.50$) between fat estimates using skinfolds and MRI or ultrasound (10, 11) illustrate this point. One plausible explanation may be that with specific exercise training, particularly resistance training, increased muscle hypertrophy compresses the extracellular space between fat cells (10). If the total amount of subcutaneous fat in the exercised limb (e.g., arm or leg) remains the same after training (fat cell diameter does not change) but in fact occupies less area because of muscle hypertrophy, this would result in a decrease in subcutaneous fat assessed by skinfold, but not with MRI. Thus, using only skinfold data would be misleading.

Body Areas Studied and Exercise Mode

The specific body areas addressed differ among studies and include subcutaneous fat over the triceps (7, 11) forearms (6), thigh (5, 10), and abdomen (3, 8, 9). In general, no consensus emerges to indicate that one body part is better than another for demonstrating exercise-induced spot reduction. In most studies, muscles in one body part (e.g., right arm) are trained and the muscles in the contralateral side (left arm) are not, and the size of the subcutaneous adipose tissue depot between the two sides is compared.

A recent study by Kostek and colleagues (12) illustrates this experimental approach and differences in conclusions depending on the measurement (MRI vs. skinfolds) method. The investigators studied 104 subjects (45 men and 59 women) who participated in 12-week supervised resistance training of the nondominant arm. Magnetic resonance imaging and skinfold calipers were used to test for subcutaneous fat in the nondominant (trained) and dominant (untrained) arms before and after training. Results showed that subcutaneous fat changes resulting from resistance training varied by gender and assessment technique. Measured by skinfold calipers, subcutaneous fat decreased in the trained arm but not the untrained arm in the men ($p \leq 0.01$) but was similar for the two arms in the total sample and in the women ($p \geq 0.05$). In contrast, MRI measures of subcutaneous fat changes were not different between arms in the total sample or by gender ($p \geq .050$). Correlations between absolute subcutaneous fat volume change and skinfold thickness change in the trained and untrained areas were all less than $r = 0.3$ ($p \geq 0.05$). The authors concluded that based on the MRI data, spot reduction did not occur; they dismissed the skinfold data because of their inherent limitations in adequately detecting subtle volumetric changes that can be observed with use of MRI scans.

An interesting study that used typical sit-up exercise training showed no changes in skinfold estimates of fat in the abdomen or other, nonexercise sites or in fat cell volume estimates. In this study (10), subjects (N = 13 males) interval trained using multiple-repetition bouts of 10 sit-ups for 27 days; the total number of sit-ups done in the 27-day experiment was 5004. The cumulative actual time spent performing the 5004 sit-ups was 2 h and 3 min or, on average, about 4 min and 33 s per day.

Table 57.1 presents results from this study. While fat cell diameter measurements were significantly reduced, there were no differences between sites over time, indicating no exercise-induced spot reduction effect. In contrast to findings in the study by Kostek and associates (12), the fat cell diameter data were mirrored by nonsignificant and minor changes in the skinfold data.

One hypothesis to support exercise-induced spot reduction involves the idea that local muscle action will stimulate nearby subcutaneous adipose tissue (SCAT) lipolysis, hence induce spot reduction of adipose tissue and thereby modify fat distribution. Recent advances in microinvasive measurements of SCAT make it possible to study this hypothesis. Stallknecht and associates (6) evaluated the hypothesis by having male subjects perform acute one-legged knee extension exercise. ^{133}Xe washout and microdialysis techniques were used to estimate blood flow and lipolysis in femoral SCAT adjacent to contracting and resting skeletal muscle.

Ten healthy, overnight-fasted males performed one-legged knee extension exercise at 25% of maximal

Table 57.1

Pre- to Postchange in Dependent Variables Resulting From Performing 5004 Sit-Ups Over 27-Day Period

	Mean, pre	Mean, post	Percent change	Significance
CELL DIAMETER, MICRONS				
Abdomen	82.64	77.39	6.4	$p \leq 0.01$
Gluteal	90.31	85.79	5.0	$p \leq 0.01$
Subscapular	76.53	73.73	3.7	$p \leq 0.01$
SKINFOLDS, MILLIMETERS				
Triceps	10.66	10.15	4.7	ns
Scapula	12.95	12.95	0	ns
Iliac	17.89	18.01	+0.6	
Abdomen	18.62	18.62	0	ns
GIRTHS				
Waist	83.0	84.0		ns
Umbilicus	84.55	85.23		ns
Buttocks	96.78	95.37		ns
Biceps flexed	28.92	28.87		ns
Calf	36.95	36.59		ns

(N = 13 Male Subjects) Note: The interaction term (site × condition) was nonsignificant, indicating no differences in the rate of change for cell diameter between anatomic sites.

Adapted from F.I. Katch et al., 1994, "Effect of sit up exercise training on adipose cell size and adiposity," *Research Quarterly for Exercise and Sport* 55: 242-247.

workload (W_{max}) for 30 min followed by exercise at 55% W_{max} for 120 min with the other leg, and finally exercised at 85% W_{max} for 30 min with the first leg. Subjects rested for 30 min between exercise periods. Femoral SCAT blood flow was estimated from washout of ^{133}Xe, and lipolysis was calculated from femoral SCAT interstitial and arterial glycerol concentrations and blood flow. In general, blood flow and lipolysis were higher in femoral SCAT adjacent to contracting muscle than in that adjacent to resting muscle. Figure 57.1 shows the data for adipose tissue lipolysis, represented by the glycerol release (nmol · 100 g^{-1} · min^{-1}) data. Lipolysis of SCAT tissue adjacent to activated muscle at 25% W_{max} was 102 ± 19 nmol · 100 g^{-1} · min^{-1} compared to the glycerol release in adipose tissue adjacent to resting muscle (55 ± 14 mmol · 100 g^{-1} · min^{-1}, $P \leq 0.06$). At 55% W_{max}, glycerol release was 86 ± 11 nmol · 100 g^{-1} · min^{-1} versus 50 ± 20 nmol · 100 g^{-1} · min^{-1}, $P \leq 0.05$; and at 85% W_{max}, it was 88 ± 31 versus −9 ± 25 nmol · 100 g^{-1} · min^{-1}, $P \leq 0.05$.

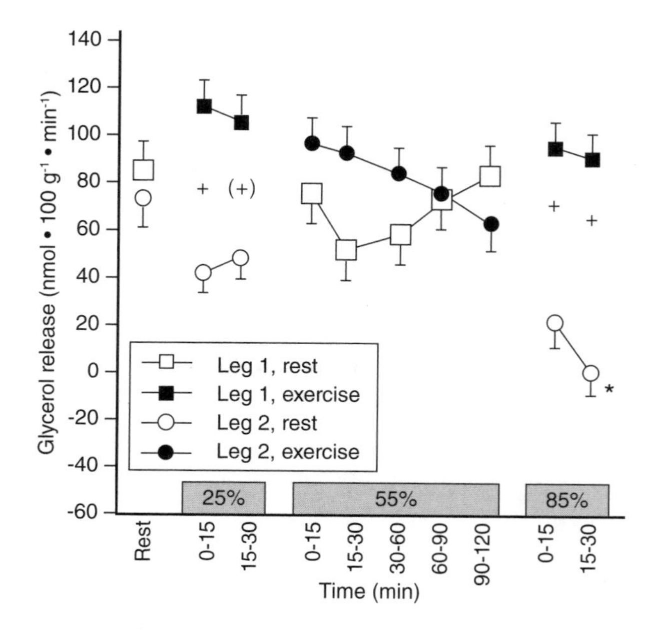

Figure 57.1 Adipose tissue lipolysis. Ten healthy, overnight-fasted males performed one-legged knee extension exercise at 25% of W_{max} for 30 min, followed by exercise at 55% W_{max} for 120 min with the other leg, and finally exercised at 85% W_{max} for 30 min with the first leg. Subjects rested for 30 min between exercise periods. Femoral subcutaneous adipose tissue lipolysis of both legs was calculated from the interstitial and arterial glycerol concentrations and adipose tissue blood flow. *$P \leq 0.05$ versus rest; +$P \leq 0.05$ between legs; (+)$P \leq 0.1$ between legs.

Adapted from B. Stallknecht, F. Dela, and J.W. Helge, 2007, "Are blood flow and lipolysis in subcutaneous adipose tissue influenced by contractions in adjacent muscles in humans?" *American Journal of Physiology-Endocrinology and Metabolism* 292: E394-E399.

These data show that blood flow and SCAT lipolysis are higher adjacent to contracting muscle versus adjacent to resting muscle, respective of exercise intensity. These data clearly support the hypothesis that specific exercises can induce "spot lipolysis" in adipose tissue. Whether this translates to actual fat loss over time at a particular site remains unknown.

Summary

In summary, confusion exists regarding the extent to which exercise-induced spot reduction is possible. It seems likely that to reduce fat content at a given body site, total body composition must be altered. More data in this area are needed.

58

Physical Activity, Visceral Fat, and Ectopic Fat Deposition

Bret Goodpaster, PhD

Physical inactivity and obesity are both associated with an increased risk for type 2 diabetes and cardiovascular disease. Exercise and weight loss induced by energy restriction both reduce those risks. What is less clear is whether the effect of physical activity or exercise on cardiometabolic risk can be partly reconciled through its effect on body weight or body fat. Although moderate exercise does not seem to promote large changes in body fat in the short term, it can be important to help induce weight loss or perhaps help in maintaining healthy weight. An alternative hypothesis is that at least some of the benefits of physical activity are due to selective reduction in visceral abdominal adipose tissue and other region-specific fat depots. This concept can be extended more generally to ectopic fat, or fat depots contained within tissues or organs, because of its potential role in type 2 diabetes, hepatic liver disease, cardiomyopathy, and aging. This chapter explores the evidence for the effects of physical activity on ectopic fat deposition and whether any such effects influence insulin resistance and other cardiometabolic risk factors in obesity.

Effects of Physical Activity on Visceral Abdominal Fat

Specific health risks have been associated with specific patterns of body fat distribution. A notable example is that of upper body, central, or abdominal obesity, which has been related to the risk for type 2 diabetes mellitus and cardiovascular disease. Visceral adiposity has distinctively been suggested to be an important link between cardiorespiratory fitness and markers of the metabolic syndrome. Moreover, improved insulin sensitivity consequent to weight loss has been specifically associated with the loss of visceral abdominal fat (1). Exercise, with or without diet-induced weight loss, seems to promote greater reductions in visceral abdominal fat relative to general body fat (2). Further, exercise training–induced reductions in visceral fat have been associated with improvements in insulin sensitivity in obese subjects (3). These associations, however, have not been consistently observed (2). Thus it remains to be determined whether the specific effects of physical activity on insulin resistance and other cardiovascular disease risk factors are due at least in part to selective reductions in visceral abdominal fat independent of total fat loss.

Effects of Physical Activity on Intra- and Extramyocellular Lipids

While abdominal adiposity has received considerable scrutiny with regard to its role in insulin resistance and the etiology of type 2 diabetes, these disorders have also been characterized by an increased amount of lipid contained within skeletal muscle (4). Analogous to associations with visceral abdominal fat, several studies have demonstrated strong associations between high intramyocellular lipid (IMCL) content and skeletal muscle insulin resistance in obesity and type 2 diabetes (4). Further, diet-induced weight loss reduces IMCL concomitant with improved insulin sensitivity in obesity and type 2 diabetes (1).

Yet, despite these numerous observations, other researchers have not seen changes in IMCL with weight loss, possibly due to the apparently different effects of diet-induced weight loss and exercise training on IMCL. We described an "athlete's paradox" in which highly insulin-sensitive endurance-trained athletes have IMCL content similar to that observed in insulin-resistant obese and type 2 diabetic subjects (5). This effect of exercise training does not appear to be limited to highly trained athletes; exercise training as short as 12 weeks can increase IMCL in previously sedentary subjects. Collectively, studies raise the possibility that weight loss and exercise have counterbalancing effects on muscle lipid.

The paradoxically higher IMCL levels associated with exercise have helped stimulate a number of investigations into the roles of IMCL and ectopic fat in obesity and metabolic disease. Specifically, studies examining IMCL at the subcellular level have revealed that IMCL as triglycerides do not confer insulin resistance but rather are likely a marker for other potentially harmful lipids within myocytes, including diacylglycerol and ceramides. Recent studies indicate that while exercise training may increase intramyocellular triglycerides, these other lipid species can be decreased with training (6). These studies suggest but do not prove that the utilization of intramyocellular triglycerides as a fuel source for exercising muscle, or at least the capacity for fat oxidation, plays an important role in partitioning between good and bad lipids within muscle. Other evidence, while sparse, suggests that the physical distribution, subcellular localization, or size of lipid droplets may also play a role in the associations among obesity, physical activity, and insulin resistance.

In contrast to visceral abdominal fat and IMCL, a body fat depot that has received relatively little attention in human obesity is that of extramyocellular lipid. While there is a considerable proportion of subcutaneous fat mass in the lower extremities, it is regarded as having only a weak relation to insulin resistance. The amount of intermuscular adipose tissue, or extramyocellular lipid, however, despite being only a small proportion of thigh adiposity, has been significantly correlated with insulin resistance in obesity (7). As with visceral fat, the responsible mechanisms for this association are uncertain but could be related to a greater resistance to insulin suppression of lipolysis, or greater and differing secretions of adipokines that induce insulin resistance, a hypothesis that to our knowledge remains untested. In any case, recent evidence suggests that physical activity selectively reduces or prevents the age-associated gain in intermuscular adipose tissue, which may be linked to improved insulin sensitivity with diet and exercise interventions. Given these observations and unanswered questions, further research is obviously warranted to more fully elucidate the potential role of IMCL and intermuscular adipose tissue in the association between physical activity and obesity.

Physical Activity and Intrahepatic Fat

Fat contained within the liver not related to chronic alcohol consumption, termed nonalcoholic hepatic steatosis (NASH), is closely related to obesity and type 2 diabetes and also to the severity of insulin resistance. The literature has also consistently and fairly convincingly demonstrated that weight loss induced by energy restriction, with or without increased physical activity, reduces liver fat. Studies combining diet-induced weight loss and increased physical activity have also shown decreases in intrahepatic fat (8). There are fewer published studies, however, concerning the specific effects of exercise on intrahepatic fat content. Devries and associates (9) showed that short-term endurance training without weight loss did not alter hepatic lipid content in lean or obese subjects. Larson-Meyer and associates (10), however, demonstrated that exercise, with or without energy restriction, reduced intrahepatic fat content in overweight subjects, although these changes in liver fat were not associated with improvements in insulin sensitivity. Although an exercise-mediated link between hepatic fat accumulation and insulin resistance has not been elucidated, the benefits of exercise on hepatic and lipoprotein metabolism may relate to the potential effects of exercise on reducing or preventing hepatic lipid accumulation.

Conclusions

In addition to the increased risk of type 2 diabetes and cardiovascular disease conferred by generalized obesity, the literature firmly supports the concept that fat distribution, and more specifically ectopic fat accumulation, plays a particular role in pathophysiology related to obesity. As summarized in table 58.1, physical activity may selectively alter various ectopic fat depots, which in turn may promote improvements in insulin sensitivity or result in other favorable changes in cardiometabolic risk.

Table 58.1

Ectopic Fat Depots Associated With Obesity and Obesity-Related Disease and How They Are Selectively Affected by Physical Activity

Ectopic fat depot	Link to obesity	Link to disease risk independent of generalized obesity	Effects of physical activity
Visceral abdominal fat	Elevated	Associated with insulin resistance, type 2 diabetes, metabolic syndrome	Selectively decreased
Intramyocellular triglycerides	Elevated	Associated with insulin resistance, type 2 diabetes	Increased
Intramyocellular diacylglycerol	Unclear	Associated with insulin resistance, type 2 diabetes	Unclear
Intramyocellular ceramides	Unclear	Associated with insulin resistance, type 2 diabetes	Decreased
Intermuscular adipose tissue	Elevated	Associated with insulin resistance, type 2 diabetes	Decreased
Intrahepatic lipid	Elevated	Associated with insulin resistance, type 2 diabetes	Unclear

Our current understanding of the potential exercise- or physical activity–mediated effects, however, is far from complete. Further research is necessary to determine how the selective reduction in visceral abdominal fat with exercise may translate into improvements in insulin resistance and the risk for both type 2 diabetes and cardiovascular disease. In addition, while IMCL have been reported to be linked with insulin resistance, we still do not know whether lipids within muscle cause insulin resistance in humans. Further research is also needed to elucidate the effects of exercise on lipid accumulation in the liver in the treatment of hepatic insulin resistance. Lastly, other ectopic fat depots such as epicardial fat and intramyocardial lipid deserve more attention regarding methods for assessment and their potential role in health and disease.

Physical Activity, Weight Loss, and Maintenance of Lean Mass

Steven B. Heymsfield, MD

Increasing physical activity levels is at the center of advice provided to overweight and obese patients as a part of lifestyle management treatments. Does increasing physical activity prevent the anticipated loss of lean mass and related lowering of energy expenditure accompanying diet-induced weight loss? Lean mass is an energy-producing functional body compartment whose loss during dieting is ideally minimized.

Body Composition and Energy Expenditure Effects

Physical activity raises energy expenditure above basal levels, and each type of activity has an energetic cost, usually expressed as a ratio to body mass. The total energy expended over 24 h reflects the sum of these individual costs combined with their duration.

Physical activity also influences energy expenditure, notably resting energy expenditure (REE), through the impact of activity-related effects on body composition. The amount of energy produced at rest reflects the sum of postabsorptive energy expended by each organ and tissue. The basal heat production, expressed as the mass-specific metabolic rate, is 4.5, 13, and 240 kcal/kg per day for adipose tissue, skeletal muscle, and brain, respectively. Adipose tissue is a low energy expenditure tissue while lean tissues, as these figures show, have a much higher range of mass-specific metabolic rates. The contribution of each organ and tissue to body weight thus influences the subject's whole-body REE.

Exercise, particularly resistance exercise, leads to skeletal muscle hypertrophy and adipose tissue loss when expended energy is counterbalanced by an increase in food intake. Hence, REE typically is higher relative to body weight in highly trained athletes. Let's assume, as an example, that vigorous bodybuilding increases a 70 kg (154 lb) man's skeletal muscle mass by 2 kg (4.4 lb). The additional expended energy is perfectly matched by the subject with an increase in food intake so that zero energy balance is maintained. Skeletal muscle has an energy content of about 1 kcal/g, so adipose tissue lipids must contribute at least 2000 kcal in order for the subject's muscles to grow. Adipose tissue, which is roughly 85% fat (i.e., triglyceride), has an energy content of about 7.5 kcal/g. Our rough calculations thus suggest that the subject's adipose tissue mass decreases at least by 0.27 kg (0.6 lb) in order to provide the 2000 kcal required for skeletal muscle growth. Since the respective postabsorptive energy expenditure of skeletal muscle and adipose tissue are 13 and 4.5 kcal/kg per day, the hypothetical subject's REE will increase by 25 kcal/day from an assumed level of 1400 kcal/day. The subject's weight will also increase by 1.73 kg (3.8 lb), and REE/weight will increase from a baseline of 20 kcal/kg per day to 20.15 kcal/kg per day after training. This greatly oversimplified example is intended only to convey the relatively small magnitude of involved metabolic and body composition effects.

Such perfect energy "balance" is possible only in theoretical experiments, and substantial increases in physical activity are usually accompanied by only partial food compensation. Subjects increasing their physical activity levels therefore typically lose weight, particularly if they are participating in a weight control program. Even if subjects are asked to maintain their food intake unchanged while increasing activity levels, this in effect creates a "relative" low-calorie diet (LCD) and negative energy balance with weight loss.

Low-Calorie Diet Effects

When prescribed a LCD, subjects enter a phase of negative energy balance and, depending on the deficit, eventually come into energy equilibrium with maintenance of a new stable lower body weight. Assuming a typical LCD program, the first one to two weeks following induction of negative energy balance are accompanied by early losses of stored glycogen with associated intracellular and extracellular water. The fraction of weight loss as lean mass during this period is accordingly relatively large, with additional contributions from oxidized body proteins and to a less extent stored lipids.

Following the initial rapid weight loss phase, the subject in negative energy balance steadily loses weight for variable periods of time, depending on baseline body weight and imposed energy deficit. The composition of weight loss during this second phase is variable and depends on several factors, although we can generalize that about 60% to 75% can be accounted for by adipose tissue, with the remainder skeletal muscle and residual lean tissues.

Lean mass is increased in the obese state as excess weight creates a mechanical load leading to enlargement of the skeletal muscle compartment. Similarly, the metabolic demands of a larger body mass and greater energy flux require a corresponding adaptive enlargement of liver, kidneys, heart, and other organs and tissues. With weight loss during LCD treatment, the mechanical and metabolic loads are lowered, and appropriate remodeling accounts for negative protein balance derived from skeletal muscle and other lean tissues.

If we consider another hypothetical man, this time an obese man weighing 100 kg (220 lb), baseline skeletal muscle mass would be about 34 kg (75 lb). At a weight of 70 kg (154 lb), nonobese men usually have about 28 kg (62 lb) of skeletal muscle. Our subject would therefore experience a 6 kg (13 lb) loss of skeletal muscle with "normalization" of body weight. While the evidence is limited, available data suggest that reduced-obese and never-obese subjects who weigh the same have similar body composition.

Forbes (1-3) and more recently Hall (4) attempted to model the "companionship" of fat and lean mass changes with weight loss (figure 59.1). The developed models leading to "Forbes' rule" support the view that losses of both fat and lean mass occur with low-calorie dieting and that the relative lean loss is less in subjects with a large initial fat mass than it is in subjects with a small initial fat mass. While we can thus generalize about the composition of weight loss with dieting, the proportions of fat and lean loss are actually not constant but are strongly influenced by the subject's baseline body composition and other factors.

Energy expenditure declines as subjects lose weight with LCD (5). Several mechanisms lead to this effect: A lower body mass reduces the energy cost of physical activities; losses of metabolically active adipose and most lean tissues cause a lowering of REE; and metabolic adaptations lower REE beyond that accounted for by weight loss–induced body composition effects.

Combined Activity and Low-Calorie Diet Effects

The usual recommendation to obese subjects enrolling in a weight loss program is to reduce food intake and increase physical activity levels. How much lean mass is preserved when people increase physical activity while simultaneously decreasing food intake? What effect does the potential preservation of lean mass have on REE (see figure 59.2)?

We can gain perspective on these questions by returning to our examples. Our obese man weighed 100 kg (220 lb) at baseline and 70 kg (154 lb) following LCD treatment; skeletal muscle loss was 6 kg (13 lb). Our exercising man weighing 70 kg increased his skeletal muscle mass by 2 kg (4.4 lb). When prescribed both exercise and LCD, our obese man should thus experience a 4 kg (8.8 lb) loss of skeletal muscle rather than 6 kg (13 lb) as with the LCD alone. The obese man's REE should also be 25 kcal/day greater after weight loss than if he had not gained skeletal muscle and lost adipose tissue. These are obviously very oversimplified calculations, but they are useful as a basis for gaining insights into the questions posed in this review.

Are our hypothetical calculations for lean mass changes with weight loss combined with exercise accurate? To experimentally answer this question we need to critically consider the following: type

$FFM = 10.4 \bullet \log_e FM + 14.2$

- *Forbes developed an empirical equation linking fat mass (FM, kg) and fat-free mass (FFM, kg) in women of similar height (1).*
- *Equation predicts a loss of FFM with corresponding loss of FM.*
- *Forbes validated the equation in longitudinal cohorts from his own studies and reports of others.*

$d(FFM)/d(BW) = 10.4/(10.4+FAT)$

- *When differentiated, Forbes' equation predicts a smaller FFM loss with body weight (BW) loss in obese subjects who have a large baseline FM, as compared to lean subjects who have a low baseline FM.*

With exercise, Forbes also observed that (2, 3):

- When weight is stable, exercise produces a "modest" gain in lean mass and thus loss of body fat.
- With weight loss of several kg or more, "erosion" of lean mass usually occurs with exercise but is less in overweight or obese than in lean subjects:
  - Individuals with a large body fat mass lose ~0.25 kg lean mass/kg body weight loss.
  - Individuals with a small body fat mass lose ~0.50 kg lean mass/kg body weight loss.

Figure 59.1 "Forbes' rule" linking changes in body weight, fat-free mass, and fat mass (1) and his later observations related to exercise effects on body composition (2, 3).

and amount of prescribed physical activity; type of LCD or very low-calorie diet and prescribed energy deficit; baseline subject sex, body mass index (BMI), race, and age; method of evaluating compliance to the prescribed exercise and diet recommendations; experimental design, including study duration and characteristics of the intervention and control groups; primary study hypothesis with related power calculations; and methods of quantifying body composition and REE. Many studies exploring these questions have been reported and, except for a very few recent ones, have been collected in several published meta-analyses (6-8).

Only rarely are studies appropriately designed and powered so that it is possible to firmly evaluate the extent to which added physical activity "preserves" lean mass and REE with weight loss treatments. Weight loss is typically modest in nonsurgical studies, usually ranging from 5 to 10 kg (11 to 22 lb) in subjects with baseline weights approaching 100 kg (220 lb). Studies lasting only several weeks may be influenced by early-phase weight loss effects, and REE is in an "adapted" state if subjects are evaluated while still losing weight. Initiating and then maintaining an exercise program is accompanied by fluid balance changes that are inseparable from lean tissue protein accretion via most applied body composition methods.

Finally, REE may be elevated for many hours following intensive exercise so that measurement timing is critical.

Experimental Observations

Given these provisos, what is the available evidence in support of Forbes' rule and our hypothetical calculations? Forbes also examined the effects of exercise on body composition by collecting available data from his own research and from other investigations and by extracting information from previously published papers (2, 3). His analysis upheld Forbes' rule by showing that with dieting and exercise there is fat-free mass (FFM) loss, but the decrements are less in obese compared to lean subjects (figure 59.1).

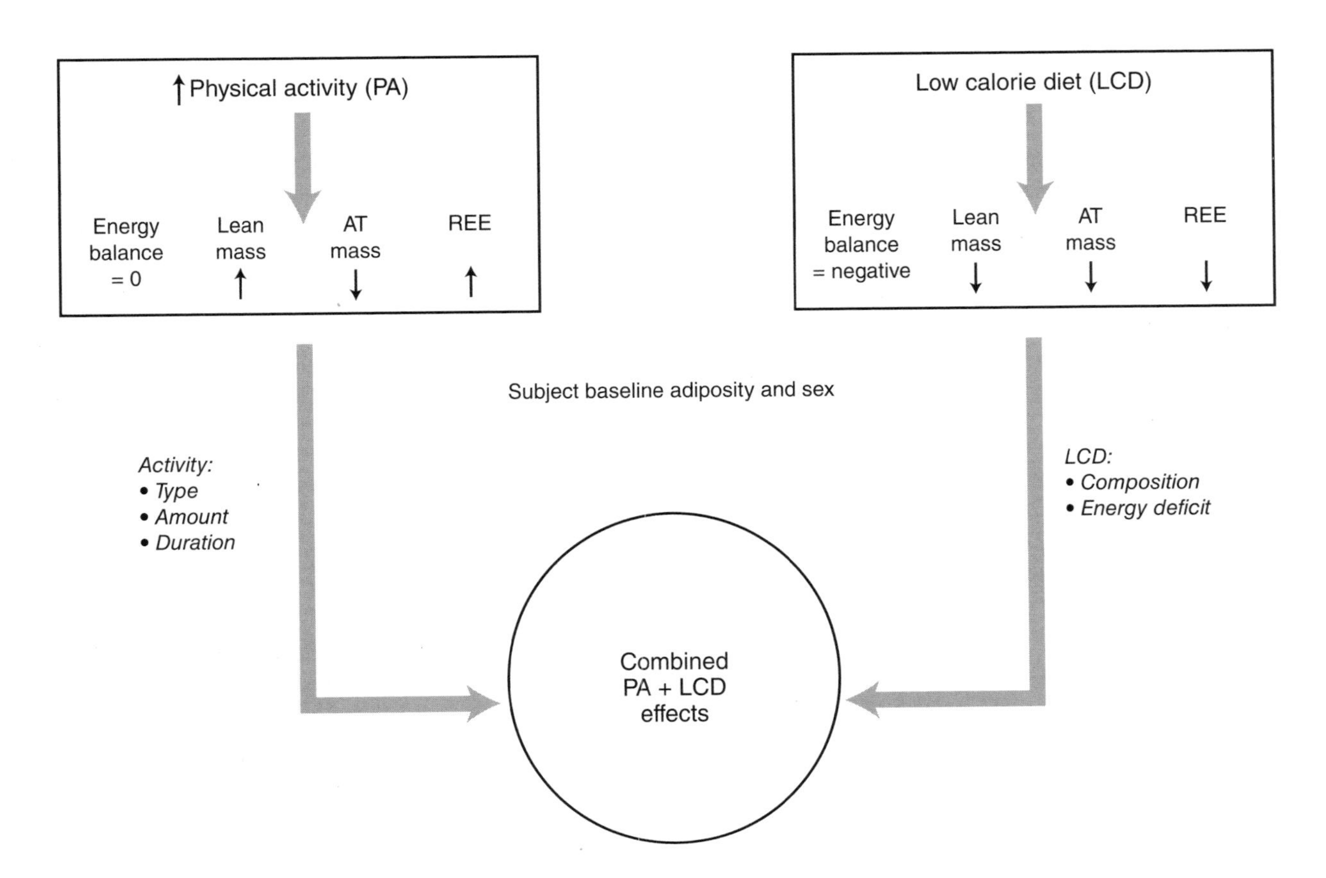

Figure 59.2 Some of the multiple factors that determine changes in lean mass and resting energy expenditure when subjects embark on a combined physical activity and low-calorie diet (LCD) weight loss program. The effects of increasing physical activity while maintaining energy balance are shown on the left, and the effects of ingesting a LCD while maintaining stable physical activity are shown on the right. AT, adipose tissue; PA, physical activity; REE, resting energy expenditure.

Three meta-analyses also explored aspects of these questions (6-8), and two are appropriate for review in the current context. Garrow and Summerbell (7) examined 28 publications that included 425 sedentary and 491 exercising subjects. Aerobic exercise without a prescribed restriction in dietary intake among men caused a weight loss of 3 kg (6.6 lb) in 30 weeks, compared with sedentary controls, and 1.4 kg (3 lb) in 12 weeks among women; but no significant effects were observed on FFM. Resistance exercise had little effect on weight loss but increased FFM by about 2 kg (4.4 lb) in men and about 1 kg (2.2 lb) in women. The authors next used regression analysis to examine the determinants of FFM changes observed in weight loss studies. For a weight loss of 10 kg (22 lb) by diet alone, the authors' model predicted a FFM loss of 2.9 kg (6.4 lb) in men and 2.2 kg (4.9 lb) in women. When an equivalent weight loss is achieved by exercise combined with a LCD, the predicted FFM is reduced to 1.7 kg (3.7 lb) in men and women.

In 2007, Chaston and associates published a meta-analysis on the changes in FFM with weight loss (8), including 26 cohorts with dietary and behavioral interventions. The degree of caloric restriction was associated with the percent of weight loss as FFM ($r^2 = 0.31$, $p = 0.006$), and no additional variance was explained by baseline BMI, magnitude of weight loss, gender, or reported exercise. The fraction of weight loss as FFM tended to be greater in men (27 ± 7%) compared to women (20 ± 7%; $p = 0.08$). Three separate randomized trials from the same research group reporting a 10 to 14 kg (22 to 31 lb) weight loss showed that aerobic exercise significantly reduced the percentage of weight loss as FFM compared to a LCD (13.0 ± 4.1% vs. 27.8 ± 6.4% in pooled subjects). Resistance training designed to increase strength reported in these studies was associated with a percent FFM loss of 16.6 ± 3.7%. A significant effect of gender was also observed in these studies, with percent FFM loss significantly ($p < 0.001$) less in women than in men.

The recent study of Redman and colleagues (9) serves as an example of research aimed at evaluating combined body composition–REE effects. Redman and associates (9), as part of the CALERIE Study, randomized men and women (age ~40 years, BMI ~28 kg/m^2) to a LCD (25% calorie deficit; $n = 12$) or LCD + aerobic exercise (each 12.5% calorie deficit; $n = 12$) for six months. Weight loss was similar and nonsignificantly different in the two groups (LCD, [X ± SEM] 10.4 ± 0.9%; LCD + exercise, 10.1 ± 0.9%). Based on dual-energy X-ray absorptiometry and data extracted from the publication, the respective losses of fat and FFM were 5.8 kg (12.8 lb) and 2.5 kg (5.5 lb) for the LCD group and 6.4 kg (14 lb) and 1.7 kg (3.7 lb) for the LCD + exercise group. These changes in body composition did not differ significantly between the groups. The changes observed from baseline in FFM were statistically significant in both groups. Sleep energy expenditure, a measure of REE, decreased from baseline by 127 kcal/day in the LCD group and by 104 kcal/day in the LCD + exercise group (P = ns for between-group comparisons).

Conclusions

The effects of increasing physical activities while maintaining energy balance and of a LCD during maintenance of stable physical activity on body composition and energy expenditure are each well established (figure 59.2). When combined in a weight loss program, are the body composition and REE effects the sum of these two interventions? Although we have no reason to think otherwise, the experiments required to critically test this hypothesis are difficult to carry out for the reasons reviewed earlier. Most of the available experimental data suggest that when prescribed a LCD alone, overweight and obese subjects enter a negative energy balance phase and lose weight with decrements in adipose tissue and to a less extent lean tissues; that the proportional loss of each compartment is related to baseline adiposity, the magnitude of energy imbalance, the LCD quality, duration of treatment, and possibly sex; and that there is a corresponding lowering of REE secondary to metabolically active tissue loss and, during the active phase of weight loss, metabolic adaptations. When an exercise program is prescribed in addition to a LCD, there is an increase in the magnitude of negative energy balance and of trophic effects on skeletal muscles that, respectively, appear to increase the proportion of weight loss as adipose tissue and decrease the proportion of weight loss as skeletal muscle and other lean tissues. The "relative" preservation of lean tissues by added physical activity may reduce the magnitude of expected REE lowering with weight loss, but this effect is likely very small and not consistently observed across studies.

60

Physical Activity for Weight Loss Maintenance

Rena R. Wing, PhD

Physical activity plays a major role in the maintenance of weight loss. Randomized trials comparing diet only, physical activity only, and the combination of diet and physical activity are quite consistent in showing that the combination produces the best weight loss—particularly in the long term. Likewise, correlational studies show that those individuals who maintain the highest levels of activity are the ones who are most successful at weight loss maintenance. The important questions facing the field have thus become (a) why is exercise so consistently related to long-term weight loss maintenance? and (b) how much exercise should be recommended for long-term maintenance of weight loss?

Relationship Between Activity and Weight Loss Maintenance

There are a wide variety of physiological and psychological mechanisms that may relate to the long-term relationship between activity and weight loss maintenance (1). These include the effects of increased energy expenditure during or subsequent to the bout of physical activity (or both), changes in lean body mass and resting metabolic rate, and changes in dietary intake or macronutrient selection. Likewise, a variety of psychological mechanisms have been suggested, including the effects of physical activity on mood, body image, self-efficacy, and stress. While each of these putative mechanisms may be operative, another possibility is that physical activity is serving as a marker of adherence to a variety of strategies associated with weight loss.

Evidence supporting this latter hypothesis comes from a variety of studies, including the Diabetes Prevention Program (2). This trial is of interest because it included over 1000 individuals in its lifestyle intervention arm. These individuals were encouraged to gradually increase their physical activity until they achieved a goal of 150 min per week of moderate-intensity activity. This goal was achieved by 74% of the participants at six months and by 67% at the end of the trial (after an average of 3.2 years). Participants who achieved the goal at six months were far more likely to achieve the goal at the end of the trial, suggesting the importance of getting participants off to a strong start. Older individuals (who may well have had more time for physical activity), men, and those who were less obese were more likely to achieve the activity goal. After adjustment for these demographic variables, frequency of monitoring intake was also strongly related to achieving the activity goal, indicating that adherence to one aspect of the intervention was related to adherence to other aspects. As expected, success at meeting the activity goal was related to success at achieving the weight loss goal (7% weight loss) at both six months and three years, showing that physical activity is related to long-term success at weight control; but the number of fat grams reported was also predictive of weight loss success. Other studies support the suggestion that individuals who make the greatest changes in physical activity during a weight loss program also make meaningful changes in their eating behavior and are most adherent to other aspects of the behavioral weight loss program (3, 4).

Amount of Physical Activity to Prescribe

Recent recommendations suggest that individuals should perform 30 to 60 min per day of moderate-intensity physical activity to prevent weight gain, but a higher level—60 to 90 min per day—to prevent weight regain. These goals for weight regain prevention are based on randomized clinical trials and on studies evaluating successful weight loss maintainers. Each of these types of studies is discussed next.

Randomized Clinical Trials

Physical activity is recognized as a critical aspect of behavioral weight control interventions. Typically, behavioral weight control programs have encouraged participants to gradually increase their activity until they achieve a level of at least 1000 kcal/week of moderate-intensity physical activity. This level of activity would equate to approximately 10 miles of brisk walking per week and would take about 150 min per week to complete. The 150 min per week exercise goal has been adopted in several large weight loss trials such as the Diabetes Prevention Program (2).

Recent studies suggest that higher levels of physical activity may be more beneficial for weight loss maintenance, raising concerns about the typical exercise prescription. Jakicic and colleagues found that women who reported doing over 200 min per week of activity (about 2600 kcal/week) at 6, 12, and 18 months had better weight losses at all time points than those reporting exercise durations of <150 min per week (about 1300 kcal/week) or 150 to 200 min per week (about 2150 kcal/week) (5).

To test this empirically, Jeffery and colleagues (6) randomized obese individuals to behavioral weight loss programs that included exercise prescriptions of 1000 kcal/week (standard behavior therapy group) or 2500 kcal/week (high physical activity group). Both groups were placed on a low-calorie, low-fat diet and participated in group classes addressing behavioral weight control strategies. There were no differences in weight loss between the 1000 kcal and the 2500 kcal exercise group at six months, but the high exercise group maintained their weight losses better from 6 to 12 months and achieved greater overall weight losses at 18 months (4.1 vs. 6.7 kg [9 vs. 14.8 lb]; see figure 60.1).

Although treatment contact was terminated at 18 months, all participants were reassessed at 30 months. At month 30, weight losses of the 1000 and 2500 kcal groups were 0.90 and 2.86 kg (2 and 6.3 lb),

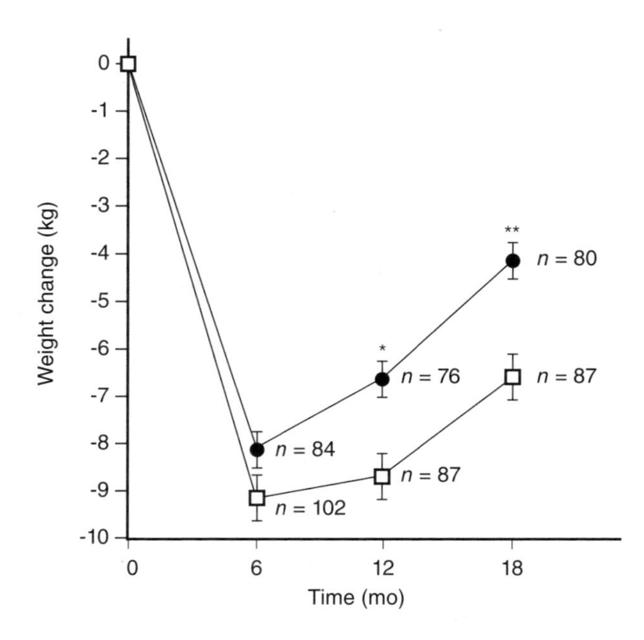

Figure 60.1 Mean (±SEM) weight change over time by treatment group (black circles, standard behavior therapy group; white squares, high physical activity treatment group). *,** Significantly different from the high physical activity treatment group (SAS GLM procedures): *P = 0.07, ** P = 0.04.

Reprinted, by permission, from R.W. Jeffery et al., 2003, "Physical activity and weight loss: Does prescribing higher physical activity goals improve outcome? *American Journal of Clinical Nutrition* 78: 684-689.

respectively, and did not differ between groups. The failure to find better maintenance of weight loss in the high exercise group resulted in large part from the failure of these participants to maintain their high doses of physical activity. Comparisons of participants who maintained physical activity levels of 2500 kcal/week or greater at month 30 with those who reported lower levels of activity (<1000 or 1000 to 2500 kcal/week) at month 30 showed that those who maintained high doses of physical activity regained only 2.3 kg (5 lb) between months 18 and 30, whereas participants with lower activity levels regained >6 kg (13.2 lb). Finally, these investigators examined the effect of consistently exercising at high levels. Those who reported >2500 kcal/week in physical activity at 12, 18, and 30 months (N = 13) were compared to all other participants. The high exercisers maintained weight loss of 12 kg (26 lb) at 30 months, compared to 0.8 kg (1.8 lb) in the other participants. While such consistently high levels of activity are associated with impressive long-term weight losses, it is clearly difficult for most obese patients to achieve and maintain these high doses of activity. Moreover, these participants reported not only high exercise levels, but also significant decreases in energy intake and dietary fat intake.

Thus, their successful weight loss maintenance probably related to their overall pattern of positive behavior changes.

National Weight Control Registry

The National Weight Control Registry (NWCR) provides further evidence that weight loss maintainers are characterized by high levels of physical activity. The NWCR was established in 1993 to investigate the characteristics of individuals who have succeeded at losing at least 30 lb (13.6 kg) and keeping it off at least one year. Currently there are over 5000 individuals in the registry. These members far exceed the minimum eligibility criteria; on average they have lost almost 70 lb (32 kg) and kept it off almost six years. Analyses of approximately 3000 registry participants indicate that the average participant reports 2691 kcal/week (7) in physical activity. Men report higher levels of activity than women (2903 vs. 2532 kcal/week), but the average level corresponds very well with current activity recommendations for weight loss maintenance. However, it should be noted that there is marked variability among NWCR participants; 25% report expending <1000 kcal per week in physical activity, whereas 35% report over 3500 kcal per week (8).

A recent analysis of registry members suggests that walking is the most commonly reported activity (8). Resistance training is the next most popular activity; cycling, use of cardiovascular exercise machines, aerobics, and running are other commonly performed activities.

When followed over a year, registry members experience an average of a 2.1 kg (4.6 lb) weight regain. One-year weight regain has been associated with demographic variables (higher age, greater weight loss, shorter duration of weight loss maintenance) and with adverse changes in eating and exercise behavior. Of particular relevance is the fact that decreases in physical activity (independent of demographic variables and other behavior changes) are significantly associated with increased risk of weight regain. Likewise, those individuals who not only decrease their activity, but also relapse in other behavior changes (e.g., both decreasing activity and increasing their television viewing), experience larger weight regains (9).

Comparison of Successful Weight Losers and Normal-Weight Controls

There are several limitations to the registry, including the use of self-report measures of physical activity and the inability to determine whether the level of activity associated with prevention of weight gain differs from the level associated with prevention of weight regain. To make this direct comparison, we (10) identified 135 women who had reduced from overweight or obese to normal weight, and who had lost at least 10% of their body weight and kept it off for five years. Thus the criteria used to define weight loss maintainers in this study were far more strict than those used in the NWCR. These individuals were compared to 102 always normal-weight women, who were weight stable and had no history of ever being overweight. Both groups were asked to wear RT3 triaxial accelerometers for one week, and comparisons were made of the number of minutes (using 10 min bouts of activity) spent in low-intensity (2 to 2.9 metabolic equivalents [METs]), medium-intensity (3.0 to 4.9 METs), or high-intensity (≥5.0 METs) activity.

The weight loss maintenance group spent more minutes per day in medium- or high-intensity activity (average of 58 vs. 52 min/day) than the always normal-weight controls, with the difference particularly apparent for the minutes of high-intensity activity (24.4 vs. 16.9 min/day, $p < .02$). Moreover, in the always normal-weight group, 56% reported 30 to 60 min/day of moderate- or high-intensity activity and 30% reported >60 min/day. This contrasted with data for the weight loss maintenance group, among whom 32% reported 30 to 60 min/day but 44% reported >60 min/day. These findings are consistent with the suggestion that 30 to 60 min/day of activity is needed to prevent weight gain, but that a higher level, 60 to 90 min/day, is required to prevent weight regain (see figure 60.2). Of note, however, these data indicate that both duration and intensity may be related to successful weight loss maintenance.

Conclusions

There is consistent evidence that weight loss maintenance is enhanced by ongoing adherence to physical activity. However, the reasons for this relationship are not clear. One possibility is that those who are maintaining high levels of physical activity are also maintaining improvements in diet and eating behaviors, and that this overall pattern of adherence (rather than exercise per se) relates to long-term weight loss maintenance.

Recent studies provide support for the current public health recommendations on physical activity and the suggestion that higher doses of activity are needed for prevention of weight regain than for prevention of weight gain. However, while successful

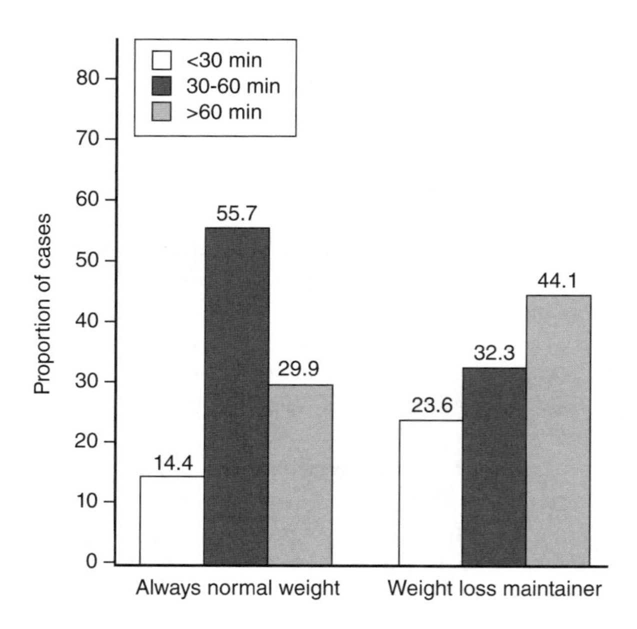

Figure 60.2 Proportion of participants spending <30 min/week, 30 to 60 min/week, or >60 min/week in moderate-intensity or higher (METs ≥3) physical activity in the always normal-weight and weight loss maintaining groups.

Reprinted, by permission, from S. Phelan et al., 2007, "Empirical evaluation of physical activity recommendations for weight control in women," *Medicine and Science in Sports and Exercise* 39: 1832-1836.

weight loss maintainers on average perform high levels of physical activity, it should be noted that there is marked variability within this group. In the NWCR, 25% of participants report <1000 kcal/week in exercise; likewise, using accelerometry, 24% of weight loss maintainers reported <30 min of activity per day. Thus, while higher levels of physical activity are common in weight loss maintainers, such high levels are neither *necessary* nor *sufficient* to achieve long-term weight control. Long-term weight loss maintenance is best produced by the combination of high exercise and reduced caloric intake, thereby most effectively altering energy balance.

61

Physical Activity and Weight Control During Pregnancy

Michelle F. Mottola, PhD

Maternal obesity and overweight during pregnancy have been increasing at an alarming rate, contributing to the obesity epidemic and increasing pregnancy complications, including gestational diabetes mellitus (GDM) and pregnancy-induced hypertension. High body mass index (BMI) in the mother is associated with high birth weight in the offspring, with maternal obesity as an independent and more reliable risk factor for large-for-gestational-age infants than glucose intolerance (1). Birth weight is directly associated with BMI later in life, represented by a "J"-shaped curve, with a slightly higher BMI in individuals born small but a larger prevalence of overweight and obesity in those born large (figure 61.1) (1).

The "J"-shaped curve shown in the figure indicates that babies born small for gestational age may have experienced undernutrition in utero, which may lead to a mismatch of fetal imprinting in that the metabolism of the developing fetus is prepared to experience low nutrition in the postpartum period. Usually, this does not occur and the infant experiences a "catch-up" growth period. The catch-up growth period is a time of rapid weight gain, which leads to abdominal adiposity and risk for obesity and cardiovascular disease later in life (1). On the other end of the birth weight spectrum, babies born large for gestational age and macrosomic are at high risk for obesity and high BMI later in life.

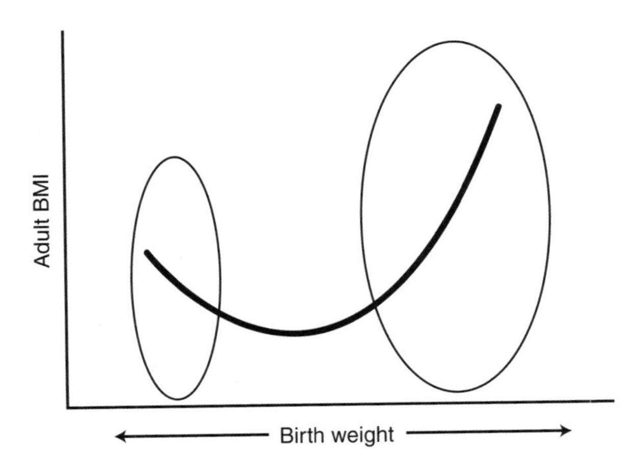

Figure 61.1 The relationship between birth weight and body mass index (BMI) development later in life

The robust link between the fetal environment and its profound influence on lifelong health and the future risk of chronic disease cannot be ignored. The maternal metabolic state may have a powerful influence on whether the emerging child develops obesity (1), which may be determined by maternal prepregnancy BMI and by the amount of excessive weight gained during pregnancy. Although the current guidelines are under review, they indicate that overweight and obese women should not gain more than 11.5 kg (25.4 lb) during pregnancy. Precluding excessive maternal weight gain may reverse and prevent undesirable effects passed on to the next generation nongenetically through the influence of the intrauterine environment. Thus, prevention of excessive weight gain during pregnancy is highly recommended to reduce the occurrence of GDM and to prevent diabetes, obesity, and hypertension in the mother after delivery, as well as these chronic disease risks in the offspring.

Healthy Lifestyle Approach for Weight Control

Pregnancy is the ideal time for behavior modification that may also promote long-term healthy lifestyle change. Although the healthy lifestyle approach is intuitive, to date only eight studies have examined the effectiveness of an intervention designed to prevent excessive weight gain during pregnancy, and not all used overweight or obese women exclusively. As shown in table 61.1, half of the studies listed (2-5) were *not* successful in preventing excessive pregnancy weight gain in overweight and obese women. Of the successful studies, one showed success in nulliparous obese women only (6) and another in obese women diagnosed with GDM (7); a third study involved small numbers and included no exercise in the intervention (8). The inclusion of mild to moderate exercise is an important part of an intervention during pregnancy, as dietary control alone may potentially reduce not only fat mass but also fat-free mass. The most recent study demonstrated prevention of excessive weight gain in overweight and obese pregnant women, with minimal weight retention at two months postdelivery (9), with use of a Nutrition and Exercise Lifestyle Intervention Program that consisted of nutrition control with pedometer step counts for walking. The ineffective studies (2-5) used education alone as an intervention, which did not prevent excess weight gain during pregnancy or minimize weight retention after delivery. Women were usually encouraged to eat from the major food groups, with suggestions of being more active, but were given no true exercise advice.

The successful interventions that used dietary control in combination with exercise as part of the lifestyle change for overweight and obese pregnant individuals included a variety of activities. Claesson and colleagues (6) offered aqua-aerobics classes (one or two times per week) designed especially for obese pregnant women. The exercise frequency was minimal and would not have improved aerobic fitness in the obese participants. Artal and colleagues (7) used a supervised moderate-intensity exercise session (walking on the treadmill or cycling on a semirecumbent bike) for obese participants with GDM, once per week, along with unsupervised exercise on the remaining six days at home. The exercise group had a mean exercise time of 153 ± 91.4 min per week, and 50% of them exercised more than 150 min per week (7). The weight gain per week was lower in the diet and exercise group than in a diet-only group.

Mottola and colleagues (9) based nutritional control on dietary intakes for women with GDM and used this in combination with a mild walking program (quantified by pedometer) for overweight and obese pregnant women. The walking program started at 25 min (an average of 2861 ± 288 steps) and then increased by 2 min every week until 40 min were reached (4484 ± 530 steps); this level was maintained until delivery. The participants had taken an average of 5678 ± 1738 steps per day before the intervention (Nutrition and Exercise Lifestyle Intervention Program). By the end of the program, the women were taking over 10,000 steps, significantly increasing their activity. A mild walking program (9) or unstructured exercise at home (7), especially with overweight and obese pregnant women, may be better than aqua-aerobics classes, as compliance and time management may be more difficult with structured classes and not everyone has access to a pool.

Activity Recommendations

Evidence-based guidelines indicate that regular prenatal exercise is an important component of a healthy pregnancy (10). In addition to maintaining physical fitness, exercise may be beneficial in preventing or treating maternal–fetal diseases. Walking is a popular activity during pregnancy and, as two of the intervention studies have shown, walking in combination with nutritional control can be effective in preventing excessive weight gain in pregnant overweight and obese women (7, 9).

The American College of Obstetricians and Gynecologists suggests that all pregnant women with low-risk pregnancies should exercise on most if not all days of the week (11). In Canada, PARmed-X for Pregnancy provides more specific guidelines for frequency, intensity, duration, and type of exercise in women who are medically prescreened and have a low obstetric risk pregnancy (10). However, the intensity represented by the target heart rate zones based on age as suggested by PARmed-X for Pregnancy may make the exercise too difficult for overweight and obese pregnant women. These target heart rate zones are provided for recreationally active women and represent approximately 60% to 80% of heart rate reserve (10). Alternate target heart rate zones (validated on 156 pregnant women) have been suggested that also take into account the fitness level of the individual. Low-risk women who are not fit should exercise at a lower heart rate, between 128 and 144 bpm (12).

Target heart rate zones specific for overweight and obese pregnant women are also available;

■ Table 61.1 ■

Summary of Studies Using an Intervention to Prevent Excessive Gestational Weight Gain

Study	Population	Intervention	Results	Outcome
Gray-Donald et al. (3)	Cree women of James Bay, 112 in intervention vs. 107 historic controls; all body mass index (BMI) categories	Goal to optimize gestational weight gain with use of exercise groups and dietary education via media campaign	No difference was seen between groups in weight gain or rate of weight gain.	Not successful, possibly because of value placed by Cree culture on weight gain
Polley et al. (2)	Low income; 120 normal-weight or overweight or obese in Pittsburgh	Subjects randomized into stepped behavior intervention or usual prenatal care; intervention group received educational materials	Among overweight or obese women, 59% had excessive weight gain in intervention group vs. 32% in usual-care group.	Not successful; many barriers to keeping up intervention identified for low-income women
Olsen et al. (4)	Low and middle/upper income in upstate New York; 179 normal or overweight enrolled in intervention compared to 381 historic controls	Goal to use two tiers: (a) health care provider information and (b) mailed materials including newsletter, postcards	Among historic controls, 45% gained excessive weight vs. 41% in intervention group, $p > 0.05$; normal-weight and overweight low-income women benefited most.	Not successful except in low-income women
Kinnunen et al. (5)	Six maternity clinics from Finland; three intervention ($N = 49$), three control ($N = 56$); control received standard care; all women primiparous; all BMI categories	Individual counseling on diet and physical activity plus information on weight gain recommendations	46% of women in intervention group vs. 30% in control group exceeded weight gain, $p > 0.05$.	Not successful; intervention had minor effects on some dietary habits but not on activity or excessive gestational weight gain
Claesson et al. (6)	Prospective case-control intervention study in Sweden; used 155 for intervention vs. 193 in control group receiving standard care; all women obese	To decrease total weight gain to <7 kg; behavioral intervention (weekly motivational talk using trained midwife) plus aqua-aerobic classes offered one or two times per week	Women in intervention group had lower weight gain and had lost more weight at postnatal checkup than control, but no difference in number of women who gained <7 kg.	Successful in controlling weight gain for nulliparous obese women; did not affect delivery or neonatal outcome
Artal et al. (7)	Obese women with gestational diabetes (GDM); self-selected intervention	Diet plus exercise ($N = 39$) vs. diet alone ($N = 57$); used standard diet for GDM in both; exercise supervised once in lab and unsupervised five more times at home	Weight gain per week was lower in intervention group; no difference in insulin usage.	Successful in limiting weight gain; less macrosomic infants
Wolff et al. (8)	Randomized controlled trial of obese, nondiabetic women, starting at 15 weeks gestation	No intervention ($N = 27$) vs. intervention (10 h of dietary counseling with *no* advice on exercise; $N = 23$)	Weight gain was about 6.7 kg more in nonintervention group; weekly weight gain of 0.26 ± 15 kg/week in intervention group vs. 0.44 ± 0.21 kg/week in nonintervention group.	Intervention successful in limiting weight gain but included no exercise
Mottola et al. (9)	Overweight and obese women, starting at 16-20 weeks gestation; prospective single arm intervention, matched by age, prepregnancy BMI, and parity	Nutrition and Exercise Lifestyle Intervention Program (NELIP) ($N = 65$); matched cohort (4:1) $N = 260$; NELIP consisted of approximately 2000 kcal/day, 200 g carbohydrate per day, and healthy food choices plus walking program using pedometers	Weight gain averaged 0.38 kg/week on NELIP; women took on average 10,000 steps per day at end of NELIP; minimal weight retention at 2 months postpartum	Successful in preventing excessive weight gain and achieving minimal weight retention at 2 months postdelivery

these, corresponding to a lower intensity (20% to 39% of heart rate reserve), were validated on 106 overweight and obese pregnant women. These target heart rate ranges for sedentary overweight and obese pregnant women are 110 to 131 bpm (20 to 29 years of age) and 108 to 127 bpm (30 to 39 years of age) (13). Low-risk pregnant overweight and obese women who use these lower intensities may have better compliance, especially when walking is used as the exercise modality. Even at these lower intensities, this group will improve aerobic fitness. Using the "talk test" will also confirm that the pregnant woman is not overexerting: The intensity is appropriate as long as she can carry on a conversation while exercising (10). Walking three or four times per week at the appropriate intensity is highly recommended to help prevent excessive gestational weight gain and other chronic disease risks for both the mother and the developing fetus.

Summary

Excessive weight gain during pregnancy may have a direct link to the obesity epidemic and pregnancy-induced diseases such as GDM and hypertension. The profound influence of the fetal environment on lifelong health and future disease risk can no longer be ignored. Excessive weight gain during pregnancy has been strongly associated with infant birth weight and the development of childhood obesity along with future disease comorbidities. Although the total weight gain recommendations during pregnancy are currently under review, it is suggested that overweight and obese women gain no more than 11.5 kg (25.4 lb). A healthy lifestyle approach to prevent excessive weight gain during gestation is intuitive; but to date, only 50% of interventions have been successful in meeting this goal. The exercise modalities used were aqua-aerobics classes, structured exercise on a treadmill or semi-recumbent ergometer, and a walking program using a pedometer to count steps. All interventions used some type of nutrition control and counseling for overweight or obese pregnant women. Suggestions for exercise guidelines (following medical prescreening) included target heart rate ranges of 110 to 131 bpm (20 to 29 years) and 108 to 127 bpm (30 to 39 years) and a frequency of three to four times per week, with walking starting at 25 min and increasing to 40 min, a duration that is maintained until delivery. Prevention of excessive weight gain during pregnancy, especially in overweight or obese individuals, is paramount, as this may avoid pregnancy-induced problems and chronic disease risk for both the mother and her unborn child, as well as prevent the development of childhood obesity.

62

Physical Activity and Postpartum Weight Loss

Cheryl Lovelady, PhD, MPH, RD

Excess weight gain during pregnancy and changes in lifestyle during the postpartum period are contributing factors to obesity among women. While many epidemiological studies indicate that the average weight retained from pregnancy is only 0.5 to 3.0 kg (1.1 to 6.6 lb), between 14% and 20% of women are at least 5 kg (11 lb) heavier at 6 to 18 months postpartum than they were before pregnancy (1).

Epidemiological evidence suggests that physical activity may help prevent postpartum weight retention. Rooney and Schauberger (2) weighed 540 women during pregnancy and also at 5 to 10 years after delivery. Women with the least weight gain during follow-up participated in aerobic exercise after pregnancy, breast-fed their infants for longer than 12 weeks, or lost all of their pregnancy weight by six months postpartum. Results from a large prospective study of 902 postpartum women showed that the odds ratio of retaining at least 5 kg (11 lb) at one year postpartum was 0.66 (95% confidence interval [CI]: 0.46 to 0.94) per daily hour of walking (3). Similarly, Olson and colleagues (4) reported that women who exercised often (approximately 30 min or more of moderate-intensity physical activity on most days) had an odds ratio of 0.22 (CI: 0.09 to 0.58) of retaining ≥4.55 kg (10 lb) at one year postpartum.

During the postpartum period, many women desire to lose excess weight gained during pregnancy. Therefore, the postpartum period may be an ideal time to implement an exercise and diet program to prevent obesity. However, there are very few reports of weight loss interventions during the postpartum period. In addition, considering the American Academy of Pediatrics recommendation that all women breast-feed their infants during the first year of life (5), the effect of exercise and caloric restriction on milk volume and composition and consequently on infant growth and health must be examined.

Exercise Effects in Lactation

Lovelady and colleagues investigated the impact of exercise on lactation performance in 16 exclusively breast-feeding women with infants aged 9 to 24 weeks in a cross-sectional study (6). The exercising women were more fit ($\dot{V}O_2max$: 46.4 ± 2.4 vs. 30.3 ± 4.7 ml O_2/kg), were leaner (21.7 ± 3.5 vs. 27.9 ± 4.7% body fat), and consumed more energy (2739 ± 309 vs. 2051 ± 335 kcal/day) than the sedentary women. They reported exercising (mainly swimming and running) an average of 88 min/day. They expended significantly more energy than the sedentary women (3169 ± 273 vs. 2398 ± 214 kcal/day, including approximately 500 kcal/day in breast milk consumed by the infants). There were no significant differences in the macronutrient composition or volume of milk produced between exercising and sedentary women.

Lovelady and associates (7) randomized 33 sedentary lactating women to an exercise or control group at six to eight weeks postpartum. The exercise program consisted of aerobic exercise (walking, jogging, or cycling), at 60% to 70% of maximum heart rate, five days a week for 12 weeks. The sessions were initially 20 min long, with increases of 5 min every three days until 45 min/day was achieved. Women in the control group did not engage in aerobic exercise more than once per week during the same period. Both groups were instructed not to restrict their dietary intake.

Women in the exercise group significantly increased their cardiorespiratory fitness level compared to the sedentary women (25% vs. 5%). Plasma triglycerides, total cholesterol, and low-density lipoprotein (LDL)-cholesterol decreased significantly in both groups over time; however, there was a trend for high-density lipoprotein (HDL)-cholesterol to increase in the exercise group and to decrease in the control group. Insulin response to a test meal decreased significantly in the exercise group but not in the control group. Both groups lost an average of 1.6 kg (3.5 lb), predominantly fat loss. There were no differences in the macronutrient composition or the volume of breast milk produced between groups. Infant weight and length gain was also similar between groups. These results suggest that sedentary women can begin an exercise program without affecting their milk macronutrient composition or volume. However, exercise without energy restriction did not promote weight loss.

McCrory and colleagues (8) examined the effects of exercise and dieting for 11 days on lactation performance. Sixty-seven exclusively breast-feeding women were randomly assigned to one of three groups at approximately 12 weeks postpartum: (1) diet group, which had an energy deficit of 35% by restricting energy intake by approximately 1000 kcal/day; (2) diet plus exercise group, which had an energy deficit of 35% by decreasing energy intake by approximately 720 kcal/day and exercising for 86 min/day on 9 of the 11 days; or (3) control group, which had no energy deficit. Women in the diet group lost 1.9 ± 0.7 kg (4.2 ± 1.5 lb), while the diet and exercise group lost 1.6 ± 0.5 kg (3.5 ± 1.1 lb) and the control group lost 0.2 ± 0.6 kg (0.4 ± 1.3 lb). Almost half of the weight loss in the diet group was fat-free mass, while the diet and exercise group lost no fat-free mass, on average. This short-term energy deficit did not affect infant growth or milk volume and macronutrient composition.

Lovelady and colleagues (9) randomly assigned 40 overweight (body mass index ≥25 to 30), sedentary, exclusively breast-feeding women to one of two groups at four weeks postpartum. The diet and exercise group restricted their energy intake by 500 kcal/day and exercised four days per week at 65% to 80% of their maximum heart rate. The exercise sessions were initially 15 min, then were increased by 2 min every day until the women were exercising (walking, jogging, or aerobic dancing) for 45 min/day. Women in the control group were instructed not to restrict their energy intake and not to perform vigorous exercise more than once per week. After 10 weeks, women in the diet and exercise group lost significantly more weight (4.8 ± 1.7 vs. 0.8 ± 2.3 kg [10.6 ± 3.7 vs. 1.8 ± 5 lb]) and fat mass (4.0 ± 2.0 vs. 0.3 ± 1.8 kg [8.8 ± 4.4 vs. 0.66 ± 4 lb]) than mothers in the control group. Cardiovascular fitness increased significantly more in the diet and exercise group compared to the control group (13% vs. 2%). The gains in infant weight and length were similar in the two groups. The results of this study suggest that overweight women may begin a program of moderate exercise and energy restriction at four weeks postpartum and that a weight loss of 0.5 kg (1.1 lb) per week will not affect infant growth.

Other Physical Activity Interventions

Kinnunen and coworkers (10) investigated the effectiveness of having nurses deliver physical activity and dietary counseling for maternal weight loss compared to usual care at the newborn's regular preventive health clinic visits (three intervention and three control) in 92 postpartum women. They reported that there was no difference in physical activity changes; however, more women in the intervention group returned to their prepregnancy weight by 10 months postpartum (50% vs. 30%, $p = 0.06$). O'Toole and colleagues (11) compared a structured diet and physical activity intervention (weekly meetings for 12 weeks and daily food and activity diaries) with a 1 h educational session on diet and physical activity for promoting postpartum weight loss in 23 overweight women. Leermakers and colleagues (12) evaluated a six-month behavioral weight loss intervention (two group sessions, correspondence materials, and telephone contact) and an informational brochure in 62 postpartum women. Both studies reported significantly greater weight loss among intervention subjects, but neither showed differences in behavior changes related to diet and physical activity. Leermakers' group found no relationship between weight loss and self-reported diet or activity but did find a relationship with the number of self-monitoring tools completed. Participants in both studies had difficulties attending the group sessions and completing the self-monitoring diaries and "homework" assignments, suggesting that the postpartum period may be too demanding a time for women to make lifestyle changes.

Conclusions

The American College of Obstetricians and Gynecologists recommends that physical activity be resumed as soon as physically and medically safe after delivery (13). Since physiological changes of pregnancy continue for four to six weeks postpartum, women should begin exercising gradually. Results of studies with breast-feeding women suggest that exercise improves aerobic fitness, plasma lipids, and insulin response. In these studies, exercise began at four to six weeks postpartum, after medical clearance had been received from the participant's physician. The exercise sessions were initially 15 min, then were increased by 2 min every day until the women were exercising (walking, jogging, or aerobic dancing) for 45 min/day, four to five days a week.

However, exercise without calorie restriction does not promote weight loss. Once lactation is established, overweight women may restrict energy intake by 500 kcal/day to promote a weight loss of 0.5 kg (1.1 lb) per week without affecting infant growth. While studies on postpartum women have determined the safety and efficacy of moderate exercise and calorie restriction during lactation, further research is needed to establish the most effective way to implement these lifestyle changes during the postpartum period.

Physical Activity and Birth Weight

Johan G. Eriksson, MD, PhD

One can focus on the topic of physical activity and birth weight from two different angles. A discussion of this topic could be based on the level of maternal physical activity during pregnancy and the effect upon offspring birth size, or could be based on the possible early programming of physical activity levels during prenatal life. This chapter addresses both aspects, but the first one only briefly. The importance of the protective effect of exercise among individuals born with a small body size is also considered.

Maternal Physical Activity During Pregnancy and Offspring Birth Size

The effect of maternal physical activity during pregnancy on offspring birth size is addressed only briefly here, as some recent excellent overviews cover this large topic (1, 2). The general conclusion is that maternal exercise during normal pregnancies is advisable and has in general not been associated with major influence on birth size of the offspring. However, there may well be ethnic differences as shown in a recent Indian study, which indicated that physical activity in the first trimester was associated with low birth weight in offspring (3).

Developmental Origins of Health and Disease

Most common noncommunicable diseases develop as a consequence of unfavorable environmental conditions in combination with a genetic predisposition for the disease. During the past decade, evidence has accumulated to suggest that several chronic diseases originate early in life. The evidence is strongest for coronary heart disease, hypertension, and type 2 diabetes (4). Large-scale epidemiological studies, as well as experimental animal work, have introduced the concept of programming: Factors present in early life, for example malnutrition and stress, can cause lifelong changes in organ structure and function, affecting later health outcomes.

Activity level, and consequently physical fitness, could be programmed early in life, as suggested by animal studies. Animal models have convincingly shown that lifestyle and exercise habits may have a prenatal origin. Offspring from undernourished mothers had a smaller body size at birth, and they were less active than offspring with normal body size at birth. This sedentary behavior was exacerbated by hypercaloric nutrition in the postnatal phase (5). However, much less is known about the association between birth size and physical fitness in humans. Regular physical activity is an important contributor to good overall health, while inactivity associated with poor physical fitness is a major health concern. Low level of physical activity is a major risk factor for cardiovascular disease and type 2 diabetes. Identifying risk groups that might be programmed to be less active is of major importance from a public health point of view.

Birth Size and Exercise

A small body size at birth is known to be associated with a smaller lean body mass or muscle mass in later life. This could potentially influence physical activity as well as willingness to exercise. Several studies have demonstrated a lower physical fitness and activity level in preterm children and adolescents born with extremely low birth weight. Prematurity is, among other things, associated with a higher risk for long-term motor sequelae potentially influencing physical performance. People who were born prematurely are certainly an important group, but because of the heterogeneity of prematurity they fall outside the scope of this chapter and are not discussed further.

Only a few studies have focused on the relationship between birth size and physical activity, and the findings have been inconsistent. The Avon Longitudinal Study of Parents and Children (ALSPAC) (6) assessed the early-life determinants of physical activity at the age of 11 to 12 years by Actigraph accelerometer for seven days in more than 5000 children. Only a few early-life factors predicted exercise levels among these children. Birth size did not predict physical activity; however, parental activity levels were associated with the activity level of the child.

A Brazilian study (7) focusing on children from the same age group showed that those with a low birth weight (<2500 g or 5.5 lb) were less active than those born with higher birth weights. However, no statistically significant differences in sedentary lifestyle were observed in the various birth weight groups, although those belonging to the lowest birth weight group tended to live a more sedentary life.

Within the Northern Ireland Young Hearts project (8), involving a representative population-based sample of schoolchildren, including adolescents, the association between birth weight and aerobic fitness was assessed at ages 12 to 15 years. This study showed a positive relationship between birth weight and fitness score at age 12 years. However, this association was no longer significant at the age of 15 years. Based on the few available studies, it seems that prenatal factors have only a limited influence on later physical activity and fitness in children and adolescents.

Likewise, only a few studies have assessed the relationship between birth size and physical activity in adult life. Laaksonen and coauthors (9) were not able to detect any association between cardiorespiratory fitness or duration of strenuous leisure-time physical activity and birth size in a prospective population-based study including 462 nondiabetic middle-aged men.

Our findings in the Helsinki Birth Cohort Study (10) have been opposite to those of Laaksonen's group. Despite the fact that a small body size at birth is strongly related to a smaller muscle mass in later life, men born thin exercised more than those not born thin when studied at a mean age of 70 years. They also exercised at a higher intensity, and their yearly energy expenditure on physical activity was higher. These associations between birth size and exercise-related outcomes were not significant among women. Research needs to focus more on the importance of prenatal factors in relation to physical activity in adult life before any general conclusions can be drawn.

Protective Effect of Exercise Among Individuals Born Small

It is well known that individuals who are physically active have a lower risk for several noncommunicable diseases including cardiovascular disease and type 2 diabetes. In general, the protective effect of exercise is strongest among those belonging to high-risk groups. Small body size at birth is a relatively recently identified risk factor for several noncommunicable diseases; consequently only a few studies have assessed the protective effect of exercise in this high-risk group.

Laaksonen and coworkers (9) have convincingly shown that better cardiorespiratory fitness and more vigorous leisure-time physical activity reduced or even abolished the negative metabolic consequences of small birth size among men studied at around 50 years of age.

Table 63.1 shows the odds ratios (ORs) for type 2 diabetes and impaired glucose tolerance according to exercise-related variables, as well as birth size, in a substudy of the Helsinki Birth Cohort Study (10). More frequent exercise, as well as exercise of at least moderate intensity, was associated with lower risk for glucose intolerance. The positive effect of exercise was strongest in subjects born with small birth size; the interaction between exercise frequency and birth weight was highly significant ($p = 0.003$) (figure 63.1). These positive effects were observed regardless of adult body mass index and gender.

Regular and moderate-intensity exercise protected against the metabolic syndrome as well as glucose intolerance in people who were at increased risk of developing type 2 diabetes because they were born thin or small. The findings in these two Finnish studies strongly suggest that small birth size is a risk factor for metabolic outcomes. However, such subjects are known to respond positively to exercise.

Conclusions

Physical activity recommendations should be targeted specifically toward people who are likely to benefit particularly from exercise. Importantly, they need first to be identified. According to the research,

■ Table 63.1 ■

Odds Ratios (95 CI%) for Type 2 Diabetes and Impaired Glucose Tolerance According to Birth Weight and Frequency and Intensity of Leisure-Time Physical Activity

	Frequency of leisure-time exercise		Intensity of exercise	
Birth weight (g)	Less than three times a week	Three or more times a week	Light	Moderate or greater
≤3000 g	5.2 (2.1-13)	1.0	3.5 (1.5-8.2)	1.0
>3000 g	3.6 (1.7-7.7)	3.1 (1.6-6.3)	3.1 (1.6-6.3)	1.9 (0.9-3.8)
P for interaction	0.003		0.12	

Adapted from J.G. Eriksson et al., "Exercise protects against glucose intolerance in individuals with a small body size at birth," *Preventative Medicine* 39: 164-167.

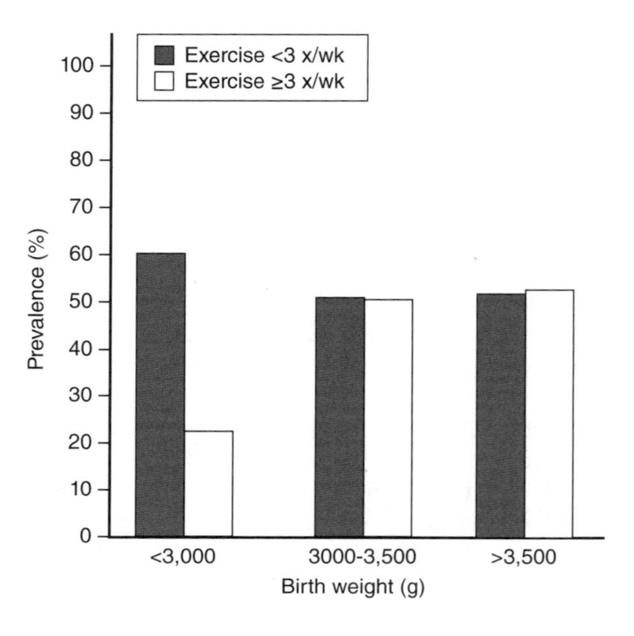

Figure 63.1 Prevalence (%) of type 2 diabetes and impaired glucose tolerance according to birth weight and weekly exercise frequency.

people who were small at birth, and thus are at high risk for the metabolic syndrome and related health outcomes, benefit from exercise and should be encouraged to maintain an active lifestyle. There is no strong evidence from human studies supporting the theory that physical activity is programmed in prenatal life. One must keep in mind that this could be due to methodological issues, and more studies applying more sophisticated protocols are needed for a better understanding of the association between birth size, physical activity, and cardiovascular fitness later in life.

64

Physical Activity and Body Composition in Children

Bernard Gutin, PhD

Because of the increasing global rates of pediatric obesity, it is necessary to implement effective preventive programs on a widespread basis. To guide the formulation of large-scale interventions in schools and communities, as well as to inform public health recommendations, it is necessary for scientists to investigate the effectiveness of various approaches in controlled settings. Unfortunately, the results of the research on preventive intervention trials have been mixed, with many failing to show clear-cut favorable impact on the indices of adiposity used as outcomes. Therefore, it is not clear which public health recommendations are appropriate or which procedures should be incorporated into large-scale interventions. This chapter describes recent research suggesting that controlled preventive trials are more likely to be successful if they use indices of body composition rather than weight as outcome measures, and if they place greater emphasis on adequate doses of vigorous physical activity (PA) rather than restriction of energy intake.

Body Mass Index as an Indicator of Intervention Effectiveness

Body mass index (BMI) has frequently been used to evaluate the effectiveness of interventions. It is easily measured and is therefore valuable for epidemiologic studies of obesity prevalence. However, BMI has important limitations as an outcome measure for intervention trials, especially when the trials include PA. From a health perspective, it is important to remember that BMI includes both fat mass and fat-free mass (FFM) in the numerator, whereas the component of body composition that is harmful to health is the fat mass. In the Danish study named Diet, Cancer and Health, the body composition of 27,178 men and 29,875 women 50 to 64 years old was measured; the people were followed up for a median time of 5.8 years. Results showed that the BMI represented the joint but opposite associations of body fat and FFM with mortality.

Both high body fat and low FFM were independent predictors of all-cause mortality (1). This is consistent with our findings that, already in adolescence, percent body fat (%BF) tends to explain more of the variance in cardiometabolic risk factors than does BMI (2). Especially noteworthy was the finding that the variance in fasting insulin (a key early marker for development of the metabolic syndrome) that was explained by %BF was 25.5% while the variance explained by BMI was 17.6%.

Another component of body composition that should be considered as an outcome measure in controlled trials is visceral adipose tissue (VAT). Already in childhood, VAT explains additional variance in some lipid risk factors beyond the variance explained by %BF (2). Measurement of VAT requires high-technology techniques such as magnetic resonance imaging, but an alternative and easily measured index of central fatness is waist

Acknowledgements
The research of our group at the Medical College of Georgia was supported mainly by grants from the National Institutes of Health (HL49549, HL55564, HL64157, HL64972, DK63391).

girth; this index explains more of the variance in some cardiometabolic risk factors than does BMI (2).

Another reason to doubt the validity of BMI as an indicant of the effectiveness of interventions that include substantial doses of PA is that fatness may be reduced at the same time that FFM is increased, with the consequence that BMI can provide misleading results (3).

Positive Energy Balance as a Cause for Pediatric Obesity

Intervention trials have generally been based on an energy balance paradigm in which youths who choose to ingest more energy, or expend less energy, than their age-mates deposit the excess energy in their fat depots and become obese. While the energy balance paradigm may apply well to adults who are no longer growing, youths are not simply little adults; they are in a dynamic phase of tissue building during which anabolic hormones drive them to maintain a positive energy balance in order to ingest sufficient energy (and accompanying nutrients) needed for development of healthy bodies. The key is for young people to partition the surplus ingested energy into FFM—muscle and bone—rather than fat mass. Because vigorous PA stimulates the development of FFM, the best scenario for a youth wishing to develop a healthy body composition is to engage in a relatively large amount of vigorous PA while ingesting sufficient energy and nutrients to support the tissue-building process. This nutrient partitioning paradigm suggests that vigorous PA stimulates the development of FFM and that ingested energy and nutrients are then "pulled" into lean tissue rather than "pushed" into fat tissue. Because the synthesis of lean tissue is metabolically expensive, and because lean tissue has a higher metabolic rate than fat tissue, youths who partition energy into lean tissue would be expected to have a *higher* energy intake than youths who have a greater proportion of fat mass.

This principle is illustrated by the results of a study recently reported by our group (4). This cross-sectional study, which investigated the relations among diet, PA, and body composition, is noteworthy for several reasons: (1) It had a relatively large sample size (661 adolescents); (2) it employed high-quality measurements of %BF with dual-energy X-ray absorptiometry and of VAT with magnetic resonance imaging; and (3) diet and PA were measured with four to seven separate 24 h recalls over a two- to three-month period (52% of the youths had recalls for all seven days). Special care was taken to train the subjects in the diet and PA recalls, using a multiple-pass approach in order to minimize underreporting. At the outset of the project, in keeping with the energy balance paradigm, we hypothesized that youths who were relatively fat would report that they ingested more energy than the leaner youths, and that total energy expenditure would be relatively low in the fatter youths. To our surprise, we found that energy intake and vigorous PA were positively correlated with each other and were both negative predictors of %BF; that is, those youths who did the most vigorous PA and ingested the most energy were the leanest. Figure 64.1 illustrates these results. When VAT was used as the outcome measure, energy intake was the sole and negative predictor of VAT; that is, the youths who ingested the most energy had the least VAT. With respect to PA, we found that lower levels of %BF were associated with greater amounts of vigorous PA, but not with moderate PA. Moderate PA includes activities like walking, while vigorous PA includes sports, games, and dance activities. Studies using accelerometry to measure PA have

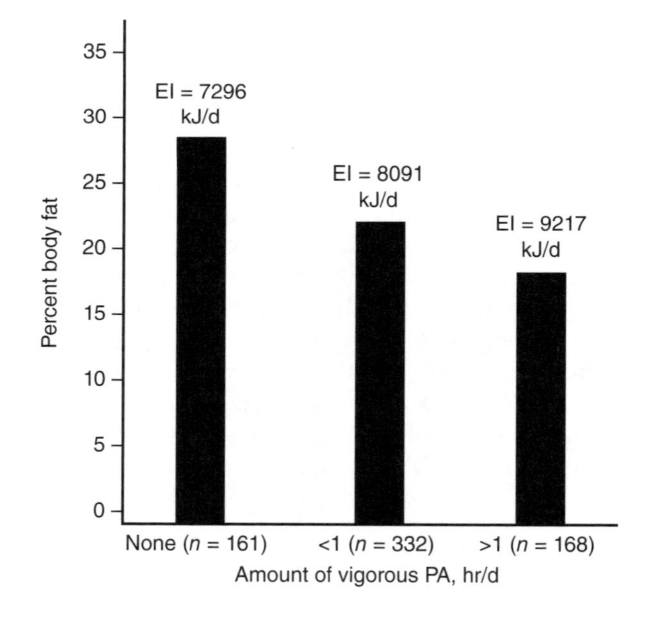

Figure 64.1 Percent body fat for 661 adolescents in relation to hours per day of vigorous physical activity (VPA), adjusted for age, race, and sex. The mean energy intake (kJ/day) of each group is shown above the bars. For both %BF and energy intake, the middle group was not significantly different from the other groups, while the group that had no VPA was significantly different from the group that had >1 h/day ($p < 0.01$).

Adapted from I. Stallmann-Jorgensen et al., 2007, "General and visceral adiposity in black and white adolescents and their relation with reported physical activity and diet," *International Journal of Obesity* 31: 622-629.

supported the idea that vigorous PA, to a greater degree than moderate PA, is linked with lower body fatness (e.g., [5]).

Because it is commonly believed that fat people ingest more energy than leaner people, one way to explain these results is to assume that fatter youths underreport energy intake to a greater degree than do leaner youths. This explanation requires that the data must be disbelieved unless they agree with the preconceived notion that fatter youths *must* ingest more energy than leaner youths. In fact, there seems to be no evidence in the pediatric literature that *fatness* (as opposed to weight, which includes both fat mass and FFM) is positively correlated with free-living energy intake, while it is possible to find studies showing results similar to ours. Thus, the available literature seems more consistent with the nutrient partitioning paradigm that emphasizes vigorous PA than with the energy balance paradigm that emphasizes restriction of energy intake (6).

It must be noted that accurate measurement of free-living diet is very difficult; as better techniques are developed, this theoretical formulation may need to be reconsidered. Moreover, the nutrient partitioning paradigm was derived retrospectively from observational studies and needs to be confirmed in prospective and experimental studies.

Experimental Trials of Physical Activity and Body Composition

Drawing causal conclusions from nonexperimental studies is complicated by the possibility of reverse causation; that is, PA may influence %BF, but it is also possible that %BF influences how much PA the youths can do. Indeed, it is likely that the relationship is reciprocal. To guide large-scale preventive interventions, we need to see what happens when youths are randomized to PA and control groups and then followed over time.

In studies in which the youths were chosen such that they were obese at baseline, the projects could be construed as obesity treatment or secondary prevention rather than primary prevention. In such youths, we and others have found that vigorous PA, without restriction of energy intake, has favorable effects on %BF, VAT, bone density, aerobic fitness, and some cardiometabolic risk factors, with PA doses of 155 to 180 min/week of moderate-vigorous PA (7).

To determine the value of PA in *preventing* fat accretion, it is necessary to study broader samples of youths who are not necessarily obese at baseline. Projects assessing the effectiveness of exercise doses that worked well for obese youths have generally shown these doses to be ineffective in changing the body composition of nonobese youths (e.g., [8]). Thus, our research group implemented a preventive intervention that employed a PA dose of 80 min/day of mostly moderate-vigorous PA, offered five days a week (i.e., as much as 400 min/week) for a 10-month period. The subjects were 8- to 11-year-old black girls ($n = 201$) who varied over the spectrum of fatness and fitness. The PA sessions took place after school in the same schools the girls attended during the school day; thus, the intervention could be incorporated into the girls' regular schedules. Healthful snacks were offered to the girls prior to the PA period, but no attempt was made to restrict their energy intake. We found favorable effects for %BF, VAT, bone density, and aerobic fitness. The total body composition results are shown in figure 64.2. Note that compared to the control group, the intervention group reduced fat mass but not FFM; thus body composition was improved. Within the intervention group, those children who attended most often and maintained the highest heart rates during the exercise sessions exhibited the greatest decreases in %BF and the greatest increases in bone density (9).

Using these results as a foundation, a three-year intervention was implemented on a larger basis (the MCG FitKid Program). Nine schools, starting

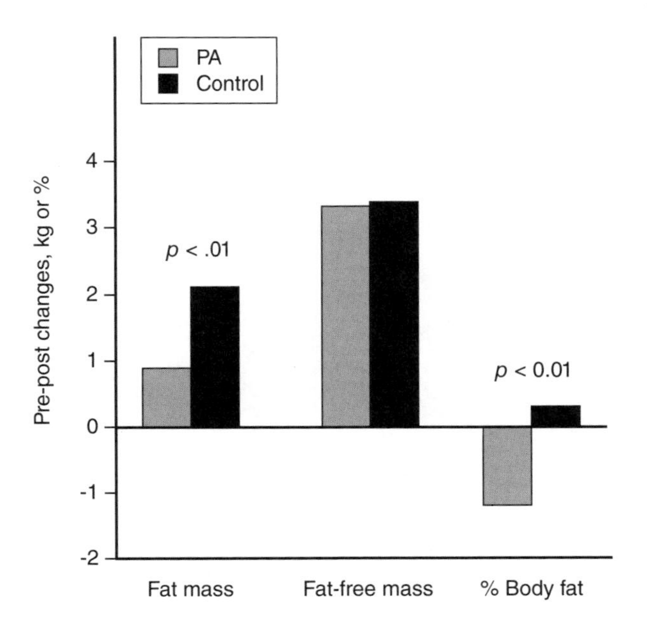

Figure 64.2 Changes in total body composition in black girls who participated in a 10-month physical intervention versus a randomized control group.

with third graders, were randomized to receive the after-school intervention and nine schools served as controls. The preliminary analysis of the three-year data showed that the favorable effects on fatness and fitness obtained during the school years, when the children were exposed to the intervention, were lost during the summers following the first and second years of intervention (9). This shows the importance of maintaining exposure to vigorous PA during vacation periods. Over the three-year period, the intervention group increased more than the controls in FFM and BMI. Thus, if we had used BMI as our index of fatness, we would have derived misleading results.

Our results are consistent with a recent review concluding that the main factor distinguishing effective from ineffective pediatric obesity prevention trials was the provision of supervised moderate-vigorous PA (10).

Conclusion

It is noteworthy that the PA intervention studies at the Medical College of Georgia, which all emphasized vigorous PA (average heart rates of ~160 bpm were maintained for 40 to 60 min periods) showed favorable effects on bone density and VAT, as well as %BF and aerobic fitness. This supports the idea that vigorous PA can help youths to develop healthy bodies without their needing to restrict their energy intake. Of course, diet is an important contributor to health and may contribute to obesity in some youths; the high intake of nonnutritive sugar-sweetened beverages is an especially important behavioral target for interventions. More controlled studies are needed in which different doses and types of PA are compared, along with their interactions with different types of diet. In youths who are already obese, it may well be necessary to restrict energy intake to impose a negative energy balance, leading to weight loss.

However, given current knowledge, it is reasonable to hypothesize that intervention trials designed to *prevent* pediatric obesity will be more effective if we shift away from the paradigm that assumes obesity to be *caused* by an excessive ingestion of energy. In fact, attempts to have youths restrict their energy intake may actually be working against the biologic demands of growth. Recent evidence suggests that a nutrient partitioning paradigm is likely to be more effective in guiding preventive intervention trials. This paradigm suggests that participation in vigorous PA drives the body to build lean rather than fat tissue, which in turn causes lean youths to ingest more energy and accompanying nutrients. Thus, we might consider the juvenile years as a period in which a healthy body composition can be most effectively developed through maintenance of high levels of both energy expenditure and intake—a high-energy lifestyle.

part VIII

Clinical Implications

A whole section of the book is devoted to a review of the clinical implications of physical activity and inactivity in obesity. The topic is of considerable importance, given the high prevalence of excess body weight in the populations of developed countries, and is covered across 18 chapters. Chapter 65 deals with the important issue of physical activity and mortality among obese individuals. A chapter is devoted to each of the following topics: impaired glucose tolerance (chapter 66), type 2 diabetes (chapter 67), hypertension (chapter 68), heart disease (chapter 69), and stroke (chapter 70). Endothelial dysfunction (chapter 71), inflammation (chapter 72), and depression (chapter 73) are then introduced. These discussions are followed by three chapters on cancer: breast cancer (chapter 74), colon cancer (chapter 75), and other types of cancer (chapter 76). One chapter addresses the issue of physical activity and metabolic syndrome (chapter 77). A chapter has also been set aside to deal with musculoskeletal disorders (chapter 78). This is followed by chapters on risk of falls in obese elderly people (chapter 79) and physical activity and the risk of adverse events in obese persons (chapter 80). Finally, two pediatric issues are reviewed: physical activity and the cardiovascular disease risk profile in obese children (chapter 81) and risk of diabetes in obese children (chapter 82).

Physical Activity and Mortality Rates in Obesity

Kevin R. Fontaine, PhD

Obesity and physical inactivity are independent risk factors for the development of serious illness and elevated mortality. Since, in many cases, obese individuals are also physically inactive, it is a challenge to disentangle the individual and combined contributions of weight and physical activity to mortality. That is, higher mortalities observed in individuals with elevated body mass index (BMI) levels could occur because of the obesity, the physical inactivity, some combination of these factors, or of course some unknown factor(s). In this chapter, I highlight a set of hypothetical causal models developed previously (1) that describe possible associations between obesity, low physical activity, and mortality rate. I then briefly review a selection of studies that evaluate whether physical activity among obese people influences mortality rate. Finally, I discuss selected methodological issues and offer suggestions for future research.

Conceptual Models

Figure 65.1 presents five hypothetical causal models that we proposed previously (1) between obesity (O), low physical activity (LPA), and mortality (M). Briefly, in model A, O causes the adoption of a sedentary lifestyle, and the resulting LPA increases M. In this model, LPA should be a stronger predictor of M than O and, after controlling for LPA, O should have no independent association to M. Thus, LPA is a mediating variable, and controlling for LPA is inappropriate to determine the total effect of O on M. In this situation, stating that O is not associated with M after controlling for LPA is like saying that falling from the roof of a multistory building is not a risk factor for dying if one controls the impact with the sidewalk below. This is a true statement, but impractical, because people who fall from a building tend to hit the sidewalk. In model B, LPA causes O, and O increases M. This is the reverse of model A in that O is a stronger predictor of M than is LPA and, after controlling for O, LPA would have no independent association with M. O is the mediating variable, and controlling for LPA will not reduce the expected value of the apparent effect of O on M. Again, such a model is impractical. In model C, O and LPA are correlated; however, LPA is the only factor that produces variation in M.

Model D is the reverse of model C in that O and LPA are correlated with each other, but only O causes variation in M. Models C and D are examples in which confounding is present; and, under such circumstances, not including both O and LPA would produce biased estimates. Finally, in model E, O and LPA are strongly associated in that O is a cause of LPA, LPA is a cause of O, and both increase M. Controlling for O should attenuate the estimated effect of LPA on M, while controlling for LPA should attenuate the estimated effect of O on M. Due to the expected high correlation between O and LPA,

Acknowledgements

I thank Shawn Franckowiak, MBA, for his assistance in the preparation of this chapter. The writing of this chapter was supported in part by NIH grants: RO1ARO53168, RO1DK076671, and K23ARO49720.

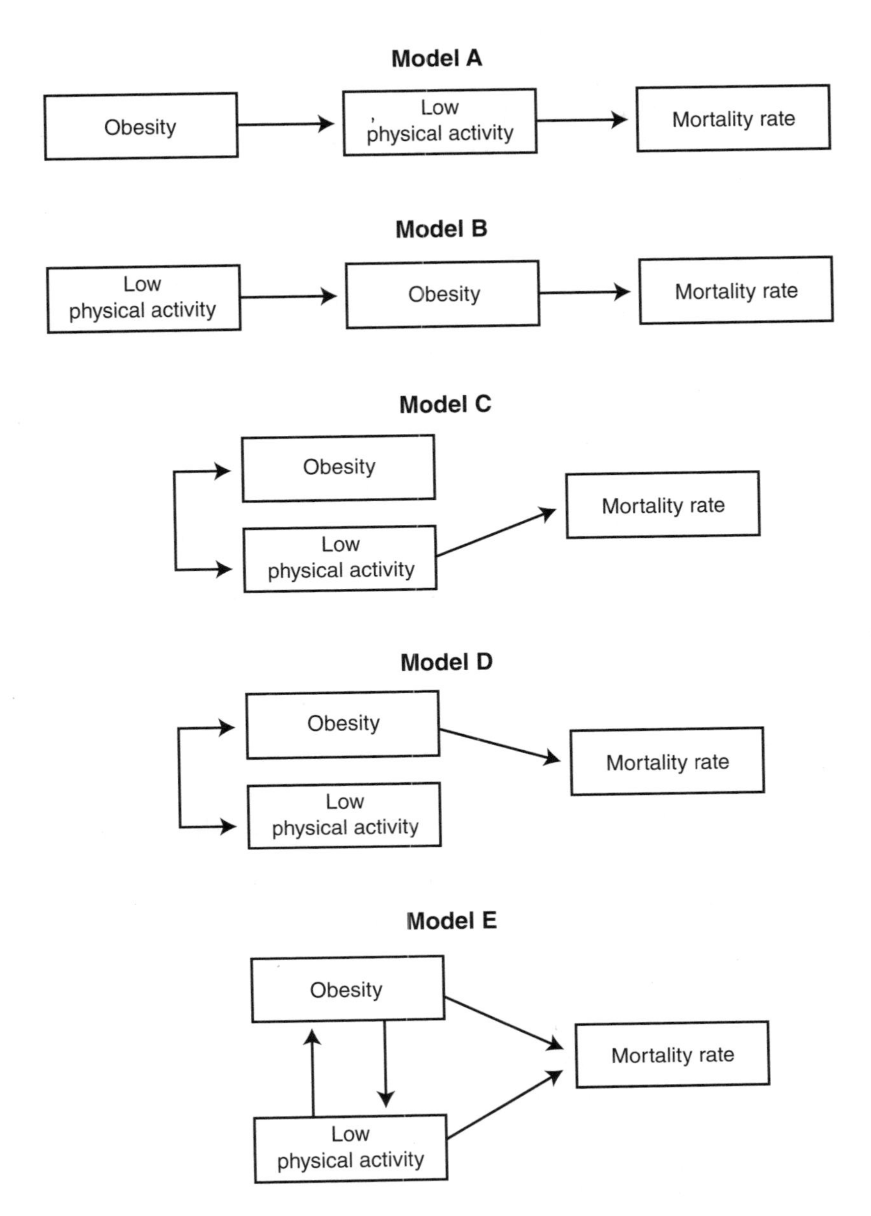

Figure 65.1 Hypothetical causal models between obesity, low physical activity, and mortality.

Adapted, by permission, from K.R. Fontaine and D.B. Allison, 2003, Obesity & mortality rates. In *Handbook of obesity,* 2nd edition, edited by G. Bray and C. Bouchard (New York: Dekker), 772.

it is hard to disentangle their independent effects on M. Model E, though difficult to translate into a testable form for statistical analysis, is most commonly thought to be the best representation of the reciprocal interplay between obesity, low physical activity, and mortality rate (1).

Overview of Selected Studies

Large-scale, randomized trials pertaining to the association of physical activity with mortality among obese adults are relatively rare (2). However, among nonobese, apparently healthy individuals, and those with or at high risk of cardiovascular disease (CVD), a growing number of studies suggest that physical activity producing a weekly energy expenditure of about 4200 kJ (~1000 kcal) significantly reduces the risk of all-cause mortality in younger and older men and women (3).

With regard to overweight and obese adults, the few epidemiologic studies that have been conducted strongly suggest a reduced risk of mortality among those who are physically active. As noted in recent review articles (2, 4), a few longer-term prospective studies have recently been published that aimed at articulating the independent and collective associations of physical activity and obesity to mortality.

In a representative example of these studies, Hu and colleagues (5) examined the association between BMI and physical activity with mortality among nearly 120,000 women, followed for 24 years, who participated in the Nurses' Health Study. Among all respondents, BMI predicted a higher risk of death than did physical activity. However, higher levels of physical activity appeared to be beneficial at all BMI levels, though physical activity level did not eliminate the risk of death associated with obesity. Specifically, compared to lean (BMI <25) and physically active women (≥3.5 h of exercise each week), lean and physically inactive women had a multivariate relative risk of death of 1.5, obese physically active women of 1.9, and obese physically inactive women of 2.4. Finally, the authors estimated that 31% of all premature deaths are driven by the joint effects of elevated BMI and physical inactivity.

In another study (6) done by a different group of investigators, among nearly 50,000 Finnish adults followed for an average of 18 years, the hazard ratios for all-cause mortality of nonobese physically inactive men compared to physically active and physically inactive obese men were 1.2 and 1.8, respectively. The hazard ratio for physically active obese women and physically inactive obese women compared to nonobese physically active women was 1.1 and 2.1, respectively, suggesting that both obesity and physical activity have strong independent associations and a joint association with mortality.

Crespo and associates (7), using data from 9824 men followed for an average of 12 years from the Puerto Rico Heart Health Program Study, investigated the associations of obesity and physical activity to all-cause mortality. After classifying the men into quartiles based on self-reported physical activity, they found that mortality was increasingly lower among those who were most physically active compared to those classified as sedentary. Moreover, within each of four weight categories, those who were most physically active had significantly lower odds of all-cause mortality.

Finally, a recent study by Sui and colleagues (8) of 2603 postmenopausal women aged 60 years or older, who completed a maximal treadmill exercise test at baseline and were followed for an average of 12 years, showed reduced death rates across incremental fifths of cardiorespiratory fitness (death rates ranged from 32.6 to 8.1 for the lowest to highest levels of fitness, respectively), independent of obesity and abdominal adiposity. It is important to note that, in an effort to pinpoint the predictive value of fitness in relation to mortality, measures of adiposity (assessed by BMI, percent body fat, and waist circumference) were conceptualized as potential confounding variables (see model C).

In sum, it is established that both obesity and physical inactivity are risk factors for elevated all-cause mortality (2, 3). Moreover, some studies (see [9] for a review) suggest that elevated BMI appears to play a more prominent role in the development of diabetes than physical activity, whereas physical inactivity may be more important than BMI in increasing the risk of CVD. Indeed, in at least one study (10), the CVD mortality risk among obese active men was even lower than that observed among inactive nonobese men. Efforts to articulate the independent and joint contributions of obesity and physical activity to mortality suggest that in most, but certainly not all studies, mortality risk is somewhat lower among obese adults who are physically active compared to inactive obese adults (4). Overall, therefore, the evidence suggests that increased physical activity can attenuate the mortality risk of obesity.

Methodological Issues

Although the literature indicates that increased physical activity has an important influence on the elevated mortality associated with obesity, several methodological issues preclude making definitive causal statements about the interplay between these factors. First, studies vary widely with respect to the methods used to assess physical activity and adiposity. Some studies (5-7, 10), though by no means all, relied on self-reports of physical activity, and some used self-reported weight and height. The use of such crude measures makes it difficult to be convinced of the validity of the findings. A related issue concerns the absence of a distinction between physical activity and cardiorespiratory fitness. That is, a number of studies use the assessment of physical activity, again typically via self-report, as a proxy for fitness. However, it is probably important to distinguish physical activity from fitness, since it is probable that increasing the level of physical activity may not confer the health benefits likely to come with a quantifiable increase in fitness. Also, some studies used BMI as a proxy for the real variable of interest, adiposity. Moreover, because BMI, adiposity, fitness, and physical activity are likely to be moderately to highly correlated, it is difficult to isolate the independent effects of these variables on mortality.

Conclusions

Overall it appears that at least some of the excess mortality risk associated with excess weight may be offset among those who are physically active. Many studies also suggest a dose–response association in that higher levels of physical activity produced a greater reduction in mortality risk among overweight and obese adults. This implies that modifying physical activity might attenuate at least some of the mortality risk associated with obesity. Moreover, overwhelming evidence indicates that physical activity produces both health benefits and reduced mortality risk compared to those of adults who are sedentary, regardless of body weight (2).

Though the aforementioned conclusions are consistent with the available data, a major drawback concerns the quality of the data. As already noted, the reliance on self-report to characterize physical activity levels, and to a lesser extent on weight status, introduces a level of imprecision that precludes definitive statements regarding the interplay between weight, physical activity, and mortality. That is, the use of crude measures opens the possibility for misclassification, which might in turn lead to faulty estimates of the association between the relevant variables. Thus, despite the cost and relative inefficiency, newer studies should be conducted that employ laboratory-based assessments of cardiorespiratory fitness and body composition. Such studies should also use objective assessments of physical activities, using accelerometers for example, rather than relying on self-reports of physical activity. In other words, more direct serial measures of the variables of interest should help us to better understand the independent and joint effects of physical activity, fitness, obesity, and body composition on morbidity and mortality.

Physical Activity and Impaired Glucose Tolerance in Obesity

Markku Laakso, MD, PhD

The prevalence of type 2 diabetes (T2DM) is rapidly increasing almost in all countries, and diabetes will be one of the main threats to human health in the 21st century. Changes in the human environment and in human behavior and lifestyle, in conjunction with genetic susceptibility, have resulted in a dramatic increase in the incidence and prevalence of diabetes in the world. The rapid increase of obesity, mainly due to lack of exercise and profound changes in the diet, forms the background for the epidemic of T2DM.

As summarized in a previous review (1), high body mass index (BMI) is the strongest risk factor for T2DM. In a cohort of 51,529 U.S. male health professionals, subjects with BMI ≥35 kg/m^2 had a 42-fold relative risk of developing T2DM compared with men with BMI <23 kg/m^2. In the Nurses' Heath Study, including 84,941 female nurses, women whose BMI was at least 35.0 km/m^2 had almost a 40-fold risk of becoming diabetic compared to women whose BMI was <23.0 kg/m^2. The Nurses' Heath Study showed that exercise for at least 7 h per week reduced the risk of T2DM by 39% compared to that in women who exercised <0.5 h/week. Other longitudinal studies have also provided strong evidence that sedentary lifestyle is predictive of T2DM.

Type 2 diabetes is preceded by a long period of asymptomatic hyperglycemia. In this prediabetic state, postprandial or postglucose levels are mildly elevated, whereas fasting blood glucose can usually be maintained within the near-normal range. The elevation of postglucose levels (2 h plasma glucose levels from 7.8 to 11.0 mmol/L) is used to define impaired glucose tolerance (IGT). The majority of these individuals are obese. Impaired fasting glucose is a condition characterized by an elevation of fasting plasma glucose level (6.1-6.9 mmol/L). From the point of view of prevention, basic mechanisms leading to T2DM, impaired insulin action (insulin resistance) or impaired insulin secretion (or both), should be targeted in addition to modifiable risk factors for T2DM. Compensatory insulin secretion from the pancreas maintains glucose levels within the normal range, but in individuals destined to develop diabetes, beta-cell function eventually declines and leads to the hyperglycemic state.

Diabetes Prevention Trials

Both cross-sectional and longitudinal studies have indicated that high physical activity decreases the risk of T2DM, but evidence from intervention studies is limited. Since lifestyle intervention studies (table 66.1) have usually included a combination of weight loss and increased physical activity, it is difficult to determine the effectiveness of physical activity in the prevention of T2DM in individuals with prediabetes.

Da Qing Study

The Da Qing Study was the first trial to test the hypothesis that lifestyle changes prevent T2DM (2). An attempt was made to determine whether a change in diet or exercise habits was more effective. The study centers, but not participating individuals, were randomized. The diet intervention was associated with a 31% reduction, exercise intervention with a 46% reduction, and combined diet and exercise intervention with a 42% reduction in the

Table 66.1

Description of Randomized Controlled Trials Including Obese Subjects With Impaired Glucose Tolerance

Reference	Study	Duration	No. of subjects	Intervention	Exercise intervention, measurement of exercise
DIABETES PREVENTION TRIALS					
Pan et al. (2)	Da Qing	6 years	530	Exercise and diet	Increase of physical activity, self-report
Tuomilehto et al. (3)	Finnish DPS	3 years	522	Exercise and diet	Increase of physical activity, self-report
Knowler et al. (5)	DPP	3 years	2161	Exercise and diet, metformin	Physical activity at least 150 min/week, self-report
Kosaka et al. (6)	Japanese	4 years	458	Exercise and diet	Increase of physical activity, not reported
Ramachandran et al. (7)	Asian Indian	3 years	531	Exercise and diet, metformin	Physical activity at least 30 min/day, self-report
OTHER TRIALS					
Mensink et al. (8)	Dutch	2 years	114	Exercise and diet	Physical activity at least 30 min/day and at least 5 days a week, $\dot{V}O_2max$
Carr et al. (9)	American	2 years	62	Structured exercise and diet	Physical activity 1 h on 3 days/week, $\dot{V}O_2max$
Oldroyd et al. (10)	English	2 years	69	Exercise and diet	Aerobic activity 20-30 min at least once a week, resting pulse, self-report

DPS= Diabetes Prevention Study; DPP = Diabetes Prevention Program.

risk of developing T2DM during a six-year follow-up period. Thus, physical exercise when added to a diet treatment did not have an additive value in relation to the prevention of T2DM.

Finnish Diabetes Prevention Study

The Finnish Diabetes Prevention Study (DPS) randomized 522 middle-aged obese subjects with IGT to either the intervention group or the control group (3). Each subject in the intervention group received individualized counseling aimed at reducing weight (5% or more), making dietary changes, and increasing physical activity (>150 min/week). Endurance exercise (such as walking, jogging, swimming, aerobic ball games, or skiing) was recommended. The cumulative incidence of diabetes was 58% lower in the intervention group than in the control group without any significant difference between men and women. The difference in weight loss between the groups was 3.4 kg (6.6 lb), and the increase in physical activity was significant (36% vs. 16). Subsequent analyses have demonstrated that the increase in physical activity prevented diabetes, independent of weight loss and dietary changes (4).

The Diabetes Prevention Program

The Diabetes Prevention Program (DPP) randomly assigned 3234 nondiabetic persons with IGT to placebo, metformin, or a lifestyle modification program (5). A 16-lesson curriculum covering diet, exercise, and behavior modification was designed to help the participants to achieve these goals. The proportion of participants who achieved the goal of at least 150 min of physical activity per week was 58% at the most recent visit. The cumulative incidence of diabetes was 58% lower in the lifestyle intervention group than in the placebo group. Lifestyle intervention was associated with greater weight loss and a greater increase in leisure physical activity compared to values in the control group. No attempt was made to analyze the independent effect of physical activity on the prevention of diabetes.

Japanese Diabetes Prevention Study

Male subjects with IGT recruited from health screening examinees were randomly assigned in a 4:1 ratio to a standard intervention group (control group, N = 356) and an intensive intervention group (intervention group, N = 102) (6). In the intervention group, detailed instructions on lifestyle (weight loss by diet, increased physical exercise) were given. The cumulative four-year incidence of diabetes was 9.3% in the control group, versus 3.0% in the intervention group; and the reduction in risk of diabetes was 67.4%. Weight loss was 0.39 kg (0.86) in the control group and 2.18 kg (4.8 lb) in the intervention group. Detailed analysis on the effect of physical activity alone was not presented.

Indian Diabetes Prevention Program

In an Indian program, a total of 531 subjects with IGT were randomized into four groups (7). Group 1 was the control; group 2 was given advice on lifestyle modification; group 3 was treated with metformin; and group 4 was given lifestyle modification and metformin. The three-year cumulative incidences of diabetes were 55.0%, 39.3%, 40.5%, and 39.5% in groups 1 through 4, respectively. No significant changes in body weight were noticed. Physical activity increased, but the effect of physical activity in the prevention of diabetes was not evaluated.

Physical Exercise and Diabetes Prevention Trials

All five diabetes intervention trials reported in table 66.1 showed a considerable decrease in the incidence of T2DM. The Finnish DPS (3) and the DPP (5) showed a 58% reduction in the risk of T2DM. The Chinese Da Qing study (2) resulted in a 31% reduction of incident T2DM with diet treatment, a 46% reduction with exercise, and a 42% reduction with diet and exercise treatment. In the Japanese trial (6), the reduction of the risk of T2DM with lifestyle intervention was 67%, and in the Indian study it was 55% (7). The success of these interventions is likely to be largely explained by weight loss. The role of physical activity in the reduction of the risk of T2DM is difficult to estimate given that changes in physical activity were based on self-report, which is likely to be unreliable in a trial setting. The Finnish DPS (4) used a standardized and validated physical activity questionnaire and concluded that the preventive effect of physical activity on T2DM was independent of weight loss. Because in all trials except the Indian trial (7) a significant weight loss was reported, the role of leisure-time physical activity in the prevention of T2DM remains unclear. The only possibility for getting a definite answer on the role of physical activity in diabetes prevention is a trial in which physical activity is significantly increased in high-risk individuals without changes in weight. This is difficult given the fact that the majority of these individuals are overweight.

Other Trials

The problem of three other trials reported in this review (table 66.1) is that they were small in size (from 62 to 114 subjects) (8-10). On the other hand, two of them (8, 9) measured the success of the exercise program using $\dot{V}O_2max$, which gives objective information on the effects of exercise. However, weight loss was also reported in these trials, which makes it difficult to evaluate the independent role of physical activity in improving metabolic parameters.

Dutch Study

In a Dutch study (8), 88 obese subjects with IGT were randomly assigned to either a lifestyle intervention group (regular dietary advice, advice to increase physical activity) or a control group to measure changes in glucose tolerance. Subjects adherent to both the diet and exercise intervention showed the largest reduction in weight loss and 2 h glucose levels and increased their aerobic capacity.

Study on Japanese Americans

A total of 62 Japanese Americans with IGT were randomized to the American Heart Association (AHA) Step 2 diet plus endurance exercise (n = 30) or the AHA Step 1 diet plus stretching (n = 32) for 24 months (9). Lifestyle modifications decreased weight and central adiposity and improved insulin sensitivity, but not beta-cell function, suggesting that this degree of lifestyle modification may be limited in preventing T2DM over the long term.

English Study

Participants with IGT (n = 78) were randomly assigned to a two-year lifestyle intervention (diet, 20 to 30 min aerobic exercise at least once a week) or to a control group (10). After 24 months follow-up, BMI was significantly lower in the intervention group compared with the control group, and insulin sensitivity improved after 12 months in the intervention group.

Conclusions

The role of physical activity—independent of other lifestyle changes (particularly weight loss)—in the treatment of obese individuals with IGT, who are at high risk of T2DM and cardiovascular disease, remains unsettled on the basis of trial evidence published so far. However, an analysis of the data from the Finnish DPS suggested that an increase in moderate-to-vigorous and strenuous structured leisure-time physical exercise largely reduced the risk of T2DM even after adjustment for weight loss and diet changes (4). This suggestive evidence needs to be proven in a trial in which the effect of increased physical activity is tested in a setting where changes in body weight and diet are not confounding the results. This trial should also include the measurement of $\dot{V}O_2max$ to provide objective evidence on the effects of the exercise program.

Physical Activity and Type 2 Diabetes in Obesity

Edward S. Horton, MD

The worldwide prevalence of obesity and type 2 diabetes mellitus (T2DM) is increasing rapidly and now presents a major health problem for both developed and developing countries. This "dual epidemic" is closely related to changes in the quantity and composition of people's diets and to decreased physical activity, associated with economic growth and development, urbanization, increased availability of food, and a more sedentary lifestyle. It is now estimated that 66% of adults in the United States are overweight (body mass index, BMI ≥25) and, among them, 32% are obese (BMI ≥30). There are approximately 24 million people with diabetes, 90% to 95% having T2DM, and 57 million at high risk of developing diabetes by virtue of having impaired glucose tolerance (1). In addition, 24% of American adults have the metabolic syndrome, according to the National Cholesterol Education Program Adult Treatment Panel III criteria (2). The health and economic consequences of this dual epidemic are tremendous; and changes in lifestyle, focusing on weight reduction and increased physical activity, are fundamental to both the prevention and the treatment of T2DM and its long-term complications.

Regular physical exercise has long been considered a cornerstone in the management of T2DM. However, the goals of an exercise program, as well as the specific risks and benefits, will vary from one patient to another and have to be individualized, depending on the person's age, degree of obesity and physical fitness, and the presence of diabetic complications or concomitant illnesses. For example, a young person with T2DM may wish to participate in vigorous or prolonged exercise as part of his or her normal lifestyle, including engaging in recreational or competitive sport. In contrast, an overweight, middle-aged person may undertake regular physical exercise as part of his or her treatment program to improve glycemic control; lose or maintain body weight; and improve blood pressure, lipids, and other cardiovascular risk factors. As a rule, regular physical activity is an important component of treatment for people with T2DM and should be prescribed along with appropriate diet and oral hypoglycemic agents or insulin as part of a comprehensive treatment program. Designing an appropriate exercise program for T2DM requires a careful assessment of the expected benefits and associated risks of exercise in individual patients, as well as appropriate monitoring to avoid injuries and other complications.

This chapter reviews the effects of physical exercise on the regulation of blood glucose and insulin concentrations and insulin sensitivity in people with T2DM.

Effects of Exercise

Obesity and T2DM are associated with insulin resistance and varying degrees of hyperinsulinemia, in both the fasting and postprandial states, depending on the function of the pancreatic beta cells. Lean people with T2DM are generally more insulin sensitive and have lower plasma insulin concentrations than do obese T2DM subjects. However, with increased duration of T2DM, there is usually a progressive loss of pancreatic beta cell function in both lean and obese subjects, which results in worsening of glucose control and the need for increased use of oral antidiabetic medications or insulin therapy. Weight loss and increased physical exercise can improve insulin sensitivity, decrease hyperinsulinemia, and improve glucose control in people with T2DM. For lean subjects, the beneficial effects of exercise are related primarily to the acute glucose-lowering effects of exercise and to the long-term effects of physical training to improve insulin sensitivity and reduce multiple risk factors for cardiovascular disease (CVD). These beneficial effects of exercise also occur in obese people with T2DM; and when exercise is combined with a calorically restricted diet, exercise also plays an important role in achieving and maintaining weight loss.

The regulation of blood glucose during and after exercise is complex and involves multiple hormonal and neural mechanisms that have been reviewed elsewhere (3, 4). The energy requirements of skeletal muscle in the resting state are derived primarily from fatty acid oxidation, with only 10% to 15% coming from glucose oxidation and 1% to 2% from oxidation of amino acids. During exercise, total energy requirements increase, and there is a shift in the pattern of metabolic fuel utilization. As exercise intensity increases, glucose metabolism increases; and, during high-intensity exercise, it becomes the predominant source of energy for muscle contraction. Glucose is derived from a combination of breakdown of muscle glycogen stores by glycogenolysis and uptake of glucose from the circulation by increased muscle blood flow and activation of muscle glucose transport, the latter becoming the major source of glucose as muscle glycogen stores become depleted. In nondiabetic subjects, blood glucose concentrations are maintained in a normal range during moderate-intensity exercise by increased hepatic glucose output. However, in people with T2DM and hyperglycemia, moderate exercise usually results in a decrease in blood glucose concentration toward normal but rarely to hypoglycemic levels. In both normal and diabetic subjects during high-intensity exercise, hepatic glucose output may exceed peripheral glucose utilization, and blood glucose concentrations increase; in contrast, with prolonged, glycogen-depleting exercise, hepatic glucose output may be insufficient to match glucose utilization, and blood glucose levels may decrease into the hypoglycemic range. However, this is rarely a problem for people with T2DM unless they are treated with insulin or insulin secretagogues.

Insulin and muscular contraction are the two main regulators of glucose uptake in skeletal muscle. Both activate the translocation of the glucose transporter GLUT 4 from an intracellular location to the plasma membrane and T-tubules. In obesity and T2DM, insulin resistance is associated with decreased insulin stimulation of skeletal muscle glucose uptake, whereas the response to exercise (muscle contraction) is normal. Thus, exercise-stimulated glucose uptake and metabolism in muscle are not impaired. This is explained by the fact that the signaling mechanisms by which insulin and exercise increase glucose uptake in muscle are different. Insulin action is dependent on its binding to its receptor and the initiation of a signaling cascade that involves autophosphorylation of the insulin receptor, phosphorylation of insulin receptor substrates 1 and 2, activation of phosphoinositol-3 kinase, and subsequent activation of other downstream proteins that ultimately result in translocation of GLUT 4 to the muscle cell plasma membranes and T-tubules where glucose uptake occurs.

Muscle contraction, however, is not dependent on phosphorylation of the insulin receptor, insulin receptor substrates 1 and 2, or phosphoinositol-3 kinase for activation of glucose transport. It activates a multifactorial signaling process that involves changes in the cellular energy state, increased intracellular calcium, and possibly activation of protein kinase C and other factors that are not yet well understood. In brief, the energy-sensitive enzyme AMP-kinase and its upstream kinase, LKB1, respond to changes in the AMP/ATP ratio and increase glucose uptake. Calmodulin also activates glucose transport independently of AMP-kinase, and protein kinase C may also play a role in activating glucose transport in skeletal muscle. While downstream signaling molecules are an area of active study, it is known that AMP-kinase phosphorylates the protein AS160 and calmodulin binds to it, and it is now considered that AS160 may be the convergence point by which both insulin and muscle contraction stimulate glucose transport. The effects of physical training are also complex and involve changes in muscle fiber type, mitochondrial biogenesis, and increased muscle expression of GLUT 4, making muscle more sensitive to insulin. Factors involved in the responses to physical training include PGC-1a, calcineurin, Ca^{2+}-calmodulin, and P38 MAPK (5).

Current data suggest that both the acute effects of exercise and the effects of physical training on muscle glucose metabolism are relatively normal and that the same beneficial effects of an exercise training program can be expected in people with T2DM as are seen in normal subjects.

Benefits of Exercise

The benefits of exercise for patients with T2DM are listed ahead. Moderate, sustained exercise usually lowers elevated blood glucose concentrations and may be used by patients on a day-to-day basis to help improve their metabolic control. In addition, physical training, independent of weight loss, results in improved sensitivity to insulin and lower fasting and postprandial insulin and glucose concentrations. Several studies have demonstrated that both aerobic and strength training programs improve insulin sensitivity and glycemic control as measured by improvement in hemoglobin A1C (6). Thus, both the acute blood glucose–lowering effects of exercise and the increased insulin sensitivity produced by physical training may be of major importance in improving long-term glycemic control in people with T2DM.

Benefits of Exercise for Patients With Diabetes

Lower blood glucose concentrations during and after exercise

Improved insulin sensitivity

Lower basal and postprandial insulin concentrations

Cardiovascular conditioning

Improved lipid profile

- Slightly decreased low-density lipoprotein cholesterol
- Decreased triglycerides
- Increased high-density lipoprotein cholesterol

Improvement in systolic and diastolic blood pressure

Lower glycated hemoglobin levels

Improved sense of well-being and quality of life

Increased strength and flexibility

Increased energy expenditure

- Adjunct to diet for weight reduction
- Preservation of lean body mass
- Increased fat loss

Another benefit of regular physical exercise is that it is an effective adjunct to diet for weight reduction and weight maintenance. Physical exercise programs alone are usually associated with little or no loss of weight and must be combined with a calorie-restricted diet to achieve significant weight loss. However, numerous studies have demonstrated that regular physical exercise is an important part of the lifestyle modification program and is particularly important for achieving weight maintenance after weight loss goals have been achieved (7).

Regular exercise is also associated with a reduction in cardiovascular risk factors through improvement of the lipid profile, reduction in both systolic and diastolic blood pressure, reduced markers of subclinical inflammation, and improved endothelial function. While several population studies have demonstrated an inverse correlation between the amount of habitual physical exercise and cardiovascular events (8), prospective randomized controlled trials of lifestyle intervention have not yet been completed, so it is not known if the improvement in cardiovascular risk factors will result in reduced cardiovascular events over time. In the LookAHEAD Trial, overweight or obese adults with T2DM who participated in a program of lifestyle modification focusing on healthy diet, weight loss, and increased physical activity showed a significant improvement in glucose control, as measured by decreased hemoglobin A1C and fasting plasma glucose concentrations, as well as reduced use of antidiabetic medications after one year of treatment. Participants had a mean weight loss of 8.6% of initial body weight and a 21% increase in physical fitness as measured by submaximal treadmill exercise testing. In addition, the lifestyle modification program resulted in significant improvements in both systolic and diastolic blood pressure, serum triglycerides, and high-density lipoprotein cholesterol and urinary albumin-creatinine ratios (9). In this and other studies that have used a combination of weight loss and physical exercise, it is not possible to clearly separate the effects of exercise. However, in clinical practice these interventions are usually combined to achieve the best results.

Finally, regular exercise has the same general benefits for people with T2DM that it does for nondiabetic people. Cardiovascular fitness and physical working capacity are improved, as are the sense of well-being and quality of life.

Risks of Exercise

There are several potential risks associated with exercise for patients with T2DM. In patients with T2DM who are treated with insulin or an insulin secretagogue, hypoglycemia may occur during or following exercise, although this is a much less common problem than in people with type 1 diabetes. It is generally not necessary to take extra food prior to or during exercise or to adjust medications unless the exercise is of very long duration. However, patients should be cautioned to check their blood glucose levels periodically and to be aware of signs or symptoms of hypoglycemia if they are being treated with insulin or an insulin secretagogue and participating in prolonged exercise.

Risks of Exercise for Patients With Diabetes

Hypoglycemia, if patient is being treated with insulin or insulin secretagogues

Hyperglycemia with very strenuous exercise

Precipitation or exacerbation of CVD

Angina pectoris

- Myocardial infarction

Arrhythmias

Worsening of long-term complications of diabetes

Proliferative retinopathy

- Vitreous hemorrhage
- Retinal detachment

Nephropathy

- Increased proteinuria

Peripheral neuropathy

- Soft tissue and joint injuries

Autonomic neuropathy

- Decreased cardiovascular response to exercise
- Decreased maximum aerobic capacity
- Impaired response to dehydration
- Postural hypotension

Careful screening for underlying cardiac disease is important for all patients with T2DM before they start a vigorous exercise program involving an exercise intensity that is unusual for the patient and is greater than that of the activities of daily living. The role of exercise stress testing is controversial. For young, generally healthy people with T2DM, an exercise stress test is not needed prior to the start of an exercise program and may be a detriment because of the risk of a false positive test. However, in older individuals, particularly those who have a history of CVD or increased risk factors for CVD, an exercise stress test will be useful to help identify silent ischemic heart disease and may identify patients who have an exaggerated hypertensive response to exercise, or may develop postexercise orthostatic hypertension, or both.

Several complications of diabetes may be aggravated by exercise, and all patients should be screened before they start an exercise program. The most important of these is proliferative retinopathy, in which exercise may result in retinal or vitreous hemorrhage.

Exercises that increase blood pressure (e.g., heavy lifting or strenuous aerobic exercise) should be avoided by patients with proliferative retinopathy requiring laser therapy until cleared by an opthalmologist, usually three to six months following laser photocoagulation. Physical exercise is also associated with increased proteinuria in patients who have diabetic nephropathy, probably because of changes in renal hemodynamics. It has not been shown, however, that exercise leads to progression of renal disease, and the use of angiotensin-converting enzyme inhibitors appears to decrease the amount of exercise-induced albuminuria.

In patients with diabetic neuropathy, there are several risks of exercise that must be considered in the design of an exercise program. With autonomic neuropathy, maximum heart rate and exercise capacity are limited; and there may be problems with orthostatic hypotension and regulation of fluid balance that may be exacerbated by exercise. Peripheral neuropathy increases the risk of soft tissue or joint injuries and requires selection of appropriate types of exercise and precautions to avoid injury. In addition, degenerative joint disease is more common in obese individuals and may be exacerbated by weight-bearing exercises.

If abnormalities are found, exercises should be selected that will not pose significant risks for worsening complications or causing injuries. Young, otherwise healthy patients with T2DM do not normally require formal exercise prescriptions, although they need specific recommendations regarding strategies for managing exercise and avoiding injuries.

Guidelines for Exercise

With few exceptions, the types of exercise a patient performs can be a matter of personal preference. It is generally recommended that most people participate in moderate-intensity exercise for at least 30 min on most days of the week. However, exercise at least three times a week has been shown to result in significant improvement in cardiovascular fitness, improved glycemic control, and reduction in cardiovascular risk factors in people with T2DM. The effect of a single bout of aerobic exercise on insulin sensitivity lasts 24 to 72 h depending on the duration and intensity of the exercise. Consequently, to achieve sustained effects, the time between exercise sessions should not be more than two consecutive days. The effects of resistance exercise training on insulin sensitivity may last somewhat longer because of the associated increase in muscle mass and adaptive changes in muscle metabolism.

To achieve the best results from an exercise program, it is commonly recommended that people alternate days of aerobic exercise with days of strength training, with a goal of achieving at least 30 min of exercise on five or six days each week. Including a 5 to 10 min warm-up of low-intensity aerobic exercise and stretching to prevent musculoskeletal injuries may be helpful for older people with T2DM, but recent studies in younger people have failed to show that this is necessary. The higher-intensity portion of the exercise session should last 20 to 45 min and then gradually be increased up to 60 min or longer, depending upon the level of physical conditioning. The general rule is to start slowly and build up gradually as cardiovascular condition and strength improve. Increased amounts of time doing moderate or vigorous aerobic exercise will result in greater cardiovascular conditioning and greater CVD risk reduction compared with lower durations of activity. The current recommendation for resistance training is a minimum of 8 to 10 exercises involving the major muscle groups, with a minimum of one set of 10 to 15 repetitions resulting in near fatigue. Alternatively, fewer repetitions (8 to 10) of the heaviest weight that can be lifted 8 to 10 times to near fatigue can be used (10).

By undertaking a program of aerobic exercise equivalent to brisk walking at least three days per week and resistance training at least two days per week, one can expect improved control of glucose, blood pressure, and dyslipidemia, as well as better maintenance of weight loss if the program is combined with reduced caloric intake.

Special Considerations for People With T2DM

For patients who have not participated regularly in exercise in the past or who have significant complications of diabetes or other impediments to exercise, supervised exercise programs may be beneficial. Often cardiac rehabilitation programs will help in supervising exercise programs for people with T2DM, particularly if the patients have established CVD or are at high risk for CVD. Many diabetes treatment centers offer supervised exercise programs. However, many patients do not need formal supervision once an initial assessment has been completed and an appropriate exercise plan has been established.

Several things can be done to facilitate the patient's motivation and participation in an exercise program. These include choosing activities the person enjoys, providing variety in the exercise program, having the person perform exercise at convenient times and locations, encouraging participation in group activities, involving the patient's family and associates for reinforcement, and measuring progress to provide positive feedback. Most important is to start slowly, build up gradually, and set realistic goals. In addition, patients should be taught how to avoid injuries or other complications while participating in physical exercise.

In patients being treated with low-calorie diets, physical exercise is generally well tolerated and does not pose any additional risks, but it is important to note that increasing activity in a person with T2DM who is on a low-calorie diet and is taking a hypoglycemic agent will increase the risk of hypoglycemia, and appropriate adjustments in medications should be made.

68

Physical Activity and Hypertension in Obesity

James M. Hagberg, PhD

Substantial evidence generated over the last three to four decades has unequivocally established a strong association between obesity and hypertension. In 1978, a large epidemiologic study in about 1 million individuals showed that 20- to 39-year-olds who were overweight had a hypertension prevalence double that of normal-weight and triple that of underweight individuals (1). The same trends were evident in 40- to 64-year-olds, with overweight individuals having 50% and 100% higher rates of hypertension than normal-weight and underweight individuals, respectively. Ten to fifteen years later, the National Health and Nutrition Examination Study III showed very similar results; obese men had a hypertension prevalence nearly triple that of normal-weight men, and obese women had a prevalence about 2.5 times that of their normal-weight peers. The Behavioral Risk Factor Surveillance System also reported similar results another 10 years later in about 200,000 Americans. The prevalence of hypertension in normal-weight individuals was 16%, whereas hypertension was nearly twice as prevalent in overweight individuals—about 2.5 times higher in class 2 obese individuals and about three times higher in class 3 obese individuals. Hypertension prevalence rates and their relationships with body mass index (BMI) differ somewhat across U.S. ethnic-racial groups; but among whites, blacks, and Hispanic Americans and across both genders, all evidence indicates that individuals with the highest BMI have substantially higher hypertension prevalence rates.

In addition to these data indicating a powerful relationship between current weight and hypertension, there is substantial evidence that obesity is a risk factor for the future development of hypertension. For example, in the Nurses' Health Study (*n* = 82,473 women), BMI at age 18 and at midlife was strongly and positively associated with the risk of developing hypertension, as was weight gain from age 18 to midlife. Numerous other reports also indicate that body weight, BMI, obesity, and overweight in early life, even in childhood and adolescence, dramatically affect a person's risk of developing hypertension later in life.

Thus, there is an unequivocal and strong relationship between people's current and historical weights and weight increase over time with their risk of having or developing hypertension. Furthermore, the effect is substantial, with a doubling to tripling of risks for individuals in the overweight and obese categories that are widely prevalent in the United States and the rest of the developed world today.

Treatment of Hypertension

Clearly, as all current treatment guidelines recommend, the optimal intervention for an obese individual, whether with or without the additional comorbidity of hypertension, is to improve diet, initiate and maintain an increased level of physical activity, readjust daily caloric balance, and, as an overall outcome, lose weight (2). However, just as clearly, it is very difficult for an obese or overweight hypertensive patient, or anyone for that matter, to make such widespread changes in his or her overall lifestyle. Quite obviously, if these changes were easy to initiate and maintain, the epidemics of overweight or obesity, diabetes, and hypertension in the United States would have been reversed long ago. Attempting to make such widespread alterations in overall lifestyle is very difficult and requires a degree of motivation that most individuals do not possess. Therefore, this review first assesses the benefits of a single lifestyle intervention—increased physical activity—for the obese or overweight hypertensive individual and then examines the efficacy of a more intense lifestyle intervention that includes increased physical activity combined with a diet-induced weight loss program.

Increased Physical Activity as a Treatment

In one of the earlier reviews of publications that assessed the impact of exercise training on blood pressure (BP) in hypertensives, we found that of the six studies published by 1984 showing significant BP reductions in hypertensive individuals with exercise training, only one also showed a significant reduction in body weight, and the weight loss was 1.2 kg or 2.6 lb (3). Furthermore, this small number of studies indicated that the reduction in BP resulting from exercise training was not related to the initial BMI of the participants. Thus, we concluded that the weight loss that may result from exercise training was not contributing to the significant exercise training–induced reductions in both systolic (SBP) and diastolic (DBP) blood pressure in hypertensive individuals, many of whom were overweight or obese. A meta-analysis by Fagard 15 years later arrived at the same conclusion (4). Fagard found that even though overweight hypertensives lost weight with training, their BP reduction was the same as that of normal-weight hypertensives not losing weight with training. He also found no significant relationships between BP changes with training and either initial BMI or the BMI change with training.

In another of our reviews, we found no significant relationships between BP and body weight changes resulting from exercise training in hypertensive individuals (5). We concluded that reductions in BP with exercise training were not dependent on substantial changes in body weight. Another meta-analysis of previous randomized controlled aerobic exercise training trials showed no significant effects of initial BMI or weight change with exercise training on the training-induced BP reductions in hypertensives (6). This study showed that normal-weight and overweight or obese hypertensives and hypertensives who did or did not lose weight with training all decreased their BP to the same extent. And, finally, another subsequent review concluded that BP reductions with exercise training occurred independent of any exercise training–induced weight loss (7).

Thus, the available literature is clear and generally consistent in showing that overweight or obese hypertensive individuals can elicit the same SBP and DBP reductions with exercise training as normal-weight individuals. As most hypertensive individuals are overweight or obese, such results provide strong evidence that these high-risk individuals are not at any disadvantage in terms of training-induced improvements in this important cardiovascular (CV) disease risk factor compared to their leaner counterparts. In addition, the published literature also quite clearly and consistently indicates that overweight or obese hypertensives do not have to lose weight for exercise training to significantly reduce their SBP and DBP. This is especially important because weight loss is generally minimal with exercise training programs not also specifically focused on weight loss.

Previous reviews quantifying the magnitude of BP reductions in hypertensives with exercise training present ranges from –3.2 to –10.6 and –2.4 to –8.2 mmHg for SBP and DBP, respectively (3-7). The relatively wide ranges for these effect sizes are a function of whether the training-induced reduction was corrected for the BP change observed in the control group in the same study. Reviews in which the training-related BP reductions were corrected present the lower effect sizes in the ranges cited. Blood pressure reductions at the higher end of these ranges were from reviews in which the generally nonsignificant BP changes observed in the control group were not corrected for. Also, those reviews often showed that hypertensives had larger training-induced BP reductions, although these differences were not always significant. Another important point is that of all the hypertensive groups that underwent endurance exercise training in previous studies, the majority experienced significant BP reductions with training, and virtually all of the remaining groups reduced their BP but not significantly with training. This review could identify from all previous studies only one hypertensive group in which endurance exercise training significantly increased BP. Thus, at least in terms of BP changes with exercise training, clinically and statistically significant negative responses are almost nonexistent.

Blair and colleagues, in their already classic studies from Dallas, Texas, make another important point relative to the benefits of increased levels of CV fitness for hypertensive and overweight or obese individuals (8). Their widely quoted primary results show very clearly that having a CV fitness level above that of the lowest 20% of the population substantially reduces all-cause and CV mortality. However, they also found that a fit individual who still has hypertension is at about 25% lower mortality risk than normotensives with low CV fitness. Similarly, fit individuals who remained overweight or obese had a 60% lower mortality risk than normal-weight men with low CV fitness. Thus, at least in this cohort, an above-average CV fitness level completely overcame and reversed the mortality risks associated with hypertension and overweight or obesity, indicating that a lack of CV fitness had more impact than either of these two CV disease risk factors on mortality risk.

Increased Physical Activity Plus Dietary Weight Loss as a Treatment

Another treatment option for overweight or obese hypertensives is diet-induced weight loss with or without the addition of exercise training. It is beyond our scope here to review the available literature on the effects of dietary weight loss interventions on BP in overweight or obese hypertensives. However, a recent meta-analysis of randomized controlled diet-induced weight loss trials in hypertensives concluded that each 1 kg (2.2 lb) weight loss was associated with 1.05 and 0.92 mmHg reductions in SBP and DBP, respectively, with individuals who lost >5 kg (11 lb) eliciting greater SBP and DBP reductions than those losing less body weight.

Only a few studies have compared the effects of diet-induced weight loss plus exercise training to those of exercise training in overweight or obese hypertensives. In 1997, Gordon and colleagues studied obese individuals with either high normal BP or stage 1 hypertension (9). Their 12-week-long interventions consisted of exercise training without weight loss, a diet-induced weight loss group, and a combined diet-exercise group. All three groups significantly reduced SBP and DBP with the interventions. However, more importantly, all three intervention groups decreased SBP and DBP to the same degree. The authors concluded that the independent effects of these two interventions for overweight or obese hypertensives—diet-induced weight loss and exercise training—are not additive; each had its own independent effect on BP, but these independent effects did not "add up," as the combined intervention did not result in a further reduction in BP.

We studied hypertensive obese older men assigned to one of four groups: a control, an exercise training group, a diet-induced weight loss group, or a combined exercise training and diet-induced weight loss group (10). The interventions lasted nine months to enable us to determine whether differential effects might appear after the 12-week intervention duration of the Gordon study. Also, participants' sodium intake was kept constant at 3 g per day, as the reduction in sodium intake associated with reducing caloric intake could have its own independent effect on BP. However, even with these additional controls in place, while both SBP and DBP decreased significantly in all three intervention groups compared to the control group, there were again no significant differences in SBP or DBP reductions elicited among the three interventions. Thus, the conclusion of this and the Gordon study is that diet-induced weight loss and exercise training without weight loss each independently and significantly reduce BP in overweight or obese hypertensives; but when the two interventions are combined, the independent effects of these interventions are not additive and do not result in any further significant BP reduction.

Since these two initial studies, three additional reports have assessed the impact of diet-induced weight loss and exercise training versus exercise training on BP. The first of these studies also showed that diet-induced weight loss reduced BP to the same degree as a combined diet and exercise training intervention in obese hypertensives. The second indicated that in overweight or obese hypertensives, diet-induced weight loss reduced BP to a greater extent than exercise training, and the final study showed that obese hypertensives reduced BP somewhat more with combined exercise training and diet-induced weight loss compared to exercise training. Thus, these results are also generally consistent and support the conclusion that both diet-induced weight loss and exercise training have independent and significant BP-reducing effects in overweight or obese hypertensives. However, these two independent effects are not "additive," as the combination of the two interventions does not reduce BP further. It may also be important to keep in mind that the magnitude of weight loss in these studies is generally not substantial for overweight or obese individuals (<7 to 8 kg [15.4 to 17.6 lb]);

therefore, it is not clear how the effects on BP of diet and exercise training interventions that elicit substantially greater amounts of weight loss might compare.

Conclusion

One can conclude that increased levels of physical activity or exercise training can have significant benefits for the overweight or obese hypertensive in terms of BP reductions, and these BP reductions can occur independent of any significant weight loss. However, adding weight loss to an increased physical activity and exercise training intervention does not appear to reduce BP to any greater extent than either intervention alone, at least across the minimal weight losses that have been studied to date.

69

Physical Activity and Heart Disease in Obesity

Timothy Church, MD, MPH, PhD

In work published in 1953, Dr. Jeremy Morris and colleagues examined the risk of having a myocardial infarction in sedentary bus drivers compared to physically active bus conductors (1). The buses were double-decker, and while the bus driver essentially sat throughout the work day, the conductors accumulated a substantial amount of physical activity as they climbed up and down the bus stairs to check tickets. It was observed that the active conductors had about one-half the number of coronary heart disease events than the sedentary bus drivers experienced. Interestingly, in what may have been the first paper examining fatness in the context of physical activity and heart disease, Morris and colleagues in 1956 noted that the physically active conductors wore smaller trousers (i.e., had smaller waists) than the sedentary drivers (2). In 1975, Dr. Ralph Paffenbarger reported an inverse relation between risk of coronary heart disease death and work-related caloric expenditure in longshoremen, providing further evidence that work-related physical activity was associated with lower coronary heart disease risk (3). These two researchers provided the first epidemiological evidence suggesting that physical activity provided protection against the development of cardiovascular disease (CVD).

In recent decades, as physical labor and thus physical activity have been largely removed from the workplace, the focus has shifted to self-reported recreational physical activity. There is conclusive evidence from numerous well-established large prospective epidemiological studies that physically active individuals have substantially lower risk of CVD incidence and mortality compared to sedentary individuals. For example, findings from the Nurses' Health Study and the Women's Health Initiative demonstrate that individuals who perform 120 min or more of moderate-intensity physical activity per week are at 30% to 50% lower risk of developing CVD than sedentary individuals.

Cardiorespiratory Fitness and Cardiovascular Disease

Starting in the late 1990s, Dr. Steve Blair and his colleagues published a series of papers examining the risk of morbidities and mortality, including CVD, across levels of cardiorespiratory fitness as assessed by a maximal treadmill test, which serves as a marker of regular physical activity. The Aerobics Center Longitudinal Study (ACLS) is composed of men and women who came to the Cooper Clinic for a preventive medicine examination and underwent a maximal treadmill test. Based on the time on the treadmill, age, and gender, each individual's fitness was categorized as low, moderate, or high. For both men and women, there was an inverse relation between fitness level and risk of cardiovascular death; the risk of cardiovascular mortality in the moderate-fitness group was less than half that in the low-fitness group.

Fitness, Physical Activity, Fatness, and Cardiovascular Disease

Though the vast majority of epidemiology studies have indicated that benefits of physical activity (and fitness) in relation to CVD remain when body weight is accounted for, there has been great interest in dissecting out the role of weight control in the benefits associated with physical activity. This has led to the question whether physical activity has health benefits by itself or whether the weight control (or loss) associated with physical activity is responsible for the benefits.

Examining combinations of fitness and fatness in men from the ACLS database (figure 69.1), Wei and colleagues reported that in normal-weight, overweight, and obese individuals, low fitness was associated with substantially higher risk of CVD mortality (4). Further, it was observed that men who were overweight or obese but fit were at lower risk of CVD mortality than men who were normal weight but unfit. This finding suggested that in relation to CVD mortality risk, it is more important to be fit than to be normal weight.

In a comparison of different combinations of physical activity and fatness in 116,564 women (2370 CVD deaths), Hu and associates reported an inverse relation between self-reported physical activity and risk of CVD death in normal-weight, overweight, and obese women (figure 69.2) (5). However, in direct contrast to the ACLS data in men, in all physical activity categories there was increased risk of CVD mortality with higher body mass index. In other words, though higher levels of physical activity conferred benefit in every weight group, higher levels of physical activity did not negate the effects of excess weight. Other studies using self-reported physical activity have presented similar results. Given the sharply conflicting data from well-established, large epidemiological studies, the issue of the relative importance of physical activity, fitness, and fatness for an individual's cardiovascular health becomes quite confusing; the ongoing dialogue on this issue has been labeled the "fitness versus fatness" debate.

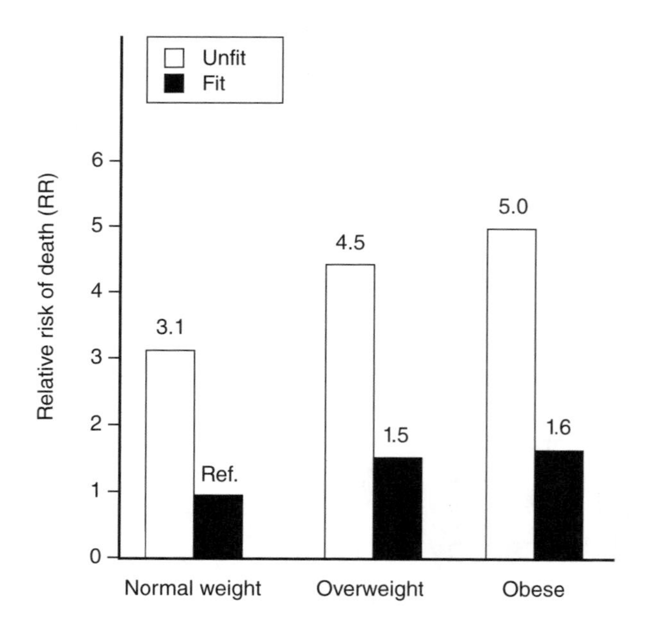

Figure 69.1 Risk of death in men by fitness and weight categories.

Adapted from R.S. J. Paffenbarger, 1999, "Relationship between low cardiorespiratory fitness and mortality in normal-weight, overweight, and obese men," *Journal of American Medical Association* 282:1547-1553.

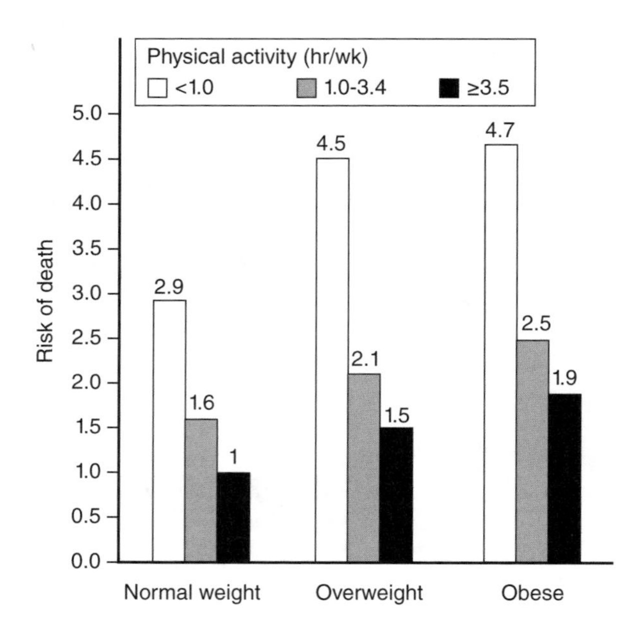

Figure 69.2 Risk of death in women by physical activity and weight categories.

Adapted from F.B. Hu et al., 2004, "Adiposity as compared with physical activity in predicting mortality among women," *New England Journal of Medicine* 351:2694-2703.

Physical Activity Quantification

A significant caveat for the fitness–fatness debate is that most of the studies not showing that fitness trumps fatness, such as the work of Hu and colleagues, did not actually measure fitness but rather measured self-reported physical activity. Self-reported physical activity questionnaires have a large degree of measurement error under the best of circumstances. Further complicating the use of physical activity questionnaires for assessing physical activity within stratifications of weight is the observation that the degree of overestimation of physical activity is positively associated with weight. In other words, the more people weigh, the more likely they are to overestimate their level of physical activity. Thus, the use of physical activity questionnaires within subgroups of weight is likely to create a disproportional amount of physical activity misclassification in the obese and overweight groups compared to the normal-weight group.

This leads one to question how many of the obese or overweight but physically active individuals are misclassified due to an overestimation of physical activity. On the other hand, fitness measurement is characterized by considerably less measurement error; more importantly, this error is not strongly associated with weight, resulting in far less opportunity for misclassification when fitness–fatness categories are created. When one creates tightly defined categories based on fitness (or activity) and fatness, the resulting analysis is strongly dependent on accurate classification of study participants.

The use of measured fitness as a surrogate measure of physical activity is not without complicating issues. Fitness has a strong genetic contribution, which is important in terms of both baseline fitness level and training response to a given level of physical activity. Many tightly controlled physical activity training trials have demonstrated that for a given increase in physical activity, although there is an increase in the group mean fitness levels, the individual changes in fitness levels are highly heterogeneous. Another complicating concept related to the use of fitness as a measure of regular physical activity is the idea that performance during a maximal exercise test represents the integrated functioning of numerous physiological systems. Thus fitness at some level is a reflection of overall health, especially of the cardiovascular system. So it should come as no surprise that fitness is a powerful predictor of premature morbidity and mortality, as poor fitness may represent the early physiological manifestation of these conditions. Given the currently available data, it is impossible to dissect the relative importance of physical activity, genes, and subclinical disease in the excess risk associated with low fitness. In summary, the methods of physical activity quantification, questionnaire versus fitness measurement, are likely to at least in part explain the discrepancy between fitness–fatness and physical activity–fatness findings.

Clinical Importance of Fitness and Fatness in Cardiovascular Health

While the "fitness versus fatness" issue has led to controversy, confusion, and debate that may never be fully resolved, the relative contribution of fitness and fatness to overall health and CVD risk actually may be of little clinical significance because low fitness and excess fatness share the common treatment of increasing regular physical activity. Increasing regular physical activity results in increases in fitness for most individuals and is a core component of successful weight loss programs, more importantly of long-term weight loss maintenance. In essence, physical activity is the common denominator for the clinical treatment of low fitness and excess weight, making the "fitness versus fatness" debate largely academic, particularly with regard to the prevention and treatment of CVD in overweight or obese individuals (6).

Physical Activity and Cardiovascular Disease Risk

Overweight and obese individuals are at increased risk of diabetes, metabolic syndrome, hypertension, and lipid disorders, all of which are strong risk factors for CVD. Any strategy to reduce the morbidity and mortality associated with these risk factors is of great clinical importance. For all of these conditions, being regularly active and avoiding low fitness has been associated with a lower risk of mortality. For example, Ardern and colleagues (7) examined the benefit of fitness within levels of Adult Treatment Panel-III-Revised (ATP-III-R) risk stratification and reported that within each level of ATP-R risk stratification of CVD mortality, there was at least twofold higher risk for unfit individuals compared to fit individuals. In other words, even in individuals whose low-density lipoprotein (LDL) and risk factor profile qualifies them for cholesterol-lowering medication, living an active lifestyle and avoiding being unfit greatly reduces the risk of CVD mortality. Thus, despite the minimal benefit (if any) of regular physical activity to LDL-cholesterol, those with elevated LDL in a relative sense stand to benefit the most from being physically active. Excess weight is a powerful risk factor for type 2 diabetes and is a primary driving force behind the recent explosion in the prevalence of diabetes. Both fitness and physical activity have been shown to dramatically reduce the risk of CVD in individuals with diabetes, and this protection is observed across the weight classifications (8). Similar survival benefits associated with physical activity without reversal of the risk factor have been shown in individuals with obesity-associated conditions such as hypertension and metabolic syndrome.

Summary

In summary, regular physical activity is associated with substantially reduced risk of CVD mortality independent of weight; and even in overweight or obese individuals, being physically active is associated with reduced CVD risk. Further, in individuals with obesity and CVD-related risk factors such as diabetes, being physically active reduces the risk of CVD. Thus physical activity should be a priority for individuals of all weights and every CVD risk factor status.

70

Physical Activity and Stroke in Obesity

Janice Eng, PhD

Stroke is a leading cause of neurological disability and serious long-term disability. Obesity, measured by body mass index (BMI), increases the risk of stroke in both men and women (1). Abdominal obesity (e.g., waist circumference, waist/hip ratio) may be a better predictor of stroke, particularly in men (1). Other well-known risk factors for stroke include increasing age, hypertension, elevated blood cholesterol, carotid stenosis and atrial fibrillation, cigarette smoking, diabetes mellitus, and heart disease. These risk factors for stroke also apply to people with obesity, although the interactions of obesity with these factors are not well established.

Physical activity is a recently established risk factor for stroke; risks of stroke are 20% and 27% lower with moderate and high physical activity, respectively, compared to low activity (2). A combination of physical inactivity and obesity increases the risk of cardiovascular disease (including stroke) compared to each factor alone. Although logically weight loss, physical activity, or both would reduce the risk of stroke, risk reduction with such treatments has not been evaluated yet.

Given the sedentary lifestyle that follows a stroke, together with premorbid conditions, it is not surprising that cardiovascular disease is the leading prospective cause of death in individuals living with chronic stroke. In addition, these individuals are at high risk for experiencing another (i.e., recurrent) stroke. Almost 30% of new cases of stroke are recurrent strokes (3).

Impact of Obesity in Stroke

Obesity has important implications for people with stroke. The prevalence of obesity in people with stroke is at least as high as, if not higher than, that in populations of similar age who have not had a stroke. Greater BMI in people with stroke is associated with poorer control of blood pressure and glucose level, as well as an increased risk of metabolic syndrome and diabetes mellitus. Obesity also negatively affects rehabilitation efforts. Obese patients with stroke commence their rehabilitation with a lower functional score and then show less improvement (4). A number of factors could contribute to these findings. Those with greater weight may have had poor premorbid health, which may be reflected in lower function. In addition, the excessive weight may cause difficulties in regaining the mobility to perform daily activities of living, and rehabilitation staff may have difficulty assisting the overweight client with basic functions such as transfers or walking.

Lack of Intensive Activity in Stroke Rehabilitation

Stroke rehabilitation can be criticized for its lack of emphasis on lifelong physical activity, which ideally should commence from the early rehabilitation phase. Although now proven incorrect, previous thinking was that strenuous activity would permanently increase spasticity and reinforce abnormal movement. Thus, stroke rehabilitation has historically involved moderate exercise programs in the hospital, with few formal programs that extend into the community setting. This is in stark contrast to the well-established intensive cardiac rehabilitation programs that extend into the community and that long ago demonstrated their cost effectiveness.

Typical therapy sessions for stroke inpatient rehabilitation do not elicit a cardiovascular stress that would induce a training effect. One stroke rehabilitation study showed that heart rate measures were within a training zone (40% to 85% of heart rate reserve) for only 2.8 min of the 55 min physical therapy sessions and for 0.7 min of the 41 min occupational therapy sessions (5). Of particular concern, there was not a meaningful increase in exercise intensity between the 2nd and the 14th week of rehabilitation. Given that many patients will not receive further therapy after inpatient stays, such low intensities will not prepare individuals to return to physical activity. The use of simple devices, such as heart rate monitors, during therapy sessions could assist clinicians in gauging their patients' exertion and progress in exercise intensity.

Inactivity in Chronic Stroke

The American Stroke Association Council and American Heart Association recommend that people with stroke engage in at least 30 min of moderate-intensity exercise on most days (3). In a population-based survey of over 24,000 adults aged 65 years and older, the group with stroke included the highest proportion of individuals who did no leisure-time physical activity (27%) compared to groups with other chronic conditions (6). Only 38% of respondents with stroke used walking for physical activity compared with 50% to 56% in all other chronic disease groups and 61% in those without chronic conditions. Hip accelerometers have been used with community-dwelling people with stroke to provide real-time estimates of the frequency, intensity, and duration of physical activity. Even in people with mild stroke severity, few met recommended levels of physical activity (7). Similarly aged healthy adults participated in twice the amount of physical activity that the stroke participants engaged in over a three-day period. Of particular concern, many stroke participants may have had sufficient ability (e.g., could complete more than 400 m [437 yd] in a 6 min walk test) but did very little physical activity at home as recorded by the accelerometer (7). This observation may be similar to the phenomenon of upper extremity "learned nonuse," seen in stroke patients who demonstrated the ability to use their arm in the clinic setting but did not use the arm at home. This is an important consideration, because if individuals with stroke do not utilize their current abilities to engage in physical activity, they are at risk for further compromise of health.

Challenges of Regular Exercise

Inactivity is a modifiable risk factor for recurrent stroke, heart disease, and obesity. However, there are numerous challenges to increasing the physical activity in those with stroke. Stroke patients have mobility difficulties, cognitive impairments, and cardiovascular conditions that may require specific screening and supervision during exercise to minimize the risk of falls and cardiac events. Transportation to community facilities is challenging, as many stroke survivors rely on caregivers or limited transit services for people with disabilities. Community membership fees can be a barrier due to financial hardships stemming from medical costs and loss of employment.

Benefits of Exercise

On the positive side, there is increasing evidence of the benefits of regular exercise in people with stroke, including a reduction of falls, increase in cardiovascular fitness, maintenance of bone density, improved mobility, greater muscle strength, and enhanced quality of life. Intensive treatment, particularly cardiovascular training, has been shown to be safe and effective regardless of the stage of stroke recovery (subacute to chronic) or modality (e.g., treadmill training, stationary bike, chest-deep water exercise, functional repetitive exercises) (8). Exercise intensities ranging from 50% to 80% heart rate reserve and durations from 20 to 40 min, three to five days per week, have been shown to improve peak$\dot{V}O_2$ or workload in addition to walking speed and walking endurance poststroke (8). Treadmill training interventions in stroke have traditionally focused on outcomes of walking ability or motor recovery. However, a recent treadmill training program demonstrated a reduction in recurrent stroke risk factors by improving glucose tolerance and reducing insulin resistance (9). The six-month program was 40 min, three times per week, at a target of 60% to 70% heart rate reserve; handrail and overhead harness support was used as modification. Although body weight, body fat, and fat-free mass did not change with treadmill training, this study is one of the few stroke trials to measure body composition. Interventions need to consider reducing the risk of cardiovascular disease and recurrent stroke, as well as traditional outcomes of motor recovery.

Strategies for Increasing Physical Activity and Healthy Eating

The management of obesity must address both activity and diet. The nutritional aspects poststroke have not received much consideration, particularly in the chronic stroke phase. Although 85% of a group of stroke patients had at least one major risk factor for cardiovascular disease, only 61% of this group reported that a doctor had told them to reduce high fat and cholesterol intake (10). Similarly, only 64% reported that a doctor had told them to exercise more (10).

Counseling through physician advice or other clinician input has been shown to be effective for promoting a healthy lifestyle. Patients with stroke who received dietary physician advice were one and a half times more likely to change their dietary behavior than if they had not received advice. In addition, those who received advice to exercise more were almost twice as likely to engage in exercise as those who did not receive such advice (10). Lastly, those who exercised had more healthy days (days when physical and mental health were perceived to be good) and fewer days on which they could not carry out their usual activities because of poor physical or mental health (10).

Physician advice is only one avenue for promoting lifestyle change. Ideally, nutrition and physical activity should be emphasized beginning early in rehabilitation and should be addressed through a multidisciplinary team approach. Only one study (over one-fourth of participants were stroke patients) has been aimed at addressing physical activity behavior during rehabilitation. This study used tailored counseling sessions on sport and physical activity during rehabilitation and up to eight weeks after rehabilitation discharge (11). The authors found that physical activity participation was higher at nine weeks and one year after rehabilitation discharge in the counseled group compared to the control group. Such a long-lasting impact demonstrates the effectiveness of a rehabilitation-based physical activity promotion program for people with physical disabilities.

Multidimensional health promotion interventions may also have benefit. A short-term (12 weeks, three times per week) health promotion program for people with chronic stroke included classes for exercise, nutrition (including cooking classes), and health behavior change (12). The health behavior curriculum included topics such as risk factors of stroke, stress, and family roles and promoted active involvement. Both the control and health promotion group improved on dietary fat intake. The health promotion group also showed improved physical function (cardiovascular fitness, strength, flexibility), reduction in total cholesterol, a small reduction in weight (average 2.8 lb [1.3 kg]; 63% were obese), and improved quality of life compared to the control group. The financial sustainability of this program is questionable, as the program and transportation were provided free to participants. However, it is noteworthy that such programs can improve the health and health behaviors of individuals with stroke who have multiple chronic medical conditions, including obesity, hypertension, depression, and diabetes.

Conclusions

Physical inactivity and obesity are major risk factors for stroke and recurrent stroke. It is recommended that a focus on physical activity and nutrition commence early in stroke rehabilitation when there are opportunities for multidisciplinary input and interaction. A goal of rehabilitation should be the gradual increase in physical activity intensity, which can be monitored and progressed with use of devices such as pedometers, accelerometers, or heart rate monitors. Outcome measures need to move beyond traditional motor recovery and include recurrent stroke risk factors such as cardiovascular fitness, physical activity, glucose tolerance, and BMI. Preliminary studies demonstrate the feasibility and efficacy of health promotion counseling to improve physical activity behavior and nutrition poststroke. Further research is required to define effective methods to promote healthy behavior upon discharge and beyond.

71

Physical Activity and Endothelial Dysfunction in Obesity

Christopher A. DeSouza, PhD

The vascular endothelium is a single-cell-thick tissue located at the interface between circulating blood and vascular smooth muscle. Once thought to be a simple barrier separating blood and the vessel wall, the vascular endothelium is now recognized as a multifunctional organ vital to cardiovascular homeostasis. Healthy endothelium functions in a vasoprotective and thromboresistant mode by contributing to the regulation of vascular tone through the synthesis and release of vasodilator and vasoconstrictor substances, in particular nitric oxide (NO) and endothelin-1 (ET-1); inhibiting monocyte and platelet adhesion; and maintaining fibrinolytic integrity. However, due to its unique anatomical position, the endothelium is susceptible to blood-borne injury, mechanical forces, and cardiovascular risk factors that can cause functional imbalances in endothelial-derived vasodilator and vasoconstrictor mediators, increased monocyte adhesion, and impaired coagulation and fibrinolysis. Collectively termed endothelial dysfunction, this systemic syndrome occurs early in the pathogenesis of vascular disease and contributes to the development of atherogenic lesions, thrombus formation, vasospasm, plaque rupture, intimal growth, and in turn coronary and cerebrovascular events (1).

Obesity is an independent risk factor for coronary artery disease, cerebrovascular disease, and atherothrombotic events. Many of the cardiovascular complications associated with obesity (e.g., hypertension, arterial spasm, myocardial infarction) are due, in large part, to endothelial dysfunction. Indeed, alterations in endothelial function associated with obesity that precede and predispose to atherosclerosis and thrombosis include impaired vasomotor and fibrinolytic regulation. In contrast to obesity, habitual aerobic exercise is associated with a reduction in cardiovascular events. One of the mechanisms by which regular aerobic exercise confers this cardioprotection is via its availing effects on endothelial function.

This chapter focuses on the negative influence of increased adiposity on endothelial vasodilator and fibrinolytic function and the beneficial effects of regular physical activity, particularly aerobic endurance exercise, on these two critical functional properties of the endothelium.

Obesity and Endothelium-Dependent Vasodilation

Impaired endothelium-dependent vasodilation is a hallmark characteristic of endothelial dysfunction and has been linked etiologically to the initiation and development of atherosclerotic vascular disease. Endothelium-dependent relaxation is mediated, in large part, by a diffusible substance initially referred to as endothelium-derived relaxing factor and later identified as NO. Derived by the degradation of L-arginine by NO synthase, NO is released by the endothelium in response to a variety of stimuli

Acknowledgements
Supported by NIH DK062061, HL068030, HL076434, HL074634, RR00051, and the American Diabetes Association.

including increased blood flow, shear stress, hormonal mediators, and several endothelial agonists. Impaired endothelium-dependent vasodilation occurs early in the atherosclerotic process, contributes to disease development and progression, and can trigger acute cardiovascular events. Moreover, endothelial vasodilator dysfunction is a strong predictor of future atherosclerotic events (2).

Increased adiposity is associated with impaired endothelium-dependent vasodilation whether assessed noninvasively via brachial artery flow–mediated dilation (FMD) or through changes in blood flow in response to intra-arterial infusions of endothelial vasodilators such as acetylcholine. Reduced FMD has been observed in obese compared with normal-weight children as young as 8 years of age (3). In obese adults, FMD is also markedly impaired in comparison with that in normal-weight individuals. In addition, increased adiposity is associated with blunted coronary and peripheral blood flow responses to a variety of endothelial agonists such as acetylcholine, methacholine, bradykinin, substance P, and isoproterenol. For example, forearm blood responses to intrabrachial artery infusion of acetylcholine are about 30% lower in obese compared with normal-weight adults of similar age (4) (figure 71.1). The mechanisms underlying obesity-related endothelial vasodilator dysfunction are complex, are not well understood, and may differ between children and adults. In general, factors such as inflammation and oxidative stress as well as increased ET-1-mediated vasoconstrictor tone are all thought to contribute to reduced vasodilator capacity with obesity. Interestingly, in obese adults, the contribution of NO to endothelium-dependent vasodilation is not impaired compared with that in age-matched normal-weight individuals, suggesting that NO bioavailability is not compromised with obesity (4). Whether this is also the case in children is unknown.

Obesity, Endothelium-Dependent Vasodilation, and Physical Activity

Regular aerobic exercise is an extremely effective lifestyle intervention strategy for improving endothelium-dependent vasodilation in both obese children and adults. Continuous aerobic physical activity (playing games such as dodge, tag, and soccer as well as jogging) for 1 h, three times per week for as little as eight weeks, can significantly improve FMD in obese children, independent of changes in body fatness (5). In obese adults, brisk walking for 12 weeks improves endothelial vasodilator capacity to levels comparable to those of normal-weight adults of similar age. Indeed, forearm blood flow responses to acetylcholine can improve by as much as 30% in obese adults with moderate-intensity aerobic exercise training (daily walking, 60% to 75% of maximal heart rate, four or five times per week) (figure 71.1). Moreover, the exercise-induced

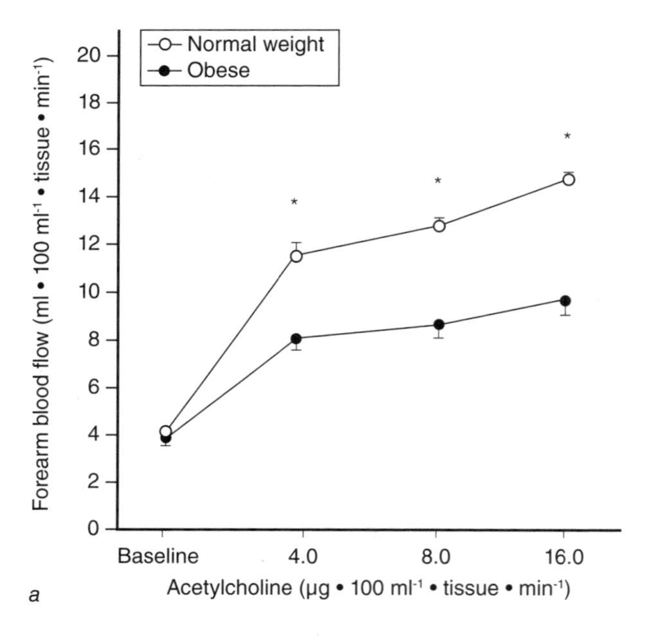

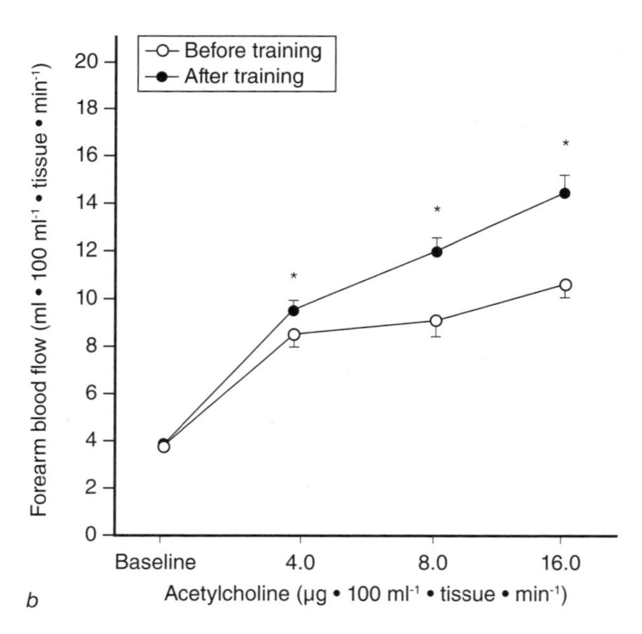

Figure 71.1 *(a)* Forearm blood flow responses to intrabrachial artery infusion of acetylcholine in normal-weight and obese adults and *(b)* before and after aerobic exercise training in obese adults. $^*P < 0.05$.

increase in endothelium-dependent vasodilation in obese adults can occur without changes in body composition, cardiometabolic risk factors, or maximal aerobic exercise capacity. Moderate-intensity resistance exercise is also associated with improvement in endothelium-dependent vasodilation in obese adults (6).

The mechanisms mediating the beneficial effects of habitual physical activity on endothelial vasodilation are not well understood. Since obesity (at least in adults) is not associated with impaired NO bioavailability, it is unlikely that the modulatory effects of exercise on endothelial vasodilator capacity are mediated by increased NO. Instead, exercise-induced reductions in inflammation, oxidative stress, and ET-1-mediated vasoconstrictor tone may all contribute to increased endothelial vasodilator ability. In older overweight adults, regular aerobic exercise reduces ET-1 system activity, alleviating endothelial vasoconstrictor tone and allowing greater vasodilator potential (7).

Obesity and Endothelial Fibrinolytic Capacity

The development of fibrous plaque and the resulting thrombus formation within the lumen of an artery is recognized as the precipitating cause of atherothrombotic events such as angina, myocardial infarction, thrombotic stroke, and sudden death. The hemostatic mechanism responsible for the proteolytic degradation of intravascular fibrin deposition is the fibrinolytic system. This enzymatic pathway maintains vascular patency by converting the inactive proenzyme plasminogen to the active enzyme plasmin, which lyses fibrin into soluble degradation products. Reduced endogenous fibrinolytic activity is a major contributor to the development, progression, and severity of atherothrombosis. The vascular endothelium plays a central role in the regulation of fibrinolysis through the synthesis and release of specific plasminogen activators and inhibitors. Endothelial cells are the principal site of synthesis and release of tissue-type plasminogen activator (t-PA), the primary plasminogen activator in fibrinolysis, and major producer of its biological inhibitor plasminogen activator inhibitor-1 (PAI-1).

Tissue-type plasminogen activator is the key enzyme in initiating an endogenous fibrinolytic response due to its ability to preferentially activate plasminogen on the surface of developing thrombi. It is the capacity of the endothelium to release t-PA rapidly and acutely from intracellular storage pools, rather than circulating plasma fibrinolytic concentrations, that determines the efficacy of endogenous fibrinolysis. Indeed, thrombolysis is much more effective if active t-PA is readily available and incorporated during, rather than after, thrombus formation. Clinically, impaired endothelial t-PA release is a strong predictor of future cardiovascular events, as it has been linked to increased coronary atheromatous plaque burden and myocardial infarction. The use of endothelial agonists such as substance P or bradykinin at subsystemic, locally active concentrations in the forearm, coupled with measures of blood flow and arterial and venous sampling, provides a reproducible method for directly assessing the rate of t-PA release from the endothelium in vivo (8).

Similar to endothelium-dependent vasodilation, the capacity of the endothelium to release t-PA is significantly blunted with increased adiposity. Net endothelial t-PA release across the forearm in response to bradykinin is about 50% lower in obese compared with normal-weight adults (figure 71.2) (9). A major underlying mechanism for the obesity-related reduction in endothelial t-PA release is oxidative stress. Oxygen free radicals inhibit t-PA release from endothelial cells and damage the fibrin-binding affinity of t-PA, reducing its thrombolytic capacity. Acute intra-arterial and chronic oral supplementation with the potent antioxidant vitamin C restores the capacity of the endothelium to release t-PA in overweight and obese adults to levels comparable with those of normal-weight adults of similar age (GP Van Guilder and CA DeSouza, unpublished data). Given the invasive nature of the protocol used to assess endothelial t-PA release in vivo, there are no data on the influence of adiposity on the capacity of the endothelium to release t-PA in children. However, there is evidence that obesity is associated with a hypofibrinolytic state; systemic concentrations of the primary inhibitor of the fibrinolytic system, PAI-1, are elevated in obese children and adolescents (10).

Obesity, Endothelial Fibrinolytic Capacity, and Physical Activity

Regular aerobic exercise confers beneficial effects on endothelial fibrinolytic regulation in obese adults. Moderate-intensity aerobic exercise training (daily walking for 12 weeks) markedly improves endothelial t-PA release in previously sedentary obese adults, independent of changes in body mass or body composition (figure 71.2). In fact, habitual

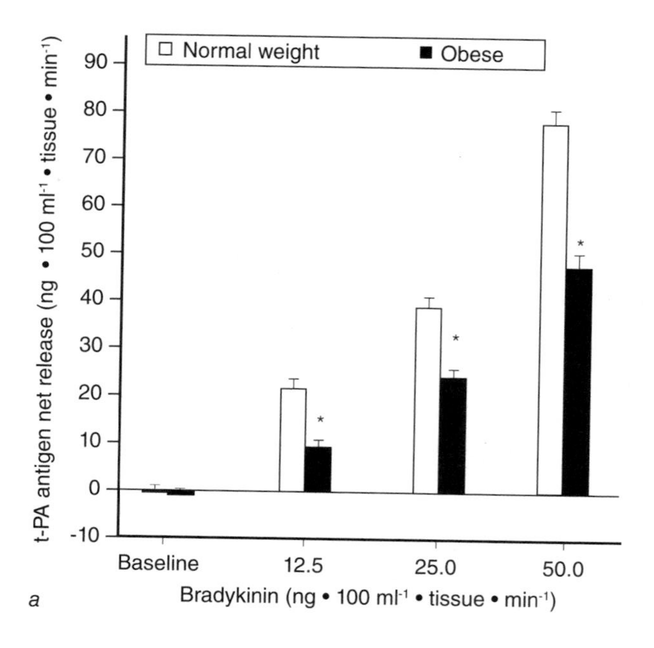

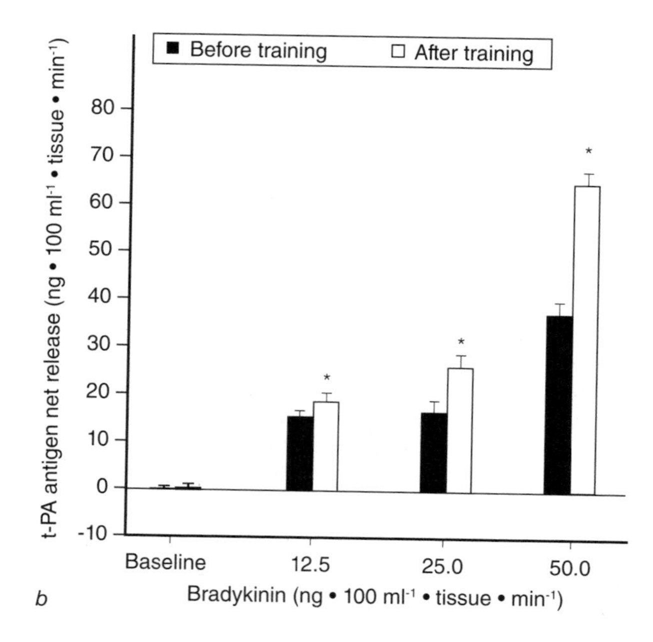

Figure 71.2 *(a)* Net release rate of tissue-type plasminogen activator (t-PA) antigen across the forearm in response to intrabrachial artery infusion of bradykinin in normal-weight and obese adults and *(b)* before and after aerobic exercise training in obese adults. $^*P < 0.05$.

Adapted from G.P. Van Guilder et al., 2005, "Endothelial t-PA release is impaired in overweight and obese adults but can be improved with regular aerobic exercise," *American Journal of Physiology-Endocrinology and Metabolism* 289: E807-E813.

aerobic exercise can restore the capacity of the endothelium to release t-PA in obese adults to levels comparable with those of their normal-weight peers. Several factors may contribute to the favorable effects of regular aerobic exercise on endothelial t-PA release, including shear stress–induced increase in t-PA transcription and protein synthesis, augmented cytoplasmic calcium levels that may potentiate t-PA release, and reduced oxidative stress (9). There are no data regarding the effects of resistance exercise on endothelial t-PA release. In contrast to the case with adults, there is a dearth of data on the effects of aerobic exercise on the fibrinolytic system in obese children and adolescents. Given the rising rates of adiposity in this segment of the population and the important role of the fibrinolytic system in cardiovascular health, future studies are needed to elucidate potential beneficial effects of habitual exercise on fibrinolysis in obese children and adolescents.

Summary

Obesity-related endothelial dysfunction is evident in both children and adults, contributing to increased risk of cardiovascular disease, acute vascular events, and premature cardiac death. Importantly, endothelial dysfunction is not an irreversible consequence of increased adiposity. Regular aerobic physical activity improves endothelial vasodilator function in both obese children and adults. Moreover, chronic aerobic exercise enhances the capacity of the endothelium to release t-PA in obese adults. As noted elsewhere in this book, obese adults who engage in habitual aerobic exercise demonstrate a lower risk of cardiovascular disease morbidity and mortality. Improved endothelial vasodilator and fibrinolytic function may be a central mechanism underlying this cardioprotection. Regular moderate-intensity aerobic exercise is the most effective nonpharmacologic intervention for improving endothelial function in obese individuals. Health care professionals need to continue to stress the importance of regular aerobic physical activity, and with renewed zeal, in an effort to improve vascular health and reduce vascular risk in a steadily growing population of obese children and adults.

72

Physical Activity and Inflammation in Obesity

Mark Hamer, PhD

The role of inflammatory processes in health and disease has gained increasing interest within the last decade. Inflammatory markers, such as cytokines and acute phase proteins, are expressed in a number of tissues, notably monocytes and macrophages, vascular endothelial cells, adipose tissue, and neurons, and have a variety of purposes within the body that range from fighting infection and wound healing to appetite regulation. A number of novel circulating biomarkers that reflect low-grade inflammation, such as C-reactive protein (CRP), interleukin (IL)-6, and fibrinogen, are now considered to be cardiovascular risk factors. Inflammation has also gained attention in relation to physical activity and obesity as a possible mechanism through which these lifestyle factors may be associated with clinical endpoints. The aim of this chapter is to examine the association between physical activity, adiposity, and inflammation.

It has been suggested that the increases in circulating IL-6 that are consistently observed after exercise promote an anti-inflammatory environment by increasing IL-1 receptor antagonist and IL-10 synthesis but inhibiting tumor necrosis factor-alpha (TNF-α) release. A large number of observational studies have confirmed an association between physical fitness or physical activity and various pro-inflammatory markers. For example, in a prospective cohort study of 4088 British men, there was a significant and inverse dose–response relationship between physical activity and a number of hemostatic and inflammatory markers (e.g., white cell count, CRP, fibrinogen) after 20 years of follow-up (1). Interestingly, participants who became active during the follow-up showed levels of inflammatory markers similar to those of participants who had remained continuously active throughout. In a prospective study of 27,055 healthy women, the inverse association between physical activity and cardiovascular disease risk after 11 years of follow-up was substantially mediated by inflammatory or hemostatic biomarkers, making the largest contribution to lower risk (32.6%) in comparison with established risk factors such as blood pressure, lipids, adiposity, and glucose control (2).

Adiposity, Inflammation, and Confounding

Given that excess adiposity is an important risk factor for low-grade inflammation, significant attention has focused on whether the association between physical activity and inflammation is really independent from obesity. There are several possible relationships between physical activity, adiposity, and inflammation. Firstly, the association between activity and inflammation may be confounded by adiposity. Secondly, adiposity may mediate the association between activity and inflammation, or thirdly, activity and adiposity may share the same causal pathways. Unfortunately, if the association between physical activity and inflammation disappears following adjustment for measures of adiposity, this is not conclusive evidence for confounding. Indeed, if adiposity mediates the relationship between activity and inflammation, then adjustment for adiposity is inappropriate.

Acknowledgements
Dr. Hamer is supported by grant funding from the British Heart Foundation (United Kingdom).

The literature relating to this issue has been difficult to interpret because although a substantial proportion of observational studies have shown independent associations between physical activity and inflammation, the randomized controlled trials are less clear. For example, in 152 female smokers, there were no changes in CRP or fibrinogen after a 12-week exercise program that was not accompanied by significant weight loss, despite improvements in physical fitness (3). In 193 sedentary, mildly obese, dyslipidemic men and women, six months of exercise training did not alter levels of CRP despite improvements in fitness and visceral and subcutaneous adiposity (4). In contrast, significant reductions in inflammatory markers were observed following three to six months of exercise training without changes in body mass index (BMI) or body fat in elderly participants (5, 6). In another trial that lasted for 18 months, a dietary weight loss intervention resulted in significant reductions in CRP, IL-6, and TNF-α compared with values in the control group, although no effects were observed for an exercise intervention group (7).

A number of factors may explain these equivocal findings, including disparity in the exercise interventions, poor adherence levels, differences between characteristics of participants, and variable follow-up times. Indeed, most intervention studies typically employ three- to six-month treatments that may not be adequate to yield significant effects. In a recent study of 176 healthy men and women recruited from the Whitehall II epidemiological cohort, we demonstrated that BMI, but not physical fitness, was independently associated with IL-6 and CRP after three years of follow-up (8). However, overweight or obese participants who were physically fit had lower odds of low-grade inflammation (CRP ≥1 mg/L) at follow-up compared with their unfit counterparts, in relation to fit and lean participants, after multivariate adjustments (see figure 72.1). These data might suggest that the anti-inflammatory effects of exercise are particularly noticeable in individuals with excess adiposity. The reason physical fitness appeared to have stronger anti-inflammatory effects in overweight or obese participants may be related to differences in central adiposity. Exercise training can result in selective loss of visceral adipose tissue even without weight reduction, and lower fitness has been associated with more visceral adipose tissue after matching of individuals with similar BMI (9). Given that visceral adipose tissue is a key production site for various inflammatory markers, this may be an important factor in the interaction between physical activity, obesity, and inflammation.

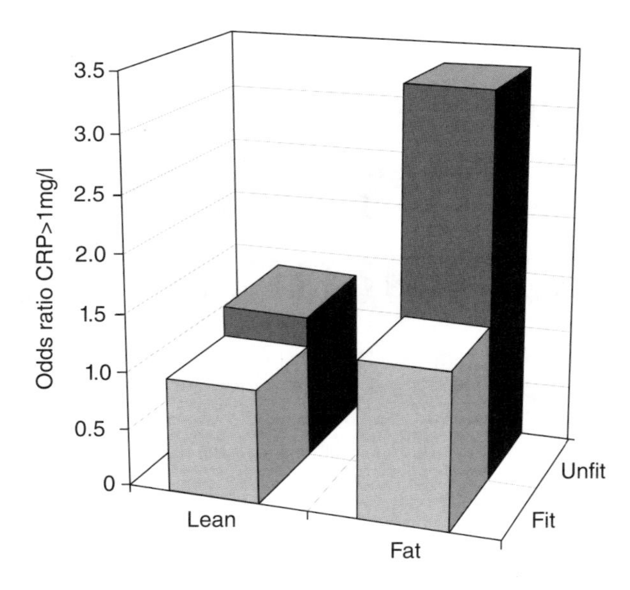

Figure 72.1 The prospective association between physical fitness, adiposity, and risk of low-grade inflammation. The reference category (odds ratio = 1.00) refers to lean–fit participants. Multivariate model includes adjustments for age, gender, smoking, blood pressure, employment grade, weight change, and baseline C-reactive protein (CRP). Physical fitness categories were derived from a medium split of the gender-specific heart rate response to a submaximal fitness test. Lean and fat were defined as a BMI < or ≥25 kg/m^2, respectively. Low-grade inflammation was defined as CRP ≥1 mg/L.

Mechanisms

A small increase in abdominal adipose tissue relative to overall body weight has been associated with significantly elevated CRP concentrations. For example, data from the Framingham Heart Study suggest that in women, an increase of 0.8 kg (1.8 lb) in visceral adipose tissue corresponds to an increase in CRP concentration of 1.8 mg/L, and in men an increase of 1.0 kg (2.2 lb) corresponds to a 0.7 mg/L rise in CRP. Several cytokines, including IL-6 and TNF-α, are produced in adipose tissue and induce hepatic production of CRP. In addition, inflammatory cells, such as monocytes and macrophages, are components of adipose tissue and accumulate in obese states. The secretion of monocyte chemoattractant protein-1 by these cells may play a key role, as there is evidence to suggest that exercise and loss of visceral fat reduces circulating concentrations of this biomarker.

Numerous mechanisms have been connected with the anti-inflammatory effects of exercise (see figure 72.2). Regular exercise has been shown to modulate nuclear factor-λB activation in animals (10), which represents a key mechanism since this is a redox-sensitive and oxidant-activated transcription factor that regulates inflammation-related gene expression. Available data suggest that aging is associated with increased nuclear factor-λB activity that may partly explain the age-associated increases in inflammation. Furthermore, Stewart and colleagues (6) recently demonstrated that a 12-week endurance and resistance exercise training program lowered CD14+ cell surface expression of toll-like receptor 4, which is thought to play an important role in inflammatory pathways. Interestingly, the authors proposed that heat shock proteins released during exercise may be a viable mechanism that could explain the training-induced changes in toll-like receptor 4. Others have speculated that the anti-inflammatory effects of exercise may be mediated by other cardiovascular risk factors such as increased insulin sensitivity, high-density lipoprotein cholesterol, and adiponectin, which are known to exhibit anti-inflammatory effects. Given that activated endothelial cells are a potent source of inflammatory cytokine release, improvements in endothelial function may also play a role. And, both weight loss and exercise are associated with reductions in reactive oxygen species, which have been implicated in inflammation.

Cause and Effect of Inflammation

Although inflammation is thought to be directly involved in key disease processes such as plaque vulnerability and rupture that are associated with cardiovascular events, the clinical utility and causal role of these novel risk markers remain widely debated. Various cardiovascular disorders, such as atherosclerosis, are widely viewed as inflammatory conditions. In several meta-analyses of large-scale prospective cohort studies, CRP and fibrinogen were found to be moderately associated with cardiovascular outcomes (11). There are, however, limited data on the predictive value of such biomarkers

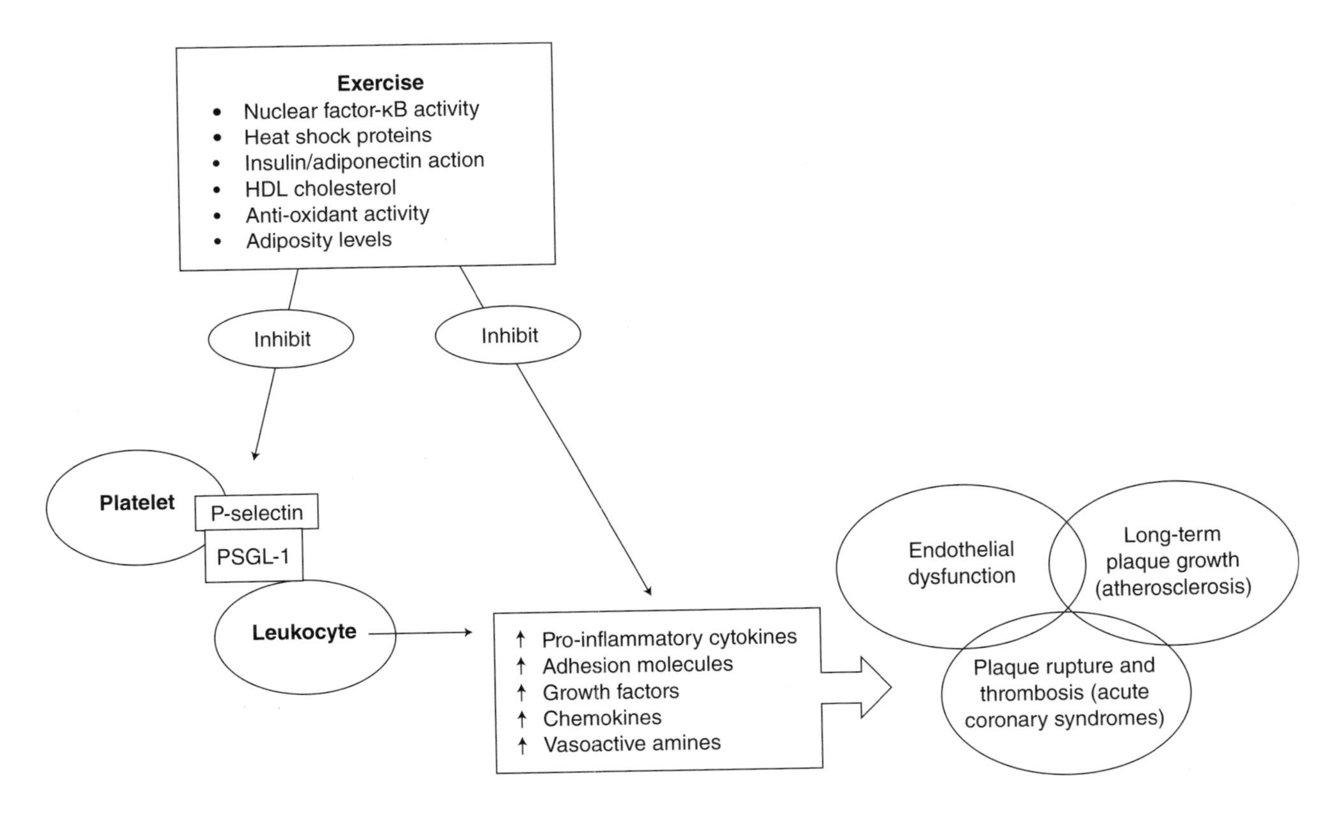

Figure 72.2 Proposed mechanisms of the anti-inflammatory effects of exercise in relation to cardiovascular risk. Platelet activation results in platelet surface expression of the adhesion molecule P-selectin, which binds to the leukocyte receptor, P-selectin glycoprotein ligand-1 (PSGL-1), leading to the formation of platelet–leukocyte aggregates. These interactions result in the release of a number of adhesive and pro-inflammatory molecules that stimulate a variety of processes promoting cardiovascular risk.

beyond that of traditional risk factors, and existing evidence is equivocal. In Mendelian randomization designs, genotypes that produce lifelong differences in various inflammatory markers are not consistently associated with coronary disease incidence. Thus, there is mixed evidence for a causal role of inflammatory risk markers in cardiovascular disease, and they may instead be a surrogate marker of the underlying disease process.

Given that a substantial amount of the evidence relating to the associations between physical activity, adiposity, and inflammation is cross-sectional, we cannot exclude the possibility of reverse causality or confounding from unmeasured variables. In particular, because elevated inflammatory responses are related to physical disability in older participants, impairment in activities of daily living could explain associations between physical activity and inflammation. In addition, several prospective cohort studies have shown that inflammation-sensitive proteins predict future weight gain at follow-up, independently of confounding factors such as obesity levels at baseline, physical activity, smoking, alcohol, or cardiovascular disease. Also, the pro-inflammatory cytokine TNF-α alters several aspects of lipid metabolism and regulates one of the major nuclear factors involved in adipocyte growth, differentiation, and function. Obesity and physical inactivity may be viewed as a marker of an unhealthy lifestyle, and the associations with inflammation may be confounded by factors such as a lack of essential nutrients, including dietary antioxidants, excess consumption of saturated fats, alcohol, and smoking. Thus, it is important to consider potential confounding and possible bidirectional associations between physical activity, adiposity, and inflammation when interpreting the data in this area.

Future Directions

The optimal exercise dose and levels of adiposity for preventing chronic low-grade inflammation are not well established. Several trials have examined the effects of different exercise intensities on CRP, although no significant findings have emerged despite a greater improvement in fitness after higher-intensity training. Few observational studies have considered the association between specific types of activity and inflammatory markers, although there is some evidence that anti-inflammatory effects are observed for participation in vigorous activities, such as sport and exercise, but not for more moderate-intensity activities that include brisk walking and domestic work. In addition, very few studies have considered the association between physical activity and anti-inflammatory markers. Trials that have examined the effects of exercise on adiponectin levels, an adipose-derived hormone that has anti-inflammatory properties, have been inconsistent and largely produced null findings, although further evidence from larger trials with a wider range of biomarkers is required to further explore this mechanism.

Conclusion

In conclusion, both physical activity and adiposity are associated with markers of inflammation, which may partly mediate the relationship between physical activity and cardiovascular risk. The precise nature of the interaction between physical activity and adiposity in explaining associations with inflammation remains poorly understood, although it is likely that both factors are partly related to inflammatory processes by independent mechanisms and partly through their relationship with one another. The anti-inflammatory effects of regular physical activity may be particularly important in lowering cardiovascular risk associated with obesity.

73

Physical Activity and Depression in Obesity

Adrian H. Taylor, PhD

This review initially considers the epidemiological link between physical activity (PA), depression, and obesity and the possible causal influences. Possible mechanisms involved in any chronic effects are discussed. Secondly the review considers the evidence for an effect of acute exercise on appetite, eating behavior, and energy intake, mediated by hedonic changes, as well as some implications for designing interventions for multiple behavior changes.

Associations Between Physical Activity, Depression, and Obesity

Academic reviews on the etiology of obesity rarely consider mental health and psychological well-being as a contributing factor (1). However, there are significant associations between measures of mental health and obesity in population studies. Only 6.5% of those with body mass index (BMI) under 25 kg/m^2 reported moderate or severe depression compared with 25.9% among those with BMI over 35, and those with moderate to severe depression were more than twice as likely to be obese compared to normal-weight individuals (2). It is less clear whether such associations are explained by confounding factors, such as PA, or whether there are causal effects. Several possible relationships are shown in figure 73.1.

Figure 73.1 shows an association between depression and obesity, with PA a confounding factor. Jorm and colleagues (3) found that, after statistically controlling for lack of PA and other associated factors (i.e., social support, education, and finance), an unexpected picture emerged: Underweight women had more depression and negative affect, and the obese and overweight tended to have better mental health than the acceptable-weight group. Depression and obesity were not related among men, even when controlling for PA. This highlights the challenge for epidemiological studies to eliminate spurious relationships.

Physical Activity, Depression, and Obesity

It is hypothesized in figure 73.1*b* that depression causes obesity. This may be explained by biological and behavioral mechanisms. For example, depression may involve a dysfunction of neuroendocrine and metabolic processes, thus lowering resting energy expenditure. Antidepressant medication may also influence metabolic processes and neurotransmitters that contribute to weight gain. Behaviorally, depression may be characterized by longer periods of sleep and rest, social isolation, and inactivity. It is important to note that the idiosyncratic nature of depression makes it difficult to consider these effects. For example, one person with depression may experience a loss of appetite whereas another may seek food to counterbalance low energy and for a sense of comfort. Figure 73.1*b* also indicates that PA reduces depression (and conversely, less PA contributes to depression).

Systematic reviews of randomized controlled trials (RCT) suggest that PA interventions reduce unipolar depression, though the evidence appears less strong among those with clinically defined depression rather than subclinical depressive symptoms (4). A small RCT (5) suggested that the adult public health guidelines of an accumulation of 30 min of moderate-intensity PA on at least five days a week was the minimal dose to reduce depression. However, it is likely that the quality of the PA

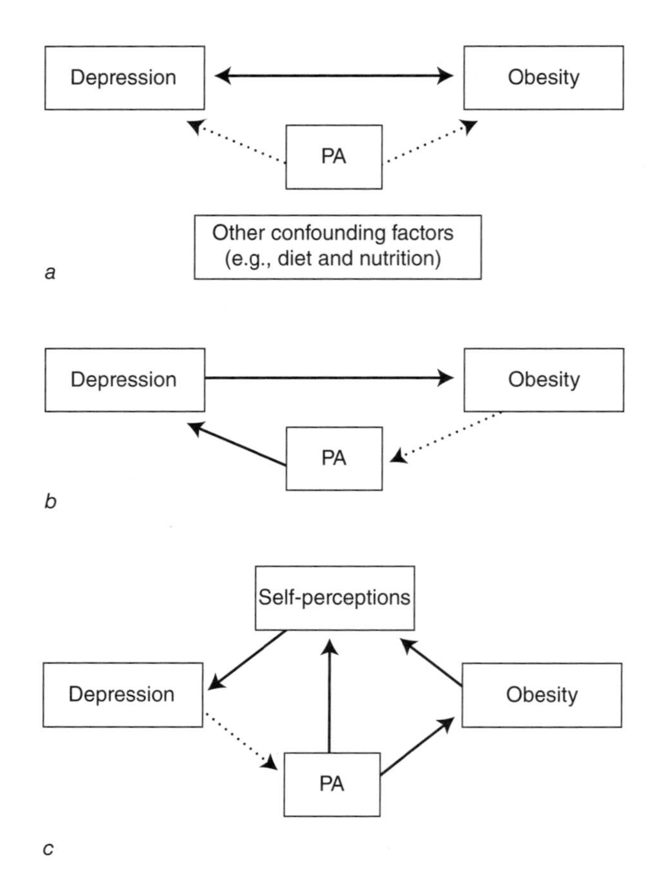

Figure 73.1 Epidemiological links between *(a)* physical activity, *(b)* depression, and *(c)* obesity.

experience (e.g., enhanced perceptions of competence, autonomy, and relatedness, from self-determination theory) is likely to be just as important as the volume of PA for reducing depression. Certainly, self-efficacy mediates the link between PA and depression. It is also not clear whether sedentary behavior contributes to depression, independently of the number of periods of exercise (e.g., three times per week). One issue also of relevance to depression is the timing of PA occurrence, since it moderates the psychophysiological responses to acute stressors.

Given that women are more commonly treated for depression than men, two special examples in which PA, depression, and obesity interact are worth noting. First, those who do more PA experience less postpartum depression and weight gain, and those with less concern for weight gain during pregnancy (who also tend to do more PA) have lower levels of postpartum distress. Secondly, weight gain and depressive symptoms are common among female smokers attempting to quit and predict nonabstinence following treatment. Importantly, PA during a quit attempt is associated with less weight gain and depression.

Finally, a dotted line is shown between obesity and PA in figure 73.1*b* This is mainly to indicate a closed loop in which obesity also contributes to less PA. This may be due to a tendency to experience greater fatigue and "need for recovery" (e.g., following work-related stress or the sleep apnea associated with obesity). It may also be challenging to perform significant amounts of PA due to physical restraints and also social physique anxiety.

It is hypothesized in figure 73.1*c* that obesity causes depression, predominantly mediated by perceptions of the self. Systematic reviews suggest that weight loss programs result in concurrent improvements in physical self-esteem, physical self-perceptions, body image, physical self-concept, and depression. Further causal evidence is required, but current theory holds that the way we feel about ourselves physically is linked to depression. It is helpful to identify the importance of hierarchical models of self-esteem in the present context. Self-esteem is a global trait-like construct, placed above contextually specific self-perceptions (e.g., social, academic, physical). In turn, the domain of physical self-perceptions (or physical self-worth) provides a more global construct than subdomain perceptions about body attractiveness, physical competence, physical strength, and sport competence. Measurement precision and conceptual clarity are important for both research and the design of interventions. For example, in a 10-week PA intervention, for adults with a mean BMI of 28.7 kg/m^2, improvements in self-perceptions of physical condition, appearance, and health were correlated with improvements in anthropometric measures after 37 weeks but more global physical self-worth was not (6). Interventions should be designed to ensure improvements in subdomain measures of self-perceptions that, only after some time, may affect more global measures of physical self-perceptions and coexistent depression.

The final link in figure 73.1*c* worthy of discussion is from depression to PA. Depression is often characterized by lower levels of PA concurrent with a loss of interest, enjoyment, and pleasure in normal activities; low self-efficacy; and reduced energy and greater fatigue. A sense of hopelessness and helplessness is indicative of more severe depression. It is difficult to demonstrate whether increases in PA cause a reduction in depression or are merely symptomatic of reduced depression. However, interventions designed to increase PA among overweight people with depression should carefully consider how the spiral of increasing depression and decreasing PA is broken.

Mechanisms

The discussion so far has focused on psychosocial factors implicated in the relationships between PA, depression, and obesity. Physical activity is a social behavior, and many regard depression as a psychosocial phenomenon. However, increasing interest, particularly with the use of animal models, has been directed toward understanding the biophysical processes by which PA influences mental health. Physical activity may affect depression through changes in cortical neurotransmitters (e.g., dopamine, brain-derived neurotrophic factor), downregulation of the hypothalamic-pituitary-adrenal stress axis (e.g., reduced noradrenaline and cortisol), and structural changes in the brain. Some evidence suggests that it may take several weeks of PA before such changes are reflected in reduced depression. This work is in its infancy, and little can be said about how obesity influences any biophysical processes involved in the link between PA and depression.

Traditionally theories of health behavior have encompassed social-cognitive constructs such as intentions, beliefs, and attitudes toward the target behavior and its expected outcomes (e.g., social-cognitive theory) to aid our understanding of why people engage in unhealthy behaviors, and also as processes to target in obesity prevention and treatment interventions. Experimental work has also given us insights into how single sessions of PA acutely influence mood and affect, engagement in further PA, and other health behaviors implicated in obesity. In the next section, this work is considered in more detail, and implications for future research and practice are outlined.

Psychological and Behavioral Acute Responses to Physical Activity

It is widely acknowledged that mood and stress play an important role in regulating a variety of lifestyle behaviors, including smoking, eating patterns (e.g., sugar snacking), alcohol consumption, PA, and sedentary behavior or hypokinesia. During temporary or more prolonged periods of negative mood, affect, and stress, there is a greater tendency to engage in behaviors that may enhance mood and affect. While *mood* is often defined (and measured) in terms of general categories (e.g., tension, depression, anger, fatigue), one way to define *affect* is in terms of a more basic state, linked to underlying physiological states (e.g., arousal and tension). Thus, circumplex affect may be described along two independent dimensions: activation and valence or pleasure. This is important because implicit urges to engage in behaviors to enhance activation and pleasure may occur independently and have distinct response profiles to different doses of PA.

During higher-intensity exercise (i.e., above the ventilatory threshold; VT), typical psychological responses reflect less pleasure compared with exercise below VT (e.g., brisk walking). In a sample of inactive obese middle-aged women, pushing the intensity to just 10% above a self-selected exercise intensity led to more displeasure than for nonobese subjects (7). Memory of pleasure or displeasure is thought to influence future engagement in PA. Exercise practitioners should encourage participants to select a preferred intensity that has been shown to be typically within the comfort zone just below the VT, rather than at a higher intensity that may lead to compensatory eating and no change in energy balance.

Perceived activation increases as intensity of PA increases, and can remain elevated for some time after exercise. In a comparison of walking and sugar snacking, Thayer (8) reported that the latter had shorter-term effects than the former in raising energy levels and perceived activation, and created a state of "tense energy" (compared with "calm energy" after exercise). However, omnipresent media messages have led to an expectancy regarding, and indeed dependence on, food and drinks that provide instant energy boosts; messages communicating that short bouts of moderate-intensity PA can provide revitalization are far less common. Smokers, during periods of boredom and low energy, also smoke to feel activated and, paradoxically, for a sense of relaxation. After smoking cessation, comfort eating often replaces smoking, resulting in weight gain of typically 3 to 4 kg (6.6 to 8.8 lb) or more.

To summarize, subconscious regulation of mood and affect by a range of lifestyle behaviors is common particularly at times of high work and social demands. Some people gain a sense of activation and pleasure from PA while others do not have the same expectancies. This may be due to experiences of PA (e.g., at school) and beliefs that exercise should be vigorous to enhance health (including feeling better). Emerging evidence suggests that we may be genetically predisposed to seek reinforcement from PA or not. From an anthropological perspective, it would seem reasonable to suggest that PA should become linked to neurobiological responses in the reward or pleasure centers of the brain: Without movement, our basic needs (i.e., food, water, shelter) would not be met. Indeed, common neurobiological mechanisms have been shown to underpin obesity (overeating) and exercise (9). Food cues are associated with increases in dopamine (DA) in anticipation of a reward in nonobese subjects, but obese subjects have reduced dopamine D2 receptors and hence search more actively for food (reinforcement) (10). Exercise may increase DA, which in turn downregulates the salience of food and food searching.

Exercise scientists have tended to focus on the effects of 20 min or more of PA on acute mood or affective benefits. Advancing technology enables us to assess the impact of sedentary behavior and multiple relatively short bouts of walking throughout the day on mood and affect. Experimental research certainly suggests that such bouts increase activation and pleasure, and logically such hedonic changes may influence engagement in other mood-regulating behaviors.

Figure 73.2 shows our multiple affective behavior change (M-ABC) model in which mood regulation (and management of acute and chronic stress) is central to a number of behaviors.

This model provides the basis for designing interventions for multiple health behavior changes based on PA promotion. Regular brief bouts of moderate-intensity PA (e.g., 10 to 15 min brisk walking) help to regulate mood and affect, which in turn subconsciously control energy intake (urges to snack, appetite), alcohol, cigarette cravings, and other behaviors commonly providing mood enhancement. We have shown that a 15 min brisk walk reduces chocolate cravings and attenuates increases in response to a chocolate cue (11) among regular but three-day-abstinent consumers.

Conclusion

The links between PA, depression (mood and affect), and obesity are complex and involve reciprocal causal effects. It is clear that a focus solely on energy expenditure to prevent and treat obesity may not be as effective as also considering the role of PA in promoting psychological well-being and reducing depression. New approaches are needed to understand how to promote multiple health behavior changes, since short bouts of PA may have effects on other obesogenic behaviors. An understanding of the neurobiology of mental health and well-being and of the effects of PA on neurotransmitters (e.g., dopamine) may enable us to promote PA more effectively.

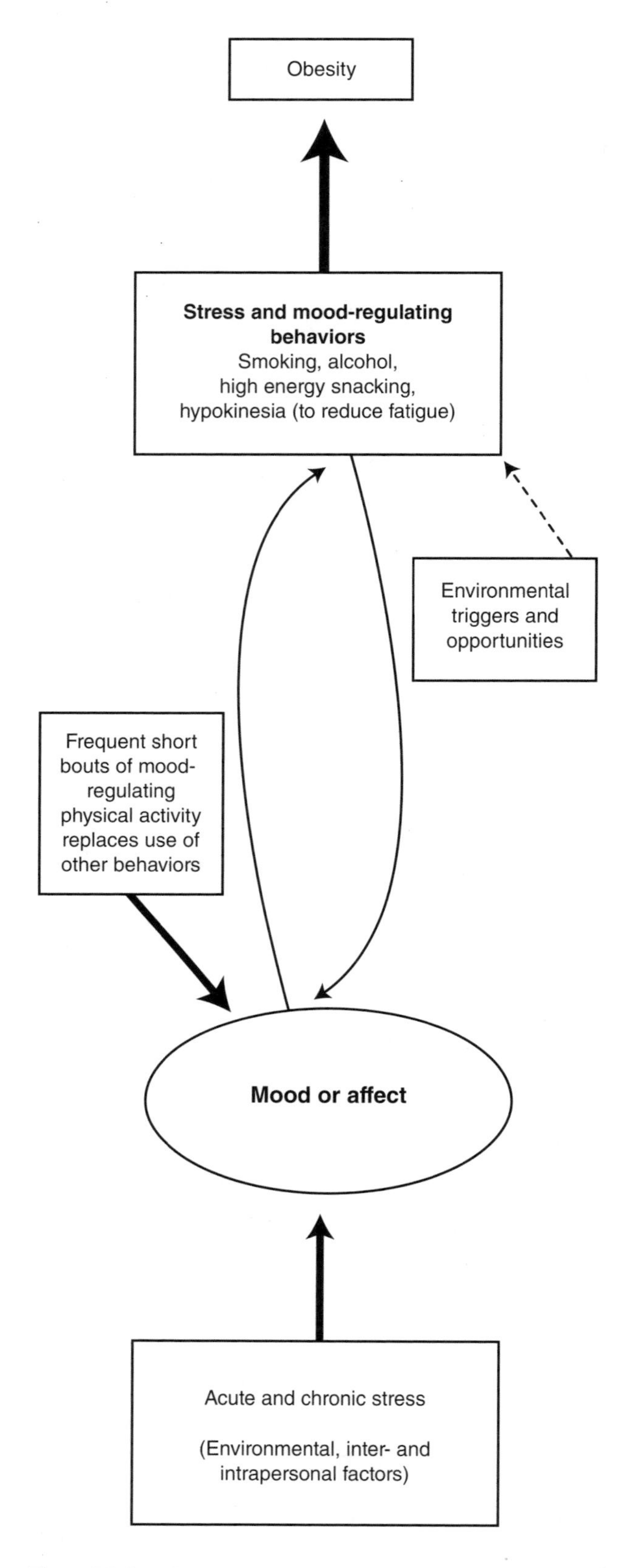

Figure 73.2 A multiple affective behavior change (M-ABC) model: promoting short bouts of moderate-intensity PA as a way of regulating mood and obesogenic behaviors.

74

Physical Activity and Breast Cancer in Obesity

Kerry S. Courneya, PhD

Breast cancer is the third most common cancer in the United States and the most common cancer in women (1). An estimated 182,460 women and 1990 men will be diagnosed with invasive breast cancer in the United States in 2008. American women have approximately a 1 in 8 probability of being diagnosed with breast cancer at some point in their lifetime. An estimated 40,480 women and 450 men will die from breast cancer in the United States in 2008. The five-year relative survival rate overall for breast cancer is 89%. If the disease is diagnosed early (localized disease), however, the survival rate is 98%, whereas if the disease is diagnosed late (distal or metastatic disease) the survival rate is only 27%. The high incidence rate and excellent survival rates have resulted in approximately 2.4 million breast cancer survivors in the United States. This growing number of breast cancer survivors has generated interest in behavioral strategies that might further reduce the risk of recurrence and early mortality and also improve quality of life. The primary purpose of this chapter is to provide an overview of the role of physical activity (PA) and obesity in the primary and secondary prevention of breast cancer and to briefly discuss possible mechanisms. Secondary aims are to (a) provide an overview of the role of PA and obesity in altering supportive care outcomes in breast cancer survivors, (b) estimate the prevalence of these risk factors in breast cancer survivors, and (c) provide examples of preliminary research on PA behavior change and weight loss interventions in this population. The chapter ends with recommendations for future research directions.

Physical Activity in the Primary Prevention of Breast Cancer

Friedenreich and Cust (2) reviewed 34 case-control and 28 cohort studies on the association between PA and the risk of primary breast cancer. They also examined possible effect modifiers, including the type of PA performed, the timing of PA, the intensity of the PA, menopausal status, body mass index (BMI), racial group, family history of breast cancer, hormone receptor status, energy intake, and parity. Overall, evidence for a risk reduction was present in 47 (76%) of 62 studies. More specifically, 30 (48%) of the studies demonstrated a statistically significant risk reduction, nine (15%) a borderline statistically significant risk reduction, and eight (13%) a nonstatistically significant reduction; 14 (23%) showed no association, and one (2%) showed a nonstatistically significant increased risk (figure 74.1). Overall, the estimated risk reduction was 25%, with a larger estimated risk reduction from case-control studies (average 30%) than from cohort studies (average 20%). A dose–response effect was reported in 28 (82%) of 34 studies that addressed this issue.

Acknowledgements
Kerry S. Courneya is supported by the Canada Research Chairs Program and a Research Team Grant from the National Cancer Institute of Canada (NCIC) with funds from the Canadian Cancer Society (CCS) and the CCS/NCIC Sociobehavioral Cancer Research Network.

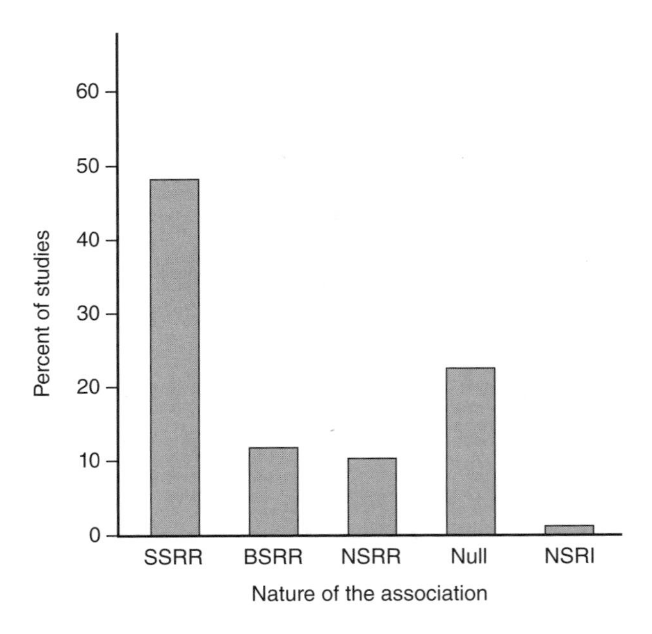

Figure 74.1 Nature of the associations between physical activity and primary risk of breast cancer in 62 studies. SSRR = statistically significant risk reduction; BSRR = borderline significant risk reduction; NSRR = nonsignificant risk reduction; Null = no association; NSRI = nonsignificant risk increase.

Adapted from C.M. Friedenreich and A.E. Cust AE, 2008, "Physical activity and breast cancer risk: Impact of timing, type and dose of activity and population subgroup effects," *British Journal of Sports Medicine* 42:636-647; and A.R. Carmichael, 2006, "Obesity as a risk factor for development and poor prognosis of breast cancer," *British Journal of Obstetrics and Gynecology* 113:1160-1166.

For the effect modifiers, stronger risk reductions were observed for recreational PA, lifetime or later-life PA, vigorous PA, postmenopausal women, lean women, nonwhite racial groups, women with hormone receptor–negative tumors, women without a family history of breast cancer, and parous women. Of particular note for the present chapter, PA was associated with a reduced risk for breast cancer among all categories of BMI except the obesity category. The authors concluded that the association between PA and breast cancer risk is convincing although it is stronger for some population subgroups than others.

The association between obesity and risk of primary breast cancer is even stronger than for PA (3). Many measures of obesity have been associated with increased risk of postmenopausal breast cancer, including BMI, percent body fat, waist/hip ratio, and weight gain (3). Several systematic reviews and meta-analyses, as well as large cohort studies, have documented these associations and indicate that high body fat, central adiposity, and adult weight gain increase the risk of postmenopausal breast cancer by 30% to 50% or more (3). Despite the stronger association with breast cancer for obesity than for PA, the PA–breast cancer association is not accounted for by obesity, indicating independent, additive, or synergistic pathways (2).

Physical Activity in the Secondary Prevention of Breast Cancer

Recent epidemiologic studies have provided promising evidence of an association between postdiagnosis PA and disease outcomes in breast cancer survivors. Holmes and colleagues (4) followed for a median of eight years 2987 women from the Nurses' Health Study who were diagnosed with stage I to III breast cancer between 1984 and 1998. Physical activity was assessed by self-report every two years. After adjustment for known prognostic factors, including BMI, analyses showed that women reporting 9 to 15 metabolic equivalent task hours (MET-hours) of PA per week (equivalent to 3 to 5 h of average-speed walking) had a 50% lower risk of breast cancer–specific mortality compared to women reporting less than 3 MET-hours/week (equivalent to 1 h of walking). At 10-year follow-up, women with greater than 9 MET-hours/week had an absolute survival advantage of 6% (92% vs. 86%) compared to women reporting less than 9 MET-hours/week. Similar risk reductions were observed for breast cancer recurrence and all-cause mortality. The association was particularly pronounced for women with stage III disease, hormone-responsive tumors, and obese women. Obese women performing 15.0 to 23.9 or >24.0 MET-hours/week had relative risk reductions for breast cancer death of 0.22 and 0.36, respectively, compared to obese women performing <3.0 MET-hours/week.

Holick and colleagues (5) examined the association between postdiagnosis recreational PA and risk of breast cancer death in 4482 breast cancer survivors enrolled in the Collaborative Women's Longevity Study. Participants were aged 20 to 79 years and had a median of 5.6 years postdiagnosis at the time of enrollment. Physical activity was assessed by self-report of six recreational activities over the past year, and women were followed for a maximum of six years. After adjustment for known prognostic factors including age, family history of breast cancer, disease stage, hormone therapy use, treatments, energy intake, and BMI, women engaging in more PA had a 40% to 50% lower risk of breast cancer death and all-cause death (figure 74.2). The association was similar regardless of age and disease stage. Although the test for interaction with BMI was not statistically significant, the relative risk reduction for lean women was 0.91 compared to 0.63 for overweight and obese women.

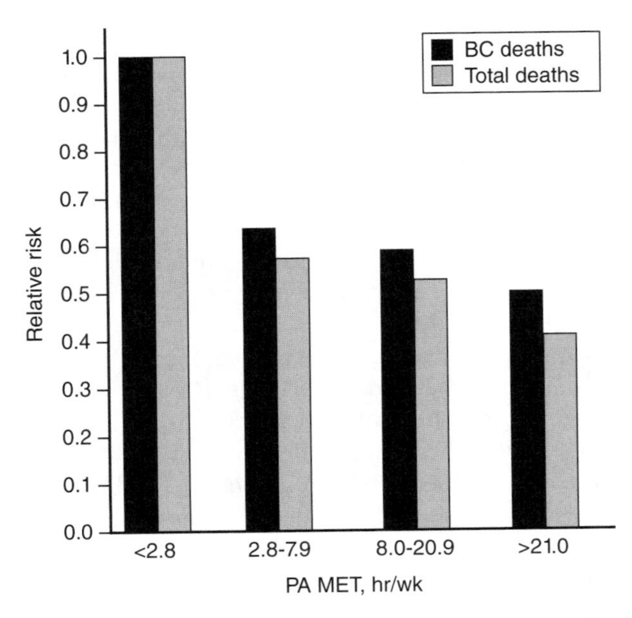

Figure 74.2 Associations between physical activity and risk of death from breast cancer and all causes in the Collaborative Women's Longevity Study. BC = breast cancer; PA MET-hours/week = physical activity metabolic equivalent task-hours per week.

Adapted from C.N. Holick et al., 2008, "Physical activity and survival after diagnosis of invasive breast cancer," *Cancer Epidemiology Biomarkers Preview* 17: 379-386.

Several systematic reviews and meta-analyses, as well as large cohort studies, have documented that obesity at the time of diagnosis and possibly weight gain over the course of chemotherapy are associated with poorer disease outcomes in breast cancer survivors (3). Obesity has consistently predicted poorer prognosis in breast cancer survivors including recurrence, breast cancer–specific mortality, and all-cause mortality even after other recognized prognostic factors such as disease stage, hormone receptor status, and treatment adequacy were controlled for (3). Again, the PA studies controlled for obesity, indicating independent or additive pathways for PA.

Possible Mechanisms

Many biological mechanisms have been hypothesized to explain the associations between PA, obesity, and breast cancer, and it is likely that no one mechanism will predominate (2). Friedenreich and Cust (2) have broadly categorized these mechanisms into (a) reduction in circulating levels of and cumulative exposure to sex steroid hormones, (b) changes in insulin-related factors and adipocytokines, (c) modulation of inflammation and the immune system, and (d) hormonal and cellular metabolism pathways (2). Moreover, different mechanisms may be operative at different stages of carcinogenesis and for different population subgroups (e.g., lean vs. obese women, premenopausal vs. postmenopausal women) (2). Finally, issues related to disease staging, tumor characteristics, treatment decisions, treatment completion rates, and treatment efficacy are additional explanations for the associations of PA and obesity with secondary breast cancer outcomes (3).

Supportive Care in Survivors

Physical activity has been examined for its effects on supportive care outcomes in breast cancer survivors, including during active treatment (e.g., chemotherapy, radiation therapy) and survivorship (i.e., after primary treatments are completed and the patient is free of disease). McNeely and colleagues (6) conducted a systematic review and meta-analysis of 14 randomized controlled trials of exercise interventions involving 717 breast cancer survivors aged 35 to 72 years. The overall pooled data from these trials showed significant positive effects of exercise on quality of life, cardiorespiratory fitness, and physical functioning. The pooled data also demonstrated a statistically significant effect of exercise on fatigue reduction, but only in the survivorship setting. There was no evidence of benefit for reducing overall body weight. Few studies have examined the effects of weight loss interventions on supportive care outcomes in breast cancer survivors.

Physical Activity in Breast Cancer Survivors

Few studies have provided population-based estimates of PA and obesity rates in breast cancer survivors and compared them to data for individuals without a history of cancer. Courneya and colleagues (7) reported such data for the Canadian population using the 2005 Canadian Community Health Survey. Participants self-reported their cancer history, height and body weight for calculation of BMI, and participation in various leisure-time activities. Fewer than 45% of Canadian breast cancer survivors were physically active, and almost 50% were overweight or obese. Moreover, obese breast cancer survivors were only half as likely to be active compared to obese women without a history of cancer. Consequently, a large proportion of Canadian women may be at higher risk for poorer disease outcomes because of their current lifestyle.

Weight Loss Interventions in Breast Cancer Survivors

Few trials have tested PA behavior change or weight loss interventions in breast cancer survivors. Vallance and coworkers (8) examined the effects of breast cancer–specific PA print materials, a step pedometer, or a combination of the two on PA and quality of life in 377 breast cancer survivors compared to a standard recommendation. At three-month follow-up, the intervention groups increased their moderate-to-vigorous intensity PA by about 40 to 60 min/week compared to the standard recommendation group. Mefferd and colleagues (9) examined the effects of a weight loss intervention on risk factors for breast cancer recurrence and cardiovascular disease in 85 overweight or obese breast cancer survivors. The intervention was designed to induce weight loss through exercise and diet modification using a once-weekly, 16-week cognitive behavioral therapy (CBT) intervention. At 16 weeks, the intervention group had lost 5.7 kg (12.6 lb; 6.8% of body weight) compared to the usual-care group's loss of 0.5 kg (1.1 lb; 0.6% of body weight). Furthermore, body fat percent, waist and hip circumference, levels of triglycerides, total cholesterol, and high-density lipoprotein cholesterol levels were also significantly reduced in the intervention group compared to the usual-care group. These results indicate that 16 weeks of a cognitive behavioral therapy program for weight management may reduce obesity and cardiovascular risk factors in overweight breast cancer survivors.

Future Directions

Indisputable evidence exists to support the adverse associations between physical inactivity, obesity, and the primary risk of breast cancer. Good evidence is available to support the contention that obesity is also a negative prognostic factor for the risk of breast cancer recurrence and early mortality. The evidence for PA in the secondary prevention of breast cancer is just beginning to be obtained, but the early research is promising. We have recently seen good epidemiologic research with valid self-report PA measures and good control of potential confounders. One major challenge to this research, however, is the difficulty of accurately quantifying exposure to short-term and long-term PA in population-based samples. All epidemiologic studies to date have relied on self-report measures of PA with their well-known measurement issues. One possible approach to this limitation is to assess cardiovascular fitness as an objective indicator of recent exercise, although this approach is also not without limitations.

The limitations of observational studies examining the associations between PA and cancer, however, can be addressed only through large-scale randomized controlled trials with disease outcomes (10). For both PA and obesity in both the primary and secondary prevention settings, there is a strong need for randomized controlled trials to examine the effects of PA and weight loss on the purported biologic mechanisms of recurrence and mortality in breast cancer survivors, as well as the disease endpoints themselves (10). These trials are warranted based on compelling evidence from (a) observational studies showing that PA and obesity are strongly associated with breast cancer outcomes, (b) intervention studies showing that PA and obesity cause changes in biologic mechanisms thought to play a role in breast cancer, and (c) behavior change studies showing that it is possible to achieve a substantial increase in PA and weight loss in the target population that can be maintained over an extended period of time.

The main challenges to such trials will be costs and the ability of investigators to sustain behavior change and weight loss over a sufficiently long period of time to affect disease endpoints. Fortunately, there are excellent recent examples of completed and ongoing behavioral trials showing the ability of behavioral researchers to induce substantial weight loss, dietary change, or exercise behavior (or more than one of these) and maintain these changes over an extended period of time. If well-conducted exercise and weight loss trials are positive, there will be a strong need to develop strategies for dissemination and implementation that reach the widest possible number of breast cancer survivors.

75

Physical Activity and Colon Cancer in Obesity

I-Min Lee, MD, ScD

In the United States, colorectal cancer is the second leading cause of cancer deaths and the third most commonly occurring cancer in both sexes (1). Approximately 6% of the population will develop this cancer within their lifetime (1). While the incidence rates of colorectal cancer increased between 1973 and 1985, in recent years and especially since the 1990s, incidence rates have been declining in both sexes (1). Over 70% of colorectal cancers in the United States arise in the colon, with less than 30% arising in the rectum.

Host factors (e.g., those causing the genetic susceptibility syndromes of familial adenomatous polyposis and hereditary nonpolyposis colorectal cancer) clearly are important in the etiology of colon cancer, but environmental factors do play a large role. Migrant studies indicate that the colorectal cancer rates of migrants change when they move to a new country. For example, U.S.-born Japanese men and women experienced incidence rates of colorectal cancer that were twice as high in men, and 40% higher in women, than in foreign-born Japanese (2). In particular, a healthy lifestyle—including physical activity, avoidance of obesity, and a healthy diet—appears to be important, with one study suggesting that as much as 71% of colon cancer may be preventable with adherence to good health behaviors (3).

This chapter discusses the roles of physical activity and obesity in the etiology of colon cancer. We focus mostly on colon cancer alone, since physical activity has consistently been associated with lower rates of colon cancer while the data for a relation between physical activity and rectal cancer have been weak and much less consistent (4).

Potential Biological Mechanisms

Several plausible mechanisms support the potential for physical inactivity and obesity to increase the risk of developing colon cancer. First, both physical inactivity and obesity are associated with insulin resistance, hyperinsulinemia, and hypertriglyceridemia and may increase the level of insulin-like growth factors (IGFs) (4). Hyperinsulinemia by itself also can produce an increase in circulating IGF-1 and a decrease in IGF binding proteins, which increases the availability of IGFs (4). Since insulin and IGFs have been implicated in the etiology of colon cancer, these represent pathways through which physical inactivity and obesity can influence cancer development.

Second, physical inactivity and obesity are associated with higher levels of systemic inflammation (5). There are some data to suggest that elevated levels of inflammatory markers such as C-reactive protein, interleukin-6, and tumor necrosis factor-alpha, and decreased levels of anti-inflammatory factors such as adiponectin may be associated with higher colon cancer rates (5); thus pathways related to inflammation may be involved in the associations of physical inactivity and obesity with colon cancer risk.

Third, the immune system may play a role in reducing cancer risk (a) by recognizing and eliminating abnormal cells or (b) through acquired or innate immune system components or both (4). A prevailing hypothesis states that the inverse relation between physical activity and colon cancer rates can be explained in part by the modulation of immune function with physical activity. The available evidence suggests that moderate levels of physical activity can enhance the number or function (or both) of various cells of the immune system (4). However, prolonged and intense exercise (e.g., running a marathon) may have immunosuppressive effects instead, leading to the hypothesis of a "J"- or "U"-shaped relation between immune function and level of physical activity (4).

Finally, physical activity may increase colonic motility and decrease intestinal transit time, potentially decreasing exposure to carcinogens, cocarcinogens, or promoters in the fecal stream (4). While this has remained a commonly cited mechanism, the evidence regarding faster transit time among physically active persons has been mixed; also, colonic motility has not been definitely linked to colon cancer risk. Some of the inconsistency in the data may be partly due to different methods of measuring transit time that have varying degrees of precision (4).

The following section summarizes the epidemiologic data on the independent and joint associations of physical activity with decreased colon cancer risk and of obesity with increased risk.

Physical Activity and Decreased Risk of Colon Cancer

At least 50 epidemiologic studies, including cohort and case-control studies, have examined the association between physical activity and the risk of developing colon cancer (4). These studies, conducted in North America, Europe, Asia, Australia, and New Zealand, consistently indicate that physically active individuals have a lower risk of developing colon cancer compared with inactive individuals. A review published in 2006 concluded that the magnitude of risk reduction with physical activity is on the order of 30% to 40% and is similar in men and women (4). In general, case-control studies have shown larger effects than cohort studies, with risk reductions associated with physical activity averaging 40% in the case-control studies and 20% in the cohort studies (4). Of particular interest to this book is the observation that many studies that did adjust for some measure of adiposity (primarily using body mass index [BMI]) also demonstrated independent reductions in colon cancer rates with physical activity (4). Another review of the literature in 2008, which included more recently published studies, presented very similar findings (6).

Of public health interest and importance is the amount of physical activity that is needed to lower the risk of colon cancer. The data currently are unclear because few studies have collected sufficiently detailed information on physical activity to answer this question. Based on the available data, it appears that at least 30 to 60 min/day of moderate- to vigorous-intensity physical activity is required to significantly lower the risk of colon cancer (4, 6). Also of public health interest is whether a dose–response relation exists between physical activity and colon cancer risk. Again, the data are limited: It appears that an inverse dose–response relation does exist between physical activity and colon cancer risk (i.e., increasing levels of physical activity are associated with decreasing colon cancer rates), but the available data preclude describing the precise shape of the dose–response curve (4, 6).

Obesity and Increased Risk of Colon Cancer

There is a large body of literature on the association of adiposity with the risk of developing colon cancer, with cohort and case-control studies on this topic conducted in North America, Europe, Asia, and Australia (5). Adiposity in these studies has primarily been assessed by means of BMI, and most studies show increased colon cancer rates at higher levels of BMI; this increase in risk is statistically significant in about half of the studies (5). The association appears to be more consistent and marked in men than in women. In a recent meta-analysis of 22 prospective cohort studies conducted in men, the elevation in risk of colon cancer associated with a 5 kg/m^2 increase in BMI was 24% (95% confidence interval, 21% to 28%), while the combined data from 19 prospective cohort studies in women indicated a 9% (5% to 14%) elevation in risk for the same increase in BMI (7).

Data on the association of body fat distribution with colon cancer risk are more limited. Biologically, it is plausible for a stronger association to exist with adiposity that is centrally distributed, since adipocytes from visceral fat appear to be more metabolically active (e.g., in their secretion of cytokines) compared with subcutaneous fat. Some data suggest that centrally distributed adiposity is strongly related to risk of colon cancer: In one large study of men, those in the top 20% of waist-to-hip ratio (a marker of central adiposity) had almost three and a half times the colon cancer rates of those in the lowest 20% (1).

Association of Physical Activity and Obesity With Colon Cancer Risk

While many studies have examined the relation between physical activity or obesity with colon cancer risk, the data on their joint association with risk are more limited. Most studies have looked at the independent association of each factor with risk rather than their joint association (i.e., investigating effect modification). To date, relatively few studies have dealt with the question whether the increased risk of colon cancer observed with higher levels of obesity differs among persons with different physical activity levels.

With regard to the independent associations, more than half of the studies of physical activity and risk of developing colon cancer have adjusted for adiposity (primarily using BMI), and continued to observe significant inverse associations between physical activity and colon cancer rates (6). And, studies of adiposity in relation to colon cancer rates also have noted that obesity increases colon cancer risk after adjustment for physical activity (1).

With regard to whether effect modification exists, the data have not been clear. The studies addressing this question have provided mixed results, with about half reporting no significant effect modification between physical activity and obesity in their relation with colon cancer (6). For example, in a prospective cohort study of 45,906 Swedish men who were followed for an average of seven years, no significant effect modification (p, interaction = 0.33) was observed between leisure-time physical activity and BMI in relation to risk of developing colorectal cancer (in this particular study, investigators noted an inverse relation between physical activity and both colon and rectal cancer rates, so they examined colorectal cancer as a combined endpoint) (8). Figure 75.1 shows the data from this study, indicating that higher levels of physical activity were associated with decreased risk and higher levels of BMI with increased risk. Further, there was no effect modification; that is, regardless of the level of physical activity, being obese (BMI ≥30 kg/m^2) was associated with about a 30% increase in risk compared with possessing normal weight (BMI <25 kg/m^2).

In contrast, several studies have indicated that higher levels of physical activity may modify the increase in risk of colon cancer associated with adiposity. For example, among 17,595 men in the Harvard Alumni Health Study followed for up to 26 years, the increase in risk of colon cancer at higher levels of BMI appeared to be ameliorated by higher levels of physical activity (9). Among men who expended <4200 kJ/day (~1000 kcal/day) in physical activity, the relative risks for colon cancer associated with BMI of <22.5, 22.5 to <23.5, 23.5 to <24.0, 24.0 to <26.0, and ≥30 kg/m^2 were 1.00 (referent), 1.91, 1.43, 2.28, and 2.41, respectively (p, trend = 0.004). Among those who expended 4200 to 10,499 kJ/day (~1000 to 2500 kcal/day), the increase in risk with higher BMI levels was weaker, with corresponding results of 1.00, 1.53, 1.24, 1.48, and 1.96, respectively (p, trend = 0.06). Additionally, among the most active men expending ≥10,500 kJ/day, there was no longer any increase in colon cancer risk with increasing BMI (1.00, 0.74, 0.77, 0.64, 0.61, respectively; p, trend = 0.14).

A case-control study of 931 men and women with colon cancer and 1552 controls in Shanghai also showed significant interactions between commuting physical activity and BMI in relation to risk among men (p, interaction = 0.002) and women (p, interaction <0.001) (10). Figure 75.2 shows the data for women: The increase in risk of colon cancer with higher levels of BMI was observed among the least active women, but this direct relation was not observed among the most active women. The data for men (not shown) also indicated similar trends.

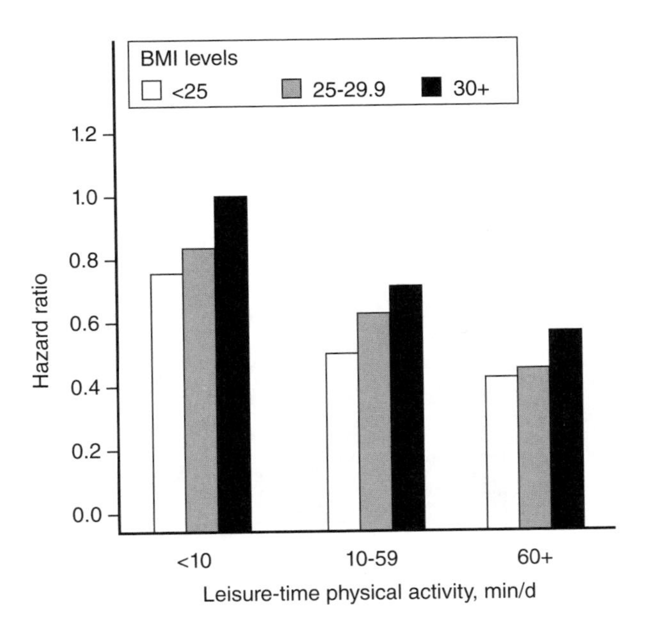

Figure 75.1 Hazard ratios for colorectal cancer by leisure-time physical activity and body mass index in Swedish men. Hazard ratios are adjusted for age, education, smoking, aspirin use, history of diabetes, and family history of colorectal cancer.

Data from S.C. Larsson et al., 2006, "Physical activity, obesity, and risk of colon and rectal cancer in a cohort of Swedish men," *European Journal of Cancer* 42: 2590-2597.

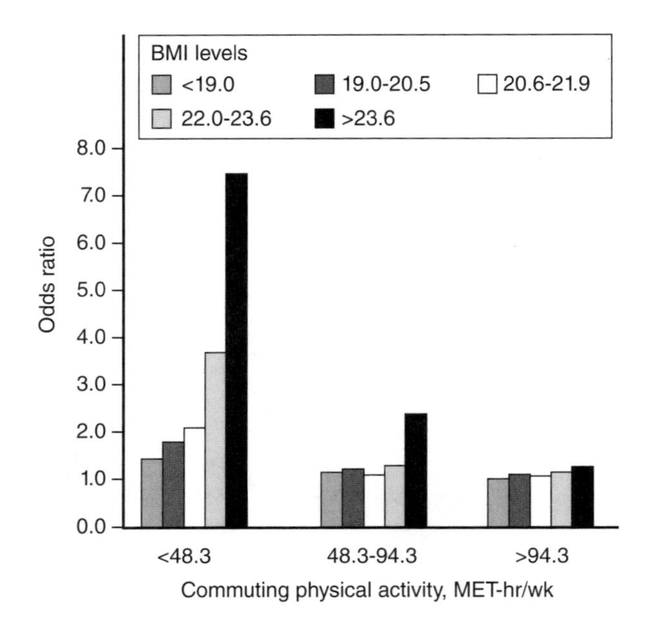

Figure 75.2 Odds ratios for colon cancer by commuting physical activity and body mass index in Chinese women. Odds ratios are adjusted for age, education, marital status, income, diet, number of pregnancies, and menopausal status.

Data from L. Hou et al., 2004, "Commuting physical activity and risk of colon cancer in Shanghai, China," *American Journal of Epidemiology* 160: 860-867.

Conclusion

Colon cancer is a leading cause of cancer morbidity and mortality in the United States. Epidemiologic studies suggest that lifestyle habits, including physical activity and avoidance of obesity, are important in preventing this type of cancer. Physical activity is associated with an approximately 30% to 40% decrease in risk, while a 5 kg/m^2 increase in BMI is associated with an approximately 9% to 24% increase in risk. Relatively few studies have examined the joint association of physical activity and obesity with colon cancer risk, and the data are inconsistent. While some studies show that possessing higher BMI increases risk at all levels of physical activity, others suggest that physical activity can ameliorate the increase in risk with higher BMI. More research is needed to clarify the joint association of physical activity and obesity with colon cancer risk.

76

Physical Activity and Other Cancers in Obesity

Roy J. Shephard, MD, PhD

For the years between 1996 and 2007, the Ovid search engine lists 2853 citations that include the key words "obesity" and "cancer/neoplasm." Many of these articles refer to the associations with obesity that are observed in breast cancer (chapter 74) and colon cancer (chapter 75). Other papers discuss incidental consequences, such as the impact of obesity on the early diagnosis of various forms of cancer by physical examination and prostate-specific antigen testing and imaging; the influence of excessive adipose tissue upon the required dose and efficacy of therapeutic irradiation; the problems of undertaking safe and effective surgery in those who are obese; and the development of obesity as a side effect of irradiation, surgery, and other treatments of cancer. A further group of reports document generally accepted associations between obesity and the mortality from endometrial, renal, pancreatic, and esophageal tumors (1). However, the increasing scale of population research has recently revealed significant associations between obesity and the risk of developing other types of neoplasm, including hepatic, biliary tract, prostatic, and ovarian tumors; leukemia; non-Hodgkin's lymphoma; and multiple myelomata. This chapter provides a brief overview of recent research; because of limitations of space, the main focus is epidemiological. Where information is available, interactions between physical activity, obesity, and the risk of the site-specific tumors are explored.

Endometrial Tumors

There are consistent reports that obesity increases the risk of endometrial cancer, possibly with a small independent effect from a low level of habitual physical activity. A case-control study from Italy and Switzerland compared 777 cases of endometrial cancer with 1550 controls. The data revealed a small univariate influence of physical activity, with an odds ratio of 1.4 (95% confidence limits: 1.1 to 1.8). However, effects for diabetes mellitus were larger, the odds ratio being 1.7 (1.2 to 2.5) for nonobese and 5.1 for obese diabetic women relative to nonobese nondiabetic women; the latter risks were apparently independent of the individual's level of habitual physical activity (2). A similar study of a 36,773 Swedish mammography cohort showed a relative risk of 6.39 (3.28 to 12.06) when obese diabetics were compared to nonobese diabetics; the relative risk was further increased to 9.61 (4.66 to 19.83) if the individuals also had a low level of physical activity (3). A case-control study from the Czech National Cancer Registry indicated that in obese women, the odds ratio of developing endometrial cancer was 3.25 (1.65 to 6.37) (4). A 21-year case-control record linkage study of some 99,000 Danish women showed a relative risk of 2.05 (1.40 to 3.00) in those who were obese (5). A 5.4-year follow-up of 1.2 million women in the United Kingdom demonstrated that the increased relative risk of incident endometrial carcinoma for a 10-unit gain of body mass index (BMI) was 2.89 (2.62 to 3.18) (6). Finally, a 6.4-year prospective study of 223,008 European women examined the incidence of cancer, finding associations with obesity (relative risk 1.78 [1.41 to 2.26]), a waist circumference >0.88 m (relative risk 1.76 [1.42 to 2.19]), and a weight gain >20 kg (44 lb) (relative risk 1.75 [1.11 to 2.77]) (7). In general, obese individuals seem likely to present at a younger age than the nonobese and to be negative with respect to microsatellite instability.

Renal Tumors

Several studies indicate a substantial effect of obesity upon the risk of renal tumors, and again there is a weak suggestion that risks are augmented by a low level of physical activity. A seven-year prospective study of 140,057 U.S. women initially aged 50 to 79 years noted that the waist/hip ratio was strongly predictive of incident renal carcinoma (relative risk in the highest vs. the lowest quartile 1.8 [1.2 to 2.5]) (8). The risk ratio was increased further in individuals with a history of weight cycling. An 8.3-year follow-up of 161,126 individuals from the Hawaii-Los Angeles cohort showed a multivariate relative risk of 1.76 (1.20 to 2.58) in obese men and 2.27 (1.37 to 3.34) in obese women (9). In this study, there was a slight trend for habitual physical activity to reduce the risk for female participants. A 5.4-year follow-up of 1.2 million British women noted a 1.53 (1.27 to 1.84) increase in the relative incidence of renal cancer for a 10-unit increase of BMI (6). Finally, after control for potential confounders, a case-control study from the Czech Republic showed that obese men had an odds ratio of 1.92 (1.14 to 3.24) for developing renal cancer (4).

Pancreatic Tumors

The 5.4-year follow-up of 1.2 million British women indicated that a 10-unit increase of BMI augmented the relative risk of pancreatic tumors by 1.24 (1.03 to 1.48) (6). A prospective study of 110,972 Japanese emphasized that the carcinogenic risk associated with obesity was present from early adulthood; thus, men with a BMI >30 kg/m^2 at an age of 20 years had a 3.5-fold increase in the risk of dying from pancreatic cancer (10). However, this study underlined that there was no significant independent partial correlation with daily walking distance or involvement in sport.

Esophageal Tumors

Two large recent studies show a strong relationship between obesity and neoplastic lesions of the esophagus and gastric cardia. A 13.3-year follow-up of 120,852 Dutch adults set the relative risk of incident esophageal adenocarcinoma at 1.40 (0.95 to 2.04) for those who were overweight (OW) and 3.96 (2.27 to 6.68) for those who were obese (OB). For lesions of the gastric cardia, the corresponding figures were 1.32 (0.94 to 1.85) for OW and 2.73 (1.56 to 4.79) for OB (11). The 5.4-year follow-up of 1.2 million British women estimated that the relative risk of incident adenocarcinoma of the esophagus was increased by 2.38 (1.59 to 3.56) for a 10-unit increment of BMI (6).

Prostatic Tumors

The respective impacts of obesity and physical activity upon the risk of prostatic cancer remain unclear. Although obesity appears to increase the risk in older adults, it may be protective in younger men, perhaps because of a greater production of estrogens or a lesser secretion of testosterone. One reason for conflicting findings may be that prostatic enlargement hampers the detection of neoplastic cells by the usual sampling technique; thus, one recent study showed that prostatic volume modulated the relationship between prostatic neoplasia and obesity, a significant association being seen only in patients with a low prostatic volume (12).

Nevertheless, it is recognized that the proliferation of tumor cells is facilitated by the presence of vascular endothelial growth factor, and animal experiments suggest that concentrations of vascular endothelial growth factor are higher in obese than in lean Zucker rats (13).

Ovarian Tumors

A 21-year record linkage project showed little relationship between obesity and ovarian tumors in a sample of some 99,000 Danish women (5). However, two larger studies have shown weak but statistically significant associations. The 5.4-year follow-up of 1.2 million British women demonstrated that the relative incidence of ovarian cancer increased by 1.14 (1.03 to 1.27) for a 10-unit increase of BMI (6). A meta-analysis identified 28 studies reporting incidence, relative risk, or odds ratios for the development of histologically confirmed ovarian cancer; a positive association with obesity was seen in 24 studies, although this was statistically significant in only 10 reports (14). The pooled assessment of risk was 1.2 (1.0 to 1.3) for OW and 1.3 (1.1 to 1.5) for OB, associations being stronger for case-control (odds ratio 1.5) than for cohort studies (odds ratio 1.1).

Hepatic and Biliary Tract Tumors

There are reports of substantial associations between obesity and lesions of both the liver and the biliary tract, although with no indication of independent influences of physical activity. A meta-analysis of hepatic tumors conducted between 1996 and 2007 identified 11 cohort studies, with 5037 cases of OW and 6042 cases of OB; the respective risks relative to those in the normal body weight range were 1.17 (1.02 to 1.34) and 1.89 (1.51 to 2.36) (15).

Researchers in a case-control study interviewed 153 patients with extrahepatic biliary tract carcinoma; the odds ratio for carcinoma in those who were obese was 2.49 (1.32 to 4.70). The added risk of obesity was particularly severe for those with gallbladder tumors, where the odds ratio was 4.68 (1.85 to 11.84) (16); however, lifestyle factors such as sedentary behavior had no influence on risk. A meta-analysis of studies conducted between 1966 and 2007 showed a relative risk of 1.15 (1.01 to 1.30) for the OW and 1.66 (1.47 to 1.88) for the OB, both risks being stronger in women than in men (17).

Multiple Myelomata

A meta-analysis of both case-control and cohort studies suggests some increase in the relative risk of multiple myelomata in those who are overweight or obese, respective values being OW 1.43 (1.23 to 1.68) and 1.12 (1.07 to 1.12) and OB 1.82 (1.47 to 2.26) and 1.27 (1.15 to 1.41) (18). The 5.4-year follow-up of 1.2 million British women suggested a similar effect, the relative incidence increasing by 1.31 (1.04 to 1.65) for a 10-unit increase of BMI (6).

Leukemia

The 5.4-year follow-up of 1.2 million British women demonstrated a significant effect of obesity upon the incidence of leukemia with a relative risk of 1.50 (1.23 to 1.83) for a 10-unit increase of BMI (6).

Non-Hodgkin's Lymphoma

Several recent case-control and prospective studies suggest a weak impact of obesity upon the risk of non-Hodgkin's lymphoma. Further, there is some suggestion that the risk is less if the individual is also physically active; this is statistically significant for the diffuse, large B-cell type of lymphomas (odds ratio 0.60; 0.40 to 0.88) (19).

A small case-control study of 387 cases of non-Hodgkin's lymphoma showed an odds ratio of 1.40 (0.9 to 2.0) (20); in this study, telephone assessments of habitual physical activity did not identify any noteworthy influences. In the much larger 5.4-year follow-up of 1.2 million British women, obesity again brought only a small if statistically significant increase in the risk of non-Hodgkin's lymphoma, a 10-unit increase of BMI increasing relative incidence by 1.17 (1.03 to 1.34) (6). A meta-analysis covering studies of incidence or mortality for the period 1966 to 2007 identified 10 case-control and 6 cohort studies; again, small increases in relative risk were demonstrated, 1.07 (1.01 to 1.14) for the OW and 1.20 (1.07 to 1.34) for the OB. Again, it was noted that obesity contributed particularly to the risk of diffuse large B-cell lymphomas (21).

Reasons for the Observed Associations

This brief review demonstrates that there are statistically significant associations between obesity and the risk of many types of cancer, although in some organs the associations are relatively weak. One must weigh several issues when interpreting the nature of these associations:

1. Many studies, rather than following nonobese and obese individuals prospectively, have looked for associations after the diagnosis of cancer or even at death; the lesion may thus have modified both body mass and habitual physical activity relative to the adult norm for the individual concerned. In particular, where a tumor is detected at a late stage (e.g., many carcinomas of the pancreas), both body mass and fat reserves may have been depleted substantially relative to values typical of the 10 to 20 years over which the tumor has developed. Moreover, if the endpoint is a registry report of death, the observed associations may reflect influences of physical fitness or obesity (or both) upon early detection, virulence of the neoplasm, or the effectiveness of treatment, rather than a primary effect upon the process of carcinogenesis.

2. The extent of an individual's physical activity and obesity have usually been assessed by relatively crude epidemiological measures, for instance, a physical activity questionnaire and determinations of the body mass index, the waist/hip ratio, or both; inevitably, this attenuates the strength of underlying relationships.

3. Where an association with obesity has been established, the variance attributed to an accumulation of body fat may be serving as a marker of some other etiological factor, particularly a lack of regular physical activity.

4. Where a positive association between cancer and obesity is *not* seen, the relationship may have been obscured by a negative association between cigarette smoking and body mass.

Conclusions

Obesity increases the risk of neoplasms in a variety of sites, including not only the colon and breast, but also the uterus, kidneys, pancreas, esophagus, liver and biliary tract, prostate, ovary, bone marrow (leukemia), lymph glands (non-Hodgkin's lymphoma), and fatty tissue (multiple myelomata). The underlying mechanisms predisposing to cancer remain to be clarified. Habitual physical activity does not generally emerge as a strong independent risk factor. Nevertheless, the substantial burden of neoplastic disease offers a strong reason to reduce excesses of body fat through a combination of appropriate regular physical activity and a restriction of food intake.

77

Physical Activity, Obesity, and Metabolic Syndrome

Mercedes R. Carnethon, PhD

The metabolic syndrome is a clustering of cardiovascular and metabolic abnormalities that places an affected individual at a twofold higher risk of developing clinical coronary heart disease and a three- to fivefold higher likelihood of developing type 2 diabetes. The syndrome initially received widespread attention when Dr. Gerald Reaven described a curious clustering of elevated blood pressure, insulin resistance, and dyslipidemia *in the absence of obesity* during his 1988 Banting lecture to the American Diabetes Association (1). He labeled this clustering "syndrome X" and suggested insulin resistance as the central feature of the disorder. By contrast, the definition proposed in 2001 by the National Cholesterol Education Program (NCEP), Adult Treatment Panel III, included abdominal adiposity as one of its components (2). Subsequent reports describe a syndrome that is approximately 4.5 to 13 times more prevalent in overweight or obese persons as compared with normal-weight persons (3). The robust correlation of metabolic syndrome with obesity across numerous study populations and using a variety of metabolic syndrome definitions (table 77.1) has led to speculation that obesity drives the development of insulin resistance and the other components of metabolic syndrome. As a result, strategies for prevention and treatment have centered on maintenance of a healthy weight in normal-weight persons or weight loss among the overweight and obese.

Physical activity is one arm of the energy balance equation and is recommended as a therapy for preventing the metabolic syndrome or reversing the syndrome once it has already occurred (4). Given the contemporary worldwide epidemic of overweight and obesity, an intriguing question is whether physical activity can prevent or reverse metabolic syndrome in individuals across the spectrum of body fatness. In the present brief review, we consider existing data that describe the associations among physical activity, obesity, and metabolic syndrome and discuss whether physical activity must produce either weight loss or improved cardiorespiratory fitness in order to lower the risk of metabolic syndrome.

Physical Activity and Metabolic Syndrome

In 2006, Ford and Li conducted an extensive review on the association of activity and fitness with metabolic syndrome (5). Because etiology is of greatest interest, I will focus attention on prospective studies—defined as those studies that measure physical activity levels prior to the onset of metabolic syndrome. At the time of Ford and Li's review, there were five prospective observational epidemiology studies on the relationship between activity and metabolic syndrome; through April 2008, we identified two more prospective studies.

The majority (four of seven) of published prospective studies show an inverse association of activity with the development of metabolic syndrome. Higher levels of physical activity are associated with a lower likelihood of developing metabolic syndrome in a dose-dependent manner. The inverse association between physical activity and metabolic syndrome is present among young, middle-aged, and older adults; in men and women;

■ Table 77.1 ■

The Three Most Common Definitions of the Metabolic Syndrome

Components	World Health Organization, 1999*	National Cholesterol Education Program, Adult Treatment Panel III, 2001†	International Diabetes Federation, 2003‡
Glucose dysregulation	Fasting insulin >25th percentile of the nondiabetic distribution or diabetes or fasting glucose ≥6.1 mmol/L	Fasting glucose >5.6 mmol/L or treatment for diabetes	Fasting glucose >5.6 mmol/L or treatment for diabetes
Adiposity	Waist-to-hip ratio >0.90 (male), >0.85 (female)	Waist circumference >102 cm (male), >88 cm (female)§	Waist circumference >94 cm (male), >80 cm (female)ll
Blood pressure	SBP ≥140 or DBP ≥90 mmHg or antihypertensive medications	SBP ≥130 or DBP ≥85 mmHg or antihypertensive medications	SBP >130 or DBP >85 mmHg or antihypertensive medications
Triglycerides	≥1.7 mmol/L	≥1.7 mmol/L	≥1.7 mmol/L or treatment
HDL-C	<0.9 mmol/L (male), <1.0 mmol/L (female)	<1.03 mmol/L (male), <1.29 mmol/L (female)	<1.03 mmol/L (male), <1.29 mmol/L (female)
Albuminuria	Urinary albumin excretion ≥20 g/min or albumin-to-creatinine ratio ≥30 mg/g	–	–
METABOLIC SYNDROME DEFINITION			
	Glucose dysregulation plus two or more of the other components (triglycerides and HDL counted as a single dysregulation)	Any three of the traits listed	Adiposity required plus two or more of the other components

Abbreviations: HDL-C = high-density lipoprotein cholesterol; DBP = diastolic blood pressure; SBP = systolic blood pressure.

**Definition, diagnosis and classification of diabetes mellitus and its complications. Report of a WHO consultation. Part 1: diagnosis and classification of diabetes mellitus*. Geneva: World Health Organization Department of Noncommunicable Disease Surveillance, 1999.

†Modified from the initial NCEP-ATPIII report following the 2003 American Diabetes Association recommendations to lower the impaired fasting glucose cut point from 6.1 mmol/L to 5.6 mmol/L. Grundy SM, Brewer HB Jr., Cleeman JI, Smith SC Jr., Lenfant C, for the Conference Participants. Definition of metabolic syndrome: report of the National Heart, Lung, and Blood Institute/American Heart Association Conference on Scientific Issues Related to Definition. *Circulation* 2004 Jan 27; 109(3): 433-438.

‡Alberti KG, Zimmet P, Shaw J. The metabolic syndrome—a new worldwide definition. *Lancet* 2005 Sept 24-30; 366(9491): 1059-1062.

§Frequently modified to cut points of overweight (body mass index [BMI] ≥25 kg/m²) or obesity (BMI ≥30 kg/m²) if only BMI measurements are available.

llCut points designated for Europid populations; ethnicity-specific cut points should be used where appropriate.

across different racial-ethnic groups; and in study populations in the United States and abroad. Longitudinal studies that included multiple measurements of activity over time showed improvements in individual components of the syndrome with increasing activity as well as a lowered incidence of the syndrome.

The two prospective studies we identified that were published after the Ford and Li review (5) were conducted in Norwegian samples of men (Oslo Study [6]) and men and women (Tromso Study [7]). In the Oslo Study of 6410 men aged 40 to 59 years at baseline, leisure-time physical activity reported as vigorous, moderately vigorous, or moderate was associated with an odds ratio of developing metabolic syndrome over 28 years that was 0.46 (95% confidence interval [CI]: 0.28, 0.74), 0.65 (95% CI: 0.54, 0.80), and 0.83 (95% CI: 0.71, 0.98), respectively, that of men who engaged in sedentary or light activities. Findings from the Tromso Study were similar, though the inverse association between physical activity intensity and incident metabolic syndrome was weaker in women as compared with men. Neither study specifically addressed the role of obesity in the association between physical activity level and incident metabolic syndrome, though authors from the Tromso Study reported the association between activity and individual components of the syndrome. As men's level of activity increased in intensity from sedentary through moderate, regular, and hard training, the hazard of developing large waist circumference decreased 15% from least to most intense (hazard ratio [HR]= 0.85, 95% CI: 0.77, 0.93). Among women, the association was not significant, though the hazard ratio indicated an inverse relationship (HR = 0.91, 95% CI: 0.82, 1.01) (7).

Obesity, Activity, and Metabolic Syndrome

Existing published investigations have not specifically addressed whether physical activity is inversely associated with incident metabolic syndrome in obese and nonobese persons studied separately (i.e., a null result for a test of effect measure modification by obesity). While such an analysis would be informative, it is likely hampered by the truncated distribution of physical activity in obese persons—fewer have engaged in the highest levels of activity. Such an analysis is also limited by the inclusion of a marker of obesity, most commonly large waist circumference or high body mass index (BMI), as one of the components of the metabolic syndrome. Participants in the normal-weight group will, by definition, express a different phenotype of the metabolic syndrome because they do not have one of the defining criteria of the syndrome. This limitation is even more pronounced when BMI is used to define obesity instead of abdominal adiposity (typically determined by waist circumference), which can be present in the absence of overall adiposity. An often-used technique in epidemiologic studies is to statistically adjust for obesity in a multivariable model and describe any remaining association. Most studies indicate an attenuated, but still inverse, association between activity and metabolic syndrome following adjustment for obesity (whether those associations remain statistically significant is determined largely based on sample size). In sum, these studies provide a valuable but incomplete picture of the complex relationship among activity, obesity, and metabolic syndrome.

More recently, intriguing findings from clinical trials examining the association of activity interventions with metabolic syndrome components have become available. In a secondary analysis of the Diabetes Prevention Program trial, intensive lifestyle modification, which includes increased physical activity levels and dietary modification, was associated with a 41% lower likelihood of developing metabolic syndrome over follow-up as compared with the standard-care (pamphlets about the importance of weight loss, but no material support) arm (8). The goal of the lifestyle modification arm was for participants to lose approximately 7% of their body weight, and many participants—nearly all of whom were overweight to begin with—met that goal. Because the authors did not conduct a second analysis in the subset of participants who did not reach targets, we cannot determine whether weight loss alongside activity was required to lower the risk of developing metabolic syndrome. In 2003, the HERITAGE Family study reported a 30% reduction in the number of participants with the metabolic syndrome following 20 weeks of supervised exercise training (16.9% before and 11.8% after training) (9). The greatest number of participants reduced their triglycerides (43%) and blood pressure (38%), followed by 28% whose waist circumference decreased, 9% who reduced glucose, and 16% who increased their high-density lipoprotein cholesterol.

Finally, in the Dose Response to Exercise in Women (DREW) trial, sedentary hypertensive postmenopausal women were prescribed exercise doses at 50%, 100%, and 150% of National Institutes of Health consensus development panel recommendations of 150 min of moderate to vigorous exercise per week for six months to evaluate changes in blood pressure (primary outcome), cardiorespiratory fitness, and metabolic factors (secondary outcomes) (10). As compared with findings for sedentary controls, there were no differences in body fat percent after six months across intervention arms, but waist circumference was significant smaller in all three exercise groups. Changes in systolic or diastolic blood pressure values did not vary across exercise groups as compared with the control group. Findings from this particular study demonstrate that even modest levels of exercise training have an effect on waist circumference, which is one of the components of the syndrome and possibly *the* component central to the syndrome.

Is Weight Loss Required to Lower Metabolic Syndrome Risk?

Obesity is clearly involved in the etiology of metabolic syndrome, but the whether it is necessary to lose weight to lower metabolic syndrome risk is a question that remains unanswered. Existing observational studies and clinical trials have not disentangled the influence of weight loss versus increased physical activity (or dietary changes) on changes in any of the metabolic syndrome components. Whether weight loss is required depends on the plausibility of other mechanisms to mediate the association between activity and metabolic syndrome (figure 77.1).

As depicted in box *a* of figure 77.1, higher levels of physical activity are associated with increased muscle strength and weight loss (expressed as a relative redistribution of fat vs. lean mass); improved cardiorespiratory fitness; and decreased symptoms of dysthymia, depression, and anxiety. These physiologic changes are associated with favorable changes in secondary mechanisms (box *b*) such as reduced inflammation and free fatty acid mobilization, as well as improved autonomic nervous system function, insulin sensitivity, and endothelial function, which in turn can improve metabolic markers that compose the metabolic syndrome (box *c*). Some of the associations between factors and box *a* and box *b* are established (i.e., increases in muscle strength and insulin sensitivity), whereas others such as improved psychological health leading to decreased inflammation or autonomic functioning (or both) are emerging. Similar hierarchies in the level of evidence between secondary mechanisms (box *b*) and metabolic syndrome components (box *c*) exist.

The large number of mechanisms and pathways displayed in this complex but not comprehensive figure suggests many ways by which physical activity could reduce the risk of developing metabolic syndrome independent of weight loss. However, because body weight can also influence the mechanisms proposed in figure 77.1, concurrent weight loss with higher levels of activity may magnify these mechanistic changes, further reducing incidence of the metabolic syndrome. In sum, there is evidence that physical activity could lower metabolic syndrome incidence without weight loss, but empirical evidence remains sparse.

Is Cardiorespiratory Fitness Needed to Lower Metabolic Syndrome Risk?

Findings from observational studies of the association between fitness and metabolic syndrome risk

Physical activity

(a) *Established physiologic changes associated with physical activity*

↑ Muscle strength ↓ Body weight ↑ Cardiorespiratory fitness ↑ Psychological health

(b) *Mechanisms associated with physical activity and individual metabolic syndrome components*

↓ Inflammatory markers ↓ FFA mobilization ↑ Autonomic function ↑ Insulin sensitivity ↑ Endothelial function

(c) ↑ HDL-C ↓ Triglycerides ↓ Glucose ↓ Waist circumference ↓ Blood pressure

Incident metabolic syndrome

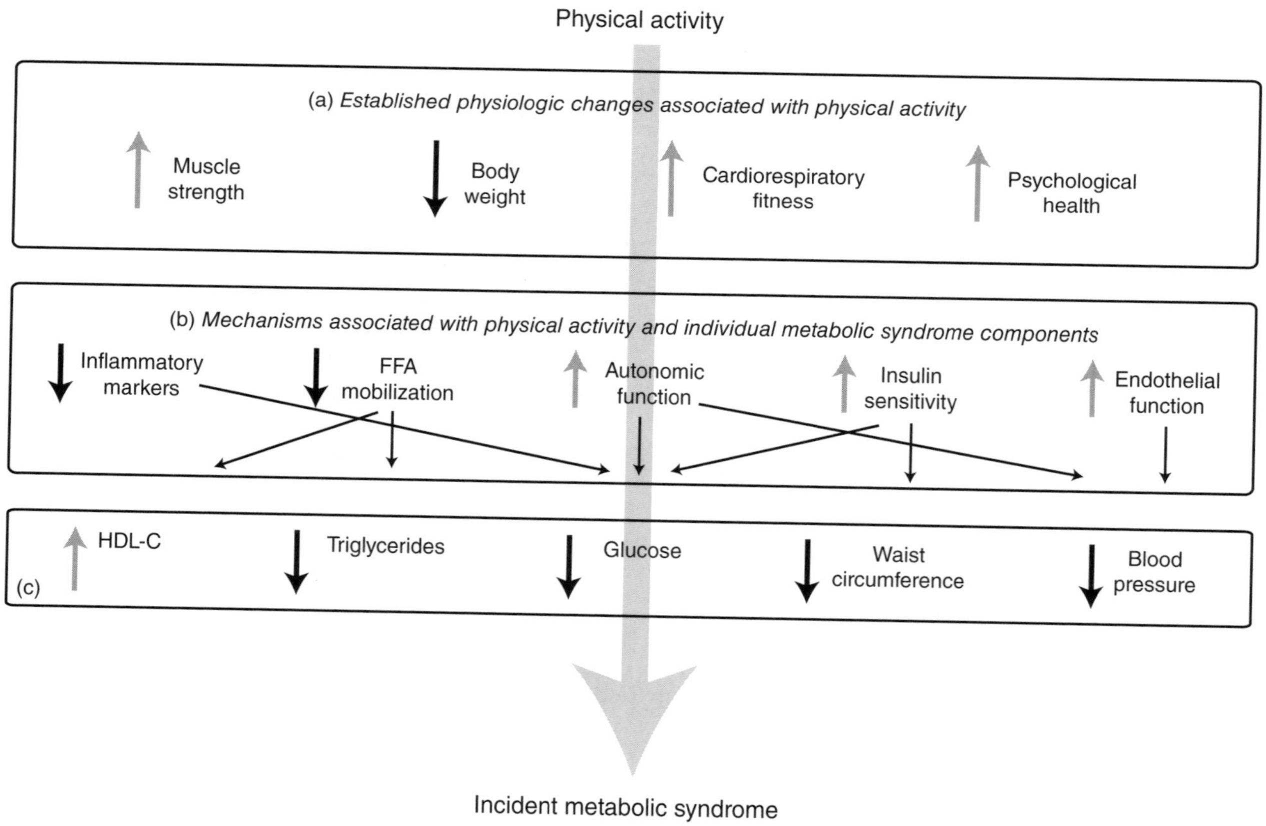

Figure 77.1 Primary and secondary mechanisms in the pathway between physical activity and development of the metabolic syndrome. FFA = free fatty acid; HDL-C = high-density lipoprotein cholesterol.

are stronger and more robust following statistical adjustment than those for the associations with physical activity (11). Part of the reason is that fitness is a physiologic trait that is measured objectively, whereas physical activity is a behavior most often measured using self-reported instruments that are prone to measurement error. Alternatively, differences in magnitude are observed because improved fitness is required to lower risk of metabolic components by some of the mechanisms proposed in figure 77.1. In the DREW study (13), any dose of activity led to short-term improvements in fitness. Despite improved fitness, the only cardiovascular disease risk factor that improved across groups over six months was waist circumference, which was lower among the exercise groups as compared with the sedentary control. From this study of sedentary postmenopausal women, it is not known whether these changes extend to men or to other persons who are at least moderately physically active already. At least based on the findings from a single study of a special population of sedentary individuals, the question whether cardiorespiratory fitness is required to lower metabolic syndrome risk seems moot if any level of activity does in fact improve fitness. Thus, the prudent answer based on available evidence is that any increase in activity levels, regardless of the potential to improve fitness, can modify certain components of the metabolic syndrome and lower metabolic syndrome risk.

Summary

In conclusion, there is a robust association between physical activity levels and risk of developing metabolic syndrome that is not completely mediated by obesity. Limited evidence from observational studies and secondary analyses of clinical trials suggests that the benefits of activity for prevention and reversal of metabolic syndrome components extend across the range of body fatness. Public health recommendations should continue to encourage physical activity as a first-level therapy for persons with the metabolic syndrome or those at risk for developing the syndrome.

78

Physical Activity and Musculoskeletal Disorders in Obesity

Jennifer M. Hootman, PhD

Musculoskeletal (MSK) disorders are common in the United States; 46 million adults have arthritis, 34 million have back pain, and 9 million have neck pain. In addition, 15 million emergency room visits per year are for unintentional injury. These MSK conditions are costly ($240 billion annually) and are among the leading causes of disability. Musculoskeletal conditions also place a large burden on the work sector, accounting for 30% of lost work days, and are the leading cause of work disability (1).

Since the human skeleton is made to produce movement, physical activity in some dose is necessary in order for the system to work properly. Disuse of the MSK system results in muscle atrophy, ligament and tendon shortening, decreased joint synovial fluid movement, and low bone mass. In contrast, excessive use (repetitive motion, physical activity) and mechanical load (body weight or resistance) of the MSK system is associated with cumulative trauma disorders (e.g., carpal tunnel syndrome, low back pain) and overuse syndromes such as tendinitis, bursitis, and stress fractures. Optimal health of the MSK system is achieved by balancing use and mechanical load (2).

Relationship Between Obesity and MSK Disorders and Disability

Excess body mass is clearly associated with incident and prevalent MSK conditions including back, neck, knee, foot, and ankle pain, as well as osteoarthritis. For example, figure 78.1 illustrates the increasing prevalence of three common MSK conditions—arthritis, neck pain, and back pain—with increasing body mass index (BMI). Knee and foot or ankle conditions are also associated with excess body mass, as is MSK-related disability. For example, in a large study of over 30,000 workers, Hagen and colleagues (3) reported a 60% increased risk of disability retirement due to back pain for the persons in the highest quartile of BMI (>28.6) even after adjusting for other known disability risk factors.

For osteoarthritis of the knee, obesity is associated with a two- to fourfold higher risk of incident disease, as well as worse disease progression. Weight loss appears to lower the risk of incident disease while weight gain has the opposite effect, and these relationships seem stronger among women. For every pound of excess body weight, 2 to 3 lb (0.9 to 1.4 kg) of excess force is exerted at large joints such as the knee (4). Thus, some experts feel that obesity is one of the most important risk factors for knee osteoarthritis, suggesting that approximately 27% to 53% of incident knee osteoarthritis could be prevented through elimination of obesity (5).

The relationship between excess body mass and MSK injuries is less consistent. Obesity has been consistently associated with occupational MSK injuries, particularly low back pain. In a study of over 30,000 men and women, persons in the highest tertile of BMI were 1.6 times more likely to report disability retirement due to low back pain, even after adjustment for a variety of known risk factors (3). Less consistent findings have been noted for the relationship between obesity and activity-related injuries or general MSK injury from all causes, although obesity has been associated with trauma-related mortality from motor vehicle accidents (6).

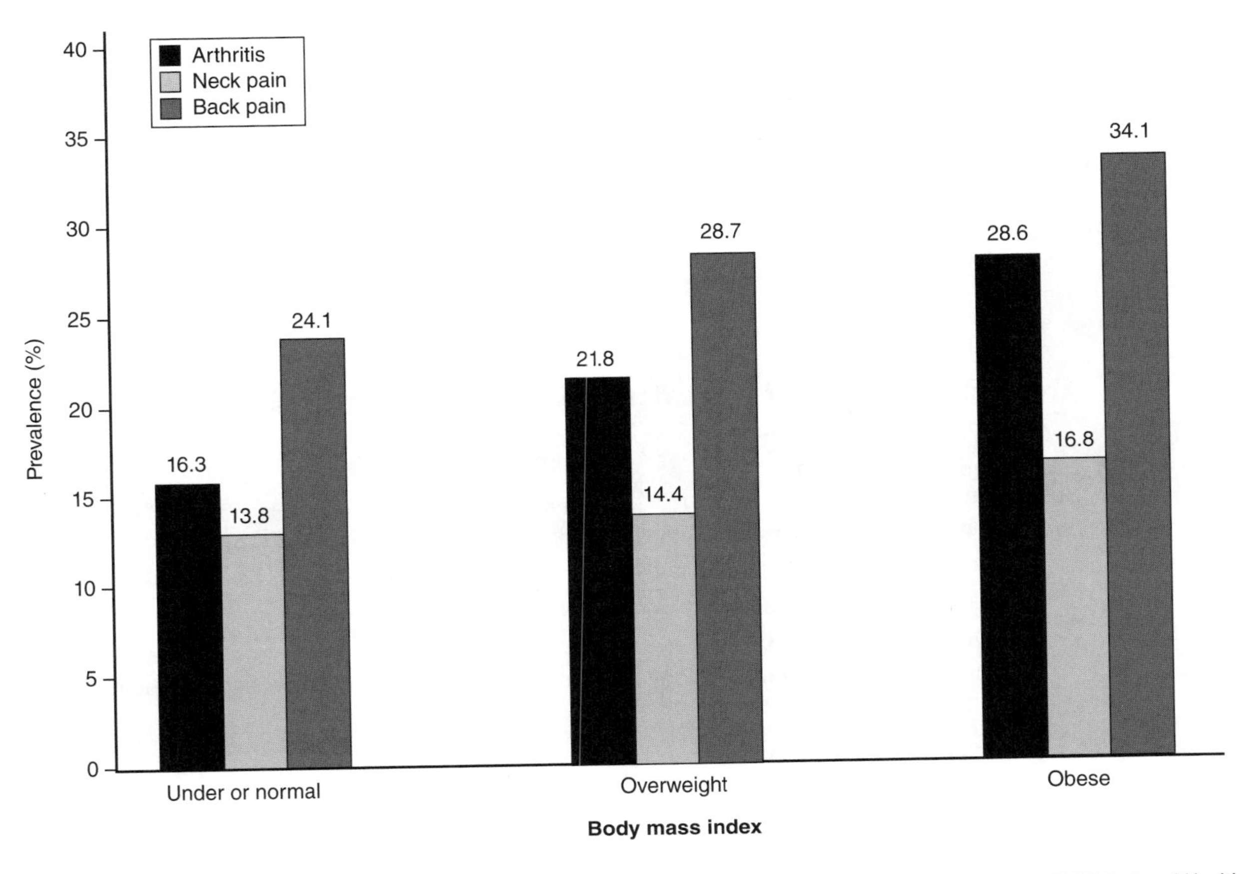

Figure 78.1 Prevalence of arthritis, neck pain, and back pain by body mass index in U.S. adults age 18+ years, 2005 National Health Interview Survey. Arthritis = self-reported, doctor-diagnosed arthritis; neck and back pain = neck or back pain lasting more than one day in the prior three months. Prevalence percent is weighted to the U.S. adult population.

Relationship Between Physical Activity and MSK Disorders

To date, no clear link has been established between participation in moderate-intensity activities, such as those recommended for public health benefits (e.g., moderate activity for 30 min/day on most days of the week), and the incidence of MSK conditions, particularly osteoarthritis. There is weak evidence, however, to suggest that regular participation in low-impact activities such as walking may provide limited protection from developing osteoarthritis (7). A possible reason is that people who engage in regular physical activity often have lower body weight. Lower body weight (reduced mechanical load) coupled with the muscle strengthening effects (improved dispersion of joint loads) and the low injury risk of walking for exercise may be a critical combination that can reduce one's risk of osteoarthritis.

In contrast, some populations may be prone to developing MSK conditions based on their occupation or recreational activities. Excessive use and overload of joints, muscles, and bones can contribute to the occurrence of MSK conditions, particularly if anatomical abnormalities (tibial varum or valgus, femoral retroversion, etc.) exist or if a severe injury such as a ligament rupture, cartilage tear, or complex fracture has occurred. For example, professional and elite-level athletes who compete for many years in high-impact sports such as football, soccer, and ballet may have some increased risk for developing osteoarthritis, particularly if they have sustained a severe joint injury during their career (7). Occupations that involve carrying heavy loads, knee bending, or squatting are also associated with increased risk of hip and knee osteoarthritis. In fact, as much as 15% to 30% of knee osteoarthritis in men may be linked to occupational activity (4). It is unclear whether excess body weight imparts additional risk on top of sport participation or occupational load, and this is an important area for further research.

Effects of Obesity on the MSK System

It is unclear whether all the effects listed in table 78.1 are purely due to the mechanical effects of obesity on the MSK system or if there are some other effects of excess body fat, such as metabolic (diabetes, lipid abnormality, etc.), hormonal, altered extremity anthropometry, or inactivity-related (4, 8). As already noted, these effects can increase the risk of incident MSK disease as well as related disability. Although there is a dearth of information in this area, weight loss seems to have beneficial effects in relation to muscular efficiency and some aspects of gait mechanics.

Weight loss also results in critical improvements in patient-oriented outcomes in those with MSK

Table 78.1

Effects of Obesity on the Musculoskeletal System

Musculoskeletal component	Effects of obesity compared to normal weight
Muscle function	Reduced muscle strength; potentially reduced muscle endurance Reduced activation of motor units Reduced oxidative capacity of muscle fibers Decreased capillary density Unbalanced muscle morphology (muscle fiber hypertrophy and increased noncontributory fat mass)
Postural balance	Increased anterior–posterior sway Altered center of gravity Greater sway area Slower sway velocity
Gait	Greater energy cost (less physiologic efficiency) at preferred walking speeds Reduced mechanical efficiency related to the following: Anterior tilt of upper body Altered step frequency Greater vertical displacement of center of mass Extraneous limb movements due to larger limb size Slower preferred walking speed and velocity Shorter step length Longer stance phase Shorter swing phase Longer period of double support Larger step width Greater hip abduction during late stance phase Reduced ankle plantarflexion More erect posture (reduced hip and knee flexion during stance) Lower knee joint torque but higher ankle joint torque Greater magnitude and rate of rearfoot eversion Higher vertical ground reaction force Increased midfoot plantar pressures during dynamic loading of the foot
Sit-to-stand task	Reduced trunk flexion Increased ankle dorsiflexion Posterior movement of the feet from initial position Reduced joint torque at hip but increased joint torque at knee

Based on S.C. Wearing et al., 2006, "The biomechanics of restricted movement in adult obesity," *Obesity Reviews* 7: 13-24.

disease. In the case of knee osteoarthritis, moderate weight loss has important effects in terms of pain reduction and improvement in physical disability. A meta-analysis of randomized controlled trials showed that a weight loss of approximately 13 lb (about 6 kg) resulted in small but significant effect sizes (0.20 for pain and 0.23 for physical disability). Weight loss of 5.1% of initial body weight strongly predicted a reduction in self-reported disability and is suggested as a potential clinical treatment goal for knee osteoarthritis patients (9). Similar benefits of weight loss on disability were noted for morbidly obese patients with low back pain who underwent surgical weight loss procedures. Almost 66% of patients with low back pain had complete resolution of symptoms two years after vertical banded gastroplasty, and the remaining patients reported less frequent bouts of pain and reduced dosages of pain medications (10).

Physical Activity and Diet Interventions for Obese Patients With MSK Disease

Both participation in aerobic or resistance exercise (or a combination of these) and dietary weight loss have proven benefits for persons with knee osteoarthritis and are nonpharmacological interventions recommended by the American College of Rheumatology. To date, only one large randomized study (11) has investigated the independent and combined effects of caloric restriction and exercise on weight loss, function, and mobility disability among obese adults with knee osteoarthritis. The Arthritis, Diet, and Activity Promotion Trial (ADAPT) included 316 subjects who were randomized to four intervention arms: healthy lifestyle (attention control and health education), diet only, exercise only, and diet + exercise combined with a goal of 5% weight loss over the 18-month trial.

Compared to the healthy lifestyle group (1.2%), the diet-only (4.9%) and diet + exercise (5.7%) groups had significant ($p < 0.05$) weight loss. In addition, the diet + exercise group showed significant ($p < 0.05$) improvement in mobility and self-reported physical function, as well as a 30% reduction in pain. These results suggest that combined diet and exercise interventions can result in clinically important reductions in body weight and disease symptoms, along with improvements in mobility and function, among older adults with a common MSK condition. Future work should focus on (1) elucidating the potential mechanisms (e.g., preserving muscle mass during weight loss) and (2) replicating these diet and exercise protocols in adults with other obesity-related MSK conditions.

Summary

Musculoskeletal conditions are common, costly, and debilitating. Excess body mass is associated with increased incidence, disease progression, and MSK-associated disability in a number of MSK conditions such as osteoarthritis, back pain, and foot and ankle problems. Physical activity can help people lose weight or maintain normal body weight and strengthen muscles and other supporting tissues, resulting in decreased and optimally dispersed mechanical loads on joints. Diet and exercise interventions are effective in reducing body weight and MSK symptoms as well as in reducing the risk of MSK-associated disability.

79

Physical Activity and Risk of Falls in Obese Adults

Teresa Liu-Ambrose, PhD

Falls are a major health care problem for older people and health care systems. To date, it has not been well established whether obesity is a significant risk factor for falls and injurious falls. This chapter presents a brief review of the epidemiology of falls among older adults, key risk factors for falls and injurious falls, evidence to date regarding the role of obesity in falls and injurious falls, and effective exercise interventions for prevention of falls.

Falls in Community-Dwelling Older Adults

About 30% of community dwellers over the age of 65 experience one or more falls every year. Older women have a higher incidence of falls than older men. The proportion of women who fall increases from about 30% in the 65- to 69-year-old age group to over 50% in those above the age of 85 years. The proportion of men who fall increases from 13% in the 65- to 69-year-old age group to approximately 30% in those aged 80 years and over.

According to the World Health Organization, falls are the third cause of chronic disability worldwide. Fatal injuries related to falls are the fifth leading cause of death among older adults in the United States, and they are the second leading cause of death due to unintentional injuries. In terms of morbidity and mortality, the most serious of these fall-related injuries is fracture of the hip. About 5% of falls result in fracture, and one-third of those fractures are hip fractures; 90% of hip fractures occur as a result of falling. In many instances, hip fractures result in death; and among those who survive, many never regain complete mobility.

Risk Factors for Falls

Falls are not random events; they occur, at least in part, due to physiological impairments such as impaired postural stability. Postural stability can be defined as the ability to maintain center of mass within limits of stability and is an essential prerequisite in daily life. Stability limits are boundaries within which the body can maintain its position without changing the base of support. The maintenance of postural stability is a complex activity, requiring input from many sensory systems, integration of this information at many levels of the nervous system, and a musculoskeletal system to implement the commands from the central nervous system (CNS). The visual, vestibular, and somatosensory systems are the primary sensory systems involved, but hearing and autonomic systems also play a role. The incidence of falls increases with age and partly due to age-related changes in the systems that contribute to postural stability. It is notable, however, that much of the decline in postural stability associated with aging is not the result of aging per se but rather the result of physical inactivity.

There is some evidence to suggest that obesity, specifically increased waist circumference, is associated with impaired postural stability (1).Thus, an abnormal distribution of body fat in the abdominal area may increase falls risk by decreasing body balance.

Risk Factors for Injurious Falls

Although falling is the strongest single risk factor for a fracture in an older person, only 1% to 2% of falls result in hip fractures, and 2% to 6% result in fractures other than of the hip in community-dwelling older adults. In addition, even older women with very low areal bone mineral density of the proximal femur have only about a 2% annual risk of hip fracture. Thus the question, "What separates those who fall without consequence from those who fall with consequences?" needs to be addressed to identify those in particular need of falls prevention. A number of studies have identified risk factors associated with injurious falls. Significant risk factors include cognitive impairment, the presence of at least two chronic medical conditions, postural instability, gait impairment, and low body mass index (BMI). Other risk factors for injurious falls are older age, Caucasian race, decreased areal bone mineral density, reduced visual acuity, and impaired neuromuscular function (such as increased reaction time and decreased muscle strength).

Risk of Falls and Injurious Falls Among Obese Adults

To date, little research has specifically focused on ascertaining the incidence of falls in older adults who are obese. Thus, it is not well established whether adults who are obese are at greater risk than others for falls or injurious falls. In fact, data from the 2001-2003 National Health Interview Surveys, conducted by the Centers for Disease Control and Prevention's National Center for Health Statistics, demonstrated that injurious falls rates for older adults in the United States did not vary significantly by level of obesity (2). Furthermore, as highlighted in the previous section, low BMI, not high BMI, has been identified as a risk factor for both falls and fall-related injuries. However, it should be noted that low BMI is related to frailty.

Nevertheless, a body of converging evidence suggests that obesity may be an important risk factor for falls and fall-related injury among older adults. Data from the National Health and Nutrition Examination Survey I suggested that high BMI is a strong predictor of long-term risk for mobility disability in older women and that this risk persists even to very old age; impaired mobility is a risk factor for falls (3).

As mentioned previously, it has been demonstrated that obesity is associated with impaired postural stability. Specifically, using mathematical modeling, Corbeil and coworkers (1) showed that older adults with increased waist circumference have a reduced capacity to maintain postural stability when exposed to perturbations. The investigators concluded that there were two main physical consequences of an abnormal distribution of body fat in the abdominal area: (1) increased mass to stabilize over the base of support and (2) anterior displacement of the body center of mass with respect to the ankle joint.

Most recently, Fjeldstad and coworkers (4) demonstrated that obese older adults, defined as those having a BMI greater than 30.0 kg/m^2, had a higher prevalence of falls than those with BMI between 18.5 kg/m^2 and 24.9 kg/m^2. Specifically, obese older adults have a falls prevalence of 27% compared with a prevalence of 15% among nonobese older adults. Furthermore, Fjeldstad and coworkers (4) reported that obese older adults had higher prevalence of stumbling during ambulation (32% vs. 14%) than nonobese older adults. High BMI was also associated with a high probability of sustaining injuries by mechanisms such as falling in a large epidemiological study that included 42,304 U.S. noninstitutionalized adults (5).

There is also evidence to suggest that obesity is a risk factor for injurious falls. The results of an epidemiological study by Owusu and colleagues (6), published in 1998, demonstrated that obesity was an important risk factor for fall-related fractures. The authors reported that waist circumference, hip circumference, and waist-to-hip ratio were all positively related to fracture incidence. Specifically, men in the highest quintile of waist circumference had a relative risk of 2.57 (95% confidence interval [CI]: 0.64 to 10.3) for hip fracture and 2.05 (95% CI: 1.06 to 3.96) for wrist fracture when compared with men in the lowest quintile. For waist-to-hip ratio, men in the highest quintile had a relative risk of 3.92 (95% CI: 1.07 to 14.3) for hip fracture and 1.50 (95% CI: 0.85 to 2.66) for wrist fracture when compared with those in the lowest quintile. It should be highlighted that BMI was not associated with fall-related fracture incidence.

In sum, recent evidence does suggest that obesity, especially when indexed by waist circumference or by the waist-to-hip ratio, is a risk factor for both falls and injurious falls.

Evidence-Based Exercise Interventions to Prevent Falls

Evidence to date has clearly demonstrated that exercise can positively influence multiple amendable falls risk factors such as postural stability, muscle strength, coordination, and reaction time in older adults. For example, Liu-Ambrose and coworkers

(7) demonstrated that both resistance training and agility training significantly reduced physiological falls risk among community-dwelling older women. Specifically, physiological falls risk was reduced by 57.3% and 47.5% with resistance training and agility training, respectively.

Exercise can also significantly reduce the incidence of falls among older adults. The Cochrane review of interventions (8) for preventing falls in elderly people showed that the following exercise interventions were likely to be beneficial:

1. A program of muscle strengthening and balance retraining, individually prescribed at home by a trained health professional (three trials, 566 participants, pooled relative risk = 0.80, 95% CI: 0.66 to 0.98)
2. A 15-week tai chi group exercise intervention (one trial, 200 participants, risk ratio = 0.51, 95% CI: 0.36 to 0.73)

The individually prescribed home program of muscle strengthening and balance retraining, also known as the Otago Exercise Program, was developed by New Zealand geriatrician John Campbell and colleagues at the University of Otago. Four trials to date have demonstrated the effectiveness of the Otago Exercise Program (9). In the first trial, 232 community-dwelling women aged 80 years and older were randomized either to an individually tailored program of physical therapy in the home or to usual care with social visits. After one year, the relative hazard for the first four falls in the exercise group compared with the control group was 0.61 (95% CI: 0.52 to 0.90). After two years, the relative hazard for all falls during the two years for the exercise group compared with the control group was 0.69 (95% CI: 0.49 to 0.97). The relative hazard for a fall resulting in a moderate or severe injury was 0.63 (95% CI: 0.42 to 0.95).

The home-based exercise program was also effective in reducing falls in men and women aged 75 years and older when delivered by community nurses. Specifically, falls were reduced by 46% (incidence rate ratio of 0.54, 95% CI: 0.32 to 0.90) after one year. The fourth exercise trial to assess the effect of this home-based exercise program on falls as observed in routine clinical practice also provided a 30% reduction in falls (incidence rate ratio of 0.70, 95% CI: 0.59 to 0.84) in both men and women aged 80 years and older. The meta-analysis of these four trials showed that the home-based exercise program was most effective in reducing fall-related injuries in those 80 and older and resulted in a higher absolute reduction in injurious falls when offered to those with a history of a previous fall.

While no falls prevention trial has been conducted specifically among obese older adults to date, it is reasonable to assume that the benefits of exercise on physiological falls risk and falls reduction would extend to this population. For example, Maffiuletti and coworkers (10) demonstrated that while postural stability was impaired among obese individuals, it could be improved with specific balance training. Aside from reducing falls risk, exercise has many beneficial effects on a number of the primary physiological systems of the human body. Specifically, regular exercise dramatically reduces the risk for many chronic diseases associated with obesity, such as cardiovascular disease and non-insulin-dependent diabetes mellitus. However, there are numerous challenges to getting older adults to take up and comply with exercise programs (11). Barriers to exercise adherence among adults who are obese can include anxiety and low motivation; factors that facilitate exercise adherence include social support, activity self-monitoring, increased self-efficacy, and appropriate exercise prescription.

Conclusions

Very little research to date has specifically focused on the issue of falls and injurious falls among older adults who are obese. However, there is some evidence to suggest that obesity increases falls risk via impairing postural stability and that obesity increases one's risk of sustaining a fracture as a result of falling.

80

Adverse Events From Physical Activity in Obese Persons

Kenneth E. Powell, MD, MPH

The purpose of this chapter is to consider whether obesity causes any unique or added risk for physical activity–related adverse events. Obesity may elevate risk either directly or indirectly. The larger mass, shift in center of gravity, smaller surface area relative to body mass, and other differences between obese and normal-weight individuals may directly increase the risk of activity–related adverse events, such as musculoskeletal injuries or heatstroke. The abnormalities in lipid profile and insulin sensitivity and other metabolic abnormalities associated with obesity increase the likelihood that obese persons have ischemic heart disease or diabetes, and may indirectly increase the risk of adverse events linked to these diseases such as sudden cardiac death during or shortly after vigorous physical activity.

The relationships between obesity and physical activity–related adverse events are complicated by the association between obesity and inactivity (1). Inactivity is one of the most important risk factors for physical activity–related adverse events (2). As a result, the relationships between obesity and adverse events may be confounded by inactivity. The causal tangle is further complicated by the fact that inactivity and obesity are independent risk factors for some diseases (e.g., ischemic heart disease) associated with risks of activity-related adverse events, as well as the fact that inactivity may cause obesity and that obesity may induce inactivity (figure 80.1). Currently available research findings have not adequately untangled the complex causal relationships among obesity, inactivity, diseases for which obesity and inactivity are causal factors, and a variety of physical activity–related adverse events.

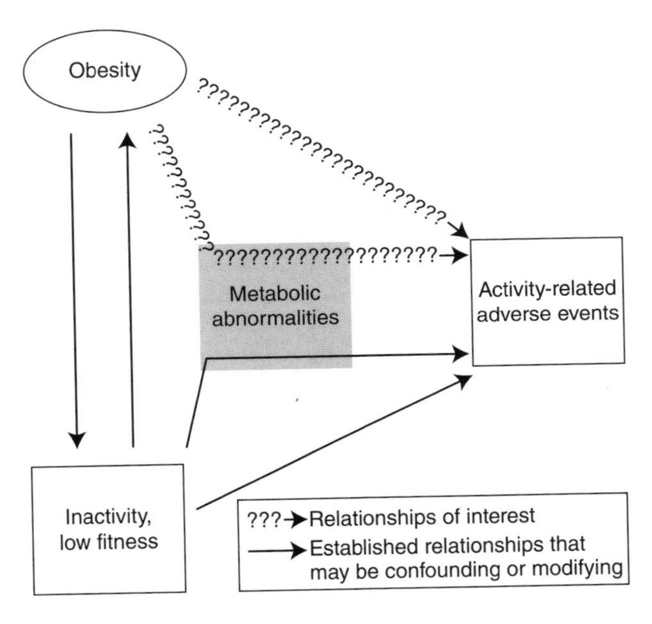

Figure 80.1 Schematic display of relationships among obesity, inactivity or low fitness, metabolic abnormalities, and physical activity–related adverse events.

The subject is further complicated by the relationship and distinction between physical fitness and physical activity. Although directly correlated, physical fitness and physical activity are different, and the terms should not be used interchangeably. However, researchers commonly measure one or the other, not both. The literature consistently demonstrates an inverse relationship between level of fitness and risk of adverse events and between level of physical activity and risk of adverse events;

rarely do studies examine or report both fitness and activity. This chapter focuses on physical activity but draws upon research utilizing fitness when that is all that is available.

Activity-Related Adverse Events and Their Risk Factors

Although the benefits of regular physical activity outweigh the risks, adverse events do occur. These "side effects" of physical activity are many, and a package insert accompanying a prescription for physical activity would have a long and varied list of misfortunes that could occur with its use.

Risk Factors for Adverse Events

The major factors that influence the risk of adverse events are related to the type and dose of the activity itself, variations in personal characteristics of people doing the activity, and the equipment or environment for the activity.

- Activity-related risk factors
  - The type of activity being performed (e.g., walking vs. rugby). Activities involving more frequent and purposeful impacts with other participants or inanimate objects have higher injury rates than activities with no or limited impacts.
  - The amount of activity performed (e.g., 500 metabolic equivalent [MET]-minutes per week of activity vs. 5000 MET-minutes per week). The risk of adverse events is directly related to the volume of activity performed.
- Person-related risk factors
  - A wide range of personal characteristics including demographic characteristics (e.g., age, sex), behavioral characteristics (e.g., physical activity habits, smoking), and health status (e.g., heart disease, obesity, previous musculoskeletal injury) (2).
  - Among personal characteristics, being inactive is one of the most important risk factors for activity-related adverse events (2).
- Equipment and environmental risk factors
  - Protective gear (e.g., bicycle helmets, football helmets, reduced-impact baseballs, mouth guards, and wrist guards for in-line skating) is known to reduce injuries.
  - Environmental conditions (e.g., the quality of playground surfaces, methods of keeping pedestrians and traffic separate, and temperature extremes) influence the risk of activity-related adverse events.

Taxonomy of Adverse Events

The spectrum of adverse events comprises a wide range of maladies or occurrences, including musculoskeletal injuries, sudden cardiac death, heat injury, infectious diseases, and respiratory disorders. No truly useful and complete taxonomy of physical activity–related adverse events exists.

Approaches for Studying Physical Activity–Related Adverse Events

The majority of studies of activity-related adverse events have focused on a specific activity or sport, such as soccer, skiing, or running; or on a specific injury, such as a medial meniscus tear. Research or surveys aimed at encompassing more activities and more types of injuries have, instead, focused on the severity of the conditions, such as death or hospitalization. All three approaches—sport specific, injury specific, or severity limited—commonly deal with adverse events associated with sport or recreation, thus omitting those associated with physically active transportation or home activities.

Another approach, the one taken in this chapter, is to select a few common or well-recognized activity-related adverse events for which obesity is a presumptive risk factor. Musculoskeletal injuries are common. Sudden adverse cardiac events (e.g., myocardial infarction, sudden death) and heat injury are severe and well recognized. Obesity may be a risk factor for all.

Obesity and the Risks of Musculoskeletal Injury

Musculoskeletal injuries are likely the most common activity-related adverse events. Assuming otherwise equivalent exposures, one might expect the greater impact forces associated with a higher body mass to place obese individuals at higher risk of musculoskeletal injuries than normal-weight individuals (3, 4). Some evidence does suggest that obese individuals are more likely to suffer injuries than are people of normal weight. But the scant available evidence suggests that the higher incidence of musculoskeletal injuries among obese persons may be due more to inactivity than to obesity.

Surveys of the General Population

For injuries of any cause, not just activity-related injuries, population-wide studies indicate that overweight and obese individuals are more likely than persons of normal weight to report a medically attended injury (5, 6). When stratified by activity level, data from one of these surveys (5) provide two pertinent suggestions (figure 80.2).

First, individuals who are obese and physically active (bar 1) are 7% more likely to be injured than individuals who are normal weight and physically active (bar 2), whereas individuals who are obese and inactive (bar 3) are 26% more likely to be injured than individuals who are normal weight and physically active (bar 2). While neither difference was statistically significant in this study, if the findings were supported by future research, this would indicate that the incidence of medically attended injuries among obese persons depends more upon their activity level than their obesity.

Second, the risk of injury appears to be directly related to body mass index (BMI) only among inactive people (figure 80.2). If this suggestion is supported by future research, it would indicate an interaction between physical activity and obesity as risk factors for musculoskeletal injuries.

Observational Studies of Selected Groups

Observational studies of injuries among highly active subpopulations, namely recreational runners and military recruits, have yielded inconsistent relationships between injury incidence and BMI (7, 8). Among both groups, however, few are obese.

Obesity and the Risks of Activity-Related Sudden Adverse Cardiac Events

As with musculoskeletal injuries, the important question pertaining to obesity and activity-related sudden adverse cardiac events (e.g., myocardial infarction, sudden cardiac arrest) is whether obesity independently raises the risk or whether obese people are at elevated risk because they are inactive.

Sudden and severe adverse cardiac events during or shortly after vigorous physical activity are rare, estimated to occur about once every 10^6 to 10^8 h of vigorous activity (2). Given their rarity, activity-related sudden adverse cardiac events have seldom been studied; a recent review citied only seven such investigations (2). These studies consistently showed an inverse relationship between regular physical activity and sudden and severe adverse cardiac events, but they included little information about the role of obesity.

Among these seven studies, only one indicated that obesity was a risk factor for sudden and severe adverse cardiac events independent of activity level. In the others, the authors stated that the risk associated with inactivity was independent of obesity but did not comment on whether obesity was an independent risk factor. Thus, limited evidence (one

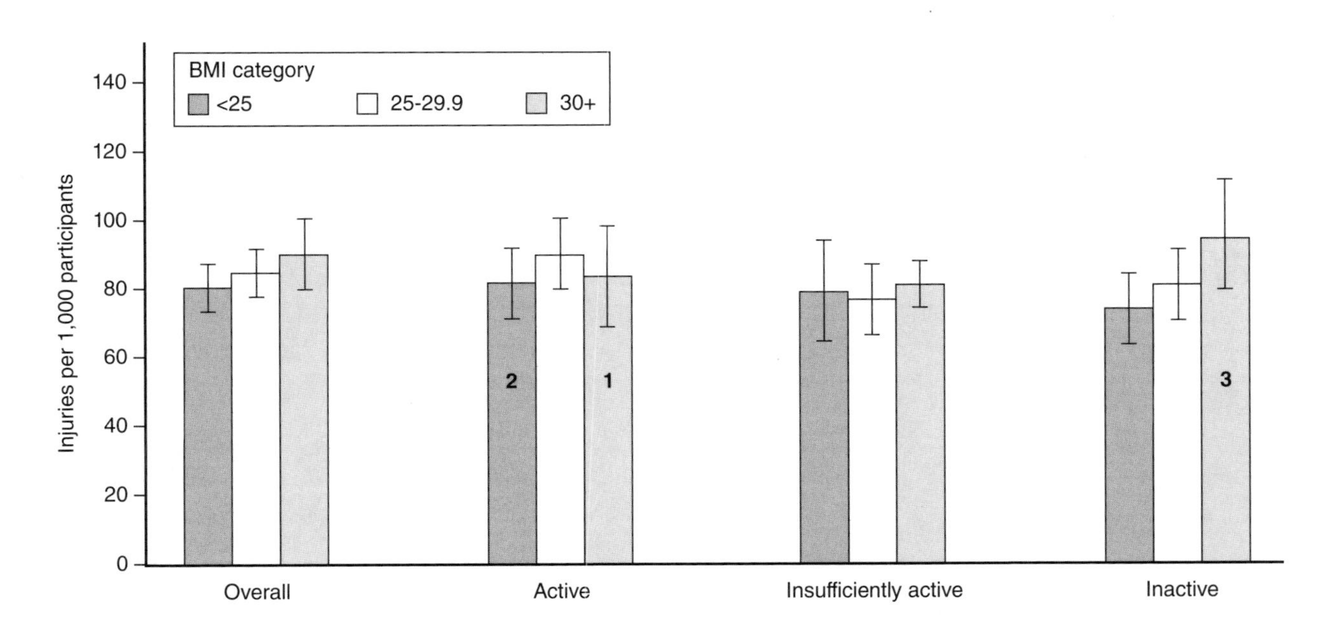

Figure 80.2 Annualized injuries per 1000 participants by activity level and body mass index, National Health Interview Survey, 2000-2002 (5). Active = 30+ min light-moderate intensity activity 5+ d/wk or 20+ min vigorous activity 3+ d/wk. Insufficiently active = some activity reported but not enough to be classified as active. Inactive = no leisure-time activity reported. Note: Bar 1 = active and obese; Bar 2 = active and normal BMI; Bar 3 = inactive and obese.

study) suggests that obesity is an independent risk factor for activity-related sudden adverse cardiac events.

Obesity and the Risk of Heat Injury

Physical activity in hot, humid environments creates a risk for heat injury among all persons, especially those who are unfit and unacclimated. Obesity is a well-established risk factor for heat injury (9), but the mechanism by which it contributes to the risk has not been determined. Reduced heat loss caused by a smaller ratio of surface area to mass is one possibility. Generation of more heat is another possibility because obese persons expend more energy than normal-weight persons while performing most tasks. One study of military recruits showed lower fitness and higher BMI to be independent risk factors for heat injury (10). However, the study included few recruits, if any, who were obese (the uppermost category of BMI was ≥26 kg/m^2), and risk did not rise steadily across the three categories of BMI (odds ratios: 1.0, 1.9, 1.6), thereby limiting the confidence with which one can draw conclusions about the independence of the relationship between obesity and heatstroke.

Conclusion

Very little can be said with confidence about whether obesity causes any unique or added risk of activity-related adverse events. One reason is that the literature addressing this potential problem is remarkably sparse. A second reason is that few studies adequately control for or stratify by activity level, an important confounder in the relationship between obesity and activity-related adverse events.

Mechanical and metabolic consequences of obesity are consistent with obesity as an important and independent risk factor for a wide range of activity-related adverse events including musculoskeletal injuries, sudden cardiac events, and heat injury. However, there are few empirical data to support this idea. Inactivity may account for a large part—in some cases possibly all—of any elevated risk associated with obesity. Future research needs to be more attentive to the confounding of obesity–adverse event relationships by inactivity.

81

Physical Activity and Cardiovascular Disease Risk Profile in Obese Children

Lars Bo Andersen, MD

The rise in childhood overweight and obesity in Western countries is apparent to all, and some studies have suggested that physical activity levels in children may also have decreased over time. The health consequences of these trends are relatively unstudied, and confusion remains about the role of overweight and physical inactivity in the development of metabolic disorders in children. There is debate around whether obesity itself is a disease in children, or whether it is the associated increase in other cardiovascular disease (CVD) risk factors that constitutes a metabolic health problem. Physical inactivity is similarly associated with high levels of other CVD risk factors (1).

It is important to assess the independent effects of physical activity and obesity on CVD risk factors, as the first is a behavior and the second is a characteristic or a condition in which large amounts of fat have accumulated in fat cells. Obesity can be changed by physical activity and diet, and many studies have looked at acute changes in adiposity related to restriction of caloric intake or performance of aerobic exercise training. The gradual changes that occur may be more interesting from a health point of view but are difficult to study in controlled settings, and most information comes from associations in observational studies. It is also important to understand the underlying reasons for the clustering of CVD risk factors in low-active or obese subjects because this may help us to form more efficient preventive strategies, which can be adapted to each individual to optimize the effect and prevent dropout.

Higher-intensity aerobic physical activity can affect cardiorespiratory fitness (CRF) over a relatively short time (months), while lower-intensity physical activity acutely affects energy expenditure and insulin levels but may not directly affect fitness per se. The insulin level may be more important than the energy expenditure itself in relation to a slow accumulation of fat, because insulin is an anabolic hormone with effects on blood pressure, blood lipids, and other CVD risk factors. However, CRF is associated with insulin sensitivity independent of the acute effect of physical activity. It has often been claimed that the intensity of aerobic exercise needs to be quite high to increase fitness. This claim could be questioned. The fitness level in normal children, those not specially trained but able to participate in ordinary play, is around 50 $ml \cdot min^{-1} \cdot kg^{-1}$ in boys and a little lower in girls depending on pubertal stage. Among children who are not able to move around freely but in other aspects are normal, such as those who are blind, fitness levels are just above 30 $ml \cdot min^{-1} \cdot kg^{-1}$ in both boys and girls (2). It is therefore likely that moderate-intensity physical activity attained through play may be important in children to preserve normal fitness. In adults, insulin may also play a role in gradual change. It is striking that appetite regulation functions well in most people but that many start to gain a kilogram of body weight or fat each year after 30 years of age. This accumulation of fat could be related to elevated 24 h plasma insulin levels, which may be caused by a gradual decline in everyday-living physical activity level or a gradual decrease in CRF.

The aims of this chapter are to describe the metabolic health consequences of low physical activity, low CRF, and obesity and to elucidate some of the physiological mechanisms responsible for the high CVD risk factor levels associated with these characteristics in children.

Physiological Mechanisms Associated With CVD Risk Factors

Acute physical activity and exercise training lead to different physiological changes, which influence CVD risk factors. Some of these mechanisms are shared with the changes caused by overweight, while others are unique to physical activity. The most important physiological change is probably in sensitivity to insulin and adrenaline; both of these hormones can double their action after a few months of exercise training. At least for insulin, the change is located in the training muscle; if only one leg is trained, the increased glucose clearance will be limited to the trained leg (3). Training increases adrenaline sensitivity and decreases adrenaline release, leading to increased glucose clearance since adrenaline blocks insulin-mediated glucose uptake. Besides this effect of training, glucose can enter the muscle cell when it contracts without the presence of insulin. This glucose transport is an important mechanism related to all intensities of activity. It causes 24 h insulin levels to decrease in the physically active, and may affect lipid storage in the fat cells and indirectly appetite regulation. Contraction-mediated glucose uptake is not blocked by adrenaline. Even during periods of stress, physical activity may help to maintain low levels of insulin. In contrast to these mechanisms located in the muscle cells, fat tissue releases cytokines such as tumor necrosis factor-alpha (TNF-α) into circulation, which can decrease insulin sensitivity systemically. Most of the common CVD risk factors are worsened by the action of insulin, which affects both metabolic pathways (blood lipids) and the sympathetic nervous system (blood pressure). The insulin sensitivity of the muscle cells is important because 80% to 90% of glucose enters the muscle cell as insulin mediated under normal conditions.

Other mechanisms related to training are increased lipoprotein lipase (LPL) activity and increased oxidative enzymes in the trained muscles. Lipoprotein lipase is situated on the inner wall of capillaries, and the number of capillaries increases in proportion to the training state of the muscle. Lipoprotein lipase catalyzes triglyceride transported from low-density lipoprotein (LDL)- or very low-density lipoprotein (VLDL)-cholesterol into the muscle cell, where it is burned, thereby improving the ratio of total cholesterol to high-density lipoprotein (HDL). Changes in oxidative enzymes related to training are substantial. After only eight weeks of training of untrained young subjects, key enzymes such as succinate dehydrogenase and hydroxyl acyl dehydrogenase have been shown to increase by 30% to 40%. This change improves the metabolism of triglycerides, which in turn improves glucose uptake.

Single CVD Risk Factors

Studies have consistently shown higher levels in all the risk factors related to the metabolic syndrome in obese versus nonobese children and adolescents (4). Blood pressure was shown to be 10 mmHg higher in the upper 10% of body mass index (BMI) compared to the lower 10% in 17-year-old adolescents, and the association between blood pressure and BMI became stronger when only the least fit part of the population was analyzed. Fitness and BMI were independently associated with blood pressure, but also with each other. Gutin and colleagues analyzed the associations between blood lipids, fasting insulin, fitness, and fatness in black and white 14- to 18-year-olds (5). Both fitness and fatness were associated with blood lipids, but when they were both entered into regression models, only fatness was significant despite a moderately strong correlation between fitness and fatness ($r = 0.69$). The authors concluded that interventions to improve lipid profile should be designed primarily to minimize fatness.

This conclusion touches an important issue related to regression analyses in cross-sectional studies. The conclusion is based on an assumption of competing risk factors, which may not be true. If risk factors (fitness and fatness) are part of the same causal pathway leading to elevated blood lipids, only the intermediate risk factor would remain in the regression when both were entered, and the real cause would disappear. For example, if we imagined that fatness was the only risk factor responsible for atherosclerosis, which also caused CVD, and adjusted the analysis between fatness and CVD for degree of atherosclerosis, fatness would no longer be significant because the degree of atherosclerosis would explain all the variation in CVD. However, in this example, fatness would still be the real cause. There is good evidence from training studies in obese individuals that as fatness decreases, blood pressure, plasma insulin, and blood lipids improve together with numerous other positive changes (6). However, it is difficult to separate out which factor is more important in these studies because it is probably impossible to train obese subjects without changing their abdominal fat.

Clustering of Risk Factors

Many studies have shown increased levels of individual CVD risk factors in overweight children, although few have analyzed clustering of risk factors. Recently, a number of studies have indicated an association between clustered cardiovascular risk factors, physical activity, fitness, and fatness. The main reason for looking at a composite score for the risk factors is that health is difficult to define in children who are free from disease but still have higher risk factor levels than their peers. Clustering of risk factors is apparent when CVD risk factors are not independently distributed but some subjects have high levels of many risk factors simultaneously (7). This happens when one or more underlying factors affect a number of risk factors at the same time. The metabolic syndrome is an example. This syndrome is thought to be caused by insulin resistance, but there may actually be more mechanisms that are responsible. Usually, scientists claim obesity to be responsible for insulin resistance, because fat tissue secretes cytokines. However, circulating cytokines may have a systemic effect on all cells, and in fact insulin sensitivity of muscle is a very local phenomenon.

Regardless of what mechanisms are responsible, clustering of risk factors is an unhealthy condition, and it makes sense to analyze a composite score of risk factors. We have recently analyzed such clustering in relation to objectively measured physical activity (1), physical fitness (7), and combinations of physical activity, fitness, and obesity (8). We found an approximately 13 times increased risk for clustering of CVD risk factors for the lower quartile of fitness compared to the upper quartile, and an approximately 10 times increased risk for the upper quartile of fatness compared to the lower. Odds ratios depend on which CVD risk factors are included in the composite score, and the fact that one exposure shows a higher odds ratio than another does not mean that it is more important. The different exposures are measured with different amounts of error variation, and physical activity variables will always have weaker associations than the more precisely assessed fitness and fatness variables even if physical activity may be just as important. However, skinfold thickness predicted clustering of risk factors as strongly as waist circumference or BMI. Physical activity exhibited a protective effect even after adjustment for both fitness and fatness in the European Youth Heart Study (see figures 81.1 and 81.2).

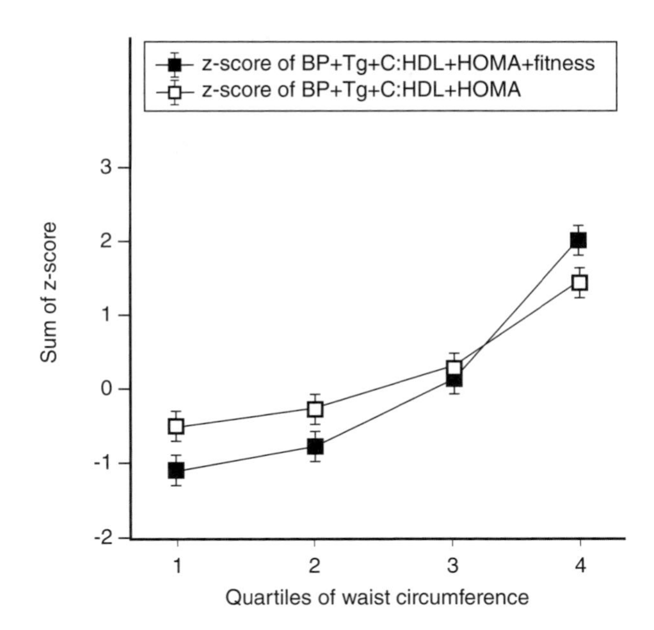

Figure 81.1 The mean sum of z-scores with standard error bars was calculated for each quartile of waist circumference. Open squares are the summed z-scores from blood pressure, triglycerides, ratio of cholesterol to high-density lipoprotein, and homeostasis model assessment (HOMA) score. Filled squares are the sum of z-scores of these risk factors plus the inverse of fitness.

Reprinted from S.A. Anderssen et al., 2008, "Fitness, fatness and clustering of cardiovascular risk factors in children from Denmark, Estonia and Portugal: The European Youth Heart Study," *International Journal of Pediatric Obesity* 3(SI): 58-66. Reprinted, by permission from Taylor and Francis. http://www.tandfco.uk/journals.

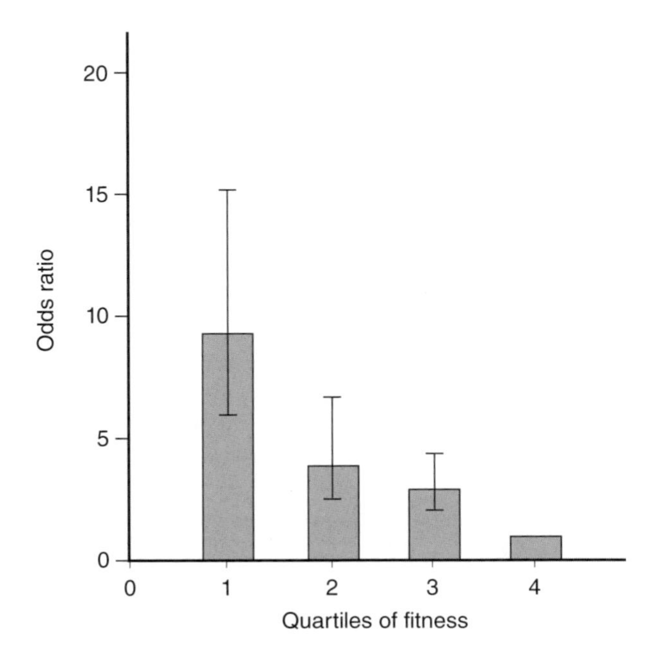

Figure 81.2 Fitness was analyzed as predictor of clustered cardiovascular risk in 2839 children from the European Youth Heart Study. Columns show odds ratios with 95% confidence intervals. Z-scores were summed for systolic BP, triglyceride, total cholesterol/HDL, and the natural logarithm of HOMA and waist circumference. 1 SD was selected as cutpoint for being at risk. A similar analysis was published earlier which included skinfold instead of waist in the clustered risk score.

The Role of Cytokines

Abdominal fat excretes a number of metabolically active cytokines. Among these are TNF-α, interleukin-6 (IL-6), C-reactive protein (CRP), and adiponectin. Tumor necrosis factor-alpha, IL-6, and CRP are inflammatory markers, which have been found to be elevated in obese subjects (9). Tumor necrosis factor-alpha has been shown to cause insulin resistance in rodents, but it is unclear whether the same mechanisms apply to humans (10). The role of IL-6 is also not clear, but IL-6 has been shown to block the effect of TNF-α and is produced in the muscles during exercise. Most studies addressing cytokines have used obese subjects with manifest severe insulin resistance. It is therefore difficult to see whether the inflammation is a result of obesity or whether the cytokines created the insulin resistance, which again could cause obesity. We analyzed IL-6, TNF-α, and CRP in a normal 9-year-old population (the Copenhagen Schoolchild Intervention Study), which included subjects with insulin resistance, obesity, and clustering of CVD risk factors. There were no associations between TNF-α or IL-6 and the homeostasis model assessment (HOMA) score, or between obesity and TNF-α or IL-6. However, we found a relatively high correlation between CRP and fitness, obesity, and HOMA. C-reactive protein is a general marker of inflammation. The preceding observation suggests that TNF-α is more a symptom of inflammation in the obese than the cause of insulin resistance. Further research may be necessary before conclusions can be drawn.

Conclusion

Cardiovascular disease risk factors cluster in children, and clustering is associated independently with obesity, low physical activity, and fitness. These observations can be accounted for by defined physiological mechanisms that are thought to be at the origin of the risk factor clustering.

82

Physical Activity and Risk of Diabetes in Obese Children

Louise A. Baur, MBBS (Hons), BSc (Med), PhD

This chapter discusses the measurement of insulin resistance, and hence diabetes risk, in children; it also reviews the relation between insulin resistance and physical activity in overweight or obese children and adolescents as evidenced in both observational and interventional studies.

Insulin Resistance and Risk of Diabetes

Insulin resistance plays a central role in the pathogenesis of several conditions associated with adverse metabolic and cardiovascular outcomes. The term "insulin resistance" refers to a reduced response to insulin-stimulated glucose uptake, an important step in the complex physiological interaction between glucose and insulin. In the healthy individual, a balance exists between glucose utilization, both insulin mediated and glucose mediated, and glucose production, primarily hepatic. With the development of insulin resistance, the individual progresses to impaired glucose tolerance and finally to type 2 diabetes mellitus when the pancreatic beta cell reserve diminishes. There are few longitudinal data on the progression from impaired glucose tolerance to type 2 diabetes in children; however, it may be more rapid than in adults, with one study showing that 24% of adolescents of mixed U.S. ethnicity with untreated impaired glucose tolerance progressed to type 2 diabetes over two years (1). Obesity, and particularly central obesity, is associated with an increased risk of insulin resistance and of type 2 diabetes in all age groups.

Measuring Insulin Action

The tools used to measure insulin sensitivity, glucose effectiveness, and glucose tolerance involve research methods that are generally invasive, labor intensive, and expensive but provide detailed dynamic information. These include the hyperinsulinemic-euglycemic clamp, traditionally regarded as the "gold standard" measure, and the frequently sampled intravenous glucose tolerance test with mathematical modeling ("minimal model method"), which assesses the physiological response to glucose and provides a rich source of information about glucose and insulin dynamics. A less invasive method is the use of a number of formulas for calculations of insulin sensitivity from insulin and glucose measurements following a standard oral glucose tolerance test. Finally, there are several surrogate measurements calculated from a single fasting blood sample, all of which correlate reasonably well in both children and adults with more gold-standard measures of insulin action. These include the fasting plasma insulin level, with high levels in the presence of normoglycemia suggestive of insulin resistance; the fasting insulin-to-glucose ratio or glucose-to-insulin ratio; the homeostasis model assessment estimate of insulin resistance (HOMA-IR = fasting insulin [mU/L] × fasting glucose [mmol/L] / 22.5); and the QUICKI score (QUICKI = 1/[log (fasting insulin in mU/L) + log (fasting glucose in mg/dL)]).

Observational Studies of Physical Activity and Insulin Resistance

Several studies have now shown that, as with adults, children with high levels of physical activity, or those who participate in physical activity, have higher levels of insulin sensitivity than less active children (2). The observational studies have generally been limited by the use of self-reported measures of physical activity, with subsequent potential for error. Likewise, the measures of insulin action have usually been derived from a single fasting blood specimen. And, finally, many studies have not adequately adjusted for body composition, so it remains unclear whether the association between physical activity and insulin sensitivity is independent of body fatness.

In a cross-sectional study, 355 U.S. nondiabetic children aged 10 to 16 years had physical activity assessed by questionnaire, body composition by skinfolds, and insulin sensitivity by the euglycemic insulin clamp; this was the first such study in children to use this gold-standard measure of action (3). After adjustment for age, ethnicity, sex, and pubertal stage, physical activity was inversely correlated with fasting insulin ($r = -0.12$, $p = 0.3$) and positively associated with clamp-derived insulin sensitivity (M_{ffm}; i.e., glucose utilization per kilogram fat-free mass per minute; $r = 0.18$, $p = 0.001$), a finding that was more marked in children with above-median systolic blood pressure ($r = -0.17$, $p = 0.03$ and $r = 0.35$, $p = 0.0001$, respectively). This association remained unchanged after further adjustment for a range of parameters, including body mass index (BMI), body fat percentage, and waist circumference.

In a subsequent large observational study from the European Youth Heart Study, 1732 randomly selected 9-year-old and 15-year-old children from Denmark, Estonia, and Portugal had fasting blood samples collected (for a range of metabolic markers, including glucose and insulin) and physical activity measured over four days via accelerometry (4). There was a small but significant negative association between physical activity (in counts per minute, adjusted for age and sex) and both glucose ($r = -0.13$, $p < 0.0001$) and HOMA-IR ($r = -0.17$, $p < 0.0001$). When cardiometabolic risk factors were clustered in a summed risk score, with systolic blood pressure, lipids, HOMA-IR, sum of four skinfolds, and aerobic fitness combined, there was a graded negative association with physical activity (either total counts or counts spent in moderate to vigorous levels of activity). The authors did not adjust for body composition in their logistic regression analyses, so it is possible that the association may be primarily mediated through level of adiposity. However, data analyses were performed separately for normal-weight and overweight children, with the odds ratios for quintiles for physical activity remaining unchanged, suggesting that this association may indeed be independent of body composition.

Results from another large cross-sectional study, the U.S. National Health and Nutrition Examination Survey (1999-2000), provide further insight into these associations (5). This study included 1783 U.S. adolescents aged 12 to 19 years and of varying ethnicity. Physical activity was assessed from a questionnaire administered during a home interview, cardiovascular fitness from a submaximal exercise test, and insulin sensitivity from fasting insulin and glucose levels (from which a value for QUICKI was then derived). In the sex-specific multiple regression analyses, in boys, after adjustment for age, ethnicity, and BMI, high levels of physical activity and high levels of cardiorespiratory fitness were positively associated with insulin sensitivity ($\beta = 0.84$, $p < 0.001$ and $\beta = 0.82$, $p < 0.001$, respectively). In contrast, in girls, physical activity was not associated with insulin sensitivity, and the association between cardiovascular fitness and insulin sensitivity disappeared after BMI was controlled for. The authors suggest that the method used to assess physical activity may be less reliable or valid in girls than in boys. Alternately, girls, who appear to have higher insulin resistance than do boys, may require higher levels of physical activity to overcome this; but such levels of physical activity are not usual in current Westernized lifestyles.

These and other studies highlight some of the uncertainties in this area. Future research will need to incorporate detailed assessments of physical activity via acceleometry (assessing duration, intensity, and amount) and more gold-standard measures of whole-body insulin action, as well as adjustment for a range of factors including sex, hormonally assessed pubertal status, and various components of body composition such as fat distribution and percentage body fat.

Intervention Studies

A range of studies have shown that exercise training (whether aerobic or resistance) can lead to improvements in insulin action in obese children or adolescents.

For example, 79 U.S. overweight children aged 7 to 11 years participated in a randomized crossover design study involving either four months of aerobic exercise training followed by four months of no training or, alternately, a no-training condition followed by the exercise training condition (6). The exercise program was offered five days per week; the children were transported to and from the gym after school and also were offered incentives for both attendance and satisfactory participation in the gym sessions (i.e., heart rate maintained above 150 bpm). The program included exercise on machines (treadmills, cycles, rowers) as well as group games. Plasma insulin concentrations decreased (by approximately 10-14%) in both groups during the periods of exercise training and increased during the no-training periods ($p < 0.001$ for group × time interaction). Thus, both training and detraining effects were observed. There was also an associated decrease in percentage body fat (via dual-energy X-ray absorptiometry) during the exercise training periods.

A Greek study has shown that the improvements in insulin sensitivity that occur with exercise training may occur independently of changes in body composition. In a before-and-after study, 19 overweight or obese girls aged 9 to 15 years took part in a 12-week aerobic training program, led by a physical education instructor, that involved three 40 min sessions each week (7). The program was of moderate intensity (heart rate >150 bpm) and included both individual and group activities. Compared with baseline values, the intervention led to significant improvements in exercise heart rate and maximal oxygen consumption. The mean insulin area under the curve (assessed from an oral glucose tolerance test) declined by 23% ($p < 0.05$), whereas glucose area under the curve remained unchanged. Interestingly, the improvement in insulin sensitivity occurred without any associated changes in body weight and percentage body fat (assessed by dual-energy X-ray absorptiometry).

A 2008 systematic review of the effects of resistance training on metabolic fitness in children and adolescents has highlighted the lack of evidence in this area (8). No studies met the full set of CONSORT criteria for the reporting of clinical trials; only four studies included data on insulin or insulin sensitivity in nondiabetic obese children, and only two of these were randomized controlled trials. Only one study (9) showed a significant change in insulin, glucose, or insulin action after resistance training when compared with values in controls or with baseline values. The authors of the systematic review highlighted the need for more rigorous study design and protocols, as well as more detailed assessment of outcome measures.

The study that showed significant changes after resistance training (9) included 22 overweight insulin-resistant Latino adolescent males who were assigned to either twice-weekly resistance training (initially low intensity and then progressively higher intensity) or a nonexercising control group for 16 weeks. At baseline and postintervention, body composition was assessed by dual-energy X-ray absorptiometry, and insulin action was assessed by the frequently sampled intravenous glucose tolerance test. The resistance training group showed a significant improvement in insulin sensitivity, both from baseline and when compared with the control group; this remained after adjustment for body composition.

In a pilot study published since the systematic review was undertaken, 14 obese Australian children (eight male) with a mean age of 12.7 years were recruited to an eight-week group circuit exercise training program, run by an exercise physiologist, for a total of 24 hour-long sessions; attendance of more than 87% was recorded (10). The exercise sessions were conducted three times per week and involved a mix of aerobic and resistance stations that progressed to 85% of maximum heart rate and ~65% of pretraining voluntary contractions, respectively. While there was no significant change in body composition as assessed by dual-energy X-ray absorptiometry, there were improvements in cardiorespiratory fitness, waist circumference, and insulin sensitivity (M value) as assessed by the euglycemic-hyperinsulinemic clamp technique. This is the only study published to date in which the gold-standard measure of insulin action has been used to assess the impact of an exercise intervention in obese children.

Conclusion

Observational studies show that, as in adults, children with high levels of physical activity or those who participate in physical activity have higher levels of insulin sensitivity than less active children. However, it remains unclear whether this effect is independent of changes in body weight or body fatness. Intervention studies show that aerobic exercise training in obese children or adolescents leads to improvements in insulin action, and there are some initial data to suggest that this is independent of body composition. As yet, there are only limited data to suggest that obese children and adolescents may have improvements in insulin action from supervised resistance training. All intervention studies to date have been of only limited duration, so the effect on long-term health risk is unknown. In addition, the impact of sex and puberty are likely to be very important but are underexplored. Therefore, in addition to incorporating more detailed measures of physical activity, body composition, and insulin action, future studies in this area will need to assess the impact of both sex and pubertal stage and also use longer durations of intervention and follow-up. The ultimate challenge is to develop and evaluate real-life interventions that are child- and youth-friendly, are sustainable outside the research setting, and have long-term health benefits.

part IX

Policy and Research Issues

The final section of the book provides a broad perspective on issues related to physical activity and obesity. It is becoming clear that multicomponent and multilevel intervention strategies will be required to stem the tide of the current obesity epidemic. Chapter 83 provides an overview of recent global policy initiatives on physical activity and obesity that encompass multiple countries and global regions. Chapter 84 describes the leadership role that governments can take in promoting physical activity at the local, regional, and national levels. Chapters 85 and 86 describe effective community-based and school-based physical activity interventions, respectively, designed to address obesity prevention and treatment. Chapter 87 highlights the potential role of private industry in the promotion of physical activity, and chapter 88 provides evidence for the effectiveness of mass media approaches in addressing the problem of physical activity and obesity. The book concludes with chapter 89, which highlights the current status of physical activity and obesity research and emphasizes the areas that require more attention in the coming years.

Global Policy Initiatives Related to Physical Activity and Obesity

Timothy Armstrong, PhD

Noncommunicable diseases (NCD), including cardiovascular diseases, diabetes, obesity, certain types of cancers, and chronic respiratory diseases, account for 60% of the 58 million deaths annually. This equates to 35 million people who died from these diseases in 2005 (1). Nearly 17% of the population are estimated to be physically inactive and an additional 41% to be insufficiently active to benefit their health, and the resultant annual death toll has been estimated to be 1.9 million deaths globally (2).

Moreover, the World Health Organization (WHO) estimates that globally, approximately 1.6 billion adults (age 15+) were overweight (BMI ≥25) and at least 400 million were obese (BMI ≥30) in 2005. Furthermore, as shown in figure 83.1, the prevalence of overweight and obesity is likely to continue to increase; by 2015, approximately 2.3 billion adults are expected to be overweight, and more than 700 million will be obese (1). Overweight and obesity also affect children, and in 2007 an estimated 22 million children under the age of 5 years throughout the world were overweight. More than 75% of overweight and obese children live in low- and middle-income countries (3).

Global Response to NCD and Physical Inactivity Burden

There is strong scientific evidence that a healthy diet and sufficient physical activity are key elements for the prevention of NCD and its risk factors, including overweight and obesity (4). In response to the growing global epidemic of NCD, an increasing effort has been made by WHO's Member States to set a favorable agenda toward the work of primary prevention of NCD and to provide a strong global mandate for WHO's efforts related to NCD prevention. Milestone resolutions are outlined in table 83.1.

Global Strategy on Diet, Physical Activity and Health

The Global Strategy on Diet, Physical Activity and Health (DPAS) should be understood as a comprehensive tool to guide the actions of WHO itself, Member States, international partners, civil society, nongovernmental organizations, and the private sector that aim to promote and protect health through healthy eating and physical activity, and therefore reduce obesity and prevent NCD.

Development of the Global Strategy on Diet, Physical Activity and Health

The DPAS was developed over a two-year period, in consultation with Member States, United Nations agencies, civil society groups, and the private sector and was endorsed in 2004 during the 57th World Health Assembly (5). The consultations carried out during the development stage were instrumental to

Acknowledgements
Special thanks to Ms. Vanessa Candeias, Technical Officer, Surveillance and Population-based Prevention Unit, Chronic Diseases and Health Promotion Department, World Health Organization, Geneva, Switzerland, for her help in the preparation of this chapter.

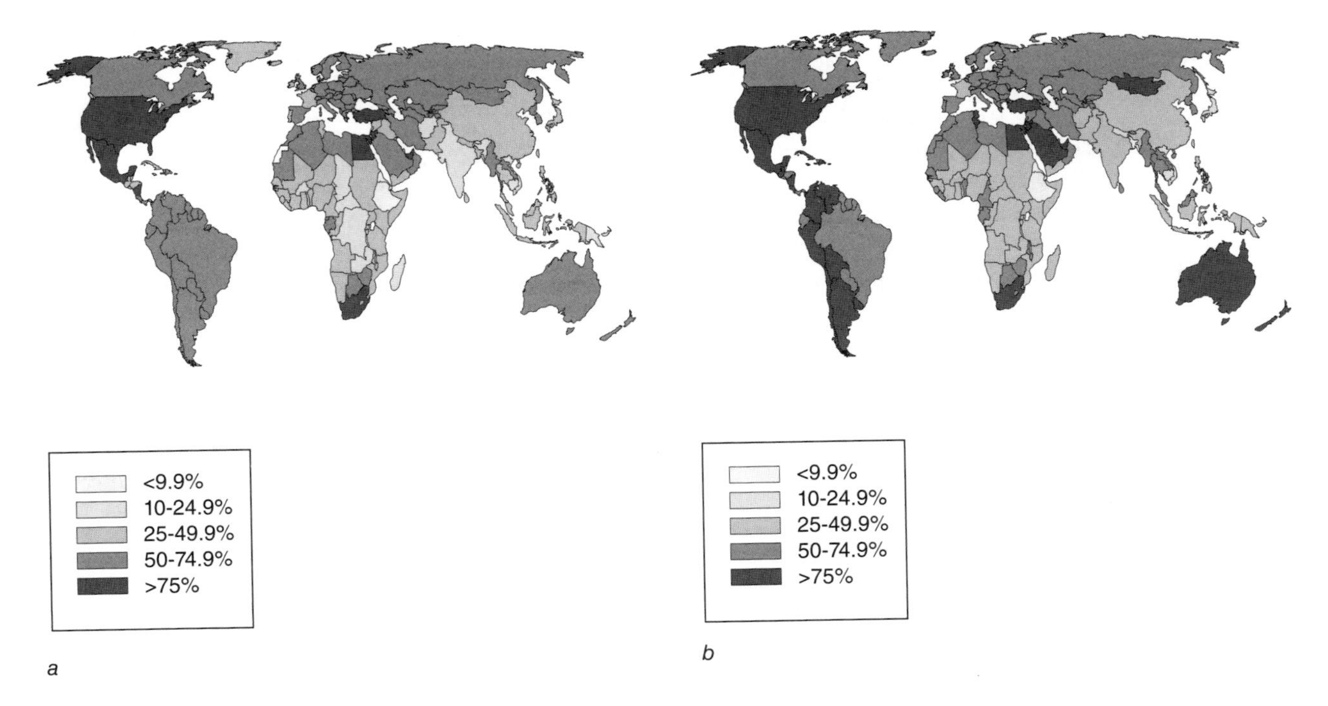

Figure 83.1 Prevalence of overweight in *(a)* 2005 and projected prevalence in *(b)* 2015.

World Health Organization, 2005, *Preventing chronic diseases: A vital investment* (Geneva, Switzerland: World Health Organization).

Table 83.1

Global Response to Noncommunicable Diseases (NCD)

Year	Action by the World Health Assembly
2000	Endorsed a resolution requesting the WHO Secretariat to give priority to the prevention and control of NCD and to provide guidance and technical support to Member States' work related to the growing epidemic of NCD (WHA Resolution 53.17).
2002	Called upon the WHO Secretariat to develop a global strategy on diet, physical activity, and health; additionally urged Member States to celebrate a "Move for Health Day" each year to promote physical activity as essential for health and well-being (WHA Resolution 55.23).
2004	Endorsed the Global Strategy on Diet, Physical Activity and Health (DPAS), providing WHO Secretariat with a clear mandate and responsibilities for work related to health promotion and primary prevention of NCD through diet and physical activity (WHA Resolution 57.17).
2007	Endorsed a resolution requesting the WHO Secretariat to prepare an action plan for the prevention and control of NCD, to be submitted to the 61st WHA (WHA Resolution 60.23).
2008	Endorsed a resolution and a six-year action plan on prevention and control of NCD (WHA Resolution 61.8). This approved action plan includes six main objectives related to the prevention and control of NCD and differentiates recommended actions for WHO, Member States, and international partners. Specifically related to physical activity, it urges Member States to - promote interventions to reduce the main shared modifiable risk factors for NCD (including physical inactivity); - implement the actions recommended in DPAS; - develop and implement national guidelines on physical activity for health; - implement school-based programs in line with WHO's health-promoting schools initiative; - ensure that physical environments support safe active commuting and create space for recreational activity by ensuring that walking, cycling, and other forms of physical activity are accessible to and safe for all; introducing transport policies that promote active and safe methods of traveling to and from schools and workplaces, such as walking or cycling; improving sport, recreation, and leisure facilities and increasing the number of safe spaces available for active play.

Resolutions available through www.who.int/gb/

validate the importance of physical activity in the agenda of all WHO regions and to highlight the significance of physical activity within DPAS. Moreover, endorsement of DPAS by the World Health Assembly represented a major step forward for the international work on physical activity.

Implementation of DPAS

The World Health Organization has progressed at the global level with DPAS implementation along the following paths:

- Provision of guidance, technical support, and tools for Member States
- Facilitation of capacity building at regional and national levels on issues related to diet and physical activity
- Direct implementation through WHO regional and country offices at the national level
- Establishment of partnerships with various stakeholders and conduct of collaborative work when and if appropriate

Provision of Tools and Guidance

The World Health Organization has produced an implementation toolbox that includes a wide range of tools to assist Member States and various stakeholders in the management and implementation of DPAS. The tools that relate to physical activity and obesity are briefly described in the next section. Further information can found on the Web site www.who.int/dietphysicalactivity/implementation/toolbox/en/index.html.

Improving Population Physical Activity Levels

As a result of the WHO Workshop on Physical Activity and Public Health held in Beijing, China, in October 2005, WHO developed a guide for population-based approaches to increasing levels of physical activity. The guide aims to assist WHO Member States and other stakeholders in the development and implementation of a national physical activity plan and provides guidance on policy options for effective promotion of physical activity at the national and subnational level (6).

Settings-Based Approach to Increasing Physical Activity Levels

Several strategies have been used to implement programs to increase physical activity levels. One approach is to design interventions in specific settings such as the workplace or schools.

Preventing NCD in the Workplace Through Diet and Physical Activity

As an outcome of an event jointly organized between WHO and the World Economic Forum in September 2007 in Dalian (China), a report on prevention of NCD through diet and physical activity at the workplace was produced. This report describes the state of knowledge regarding use of the workplace as a setting for NCD prevention; highlights the rationale for addressing diet and physical activity in the workplace; outlines key elements of successful workplace health promotion programs; and identifies potential roles for different stakeholders in the development and implementation of these programs; and components of monitoring and evaluating these programs (7).

School Policy Framework: Implementation of the DPAS

Building on various school health and nutrition programs of the United Nations System, a policy framework describing the core elements of the development and implementation of a national or subnational DPAS school policy was developed. Its purpose is to guide policy makers at national and subnational levels in the development and implementation of policies that promote healthy eating and physical activity in the school setting through changes in environment, behavior, and education (8).

Surveillance, Monitoring, and Evaluation

The DPAS recommends regular surveillance, monitoring, and evaluation of all activities by Member States, nongovernmental organizations, and private sector and research communities.

A Framework to Monitor and Evaluate Implementation

The World Health Organization developed a tool to assist Member States with their monitoring and evaluation activities in the area of promoting a healthy diet and physical activity. This tool includes a framework explaining how policies and programs and their implementation influence populations, leading to behavior, social, health, economic, and environmental changes; and it suggests how adequate monitoring and evaluation indicators can be integrated in the process of change. Additionally, it includes a series of tables of indicators that were developed according to DPAS recommendations and should be used, if and as appropriate, after adjustment to the country reality and activities (9).

Global Physical Activity Questionnaire

The development of an appropriate, valid, and reliable tool to measure physical activity is a challenging task. Several components, such as intensity, duration, and frequency, need to be taken into account. Additionally, it is desirable to assess the domain in which physical activity is performed (e.g., occupational physical activity, transport-related physical activity, and physical activity during discretionary or leisure time).

Given the increased global interest in the interaction between physical activity and health, WHO developed the Global Physical Activity Questionnaire (GPAQ) for physical activity surveillance. This instrument, designed mainly for use in developing countries, allows incorporation of variations in cultural and other national characteristics. The GPAQ has undergone a research program showing that it is a valid, reliable, standard measurement tool for monitoring and evaluating a population's physical activity level. If used consistently, GPAQ allows comparison of data collected in different countries and permits a better description of within- and between-country physical activity prevalence and changing trends in physical activity levels over time than has previously existed with existing instruments. The GPAQ is being promoted and disseminated as a part of the WHO STEPwise approach to chronic disease risk factor surveillance and as a stand-alone physical activity surveillance tool (10). Between 2001 and 2008, over 68 countries globally have repeatedly used GPAQ for their national and subnational surveillance activities.

Facilitating Capacity Building at the Regional and National Level

A part of the activities carried on by WHO to implement DPAS, which also relate to increasing physical activity levels and preventing obesity, is the organization of global or regional capacity-building workshops. These workshops bring together Member States and representatives from various relevant stakeholders. Overall, they aim to strengthen the understanding, dissemination, and utilization of the tools produced for DPAS implementation; support the development and implementation of regional or national policies and strategies related to diet and physical activity; facilitate intercountry cooperation in DPAS implementation; and foster the development of multistakeholder and multisectoral approaches.

From the endorsement of DPAS in 2004 through 2008, over 25 regional workshops were held in all WHO regions (Africa, Americas, Eastern Mediterranean, Europe, South East Asia, and Western Pacific).

Implementation at the Regional and Country Level

The 2006 WHO Global Survey on Assessing the Progress in National Chronic Noncommunicable Diseases Prevention and Control showed that of the 85 countries included in the survey, 29 had adopted multisectoral strategies and plans on healthy diet and physical activity based on the WHO Global Strategy on Diet, Physical Activity and Health.

Further information from Regional Offices indicates that several additional countries have developed policies or strategies (or both) in response to the DPAS recommendations or plan to implement DPAS, either independently or as a part of existing chronic disease prevention, nutrition, or health promotion programs.

Establishing Partnerships and Conducting Collaborative Work

Regarding the development and implementation of physical activity policies, WHO plays an important role in promoting multistakeholder interaction by organizing multistakeholder capacity-building workshops; supporting intercountry collaboration; encouraging and facilitating the development of physical activity networks to promote sharing of knowledge and information on physical activity and health; and collaborating with other agencies that have an international scope of work, such as the WHO Collaborating Center for Physical Activity and Health Promotion (U.S. Centers for Disease Control and Prevention) and the International Olympic Committee.

Conclusions

In response to the rapid global growth of the epidemic of NCD and their common risk factors, DPAS has been developed and endorsed as an international key policy instrument. This chapter provides an overview of the activities carried out by WHO, in relation to physical activity, as part of DPAS implementation.

The World Health Organization developed a set of tools to support Member States in implementation of DPAS at regional and national levels. Specifically in relation to physical activity, these implementation tools include resources on increasing physical activity at the population level, through a national school policy, and at the workplace. This chapter highlights two additional tools, the *Global Physical Activity Questionnaire* and the *framework to monitor and evaluate DPAS implementation,* that are useful for surveillance, monitoring, and evaluation activities.

In addition to the existing tools, WHO is currently developing global recommendations for physical activity. The aim of providing global recommendations on physical activity is to assist policy makers in the development of national guidelines for health-enhancing physical activity. The guidelines will provide a focal point for policy development and also provide a base from which a national plan on physical activity can be implemented and evaluated. The guidelines under preparation will focus on the amount and types of physical activity that promote health based upon best available evidence; reflect on the various types of activity (aerobic, strength, flexibility, balance) and amounts (frequency, duration, intensity); distinguish three main age groups (youth, adults, and older adults); and address the different domains of physical activity that must be explored for the achievement of recommended levels in all age groups (leisure, occupational, transportation).

The chapter has outlined how the existing tools are disseminated to different WHO regions and countries through multisectoral capacity-building workshops. Recent data indicated that 29 countries had adopted multisectoral strategies and plans on healthy diet and physical activity based on DPAS, and WHO's future activities aim to actively increase the number of countries implementing DPAS.

In 2008, the World Health Assembly endorsed an action plan on prevention and control of NCD. Overall, the objectives for DPAS implementation are integrated into this action plan, and DPAS itself is highlighted as a key strategy to address unhealthy diets and physical inactivity globally. The endorsement of this action plan reinforces WHO and its Member States' commitment to preventing and controlling NCD.

84

The Role of Government in the Promotion of Physical Activity

Art Salmon, EdD

> "There should be no doubt that maintaining a healthy weight must be the responsibility of individuals first—it is not the role of Government to tell people how to live their lives and nor would this work. Sustainable change will only come from individuals seeing the link between a healthy weight and a healthy life and so wanting to make changes to the way that they and their families live.
>
> The responsibility of Government, and wider society, is to make sure that individuals and families have access to the opportunities they want and the information they need in order to make healthy choices and exercise greater control over their health and their lives."
>
> **Prime Minister Gordon Brown**
> **"Healthy Weight, Healthy Lives: A Cross-Government Strategy for England"**
> **January 2008**

From the opening quote, one could conclude that governments do have a role in promoting physical activity as a key component of a healthy public policy platform. In support of this position, the late Dr. Michael DeBakey of the Baylor College of Medicine stated that it is important for governments to ensure the health of the people as "is strongly implied in the American policy established by our founding fathers based on the concept of natural law and natural rights emanating from the Enlightenment and establishing our 'right' to life, liberty, happiness, and property. Health is necessarily encompassing in this concept, for without health 'life', in its popular sense, and the attainment of all these 'rights' is extremely difficult, if not impossible" (1; p. 153).

In most jurisdictions, the promotion of physical activity is seen to, within the context of public health, by government, whose mandate is to improve the health of the population through health promotion, disease prevention, and health protection. Public health generally refers to the set of government-funded services whose interest is to improve and protect the health of the public (2). In the past, government's efforts in public health focused on such efforts as safe drinking water, disposal of sewage, and wide-scale immunization. Today, while efforts in the area of infectious diseases continue, public health policy is more and more focused on addressing challenges and issues related to chronic disease and injuries, healthy child development, family and community health, and environmental health.

Global Governmental Initiatives

Governments at the national, regional and state, and local level are being challenged by dramatic increases in the incidence of chronic diseases including obesity, type 2 diabetes, cardiovascular disease, and various forms of cancer. With the recognition that almost without exception, no jurisdiction is immune to the epidemic of chronic disease, it has become a global imperative that governments develop effective policies, strategies, and programs to address this pressing issue.

In May 2004, the 57th World Health Assembly endorsed the World Health Organization (WHO) Global Strategy on Diet, Physical Activity and Health. In releasing its strategy, the WHO noted that unhealthy diets and physical inactivity are among the leading causes of the major noncommunicable diseases and that "governments have a central role, in cooperation with other stakeholders, to create an environment that empowers and encourages behavior changes by individuals, families and communities, to make positive, life-enhancing decisions on healthy diets and patterns of physical activity" (3; p. 38).

The WHO Global Strategy has four primary objectives that point to appropriate actions by governments at all levels (3):

1. Reduce the risk factors for noncommunicable diseases that stem from unhealthy diets and physical inactivity by means of essential public health action and health-promoting disease prevention measures.
2. Increase the overall awareness and understanding of the influences of diet and physical activity on health and of the positive impact of prevention initiatives.
3. Encourage the development, strengthening, and implementation of global, regional, national, and community policies and action plans to improve diets and increase physical activity that are sustainable, comprehensive, and actively engage all sectors, including civil society, the private sector, and the media.
4. Monitor scientific data and key influences on diet and physical activity; support research in a broad spectrum of relevant areas, including evaluation of interventions; and strengthen the human resources needed in this domain to enhance and sustain health.

Objective 3 includes the idea that collaborative action is required at multiple levels by numerous partners and that coordinated action of governments at all levels is critical if a positive impact is to be achieved in public health. The WHO suggests that governments have a primary leadership role to play in moving a physical activity strategy forward to a successful conclusion. The development of an overarching national strategy is often an effective way to harmonize efforts within a country and ensure that a focused approach is taken across various jurisdictions. As an example, in Canada in 2002, the need for a national healthy living strategy was expressed by federal and provincial–territorial ministers of health who were seeking a coordinated approach to reduce noncommunicable diseases. As a result, the Integrated Pan-Canadian Healthy Living Strategy was announced in 2005 (4). Similarly, in 2008, the British government announced a comprehensive national obesity strategy, "Healthy Weight, Healthy Lives: A Cross-Government Strategy for England." The aim of the strategy is to address "a growing problem of lifestyle diseases of which obesity is the foremost, creating a future of rising chronic disease and long-term ill health" (5; p. iii).

In both documents, there is recognition that the national government alone cannot resolve the issues surrounding healthy eating and physical inactivity but rather must work in partnership with many other stakeholders. Even within governments, there is now widespread recognition that it is not solely the responsibility of the department of health or the ministry of sport and recreation to implement a strategy to promote active living. Other government departments with key roles to play include agriculture and food, social services, education, children and youth services, industry and trade-commerce, transportation, and the environment. When numerous governmental departments embrace and support a fully integrated policy and a "whole of government approach" is taken, a much greater opportunity exists to achieve success.

Governmental Roles, Strategies, and Tools

Accomplishing the objectives outlined by the WHO will require a comprehensive and coordinated effort by national, state and provincial, and local governments in partnership with the private sector, nongovernment organizations (NGOs), the mass media, and, very importantly, the general public. The formation of sound public policy that is embraced by the general population is an important role of government. Fundamentally, it is the role of government in policy development to

- examine, through surveillance and monitoring, the severity of a problem (in this case the rise in the prevalence of obesity and the level of physical inactivity within the population);

- through research, determine the certainty with which the causes and solutions for the problem are known;
- determine the amount of public funds to be expended; and
- determine the strategies or methods of implementation to be undertaken to resolve the problem (6).

Often governments seek to examine the severity of a particular issue or problem through systematic consultations in order to determine an appropriate course of action. In Canada and the United Kingdom, for example, House of Commons committees were established earlier in this decade to examine and recommend strategic approaches to address childhood and adult obesity. The Canadian Standing Committee on Health sought to determine trends in childhood obesity; understand the determinants that influence childhood overweight and obesity rates; and identify existing and potential roles for the federal government in addressing this issue (7). When governments take this type of approach, they typically invite individual experts and stakeholder organizations to share their views on effective approaches that the government should take to address the particular issue under examination. The resulting government report is often viewed as a "blueprint" for future action and establishes a link between proposed government policy and interested stakeholders who have the capacity to assist in implementing strategies at the local level. In Great Britain, after consultations and a comprehensive scientific literature review, the government announced that it will focus its efforts to combat obesity in five areas: promoting children's health; promoting healthy food; building physical activity into daily life; supporting healthy workplaces; and providing effective treatment and support when people become overweight or obese (5).

As governments seek informed perspectives in the formation of a policy direction, it is critical that any action be based on sound scientific evidence. In order to ensure that an evidence-based scientific foundation exists for policy formation, a key role of governments is to support and nurture the research community so that an accurate assessment of the causes of and potential solutions to an issue can be developed. As an example of research forming the basis of a government policy statement, several countries, including the United States, Australia, and Canada, have developed national physical activity guidelines. The national government in each case saw the need to lay out a specific set of recommendations that, when adopted by sport and physical activity organizations, would reduce the possibility of promulgating confusing messages that might undermine efforts to promote physical activity. Often governments see one of their primary roles as communicating simple, clear, and concise health messages and as such will develop national guidelines or adopt internationally respected recommendations from organizations like the WHO or the Centers for Disease Control and Prevention in the United States. In conjunction with the establishment of guidelines or recommendations, governments often undertake to establish and coordinate a national social marketing campaign. For over 30 years, the government of Canada supported the national ParticipACTION campaign (www.participaction.com), which encouraged Canadians to "Walk a Block a Day" or "Just Do It" (8). Similar national campaigns have been undertaken in Brazil (Agita Sao Paulo), Australia (Active Australia; www.healthyactive.gov.au), and the United States (The President's Council on Physical Fitness and Sports; www.fitness.gov/).

There are undoubtedly many challenges confronting any government as it attempts to improve public health. Thankfully there are also many policy levers, resources, and tools available to governments to influence the decisions and actions of individuals, groups, and organizations. One such lever involves tax policy. Governments can help create environments or conditions that support physical activity by implementing "physical activity–friendly" tax policies. For example, employers may be given tax incentives for establishing workplace physical fitness initiatives; parents may be able to reduce their personal income tax through a system of tax credits for enrolling their children in sport and physical activity programs; and corporations may be influenced to invest in community sport and recreation facilities if tax incentives are in place to support private sector infrastructure investment.

Establishing national policies and strategies is an important first step, but it is only a first step. The role of municipalities or local governments in combating obesity through increasing physical activity cannot and should not be overlooked. Dimitris Avramopoulos, Minister for Health and Social Solidarity of Greece and the former mayor of Athens, noted that "no country can tackle the challenge of obesity and create favorable conditions for physical activity without recognizing the important role of local government. I am convinced that this is an area that Ministers of Health should find Mayors as natural allies in their efforts to promote healthy living and to prevent and reduce obesity in the young and adult population" (9; p. 2).

Governments at the local level play a critical role in creating environments that promote safe and attractive opportunities for physical activity. Senior government officials and bureaucrats can provide leadership, legitimacy, and an enabling environment for developing and implementing policies that support active living for all citizens (9). For example, a considerable number of studies have shown a relationship between physical and social environmental factors and physical activity. Local governments play a critical role in ensuring that the design elements in the built environment, such as street layout and lighting, land use, and the location of recreation centers and parks for recreational and competitive sport, meet the needs of their citizens. Other factors, such as personal safety (crime, road safety, etc.), environmental pollution, and public transportation planning that encourages walking and cycling, are also key elements that local governments must address if they are to be successful in encouraging higher levels of physical activity among children and adults.

Conclusion

Any government that wants to change the behavior of its citizens in response to a perceived health risk can take a number of different approaches. It can undertake a mass media campaign to encourage change in individual behavior. It may also introduce enabling measures that promote change within the population, such as increasing or decreasing taxes to encourage particular behaviors. Finally, governments can introduce laws that restrict or regulate behavior such as recently announced smoking bans in public places. Whatever the approach taken, governments at all levels must work together to increase participation in health-enhancing physical activity. No one level of government has the exclusive responsibility to take on this task; what is needed is a multilevel approach with coordinated action to ensure success.

85

Community-Based Physical Activity Programs to Address Obesity

W. Kerry Mummery, PhD

> "It makes little sense to expect individuals to behave differently from their peers; it is more appropriate to seek a general change in behavioral norms and in the circumstances which facilitate their adoption."
>
> **Geoffrey Rose, 1992**

The etiology of overweight and obesity at the level of the individual is complex, and not entirely understood. At the population level the picture is somewhat clearer: Energy intake is outpacing energy expenditure. The resulting energy imbalance can simplistically be considered a cause of the current obesity problem in developed societies. The prevalence and costs of adult and childhood obesity are discussed elsewhere in this book, but it is safe to say that obesity is a societal problem of the largest order. If one is to accept this argument, it is apparent that the next logical step would be to identify approaches to address the problem with the explicit aim of reducing the prevalence of obesity in modern society. This chapter discusses community-based physical activity programs and their role in addressing obesity. A rationale for community-level physical activity interventions is presented; key issues relating to the implementation and delivery of community-based programs are discussed; and examples are given from an Australian whole-of-community physical activity study, the 10,000 Steps Rockhampton project (1-3).

There is no need at this point in the book to define physical activity or obesity, but it is likely necessary to consider what is meant by a community, at least for the purposes of this discussion. In its broadest sense, a community could be thought of as almost any amalgamation of individuals who are bound together by some common thread; but when it comes to the logistics of implementing a planned program of intervention or change, it is much more convenient to think of a community in terms of people, place, and governance. A community, by this definition, would be a definable group of people who reside in a specific locality that has a level of government charged with the responsibility of servicing the collective. In Australia, by example, there are three levels of government: federal, state, and local (municipal) governments. The local government generally constitutes the community level, which could be a city, a town, or perhaps a suburb, although suburbs often lack a unique system of governance. The community is the entity that people may define themselves by, at least geographically. Communities generally have a diverse range of constituents and stakeholders, from places of work to places of worship and recreation.

Importance of Community-Level Interventions for Physical Activity

Population-based concerns relating to physical inactivity as a behavioral determinant of overweight and obesity are, in fact, related to the sum of individual outcomes. Current approaches to the problem range from individual-level interventions—either driven by a motivated individual or initiated by a supportive clinician— to population-based actions normally founded in promotional, policy, and environmental activities by national or regional (state, territorial, provincial) governments or both.

Traditional individual-centric approaches to the treatment and prevention of obesity by means of increased physical activity (exercise) aim to identify high-risk individuals and apply some form of protective, preventive, or ameliorating treatment. This "high-risk" approach generally entails an intervention targeted at the individual in question and requires a high level of participant and clinician motivation along with a cost-effective use of resources (4). In what could be called the "biggest loser" approach to overweight and obesity, the current societal norm would likely be for individuals to utilize the services of medical doctors or exercise or related allied health professionals to provide a tailored program of activity or exercise aimed specifically at reducing or maintaining body mass. The advantages of an individually based approach are apparent. First, the program for the self- or other-identified participant can be tailored and theoretically appropriate to the individual, which frequently leads to (at least) short-term enhanced participant motivation. The individual approach also can be motivating for the professional service provider as participants are seeking help and the professional is providing it, which is often preferable to dispensing physical activity advice or exercise prescription to a person who is not contemplating change. Finally, individually based exercise or activity prescription offers an often favorable promise of some form of measurable outcome, which helps to complete and enhance the cycle of motivation and behavior.

There are, however, notable limitations to the individual-centric approach. The approach is cost intensive and often not generalizable to the population. It also treats the symptom rather than the cause of inactivity, which exacts an ongoing motivational toll on the participant. The endpoint of this toll can be relapse to previous low levels of activity. Finally the individual-centric approach to promoting activity may be behaviorally inappropriate. As suggested in the introductory quote from Geoffrey Rose (4), individual behavior is constrained and influenced by social norms, making it more appropriate—and ultimately more beneficial at the population level—to change societal norms in terms of physical activity and exercise than to deal only with the individuals who suffer from the detrimental effects of a lack of activity.

By contrast, population-based approaches seek to positively affect the determinants of physical inactivity or obesity and address them in the whole of the population. Once again, the benefits are logically apparent. Effecting a positive shift in societal norms regarding physical activity offers the promise of large-scale returns in terms of the prevalence of overweight and obesity, whereas a small shift in the percentage of the population accumulating sufficient levels of physical activity can result in notable reduction in levels of overweight and obesity (5). More importantly, the approach, to use Rose's terminology, is "behaviorally appropriate." Through addressing the causal determinants of inactivity and effecting change in societal norms, a subsequent reduction in motivational toll is achieved. This reduction in motivational toll offers the promise of ongoing behavioral change with reduced need for intervention, thus achieving a motivationally sustainable behavior at the population level.

Still, the population-based approach is not without its limitations, principal among which is the "prevention paradox" (6) whereby approaches that may achieve large benefits at the level of the population result in little noticeable benefit to the individual. This can be noted in the ongoing debate relating to the evolution of physical activity recommendations (7); here, a public health approach to the problem aims to move more of the population a little (i.e., 30 min of moderate physical activity on most days of the week) rather than high-risk individuals a great deal more. These resultant generic, often subtherapeutic recommendations can leave individuals in need of specifically tailored physical activity prescriptions confused and their related clinical provider at apparent odds with public policy.

A desire to blend the benefits of the individual-centric approach with that of a broader population view suggests the community as a feasible option in terms of both the level and type of intervention aimed at increasing physical activity to address problems with overweight and obesity. The community is the convergence of individual, local, regional, and national influences, including aspects such as public transport and town planning, health care and recreational services, government and nongovernment agencies, and print and broadcast media, as well as the individuals' places of work, education, and worship—all of which may be brought to bear on individual and community levels of physical activity. The community can, therefore, potentially obtain the best of both worlds in terms of individual-centric and population-based approaches to promoting increased levels of physical activity. Such community-level approaches often operate effectively in the form of "top-down, bottom-up" strategies (8) that strike a balance between the broader needs of the population (top-down) while engaging community-level coalitions that can develop local solutions to local problems (bottom-up). Although this convergence offers much promise, there remains a need to coordinate and drive the initiatives effectively across multiple stakeholders in order to leverage the success that single groups or partners could expect to accomplish on their own. This coordination normally takes the role of some form of community coalition, network, or steering committee, which actively seeks to mobilize a broad spectrum of community partners to improve the overall health of the community.

Role of Community Coalitions in Physical Activity Promotion

Work at the community level is driven by a series of collaborations between stakeholders with a direct interest in physical activity and health and other partners or parties who have the potential to benefit from the outcome of the collaborative efforts. The coalition becomes the active agent of change—engaging the community and its members, seeking local answers to decreased levels of physical activity and increased levels of obesity, representing the varied interests of its members, and offering some promise of sustainability through diversity and ongoing engagement. Certainly it is unlikely that any community-based physical activity program has any chance of success without an organized collaboration among stakeholders and benefactors. Of course not all coalitions or networks achieve their objectives. Sometimes this is a result of inappropriate strategies or insufficient resources. In other instances, it may just be the failure of the coalition itself to effectively facilitate the promotion of activity at the community level. Recent research indicates that coalition effectiveness may be enhanced by the development of formal governance structures, the presence of strong leadership, the active participation of members, the nurturing of diverse group membership, the promotion of interagency collaborations, and the facilitation of group cohesion (9).

10,000 Steps Rockhampton

Rockhampton is a major regional center in Queensland, Australia. Approximately 700 km north of the capital, Brisbane, Rockhampton had a population of almost 60,000 people and serviced a region of over 180,000 persons at project start in 2001. Although Rockhampton is small by contrast to regional communities in many countries, its size, the existence of local media (print, radio, and television), and the geographic distance from major metropolitan centers made it an excellent and economically efficient "test market" for the development and evaluation of a whole-of-community approach to the promotion of physical activity.

The 10,000 Steps Rockhampton project began life as a research grant submission made by a community consortium led by academics from three Australian universities (1). Guided by existing evidence regarding effective approaches to the promotion of physical activity (10), the project utilized a local physical activity task force (community coalition) to implement a series of preconceived and concurrently developed strategies aimed at

1. promoting physical activity through general practices (medical doctors) and other health services,
2. improving social support for physical activity among disadvantaged groups, and
3. affecting policy and environmental change to support physical activity within the community.

In order to ensure a bottom-up approach to community engagement, a grass-roots microgrant scheme was also implemented to support physical activity initiatives from informal groups within the community who required small amounts of start-up funding normally unavailable to them. The project's overall objective was to increase levels of incidental physical activity by promoting the accumulation of movement (10,000 steps) throughout the day as monitored by the use of a pedometer. The entire project was promoted under the umbrella of a local media campaign using print, radio, and television to publicize the distinctive logo and "brand" of 10,000 Steps Rockhampton (figure 85.1).

10,000 Steps Rockhampton ran as a community-based intervention project from 2001 through 2003; subsequent evaluation showed high levels of project awareness and increases in physical activity levels, especially in women, compared to those in a matched comparison community (2). Results of regular population surveillance studies in the city have shown a small, nonsignificant decline in the percentage of the population classified as overweight or obese during that time frame. Since 2003, the project has evolved into a program of support that works to promote physical activity at the

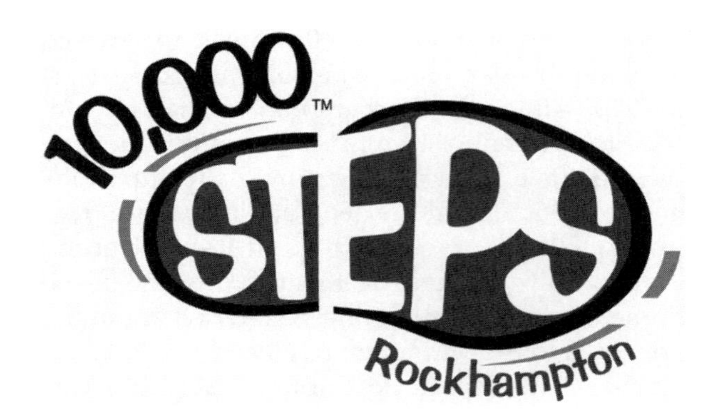

Figure 85.1 Logo from the 10,000 Steps Rockhampton Project.

© Kerry Mummery

community and individual level under the 10,000 Steps brand (3). Five years after the completion of the initial intervention study, awareness of the program remains high in Rockhampton. The ongoing existence of a multiagency regional physical activity task force (Physical Activity Capricornia Taskforce or PACT) and the presence of 10,000 Step–themed walking trails and signage, as well as support of the local council, remain as legacies of the original program. As of submission of this chapter, more than 112,000 registered members were using the existing program Web site (www.10000steps.org.au), which arose from the original community-based intervention and provides support and resources for the promotion of physical activity at the individual, workplace, and community levels (figure 85.2).

Figure 85.2 A screenshot from the Web site developed for the 10,000 Steps Rockhampton Project.

© Kerry Mummery

Conclusion

The community, however defined, offers a unique setting to address the ills of modern society in terms of physical activity and obesity. As a natural convergence between individual-level attributes and population-level factors, the community can be a vehicle for multilevel, multistrategy approaches to improving levels of physical activity in the person and the population. Community-based interventions to promote physical activity involve the development of strong strategic partnerships and the formation of mutually beneficial coalitions and networks to promote physical activity. If successful, they offer the promise of establishing behavioral norms for physical activity whereby an active lifestyle becomes the rule, rather than the exception, in society. Existing physical activity research at the community level is sparse, however, and issues relating to units of analysis and identification of best-practice strategies remain—which means that more work is required to strengthen the evidence base in terms of community intervention. Even so, the practitioner should not be dissuaded from engaging in collaborative action at the community level with the intent of facilitating the adoption of physically active lifestyles by the majority of the population.

86

School-Based Physical Activity Programs to Address Obesity

Chantal Simon, MD, PhD

Many of the societal and cultural changes of the last decades that we view as progress reduce the amount of energy expended by children in everyday life, during leisure-time activities and in schools, and so collectively contribute to the growing levels of obesity in youth. Consequently, physical activity (PA), which is inversely related to overweight and to the associated metabolic and cardiovascular risks, both in youth and in later life, has been thought to be a key element in addressing this epidemic. Since PA patterns and obesity track from childhood to adulthood, promotion of PA in youth may be especially efficient and potentially cost-effective if designed for the long term.

Although out-of-school PA is an important part of total PA, schools play a central role in educating youth toward PA and providing them with opportunities to practice. Recent trends, however, indicate that PA as traditionally provided in schools is no longer sufficient to compensate for the decline in everyday-life PA. Even more, data in some countries show a decrease in compulsory physical education (PE) classes and specifically in time devoted to activities of moderate to vigorous intensity.

Nevertheless, schools are an important avenue for delivery of preventive PA programs at a population level, for at least three reasons. First, schools are a unique way of reaching and educating all the children of a country, including lower socioeconomic sections of the community known to be particularly vulnerable to increases in prevalence of inactivity and of obesity. Second, youngsters spend most of their waking time in schools. Third, schools offer several avenues, not limited to academic curriculum and PE classes, for learning and making decisions about PA behaviors and for promoting PA (1). These include health curricula but also PA during breaks, recesses, lunch, and after-school time. The potential to increase the energy expenditure goes beyond the school day and includes walking and biking to and from school or the use of school facilities as a community resource. Lastly, schools can also be a means to reach families, an important facet of obesity prevention.

Today, the question is not whether we should target PA to address overweight and obesity in youth, but rather what preventive measures to use. As underlined by the latest systematic reviews of controlled interventions for obesity prevention (2, 3), the evidence base documenting effectiveness of programs remains limited, whatever the setting of the intervention. Nonetheless, new well-designed intervention studies to promote PA (aimed at preventing overweight or not) suggest that school-based interventions with specific involvement of the family or community, and multilevel interventions combining educational and environmental approaches, can indeed increase PA (4). In this chapter, it is not my purpose to update the systematic reviews already available but rather to explore the extent to which the PA interventions are associated with favorable effects on weight status. The chapter (1) identifies and discusses the elements of school-based PA interventions that are likely to succeed in addressing overweight, with a special focus on recent multilevel studies that target environmental factors; and (2) highlights unresolved issues, offering guidance for future investigations.

Summary of Evidence Base

This section reviews the available evidence base on existing school-based PA interventions in children and youth.

Sources of Data

Controlled school-based PA intervention studies that have examined weight status as an outcome measure were identified from recent systematic reviews on PA promotion (4) or obesity prevention (2, 3), with the following inclusion criteria: a control group, follow-up for at least one academic year, nonclinical groups of subjects, a minimum of 60 subjects per group, and at least one anthropometric measure of weight status. Recent additional studies were located through a Medline search from January 2007 to April 2008. The one-year follow-up limit was chosen because the literature consistently shows that short-term lifestyle changes are relatively easy to obtain but far more difficult to sustain. Moreover, long-term behavior changes may be necessary to allow observation of a significant effect on body weight or body composition.

Characteristics of the Studies

Table 86.1 details the 21 identified studies (17 referred to in previous systematic reviews and four additional recent studies). Two-thirds took place in the United States and the others in Europe. Six studies targeted PA alone, while the others targeted both PA and nutrition.

Enhancing Physical Activity Tends to Prevent Overweight

About 50% of the studies (10 of 21) reported at least one statistically significant favorable effect of PA on body mass index (BMI) or body composition. Interestingly, more than half of the studies were published over the last three years; and 72% of them showed a significant effect on weight status.

Due to the diversity of the methods used, the variety of the populations targeted, and the multifaceted nature of the designs, evaluation and comparison of the literature on school-based interventions addressing overweight are difficult. While some studies focused on PA, others addressed various health behaviors, including different weight-related behaviors such as food choices. Moreover, differences exist in the approaches used (informational, behavioral and social, environmental, or a combination of these) and in the number and types of PA targeted. It is often impossible to determine which components were effective. Also, the process evaluation, the quality of program implementation, and subjects' attendance and compliance are incompletely or inaccurately reported.

That said, observation of a beneficial intervention effect on weight status in four out of the six studies that focused on PA alone confirms the potential for PA programs to effectively prevent excessive weight gain.

Physical Activity Promotion Strategies

This section reviews interventions according to the main approach employed. Studies either used a health curriculum–based approach, implemented social and behavioral strategies mainly directed at modifying PE classes, or used a multilevel approach associating extracurricular PA and environmental targets with the other two approaches

Health Curriculum–Based Strategies

Three studies evaluated health curriculum–based interventions used alone. Two of these (Walters 1988; Bush 1989; see table 86.1) were long-term middle school interventions focusing on risk factors associated with coronary heart disease; the third one, conducted in elementary schools, targeted diet, PA, and overweight through lunchtime classes involving parents (Warren 2003; see table 86.1). None of these studies showed any significant effect on PA level or BMI. In general, health education–based strategies that focused on the individual have improved health knowledge but have shown their limits with regard to permanent changes in behavior and impact on weight status. An exception is interventions that included a component aimed at reducing TV viewing. Four school-based interventions included a curricular component to reduce sedentary behavior (5), either as part of a broader program (Gortmaker 1999; see table 86.1; [6]) or used alone (two short-term studies). All four studies succeeded in reducing TV viewing and preventing excessive weight gain as compared to values in control schools. Although the combination with other intervention components precludes definitive conclusions, the consistent effect across studies strongly suggests that efforts to reduce sedentary behaviors may be an effective means to reduce overweight.

Physical Education Programs

Seven studies in elementary schools and two in middle schools essentially modified curricula to

Table 86.1

Selected Controlled Long-Term School-Based Physical Activity Programs to Prevent Obesity

Authors, name or acronym, and country	Participants*, follow-up (spec. target)	Intervention components						PA outcomes**				Obesity outcomes**		
		Nut	Ed	PE	Other PA	Fam	Env	PE	Other PA	FIT	TV	BF	Obesity	BMI
ELEMENTARY SCHOOLS														
Luepker 1996[a], CATCH, U.S.	*n* = 5106, 3 years	x		x		x		x	x			ns		ns
Sallis 1997[a], SPARK, U.S.	*n* = 955, 2 years		x	x				x	ns	x		ns		
Sahota 2001[a], APPLES, U.K.	*n* = 634, 1 year	x	x		x				ns		ns			ns
Warren 2003[a], Be Smart, U.K.	*n* = 213, 14 months	x	x			x			ns				ns	
Caballero 2003[a], Pathways, U.S.	*n* = 1704, 3 years (ethnic minority)	x	x	x		x			x			ns		ns
Fitzgibbon 2005[a], Hip-Hop to Health, U.S.	*n* = 409, 2 years (ethnic minority)	x	x		x	x			ns		ns			x
Coleman 2005[a], El Paso CATCH, U.S.	*n* = 896, 3 years (ethnic minority; low SES)	x	x	x				x		x			x	ns
Manios 2005[a], Greece	*n* = 1046, 10 years	x	x	x				x	x				ns	x
Fitzgibbon 2006[a], Hip-Hop to Health, U.S.	*n* = 420, 2 years (ethnic minority)	x	x		x	x			ns		ns			ns
Graf 2008[10], CHILT Project, Germany	*n* = 615, 4 years		x	x						ns			ns	
Sollerhed 2008[11], Sweden	*n* = 132, 3 years			x						x				x
Gutin 2008[8], FitKid Project, U.S.	*n* = 525, 3 years	x			x					x		x		Higher

(continued)

Table 86.1 *(continued)*

Authors, name or acronym, and country	Participants*, follow-up (spec. target)	Intervention components						PA outcomes**				Obesity outcomes**		
		Nut	Ed	PE	Other PA	Fam	Env	PE	Other PA	FIT	TV	BF	Obesity	BMI
MIDDLE AND HIGH SCHOOLS														
Walter 1988[a], U.S.	*n* = 3388, 5 years	x	x						ns					ns
Bush 1989[a], U.S.	*n* = 431, 4 years	x	x						ns			ns		ns
Donnelly 1996[a], U.S.	*n* = 338, 2 years	x	x	x				x	Lower	ns		ns		ns
Gortmaker 1999[a], Planet Health, U.S.	*n* = 1560, 2 years	x	x	x							x		x in girls	
Sallis 2003[a], M-SPAN, U.S.	*n* = 26,616; 1434 with BMI, 2 years	x		x	x		x	x						x in boys
Pate 2005[a], LEAP, U.S.	*n* = 2744, 1 year (girls)		x	x	x	x	x		x				ns	
Haerens 2006[a], Belgium	*n* = 2840, 2 years	x	x		x				x					x in girls
Simon 2008[6], ICAPS, France	*n* = 954, 4 years		x		x	x	x		x		x	x		x
Martinez Vizcaino 2008[12], Spain	*n* = 1119, 1 year				x				x			x		ns

Nut, nutrition; Ed, education; PE, physical education; PA, physical activity; Fam, family; Env, environment; FIT, fitness measures; TV, television; BF, body fat; BMI, body mass index.
[a]Complete references are to be found in the systematic obesity and PA reviews[2-5]; *participants at inclusion; **x = significant beneficial effect; ns: nonsignificant.

increase the amount of time students spent in moderate or vigorous activities during PE classes, either by providing additional classes or by changing the content of the PE classes. All seven of these interventions also included a health education–based curriculum, and six targeted both PA and nutrition. Physical education programs showed consistent increases in PA during PE classes, with, in almost all the studies, an improvement in fitness, endurance performance, motor skills, or more than one of these. However, their effectiveness in relation to out-of-school PA and weight status has been questioned; some studies even showed a compensatory decreased PA outside of the classes (Donnelly 1996; see table 86.1). Similarly, two studies designed to reduce gains in BMI in preschool minority children, which included three 20 min PA sessions weekly in association with an educational component, had no effect on PA and sedentary behavior at years 1 and 2 of follow-up (Fitzgibbon 2005, 2006; see table 86.1).

Only four of the studies discussed here had a significant impact on weight status. In one (Gortmaker 1999; see table 86.1), the reduction of obesity prevalence was attributed to a reduction in TV viewing rather than an increase in PA. Two others were studies that proposed the highest levels of compulsory PA (90 to 120 min in addition to the 2 h of PA in the normal curriculum). Taken together, these data agree with findings in a recent Australian cohort study (7) that the amount of compulsory PA reported by schools is not associated with total PA and that, if used alone, more than 220 min of compulsory PE weekly might be required to decrease the prevalence of obesity in childhood.

Multilevel Programs

Recently, the importance of taking into account the sociocultural, physical, and policy environment in which students live has been emphasized. According to the socioecological model, education-based interventions associated with social support and environmental changes that reduce the barriers to adopting an active lifestyle have a higher potential for changing PA habits. In the context of schools, apart from PE classes, the main targets of environmental components have been leisure PA within the school and active commuting to school. Five controlled studies based on this approach have examined anthropometric measures as an outcome, one in elementary schools (8) and four in middle or high schools (Sallis 2003; Haerens 2006; Pate 2005; Simon, 2008; see table 86.1; [6]). In all studies, the intervention was associated with an increase in moderate to vigorous PA as compared to values in the control group. All but the study conducted in high schools (which had a follow-up less than two years) (Pate 2005; see table 86.1) had a significant impact on weight status.

An example of interventions based on a socioecological perspective is the Intervention Centered on Adolescents' Physical activity and Sedentary Behavior (ICAPS) study (6), a four-year randomized controlled trial conducted in French middle schools. This study emphasized the dynamic interplay among personal factors, behaviors, and social or physical environmental influences and postulated that interventions should be most effective when targeting each level. The theory-based multilevel program, designed to enhance in- and out-of-school PA, involved not only the school settings but also numerous partnerships, with three objectives: (1) changing attitudes and motivation toward PA, (2) promoting social support by parents and educators, and (3) providing environmental and institutional conditions that encourage adolescents to use the knowledge and PA skills they have acquired.

The program that complemented the usual PE classes included an educational component focusing on PA and sedentary behaviors. New opportunities for extracurricular supervised PA were offered during leisure time throughout the school day, either on or near the campus. The focus was on enjoyment of participation to encourage less confident children to develop the competencies needed to adopt an active lifestyle. Cycling-to-school days were organized to affect active commuting to school. Parents and educators were encouraged to provide support to enhance the adolescents' PA level. Policy makers in local communities were asked to help in providing a supportive environment that promotes enjoyable PA. At four years, compared with controls, intervention adolescents had an increase in supervised PA of about 1 h weekly and a decrease in TV or video viewing of 16 min per day. Furthermore, the program had a sustained preventive effect on excessive weight and fat mass gain in initially nonoverweight adolescents, with a 50% reduction in four-year overweight incidence, but did not produce a long-term weight loss in initially overweight students.

The four other studies were designed to increase more specifically extracurricular leisure-time PA in the school setting (FITKID [8], M-SPAN [Sallis 2003]; Belgian Study [Haerens 2006]; LEAP [Pate 2005]; see table 86.1). Targeted environmental changes were generally aimed at increasing supervision, equipment, and organized activities. The FITKID Project (8) tested an after-school program focusing on PA in children from elementary schools. Preliminary results showed that the youth who attended at least 40% of the sessions in each of the three years of intervention improved fitness and reduced body fat during the school years but returned to levels similar to those of the controls during the summers.

One study in Belgian middle schools evaluated the effects of a PA and food intervention, including an environmental and computer-tailored component (Haerens 2006; see table 86.1). The schools were assigned to three conditions: an intervention with parental support group, an intervention-alone group, and a control group. After two years, children in the intervention groups had an increase in leisure-time PA compared with control children. The intervention with parental support had also a beneficial effect on BMI in girls (–0.55 kg/m^2 vs. controls), but not in boys. The final two studies aimed at increasing PA both during leisure time on campus and in PE classes. M-SPAN (see Sallis 2003; table 86.1), a large randomized controlled trial conducted in U.S. middle school students, associated a nutritional with a PA intervention. There was no classroom health education, but both key school staff and students were engaged in policy change efforts, and there was also a family component. A significant reduction in BMI (–0.64 kg/m^2 vs. controls) in boys along with a significant increase in PA was reported at two years follow-up, while there was no significant effect for girls. The LEAP study (see Pate 2005; table 86.1) targeted U.S. high school girls, focusing on PA and including education classes as well as family and community components. At year 1, the intervention induced a significant increase in participation in vigorous PA but had no significant effect on overweight prevalence.

Although a definitive conclusion cannot be drawn because of the very few studies available to date, taken together these data provide promising evidence that school-based multilevel interventions integrating environmental components and targeting extracurricular PA have a high potential to increase total PA and to address overweight, at least in adolescents.

Keys to Success and Mediators of Effectiveness

Inevitably the design of the school-based studies makes it difficult to determine precisely the keys to success. However, it is possible to draw conclusions relating to the type of PA targeted, length of follow-up and outcome measures, parental involvement, and subject sociodemographic characteristics and weight status, as well as to identify questions that remain open.

Type of Physical Activity Targeted

A common objective of the multilevel interventions was to go beyond PE classes to directly affect noncurricular PA. Offering PA opportunities in the vicinity of the schools, as compared to attempting to increase leisure PA outside the structured school environment, probably contributed to increasing attendance in the sessions and to overcoming the most frequent barriers to PA (e.g., transportation, time constraints, and safety concerns). Setting the adolescents in motion through attractive, convenient activities was doubtless crucial for improving their attitudes and skills in relation to PA, which in turn favored their adherence to the activities. The development of strategies adapted to the age of the pupils and to the various environmental contexts—meeting specific needs of participants—was another important element. There is a need to further explore the superiority of practice supervision by qualified educators versus regular classroom teachers to guarantee that sessions meet the objectives (Sallis 1997; see table 86.1).

Other approaches described in reports on programs that did not meet our follow-up or outcome inclusion criteria, such as redesigning and painting school playgrounds (4), have yielded encouraging results and should be evaluated in future long-term studies. Active travel to school was targeted by only one of the studies reported here (6), with limited success. However, a recent intervention involving safe walking and biking to school (9) indicates that provision of active travel plans and creation of safe routes to school, associated with information, education, and motivation strategies, can favorably modify school travel patterns; this suggests that such a strategy might be usefully added to a multilevel school intervention.

Finally, although interventions on compulsory PE used alone have shown little effect on long-term total PA and weight status, PE represents an important part of moderate to vigorous PA in youth. Thus interventions targeting PE, extracurricular PA, active transport, and also sedentary behavior might be required to successfully address overweight in youth. It is clear that such broad interventions cannot be conducted by schools alone but require cooperation with several partners, including communities and sport clubs. Links with communities will also be essential to the implementation of year-round programs, as it has been suggested that school-based interventions lose part of their effectiveness during the holidays, and to the sustainability of programs.

Follow-Up Length and Outcome Measures

Implementation of programs takes time, especially those that include environmental actions or require the involvement of several partners. Moreover, data from different studies (6) indicate that interventions and changes in behavior must be sustained in order to affect body composition, and that even a one-year intervention and follow-up period may be too short to show any significant effect on BMI or overweight incidence.

The absolute magnitude of improvement in weight parameters in these studies is often modest. We should keep in mind that in growing nonoverweight children, even a small reduction in the rate of BMI increase as compared with that in controls may result in a substantial reduction in overweight incidence over time and have public health consequences. Also, since interventions that include a PA component may increase lean mass, BMI may be a less sensitive marker of the intervention effect than body composition measurements. All this could explain why some studies failed to demonstrate a significant effect on weight status despite a real improvement in PA level. Future studies should incorporate long-term measures of body fat as an outcome, also at distance from the intervention, so that retention of changes can be evaluated. Furthermore, researchers should aim to strengthen the evidence with use of objective measures of overall PA and energy expenditure.

Parental Involvement

Parents are important role models for health behaviors. An increasing number of school-based interventions incorporate a family component. Multilevel interventions that include some parental participation seem more successful than others, and it has been suggested that parental support is essential for the sustainability of school-based programs. However, the real level of parental involvement from one study to the next is often difficult to evaluate. The extent to which caregiver involvement contributes to effectiveness should be a focus in future research, as well as the strategy to be used.

Sociodemographic Characteristics of Subjects

Only 42% of the studies conducted in elementary schools were effective as compared to 62% of those conducted in middle schools. However, it should be noted that studies in children have mainly focused on PE and that only one multilevel study including an environmental component took place in elementary schools; this may partly explain the ineffectiveness of the interventions in the elementary schools. On the other hand, both cognitive and physiological development are likely to influence the impact of specific interventions on a given child. Interventions may be more effective in adolescents due to their increased control over leisure activities, greater concerns about appearance, or peer group–mediated changes in behavior. Also, since the amount of PA has been shown to dramatically decrease during adolescence, targeting these slightly older children as they begin to establish long-term health behavior may be more appropriate than targeting younger children.

Another important point relates to gender. Although some studies have shown differing results by gender, others were effective in both boys and girls. The need to tailor school-based interventions by gender warrants further research.

Lastly, obesity and physical inactivity prevalence differ according to socioeconomic and cultural backgrounds. Multilevel school-based interventions have the potential to have an effect in all children (6), including those of low socioeconomic status. However, interventions are just beginning to point to this issue, and only few studies have taken place in low-income populations or in ethnic minorities. It is therefore important that future studies verify whether the strategies used work at least equally well in less advantaged subjects or ethnic minorities in order to avoid an increase in health inequalities.

Weight Status

While effective in preventing excessive weight gain in nonoverweight students, several interventions were unable to induce a significant remission in obesity or a long-term weight loss in overweight subjects despite significant changes in PA levels. Although this result may be attributable to a smaller effect on out-of-school PA in overweight subjects, it suggests that long-term weight loss requires more PA than does prevention of weight gain, necessitating the use of specific complementary strategies, including dietary counseling. These findings also further support the idea that prevention of excessive weight in children who are not yet overweight may be a more effective approach than weight management in overweight and obese youth.

Conclusion

The few studies available provide encouraging data regarding the effectiveness of comprehensive school-based PA interventions that simultaneously address the school environment, the perception of the school environment and motivation toward PA, to prevent the development of overweight. It may be key for interventions to incorporate multiple facets and levels in order to increase PA levels (both inside and outside the school) in a large proportion of the targeted population and to achieve long-term maintenance of behavior and body composition changes. Future studies are necessary to evaluate to what extent school actions reinforce each other, not only within the school but also in the wider community. There is a special need for trials exploring whether such strategies apply to the youngest children, to other cultural contexts, and to countries with different school organization systems (most studies were conducted in the United States and in a few European countries). It is also crucial that studies include economic data and address cost-effectiveness, which are essential from a public health and decision-making perspective.

87

The Role of Industry in the Promotion of Physical Activity

John C. Peters, PhD

In much of the developed world, the majority of the workforce is employed by the private sector (referred to as "industry") in all manner of enterprises large and small. For example, in the United States, nearly 75% of the workforce is employed by the private sector (1), and over half of these people work for companies having more than 100 employees. The landscape is similar in Europe; and the largest country in Asia, China, is rapidly shaping up to fit a similar mold. Almost regardless of what minimum size one might select to qualify a private enterprise as being part of "industry," this employer segment as a whole is responsible for the livelihood of a large fraction of the population in developed countries. The productivity of these employees, and hence of industry, is the fuel for the global "prosperity" engine, the sum total of all goods and services produced by the global economy. It is this prosperity engine that is largely responsible for the good fortunes of each of us as individuals and as a collective global society. Maintaining the health of the prosperity engine is dependent on maintaining the health of the workforce, and an impressive body of evidence supports the positive role of physical activity in promoting good health, including the prevention and management of obesity (2).

Given this connection between employee health and the industrial economy, one big question is; what is an appropriate role for industry in promoting physical activity? That is, should industry focus only on those activities that can be tied directly to traditional measures of business success, like sales and costs? Or, should the focus extend beyond the business into activities that touch the lives of non-employees and nonconsumers of its goods? And, why would industry do this?

How Industry and the Public Are Connected

When we examine the potential role of industry in physical activity promotion, there are at least two perspectives to consider. First, what role does industry play in the direct service of its own interests? Secondly, what other role does industry play in contributing to the greater public good, sometimes called social responsibility?

Because industries have such a large presence in people's lives, from employing a large proportion of the population to shaping lifestyle behaviors through the marketplace, I would argue that there really is only one major role that they play and that serves both their business interests and their social interests. It begins with fostering good employee health. This leads to healthier business, which then leads to greater economic prosperity for more people, enabling communities to invest more in programs and facilities that support better health in the community (e.g., sidewalks, parks, trails). And, because these activities directly touch the lives of the greater public, a benefit accrues to the industries involved, which completes the cycle. In the big picture, what is good for the public is good for the industry, and what is good for industry is good for the community.

I am sure that average consumers do not think they are connected to private sector industries and likely do not believe that they have much say in what the industry produces or promotes. However, as in the political process in many countries, individuals do exert power in the workforce and in the market through their collective voice and behavior; and it is this power that we need to tap into in order to improve physical activity patterns in the general population.

Should Industry Promote Physical Activity?

It may be useful to break the business interests of industry into several components in order to understand what role they might play in physical activity promotion: (1) reducing expenses (e.g., health care costs), (2) improving income (sales), and (3) improving corporate equity. All three of these elements directly benefit an industry because they affect overall profitability. And, in this highly competitive, short-term–gain driven market, profitability is essential to remaining in business whether a business is publicly or privately held.

Perhaps one of the most desirable direct benefits to industry of physical activity promotion would be reducing employee health care costs (whether in a country with nationalized health care or not). Evidence is convincing that physical inactivity contributes to greater health problems and higher costs (2, 3). In addition, many studies have provided evidence that comprehensive health promotion programs in the workplace can reduce employee health care expenditures (4), and these programs generally have physical activity promotion as one element. Fewer studies have examined the benefits of physical activity alone in reducing employee health care charges (5), so it is difficult to know what unique contribution physical activity makes to lower costs in these settings.

What about improving worker productivity, which should have an impact (although perhaps one that is hard to measure) on company sales? Data linking physical activity levels and productivity are extremely limited and have not provided a compelling picture of benefit (6). However, there is some evidence that physical activity reduces absenteeism, which logically should relate to productivity (7). It is understandable that these and other measures linking employee costs and employee productivity would be primary drivers of industry's interest in physical activity promotion.

In addition to focusing directly on employees, most large companies today engage in specific activities to build their corporate equity, the total value of everything they stand for in the minds of their customers and their business partners (e.g., supply chain). This builds loyalty and respect; and when a company is respected for what it does, the respect gives customers one more reason to buy its products and tell friends about it. With the tremendous reach of the Internet, the explosion of information available to anyone at any time has increased the need for transparency of for-profit and not-for-profit activities alike, such that consumers increasingly care about how a company does business and not just what it sells. So, even corporate equity-building activities, which may seem somewhat remote to the primary business of the company, have some impact on sales, although it is harder to measure their fractional contribution with traditional metrics.

Cannot a business simply give money to a good cause, like physical activity promotion, and expect nothing in return? Certainly. But not many industries give money anonymously for causes, whether or not the cause is related to their business. Any support that comes with no expectations, but is not anonymous, would contribute to building corporate equity and therefore help build overall success and profitability. The point is that because everything is interconnected and interdependent, most actions of industries, whether aimed directly at fueling their primary business or not, have an impact on their profitability, one of the key markers of long-term sustainability.

Why Physical Activity Promotion Is Not the Norm in the Workplace

Most large companies have health promotion or wellness programs that include physical activity components. However, many smaller businesses do not offer such programs. One likely explanation is that the evidence base supporting a benefit is not widely recognized and that the evidence for some benefits is not yet convincing. This is especially true for smaller businesses that do not have professionals on staff managing their employee health programs and are unable to implement complex programs and monitor cost-effectiveness. Understanding the conditions necessary for an investment in physical activity promotion to be considered a good return on investment in terms of employee participation, reduced health care expenditures, reduced sick leave, decreased presenteeism, improved productivity, and turnover is likely a key ingredient for increasing industry participation. And, as has been pointed out by scientists close to the issue, many of these programs have not yet demonstrated that they actually increase the level of physical activity (8).

Missing Factors

While the case for physical activity promotion to employees is gaining some ground, what is not clear is whether industries stand to gain more by promoting physical activity outside their walls in the marketplace and the community compared to other worthy causes that they could promote like education or violence prevention. The opportunity to reach a global audience is great behind brands that today reach billions of people. And the impact on physical activity and health could far exceed that which could be achieved through employee wellness alone. What would it take for industry to become more actively involved in physical activity promotion throughout its local markets and beyond?

One factor that might help would be a clearer business model to show how improving physical activity in the general population would improve the economy and the environment for business. This idea, that healthful behaviors should be tied to certain economic gains, may sound unsatisfactory in principle; but economic forces already provide the underpinning of many current behavior patterns. If we could develop new models showing that physical activity is worth something in the community beyond its intrinsic value to the individual, then we could leverage the economic engine already present in the community to drive up physical activity levels. Think about the food industry, which has a clear business model based on a tangible benefits exchange between the food purveyor and the consumer. One could argue that the health club sector already has a business model for physical activity. One key difference is, however, that humans have to eat to survive, so there is a biological imperative to acquire food that supports the business model. Humans do not have to move much to subsist, and there is no external incentive system to stimulate and reinforce such behavior; so, compared to that for food, the current business model for physical activity is comparatively weak.

Importance of Incentives

What if physical activity were used as a means to provide incentives to people in a manner consistent with the way business is currently conducted in many countries? For example, customers might be given pedometers at the entrance to a large retail store, which would make them eligible for product promotions (e.g., discounts) through accumulation of a preset number of steps during their visit. The promotions might be good only for "healthier" products, such as nutritious food, to stimulate more trial of such goods. The store benefits because more people spend more time there shopping for all the items on promotion, which generally leads to greater purchases of both promotional and nonpromotional items. The consumers benefit because they get great deals on all sorts of things and become a little healthier as a function of being more physically active. In this case, health is a by-product of "business as usual" versus an outcome of physical activity promotion. This approach might be characterized as a sort of "behavioral judo." Rather than trying to get people to add more activity to their lives by focusing on the activity, you work with the power and momentum of existing motivations (e.g., the desire to get a good deal) and harness them in a way that increases physical activity.

Future Directions

Based on the foregoing discussion, it seems clear that industries indeed have a role in physical activity promotion, one that can directly affect the health of their business. This starts with promoting activity among employees and, by extension, their families. Although this is important, the fact is that we don't yet know how to engage large numbers of people, even when we have a structure to do so (the workplace), and how to improve their physical activity behaviors in a sustainable way. More work is desperately needed here.

The bigger opportunity for industry may be to leverage what it already does best, reach people and offer them an attractive benefits exchange, only with a twist. Rather than trying to get people to engage in active behaviors for the sake of becoming more active, the twist would be to build physical activity into many of the activities people want to engage in anyway, like the "shop for the deal" example discussed earlier. How many possibilities are there such that increased activity can be simply a by-product of some other desirable consumer behavior? As long as something is fun and engaging, consumers will likely approve.

Finally, I believe that industry can play an important role in helping to link physical activity and other healthy behaviors to a much bigger social movement whose principles encompass healthy living in everyday life. For example, there seems to be a growing recognition that behaviors that are good for the health of the planet are also good for the health (and happiness) of individuals. This is the basis of a new movement called "Blue," recently announced by the noted environmentalist Adam Werbach (9), which is about leveraging grass-roots consumer behavior to build sustainability practices into everyday life. Sustainability encompasses social, cultural, economic, and environmental elements and is intimately linked to human health. Because this movement is ultimately aimed at alleviating global climate change, which is already affecting peoples' lives today, its power to motivate and engage individuals in every walk of life far exceeds that of even the best marketing initiatives.

This movement is built on the premise that the most powerful way to change population behavior is to target the behaviors people practice every day and change them in some small but better way. Werbach invented the term "personal sustainability practice" or PSP to represent one small thing people would do each day to improve the health of the planet. These PSPs include activities like biking to work, parking a little farther away from the building, and eating more fruits and vegetables, which are all behaviors that would improve overall health as well as body weight control. Every industry is mapping out a sustainability strategy and will be, if it is not already, deeply engaged in changing its behaviors, by economic necessity. The current global economic slowdown has underscored the need for a focus on long-term sustainability and on what will be required by each of us to support a quality of life that can be enjoyed by generations to come. Why not harness the power of the emerging sustainability movement to give consumers more ways to increase physical activity motivated by an interest in the larger cause of sustaining our world?

Conclusion

In summary, the most powerful role industry could play to promote physical activity would be to help find and implement countless ways of making physical activity a natural part of everyday consumer behavior, as well as helping to sustain planet Earth. What bigger incentive could there be?

88

Mass Media Approaches to Addressing Physical Inactivity and Obesity

Adrian Bauman, PhD, MB, MPH

Physical activity and obesity are major contributors to the total burden of disease in developed and many developing countries; and yet few coordinated, well-funded programs have been devised to address these issues. An initial component of effective prevention is raising community awareness, setting the agenda for change, increasing community understanding, and influencing attitudes and social norms toward an issue. In this context, population-wide mass media and social marketing campaigns may be first steps in physical activity promotion and obesity prevention.

Media campaigns are organized, purposive interventions that communicate a specific health message through mass-reach channels to a large population. This approach is different from smaller-scale mediated interventions, which target specific individuals, and use mediated channels for intervention tailoring and delivery. These might comprise mailed materials, Internet, or short message service (SMS) text messaging as communication channels for the intervention. Effective mass campaigns usually use paid media and are expensive; the purpose is to achieve high levels of penetration and reach into the community. In addition, ideally, they should be linked to other community-wide interventions and facilities and should be persistent—that is, they should be serial repeated campaigns, with sequences of relevant messages developed under an overarching campaign theme (1). Mass campaigns do not work in isolation and should be part of long-term community-wide change approaches; they should be linked to other community interventions, trained health and other professionals, and the availability of facilities and resources to support the overall preventive program.

The primary goals of short-term mass-reach campaigns are to influence awareness, attitudes, and social norms. These are the proximal outcomes mostly likely to be influenced by a short-term campaign, and should be the primary measurable endpoints for specific campaigns (2). Longer-term communications, combined with multicomponent community-wide interventions, are required to influence the endpoint behaviors of inactivity and overnutrition (2, 3).

In terms of understanding media campaign effects, researchers have proposed a "hierarchy of effects," whereby the proximal outcomes of awareness and understanding are thought to lead to intermediate outcomes (such as attitudes, perceptions, social norms) and these in turn lead to distal outcomes (self-efficacy, behavioral intention, and actual measures of behavior). This is illustrated in figure 88.1. Although widely used in campaign planning, causal models demonstrating the validity of this sequence of changes remain a limited area of research (4).

In some cases, social marketing approaches are used, with the objective of demonstrating the relative advantages to the consumer of adopting the new "product" (healthy eating or incidental physical activity), as well as its accessibility and affordability. The principles of full-blown social marketing interventions may be more extensive than for most mass media communications campaigns, as they

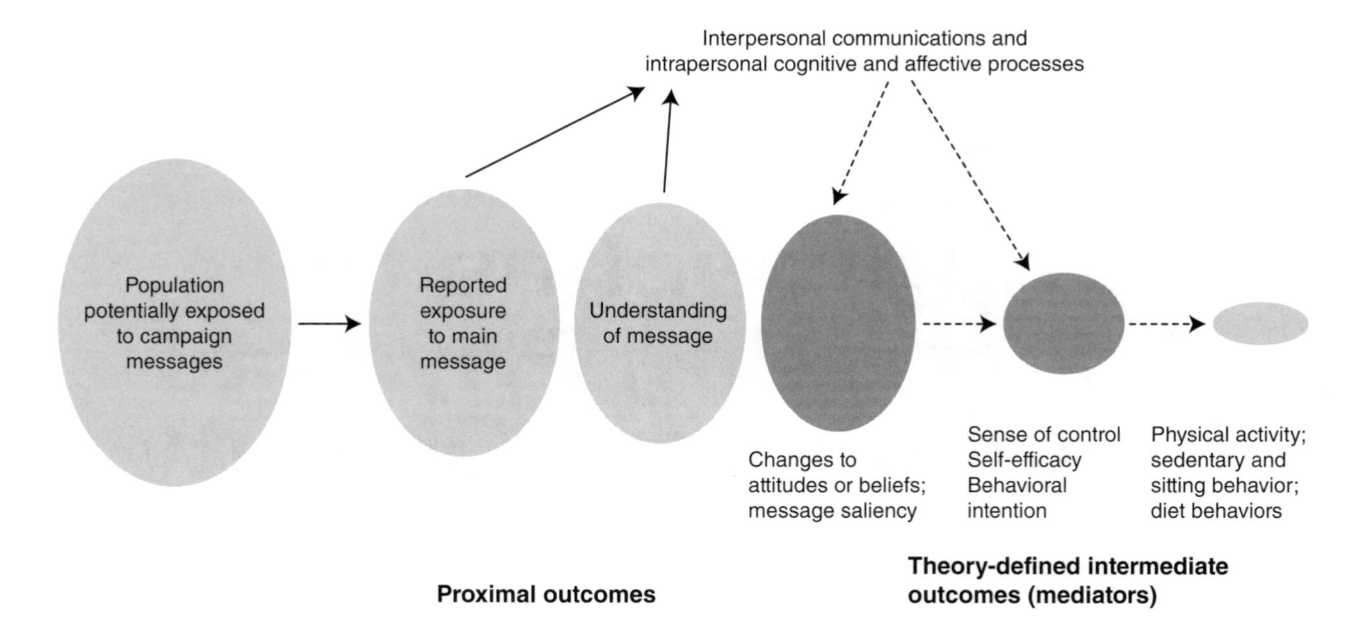

Figure 88.1 The potential processes involving mass campaign effects: mediators and moderators of behavior change. The size of the circles indicates possible numbers influenced at each stage. Potential moderator factors (environmental and ecological factors that could operate to influence campaign recall and uptake differently) are physical environment, social environment, cultural environment, and economic factors.

include subgroup targeting, consumer education, environmental change, and regulatory and policy change components (5).

The first step in campaign development is developing the right message. The communications strategy is guided by the campaign goals, as well as by whether the campaign is focusing on obesity prevention (physical activity and healthy eating) or is emphasizing physical activity or diet as separate health concerns. Here we focus on the interdependence of these two concerns in obesity prevention and point to differences in the messaging that needs to be developed based on the primary target behavior or outcome.

Increases in obesity rates are likely due to dietary changes in combination with decreases in total energy expenditure. The relevant dietary message is related to healthy choices but emphasizes decreasing energy intake. It does not focus on fruit and vegetable intake alone for their micronutrient benefits (for example in cancer prevention), but stresses their potential as healthy options to replace more energy-dense foods. Appropriate media messages would include awareness of portion size, understanding the pathology of "snacking" behaviors, and becoming aware of alternative food choices for everyday consumption.

The declines in total energy expended are due to social and cultural changes, including industrialization and mechanization, and likely to declines in daily energy expended at work and in domestic and active commuting settings (6). This means that the quantum of total physical activity required for population-level obesity prevention is more than the minimum health recommendation of moderate-intensity physical activity for health. Importantly, this implies a different physical activity message, requiring larger shifts in population (walking) behavior for obesity prevention campaigns (7) compared to the generic "30 minutes" message for general health and cardiovascular disease prevention. It is likely that 60 to 90 min per day are required for population weight loss or obesity prevention, and this is a major challenge for habitually sedentary adults.

The History of Mass Media Campaigns

A review of mass media campaigns for physical activity has identified almost 20 examples since the 1980s (3), but these have usually had a specific "leisure-time physical activity" focus. These campaigns mostly used positive and upbeat messages supporting the idea that physical activity is fun and is a part of social inclusion and community participation. Few campaigns used a disease focus or used fear or threats of developing disease as a mechanism for encouraging physical activity.

Major physical activity campaigns have been evaluated in North America, Europe, and Australasia; most have addressed physical inactivity among middle-aged adults. In the early years, physical activity campaigns were part of community-wide cardiovascular disease prevention programs, but since the mid-1990s they have been stand-alone physical activity and health initiatives. Around two-thirds emanated from the health sector alone; the remainder were conducted by departments of sport and recreation or nongovernment organizations such as the heart or stroke foundations or were achieved through physical activity coalitions and partnerships involving a range of organizations. Tracking surveys, using representative samples, demonstrated that these campaigns increased community awareness about physical activity and started to influence attitudes, but infrequently resulted in changes to leisure-time physical activity behavior (3). Only recently have campaign planners begun to include and portray incidental physical activity and "active living" behaviors as acceptable behavioral outcomes.

Specific reports on the results of obesity-focused mass media campaigns are rare. There are numerous nutrition campaigns; but these emphasize the generic health benefits of nutrition, for example, increasing servings of fruit and vegetables in adults and children. Two obesity-specific campaigns are described here, but relatively few exist. The first was the BBC "Fighting Fat and Fit" intervention, which was based on a seven-part BBC television documentary series in England (8). The impact of this documentary series was recognition by about one-fifth of adults, but less than 1% responded to the "call to action," in this case seeking further resources or registering for the weight loss and lifestyle program; nonetheless, at the population level, a 1% change reflects a large population reach.

A more recent intervention in the Netherlands (2) evaluated a three-year weight gain prevention campaign targeting young adults. Again, message awareness was high, and some attitude shifts occurred, but no impact on distal variables such as weight was noted. Other public sector obesity campaigns may be continuous, low-intensity messages that utilize PSAs (public service announcements). An example is the "Small Steps" campaign organized by the U.S. Centers for Disease Control and Prevention (9). An alternative model may involve public–private partnerships, such as the obesity campaigns generated by the weight loss industry, or, for example, by competing health insurance companies in Germany (9). These initiatives have a commercial orientation and are not evaluated for their impact on whole populations.

Future Directions

There have been few published papers on campaigns integrating both healthy eating and physical activity messages. This may be required for specific obesity prevention campaigns. Examples include the BBC *Fighting Fat, Fighting Fit* documentary series (8) and the more typical mass media campaign targeting adults in New Orleans (10). The latter study used intensive paid media to promote healthy fruit and vegetable consumption and walking among middle-aged African Americans. The campaign resulted in high awareness, as well as attitude shifts toward healthy choices that were in turn related to short-term behavioral trialing of healthy foods and walking.

To tackle obesity, innovative communications are needed. These might comprise different kinds of physical activity messages, including the promotion of active living and incidental activities and efforts to reduce sitting and sedentary time. For the energy intake side of obesity, creative efforts at advocacy campaigns, including countermarketing media (that have been used as a part of public health approaches to tobacco control) have been used. For example, messages could build community support to develop policies that regulate unhealthy food advertising to children, or even more complex messages could focus community attention on the causes of decreased energy expenditure. Serial and sustained campaigns would be necessary in order for the community to appreciate the potential health impacts of sedentary lifestyles and to understand that their genesis involves mega-industries that create sedentary choices in society, such as gasoline and automobile manufacturers and the television and domestic entertainment industries.

Conclusion

In summary, media campaigns are useful in the initial phases of large-scale obesity and chronic disease prevention efforts at the national or regional level, especially where a new message or new evidence is central to the mass communication. They do not work in isolation, and should be encompassed within a broader health promotion program composed of multiple elements targeting individuals and their social and physical environments. The limited evaluation data on the effectiveness of campaigns in influencing the community and policy agendas mean that this is an important area for future public health research.

Future Directions in Physical Activity and Obesity Research

Peter T. Katzmarzyk, PhD, and Claude Bouchard, PhD

This book presents information on a variety of topics related to physical activity and obesity. The format of the book has allowed authors to present evidence on narrowly defined topics in an authoritative, succinct manner. As such, the information is cutting edge and current as of mid-to-late 2008. However, by the time this book appears in print, some of the information, particularly in some of the faster-moving fields, will be in need of updating. Indeed, we hope to see further scientific advances in the coming years in all of the topic areas covered by this book.

The topics covered in this book span the spectrum of physical activity and obesity research, from the laboratory, to the clinic, to population, and to policy. There has been a major increase in physical activity and obesity research in recent years. Figure 89.1 presents the proportion of the published scientific literature on "physical activity" or "exercise" between 1970 and 2007 that included the key word "obesity." It is clear that there is growing interest in the topic of obesity among the physical activity research community. Although this research has yielded many advances in recent years in all areas covered by this book, it has also identified new questions that require further investigation.

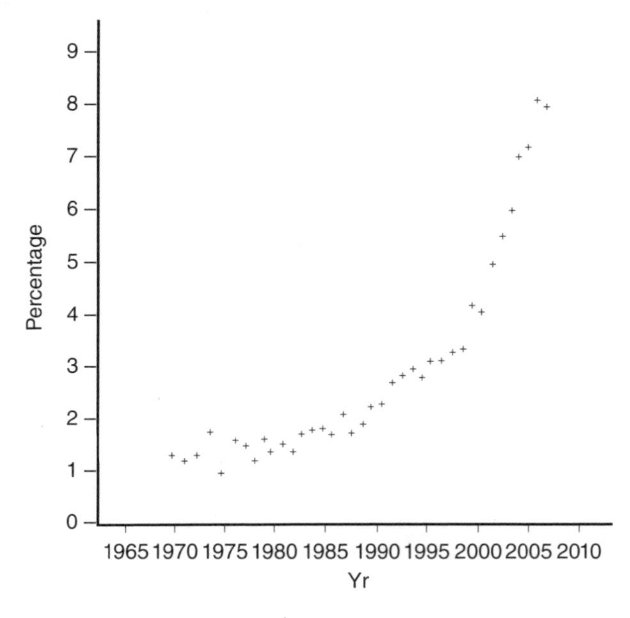

Figure 89.1 Percent of "physical activity" or "exercise" publications that included the key word "obesity," 1970-2007.

It is an exciting time to explore issues related to physical activity and obesity. As the final manuscript of this book is being prepared to be sent to the publisher, the first U.S. *Physical Activity Guidelines for Americans* has been released by the federal government (1). The release of these guidelines acknowledges the widespread recognition of the importance of physical activity to the health of society. At no other time in history has research on physical activity and obesity been more germane to the health of the world's population. Recent estimates are that more than 1 billion adults worldwide are overweight or obese, and more than 40% are not obtaining sufficient physical activity (2, 3). Research into all facets of the relationship between physical activity and obesity will contribute to addressing this public health catastrophe. Here we attempt to dissect out some of the pressing areas for future research as reflected in the sections of this book.

Definition and Assessment of Physical Activity and Obesity

Physical activity is a behavior, while obesity is a trait that is influenced by both biology and behavior. In general, markers of obesity are measured with greater precision and reliability than measures of behavior, including physical activity. However, in surveillance studies, information on both physical activity levels and obesity is often obtained from self-report questionnaires, which may introduce considerable bias into the population estimates obtained. More research is necessary to understand the extent of the measurement bias associated with self-reported measures across ethnicity, gender, and socioeconomic groups. Greater efforts should be made to incorporate direct measures of physical activity and obesity in population surveillance systems.

There is growing understanding of the health effects associated with the continuum of physical activity behavior from sedentary to highly physically active. Emerging evidence suggests that sedentary behavior may have health effects that are distinct from those associated with overall physical activity levels. More research is required to develop methods to accurately measure sedentary behavior and to quantify the daily physical activity and inactivity profiles of individuals. Current methods of physical activity assessment employed in population studies typically probe levels of leisure-time physical activity, and little is known about the amount of energy expended in the occupational, domestic, and active commuting domains of physical activity. The current methods of physical activity assessment are not accurate enough to capture small differences in activity level that may have implications for energy balance. More emphasis should be placed on developing and validating methods to quantify total daily energy expenditure and physical activity level across the whole spectrum.

Physical activity and physical fitness are both associated with the risk of obesity, chronic disease, and premature mortality. However, as already noted, physical activity is a behavior, and physical fitness is a measureable trait of an individual. Markers of physical fitness are generally more accurately measured than physical activity behaviors, and this may account for some of the differences observed in the relationship with other risk factors. However, there is more than simple measurement accuracy operating here. Fitness is determined by genetic characteristics, overall health status and robustness, and physical activity level. In other words, fitness is not a simple surrogate for activity level. Future research needs to incorporate indicators of both activity level and fitness.

Recent advances have made the laboratory assessment of adiposity very accurate. Methods now exist for the precise assessment of total body fat as well as depot-specific adipose tissue levels. However, most population-based studies rely on anthropometric markers of obesity such as the body mass index (BMI), waist circumference, and subcutaneous skinfolds. Where possible, efforts should be made to include the objective measurement of adiposity rather than relying on self-reported data. Most markers of adiposity change considerably with normal growth and maturation in children and adolescents, and more research is required to determine the best anthropometric markers of obesity-related health risk during growth.

Prevalence and Cost Issues

Physical inactivity and obesity are both highly prevalent conditions around the world. Given the existence of standard international thresholds for obesity (BMI) in children (4) and adults (5), we have a better understanding of the global prevalence of obesity than of physical activity. However, international comparisons are hindered by differences in collection methods across countries (e.g., self-report vs. measured) and the adoption of ethnic-specific anthropometric thresholds in some regions (e.g., Asia). Efforts should be made to standardize data collection methods across countries, and investigators should consistently report obesity prevalence using the anthropometric thresholds accepted by the World Health Organization, in addition to local region-specific thresholds.

Given that the surveillance of physical activity generally relies on questionnaire data, the measurement of physical activity across the globe is complicated by language differences, cultural differences in the perception of physical activity, and differences in the background levels of physical activity (6). Efforts should be made to incorporate more objective measures of physical activity, such as those obtained by pedometers and accelerometers. Collaboration between scientists from different countries would allow for the standardization of physical activity methods and eventually improve the global surveillance of physical activity.

Studies of the economics of physical inactivity and obesity are important for identifying the public health burden associated with these risk factors and for identifying cost-effective strategies to tackle these issues. Economic burden estimates rely on good estimates of prevalence, and thus the development of accurate surveillance methods is important in this regard. Very little is known about the cost–benefit ratios associated with increasing physical activity levels with respect to their impact on obesity. Future research on this topic will require that cost information be incorporated into the design of clinical and community trials from their inception.

Determinants of Physical Activity Levels

The ultimate goal of much research in the field of physical activity and obesity is to increase the physical activity levels of the population in order to reduce the personal and public health burden associated with obesity. The realization of this goal will require investment in interventions at the levels of the individual and the population. However, in order to design effective interventions, information on the determinants of physical activity and inactivity is required. This information is necessary to identify high-risk subgroups of the population and to identify potential factors to intervene on, such as parental support, sedentary behavior, school curricula, and aspects of the built environment. Given the high prevalences of physical inactivity and obesity, there is a pressing need to develop effective interventions; however, studies of the determinants of physical activity should be incorporated into the design. It is also urgent that more research be devoted to the biological determinants of physical activity behavior so that a more complete picture of the factors associated with high and low levels of activity can be drawn. This effort will require that genetic, in utero programming, and epigenetic factors be incorporated into the models.

Physical Activity and Risk of Obesity

Physical activity represents an important component of the energy balance equation, as it is a modifiable contributor to energy expenditure at the level of the individual. However, research to date using prospective designs has not demonstrated a strong, consistent link between physical activity and the risk of obesity. A major limitation with this research has been the lack of precise measurements of physical activity and inactivity. More research is required to develop and validate methods of physical activity assessment that capture a significant fraction of the variance in total daily energy expenditure. It is also important to include indicators of the level of inactivity in the research focused on weight gain and obesity.

Physical Activity and Biological Determinants of Obesity

In order to demonstrate a cause-and-effect relationship between physical activity and obesity, one must understand the molecular and metabolic mechanisms involved in the changes brought about in adiposity by decreasing inactivity time or increasing energy expenditure of activity. To date, a host of biological determinants have been linked with the development of obesity, and these determinants are being studied with respect to their association with physical activity. The evidence suggests that physical activity affects several biological pathways related to obesity, but further research is required to better delineate these relationships. Of particular interest are early-life events (in utero, during early childhood) that predispose toward obesity and the role that physical activity (of both mother and child) may play in "programming" the future health and predisposition to obesity of children.

The response to a standardized exercise training program varies from individual to individual, and part of this variation in response has been attributed to familial factors (7). Further research is required to understand the genetic contribution to individual responses to exercise in terms of adiposity, fat topography, adipose tissue metabolism, and other traits, as well as to identify the alleles influencing the ability to adapt to exercise. An important research area is that of the role of inactivity and activity on fat topography, visceral fat, and ectopic fat deposition in skeletal muscle, liver, heart, pancreas, and other tissues in normal-weight and obese people.

Physical Activity and Behavioral and Environmental Determinants of Obesity

At the most basic level, obesity is the result of an energy imbalance whereby energy intake exceeds energy expenditure. However, humans do not operate in a vacuum, and the energy balance equation must be viewed within the context of the broader behavioral and environmental milieu of the individual. Changing behavior and the environment seems to be the most logical solution to increasing physical activity and reducing obesity, although we may be able to modulate DNA methylation and epigenetic events to alter gene expression in the future. Most classic behavior change theories are hinged on changing the behavior of the individual (i.e., intra- and interpersonal). The theory of planned behavior, the transtheoretical model, social-cognitive theory, and the health belief model have been the backbone of many dietary and physical activity interventions targeting individuals or small groups (7).

However, there is a growing awareness that large-scale population changes in lifestyle behaviors such as diet and physical activity may be better served by ecological models that incorporate several levels of intervention. The social-ecological model provides a useful framework to ground this research and involves several levels of influence, including intrapersonal factors, perceived environment, behavior, behavior setting, and the policy environment (7). For both diet and physical activity, the upstream influences of policy and environmental interventions have received far less attention than the individual-level intervention, and this represents an important area of future research. A focus on the built environment has a role to play in increasing physical activity levels and reducing obesity. However, at this time, the evidence base is mainly cross-sectional and requires further development using prospective study designs in order to understand the characteristics of the built environment that are most important for diverse population subgroups, as well as the cost-effectiveness associated with their implementation.

Physical Activity in the Prevention and Treatment of Obesity

There is a clear need to develop evidence-based strategies to reduce the incidence of obesity and related disorders. However, given the currently high prevalence of obesity in many populations, a strong focus on treatment is also required. The goals of obesity treatment are to prevent further weight gain, stimulate weight loss, or maintain weight after a significant loss. The amount of physical activity required for weight loss is a function of the desired amount of the weight loss and is confounded by changes in dietary intake. More research is required on the interaction between physical activity and dietary intake. The role of compensatory behaviors such as reduced incidental activity and increased dietary intake in the context of physical activity interventions deserves further exploration. The role of physical activity as an adjunct to pharmacotherapy for obesity has been largely unexplored; however, given the increased interest in identifying drugs that influence obesity, this is an important area for future research.

Clinical Applications

Given the high prevalence of obesity, studies are needed on the effectiveness of physical activity in the management of obesity. It is anticipated that the reduction and maintenance of body weight will result in a reduced level of comorbidities and improved health. However, there is evidence that physical activity in the absence of weight loss also has health benefits. More research is required to determine which clinical conditions will benefit from increases in physical activity among obese individuals and to identify the mechanisms by which physical activity improves health status irrespective of changes in body weight. Further, it is of great interest to determine which conditions are improved by physical activity–induced weight loss in addition to the increase in physical activity itself. This information will be critical for the development of patient-centered physical activity plans for the reduction of obesity and related comorbidities.

Policy and Research Issues

Changing public policies can be a powerful strategy that can affect an entire population. It is hypothesized that a coordinated set of policies aimed at transforming our environment to support active living may have a positive impact on the prevalence of obesity. However, policy research in the field of physical activity and obesity has received less attention than other approaches. There is a role for both government and industry to play, but these sectors need to partner with researchers in order to evaluate the effectiveness of policy change strategies designed to prevent or reduce obesity from a community and population perspective.

Summary

There is no shortage of research questions to be answered in the field of physical activity and obesity. The recent advances in this area have created new directions for future research, and the potential for discovery in the coming years is very high. Many of these discoveries will be made as methodological advances allow for the improved quantification of physical activity and inactivity behaviors.

References

Chapter 1

1. Bouchard C (ed.). *Physical activity and obesity.* Human Kinetics: Champaign, IL, 2000.
2. World Health Organization. *The global burden of disease: 2004 update.* World Health Organization: Geneva, 2008.
3. World Health Organization. *The world health report 2002: reducing risks, promoting healthy life.* World Health Organization: Geneva, 2002.
4. Bouchard C. The biological predisposition to obesity: beyond the thrifty gene scenario [commentary]. *Int J Obes* 2007; 31: 1337-1339.
5. O'Rahilly S, Farooqi IS. Human obesity: a heritable neurobehavioral disorder that is highly sensitive to environmental conditions. *Diabetes* 2008; 57: 2905-2910.
6. Bouchard C, Agurs-Collins T (eds.). Gene-nutrition and gene-physical activity interactions in the etiology of obesity. *Obesity* 2008; 16 suppl. 3: S1-96.
7. Frayling TM, Timpson NJ, Weedon MN, Zeggini E, Freathy RM, Lindgren CM, et al. A common variant in the *FTO* gene is associated with body mass index and predisposes to childhood and adult obesity. *Science* 2007; 316: 889-894.
8. Andreasen CH, Stender-Petersen KL, Mogensen MS, Torekov SS, Wegner L, Andersen G, et al. Low physical activity accentuates the effect of the *FTO* rs9939609 polymorphism on body fat accumulation. *Diabetes* 2008; 57: 95-101.
9. Rampersaud E, Mitchell BD, Pollin TI, Fu M, Shen H, O'Connell JR, et al. Physical activity and the association of common *FTO* gene variants with body mass index and obesity. *Arch Intern Med* 2008; 168: 1791-1797.
10. Bray GA, Bouchard C (eds.). *Handbook of obesity: clinical applications.* 3rd ed. Informa Healthcare: New York, 2008.

Chapter 2

1. Warburton DER, Katzmarzyk PT, Rhodes RE, Shephard RJ. Evidence-informed physical activity guidelines for Canadian adults. *Appl Physiol Nutr Metab* 2007; 32: S16-S68.
2. Warburton DE, Nicol C, Bredin SS. Health benefits of physical activity: the evidence. *CMAJ* 2006; 174: 801-809.
3. Katzmarzyk PT, Janssen I. The economic costs associated with physical inactivity and obesity in Canada: an update. *Can J Appl Physiol* 2004; 29: 90-115.
4. Bouchard C, Shephard RJ. Physical activity fitness and health: the model and key concepts. In: Bouchard C, Shephard RJ, Stephens T (eds.), *Physical activity, fitness, and health: international proceedings and consensus statement.* Human Kinetics: Champaign, IL, 1994, pp. 77-88.
5. Ross R, Janssen I. Physical activity, fitness, and obesity. In: Bouchard C, Blair SN, Haskell WL (eds.), *Physical activity and health.* Human Kinetics: Champaign, IL, 2007, pp. 173-190.
6. Saris WH, Blair SN, van Baak MA, Eaton SB, Davies PS, Di Pietro L, et al. How much physical activity is enough to prevent unhealthy weight gain? Outcome of the IASO 1st Stock Conference and consensus statement. *Obes Rev* 2003; 4: 101-114.
7. Esliger D, Tremblay MS. Physical activity and inactivity profiling: the next generation. *Appl Physiol Nutr Metab* 2007; 98: S195-207.
8. Warburton DE, Nicol C, Bredin SS. Prescribing exercise as preventive therapy. *CMAJ* 2006; 174: 961-974.
9. Tudor-Locke CE, Myers AM. Challenges and opportunities for measuring physical activity in sedentary adults. *Sports Med* 2001; 31: 91-100.
10. Hamilton MT, Hamilton DG, Zderic TW. Role of low energy expenditure and sitting in obesity, metabolic syndrome, type 2 diabetes, and cardiovascular disease. *Diabetes* 2007; 56: 2655-2667.

Chapter 3

1. Katzmarzyk PT, Baur LA, Blair SN, Lambert EV, Oppert J-M, Riddoch C. International Conference on Physical Activity and Obesity in Children: summary statement and recommendations. *Appl Physiol Nutr Metab* 2008; 33: 371-388.
2. Eisenmann JC, Bartee RT, Smith DT, Welk GJ, Fu Q. Combined influence of physical activity and television viewing on the risk of overweight in US youth. *Med Sci Sports Exerc* 2008; 32: 613-618.
3. Hamilton MT, Hamilton DG, Zderic TW. Role of low energy expenditure and sitting in obesity, metabolic syndrome, type 2 diabetes, and cardiovascular disease. *Diabetes* 2007; 56: 2655-2667.
4. Healy GN, Dunstan DW, Salmon J, Shaw JE, Zimmet PZ, Owen N. Television time and continuous metabolic risk in physically active adults. *Med Sci Sports Exerc* 2008; 40: 639-645.
5. Rosenberg DE, Bull FC, Marshall AL, Sallis JF, Bauman AE. Assessment of sedentary behavior with the International Physical Activity Questionnaire. *J Phys Act Health* 2008; 5(suppl. 1): S30-S44.
6. Tremblay MS, Esliger DW, Tremblay A, Colley R. Incidental movement, lifestyle-embedded activity and sleep: new frontiers in physical activity assessment. *Appl Physiol Nutr Metab* 2007; 32(suppl. 2E): S208-S217.
7. Shields M, Tremblay MS. Sedentary behavior and obesity among Canadian adults. *Health Rep* 2008; 19: 19-30.
8. Shields M, Tremblay MS. Screen time among Canadian adults: a profile. *Health Rep* 2008; 19: 31-43.
9. Canadian Paediatric Society. Impact of media use on children and youth. *Paediatr Child Health* 2003; 8: 301-306.
10. Tudor-Locke C, Bassett DR Jr. How many steps/day are enough? Preliminary pedometer indices for public health. *Sports Med* 2004; 34: 1-8.
11. Tudor-Locke C, Hatano Y, Pangrazi RP, Kang M. Revisiting "how many steps are enough?" *Med Sci Sports Exerc* 2008; 40(suppl. 7): S537-S543.

12. Reilly JJ, Penpraze V, Hislop J, Davies G, Grant S, Paton JY. Objective measure of physical activity and sedentary behavior: review with new data. *Arch Dis Child* 2008; 93: 614-619.

13. Esliger DW, Tremblay MS. Physical activity and inactivity profiling: the next generation. *Appl Physiol Nutr Metab* 2007; 32(suppl. 2E): S195-S207.

Chapter 4

1. U.S. Department of Health and Human Services. *Physical activity and health: a report of the surgeon general.* U.S. Department of Health and Human Services, Public Health Service, CDC, National Center for Chronic Disease Prevention and Health Promotion: Atlanta, 1996.
2. Haskell WL, Lee IM, Pate RR, Powell KE, Blair SN, Franklin BA, et al. Physical activity and public health: updated recommendation for adults from the American College of Sports Medicine and the American Heart Association. *Med Sci Sports Exerc* 2007; 39: 1423-1434.
3. LaMonte MJ, Ainsworth BE. Quantifying energy expenditure and physical activity in the context of dose-response. *Med Sci Sports Exerc* 2001; 33: S370-S378.
4. Physical Activity Guidelines Advisory Committee. *Physical Activity Guidelines Advisory Committee Report,* 2008. Washington, DC: U.S. Department of Health and Human Services, 2008.
5. McKenzie TL. Use of direct observation to assess physical activity. In: Welk GK (ed.), *Physical activity assessments for health-related research.* Human Kinetics: Champaign, IL, 2002, pp. 179-196.
6. Ward DS, Evenson KR, Vaughn A, Rodgers AB, Troiano RP. Accelerometer use in physical activity: best practices and research recommendations. *Med Sci Sports Exerc* 2005; 37: S582-S588.
7. Tudor-Locke C, Bassett DR. How many steps/day are enough? Preliminary pedometer indices for public health. *Sports Med* 2004; 34: 1-8.
8. Ainsworth BE, Haskell WL, Whitt MC, Irwin ML, Swartz AM, Strath SJ. Compendium of physical activities: an update of activity codes and MET intensities. *Med Sci Sports Exerc* 2000; 32: S498-S516.
9. Pettee KK, Tudor-Locke C, Ainsworth BE. The measurement of energy expenditure and physical activity. In: Wolinsky I, Driskell JA (eds.), *Sports nutrition.* CRC Press: Boca Raton, FL, 2008, pp. 159-189.
10. Rodriguez DA, Brown AL, Troped PJ. Portable global positioning units to complement accelerometry-based physical activity monitors. *Med Sci Sports Exerc* 2005 Nov; 37(11 suppl.): S572-581.

Chapter 5

1. Dishman RK, Heath GW, Washburn RA. *Physical activity epidemiology.* Human Kinetics: Champaign, IL, 2004.
2. Brown WH, Pfeiffer KA, McIver KL, Dowda M, Almeida MJCA, Pate RR. Assessing preschool children's physical activity: an Observational System for Recording Physical Activity in Children—Preschool Version (OSRAC-P). *Res Q Exerc Sport* 2006; 77: 167-176.
3. Sallis JF. Self-report measures of children's physical activity. *J Sch Health* 1991; 61: 215-219.
4. Dorminy CA, Choi L, Akohoue SA, Chen KY, Buchowski MS. Validity of a multisensor armband in estimating 24-h energy expenditure in children. *Med Sci Sports Exerc* 2008 40: 699-706.
5. Zhang K, Werner P, Sun M, Pi-Sunyer FX, Boozer CN. Measurement of human daily physical activity. *Obes Res* 2003; 11: 33-40.
6. Zhang K, Pi-Sunyer FX, Boozer CN. Improving energy expenditure estimation for physical activity. *Med Sci Sports Exerc* 2004; 36: 883-889.
7. Troped PJ, Oliveira MS, Matthews CE, Cromley EK, Melly SJ, Craig BA. Prediction of activity mode with global positioning system and accelerometer data. *Med Sci Sports Exerc* 2008; 40: 972-978.

Chapter 6

1. Screening and Interventions to Prevent Obesity in Adults, topic page. U.S. Preventive Services Task Force. Agency for Healthcare Research and Quality, Rockville, MD. Dec 2003. www.ahrq.gov/clinic/uspstf/uspsobes.htm. Accessed 26 May 2008.
2. NHLBI Obesity Education Initiative Expert Panel on the Identification, Evaluation, and Treatment of Overweight and Obesity in Adults. Clinical guidelines on the identification, evaluation, and treatment of overweight and obesity in adults—the evidence report. *Obes Res* 1998; 6: 51S-63S.
3. World Health Organization. *Obesity: preventing and managing the global epidemic.* World Health Organization: Geneva, 1998.
4. Bray GA, Bouchard C, James WP. Definitions and proposed current classification of obesity. In: Bray GA, Bouchard C, James WP (eds.), *Handbook of obesity.* Marcel Dekker: New York, 1997, pp. 31-40.
5. Bray GA. Classification and evaluation of the overweight patient. In: Bray GA, Bouchard C, James WP (eds.), *Handbook of obesity.* Marcel Dekker: New York, 1997, pp. 831-854.
6. U.S. Department of Health and Human Services. *Clinician's handbook of preventive services: put prevention into family practice.* U.S. Government Printing Office: Washington, DC, 1994, pp. 141-146.

Chapter 7

1. Freedman DS, Wang J, Maynard LM, Thornton JC, Mei Z, Pierson RN, et al. Relation of BMI to fat and fat-free mass among children and adolescents. *Int J Obes* 2005; 29: 1-8.
2. Kuczmanski RJ, Ogden CL, Grummer-Strawn LM, Flegal KM, Guo SS, Wei R, et al. CDC growth charts: United States. *Adv Data* 2000; 314: 1-27.
3. Krebs NF, Himes JH, Jacobson D, Nicklas TA, Guilday P, Styne D. Assessment of child and adolescent overweight and obesity. *Pediatrics* 2007; 120(suppl. 4): S193-S228.
4. Cole TJ, Bellizzi MC, Flegal KM, Dietz WH. Establishing a standard definition for child overweight and obesity world wide: international survey. *Br Med J* 2000; 320: 1240-1243.
5. de Onis M, Garza C, Onyango AW, Martorell R. WHO child growth standards. *Acta Paediatr* 2006; 95(suppl. 450): 1-101.
6. Himes, JH, Park K, Styne D. Menarche and assessment of body mass index in adolescent girls. *Journal of Pediatrics* 2009; 155: 393-397.
7. Himes JH. Agreement among anthropometric indicators identifying the fattest adolescents. *Int J Obes* 1999; 23(suppl. 2): S18-S21.
8. Heymsfield SB, Lohman TG, Wang ZM, Going SB (eds.). *Human body composition.* 2nd ed. Human Kinetics: Champaign, IL, 2005.
9. McCarthy HD, Cole TJ, Fry T, Jebb SA, Prentice AM. Body fat reference curves for children. *Int J Obes* 2006; 30: 598-602.

10. Jartti L, Hakanen M, Paakkunainen U, Raittinen P, Rönnemaa T. Comparison of hand-to-leg and leg-to-leg bioelectric impedance devices in the assessment of body adiposity in prepubertal children. The STRIP study. *Acta Paediatr* 2000; 89: 781-786.

Chapter 8

1. Connor Gorber S, Tremblay M, Moher D, Gorber B. A comparison of direct vs. self-report measures for assessing height, weight and body mass index: a systematic review. *Obes Rev* 2007; 8: 307-326.
2. Behavioral Risk Factor Surveillance System. Nationwide (states and DC) prevalence data for 2004 – overweight and obesity. http://apps.nccd.cdc.gov/brfss/sex.asp?cat=OB&yr=2004&qkey=4409&state=UB. Accessed 30 April 2008.
3. Lethbridge-Çejku M, Rose D, Vickerie J. Summary health statistics for U.S. adults: National Health Interview Survey, 2004. *Vital Health Stat* 10(228). National Center for Health Statistics: Hyattsville, MD, 2006, p. 81.
4. Ogden CL, Carroll MD, Curtin LR, McDowell MA, Tabak CJ, Flegal KM. Prevalence of overweight and obesity in the United States, 1999-2004. *JAMA* 2006; 295: 1549-1555.
5. Sallis JF, Saelens BE. Assessment of physical activity by self-report: status, limitations, and future directions. *Res Q Exerc Sport* 2000; 71: 1-14.
6. Troiano RP, Berrigan D, Dodd KW, Mâsse LC, Tilert T, McDowell M. Physical activity in the United States measured by accelerometer. *Med Sci Sports Exerc* 2008; 40: 181-188.
7. Hagströmer M, Oja P, Sjöström M. Physical activity and inactivity in an adult population assessed by accelerometry. *Med Sci Sports Exerc* 2007; 39: 1502-1508.
8. Plankey MW, Stevens J, Flegal KM, Rust PF. Prediction equations do not eliminate systematic error in self-reported body mass index. *Obes Res* 1997; 5: 308-314.
9. Troiano RP, Dodd KW. Differences between objective and self-report measures of physical activity. What do they mean? *J Kor Soc Meas Eval* 2008; 10(2): 31-42.

Chapter 9

1. Brown W, Bauman A, Chey T, Trost S, Mummery K. Comparison of surveys used to measure physical activity. *Aust NZ J Pub Health* 2004; 28: 128-134.
2. Haskell WL, Lee IM, Pate RR, Powell KE, Blair SN, Franklin BA, et al. Physical activity and public health: updated recommendation for adults from the American College of Sports Medicine and the American Heart Association. *Circulation* 2007; 116: 1081-1093.
3. Sjöström M, Oja P, Hagströmer M, Smith BJ, Bauman A. Health-enhancing physical activity across European Union countries: the Eurobarometer study. *J Pub Health* 2006; 14: 291-300.
4. Armstrong T, Bull F. Development of the World Health Organization Global Physical Activity Questionnaire (GPAQ). *J Pub Health* 2006; 14: 66-70.
5. Martinez-Gonzalez MA, Varo JJ, Santos JL, de Irala J, Gibney M, Kearney J, et al. Prevalence of physical activity during leisure time in the European Union. *Med Sci Sports Exerc* 2001; 33: 1142-1146.
6. Rutten A, Abu-Omar K. Prevalence of physical activity in the European Union. *Soz Präventivmed* 2004; 49: 281-289.
7. Guthold R, Ono T, Strong K, Chatterji S, Morabia A. Worldwide variability in physical inactivity. A 51 country study. *Am J Prev Med* 2008; 34: 486-494.
8. Bassett D, Schneider PL, Huntington GE. Physical activity in an old order Amish community. *Med Sci Sports Exerc* 2004; 36: 79-85.
9. Bull F, Armstrong T, Dixon T, Ham S, Neiman A, Pratt M. Physical inactivity. In: Ezzati M, Lopez A, Rodgers A, Murray C (eds.), *Comparative quantification of health risks: global and regional burden of disease attributable to selected major risk factors*. World Health Organization: Geneva, 2004, pp. 729-881.
10. Centers for Disease Control and Prevention. Prevalence of regular physical activity among adults—United States, 2001 and 2005. *MMWR* 2007; 56: 1209-1212.
11. Bryan SN, Tremblay MS, Pérez CE, Ardern CI, Katzmarzyk PT. Physical activity and ethnicity: evidence from the Canadian Community Health Survey. *Can J Pub Health* 2006; 97: 271-276.
12. Ross J, Hamlin M. Maori physical activity: a review of an indigenous population's participation. *Health Promot J Aust* 2007; 18: 73-76.

Chapter 10

1. Katzmarzyk PT, Baur LA, Blair SN, Lambert EV, Oppert JM, Riddoch C. International conference on physical activity and obesity in children: summary statement and recommendations. *Appl Physiol Nutr Metab* 2008; 33: 371-388.
2. Wareham NJ, Rennie KL. The assessment of physical activity in individuals and populations: why try to be more precise about how physical activity is assessed? *Int J Obes Relat Metab Disord* 1998; 22: S30-S38.
3. Ness AR, Leary SD, Mattocks C, Blair SN, Reilly JJ, Wells J, et al. Objectively measured physical activity and fat mass in a large cohort of children. *PLoS Med* 2007; 4: e97.
4. Troiano RP, Berrigan D, Dodd KW, Masse LC, Tilert T, McDowell M. Physical activity in the United States measured by accelerometer. *Med Sci Sports Exerc* 2008; 40: 181-188.
5. Riddoch CJ, Bo Andersen L, Wedderkopp N, Harro M, Klasson-Heggebo L, Sardinha LB, et al. Physical activity levels and patterns of 9- and 15-yr-old European children. *Med Sci Sports Exerc* 2004; 36: 86-92.
6. Riddoch CJ, Mattocks C, Deere K, Saunders J, Kirkby J, Tilling K, et al. Objective measurement of levels and patterns of physical activity. *Arch Dis Child* 2007; 92: 963-969.
7. Webber LS, Catellier DJ, Lytle LA, Murray DM, Pratt CA, Young DR, et al. Promoting physical activity in middle school girls: Trial of Activity for Adolescent Girls. *Am J Prev Med* 2008; 34: 173-184.
8. Andersen LB, Harro M, Sardinha LB, Froberg K, Ekelund U, Brage S, et al. Physical activity and clustered cardiovascular risk in children: a cross-sectional study (The European Youth Heart Study). *Lancet* 2006; 368: 299-304.
9. Pate RR, Stevens J, Pratt C, Sallis JF, Schmitz KH, Webber LS, et al. Objectively measured physical activity in sixth-grade girls. *Arch Pediatr Adolesc Med* 2006; 160: 1262-1268.

Chapter 11

1. WHO expert consultation. Appropriate body-mass index for Asian populations and its implications for policy and intervention strategies. *Lancet* 2004; 363: 157-163.
2. James WPT, Jackson-Leach R, Ni Mhurchu C, Kalmara E, Shayeghi M, Rigby NJ, et al. Overweight and obesity (high body mass index). In: Ezzati M, Lopez AD, Rodgers A, Murray CJL (eds.), *Comparative quantification of health risks. Global and regional burden of disease attributable to selected*

major risk factors (chap. 8, vol. 1). World Health Organization: Geneva, 2004.

3. Yoshiike N, Seino F, Tajima S, Arai Y, Kawano M, Furuhata T, et al. Twenty-year changes in the prevalence of overweight in Japanese adults: the National Nutrition Survey 1976-1995. *Obes Rev* 2002; 3: 183-190.
4. Bauman A, Allman-Farinelli M, Huxley R, James WPT. Leisure-time physical activity alone may not be a sufficient public health approach to prevent obesity—a focus on China. *Obes Rev* 2008; 9 suppl. 1: 119-120.
5. Ezzati M, Hoorn SV, Lawes CMM, Leach R, James WPT, Lopez AD, et al. Rethinking the "diseases of affluence" paradigm: economic development and global patterns of nutritional risks in relation to economic development. *PLoS Med* 2005; 2: E133.
6. Chen CM. Overview of obesity in Mainland China. *Obes Rev* 2008; 9 suppl. 1: 13-20.
7. James WPT, Schofield EC. *Human energy requirements: a manual for planners and nutritionists.* Published by arrangement with the Food and Agriculture Organization of the United Nations. Oxford University Press: Oxford, 1990.
8. Allman-Farinelli MA, Chey T, Bauman AE, Gill T, James WPT. Age, period and birth cohort effects on prevalence of overweight and obesity in Australian adults from 1990 to 2000. *Eur J Clin Nut* 2007; 62: 898-907.
9. James WPT. The epidemiology of obesity: the size of the problem. *J Int Med* 2008; 263: 336-352.

Chapter 12

1. Katzmarzyk PT, Janssen I. The economic costs associated with physical inactivity and obesity in Canada: an update. *Can J Appl Physiol* 2004; 29: 90-115.
2. Yates J, Murphy C. A cost benefit analysis of weight management strategies. *Asia Pac J Clin Nutr* 2006; 15: 74-79.
3. Great Britain Parliament House of Commons Health Committee. *Obesity: third report of session 2003-04. Report, together with formal minutes* (vol. 1). London: Parliament. HMSO; 2004.
4. Wolf AM, Colditz GA. Current estimates of the economic cost of obesity in the United States. *Obes Res* 1998; 6: 97-106.
5. Zhao W, Zhai Y, Hu J, et al. Economic burden of obesity-related chronic diseases in Mainland China. *Obes Rev* 2008; 9: 62-67.
6. Raebel MA, Malone DC, Conner DA, Xu S, Porter JA, Lanty FA. Health services use and health care costs of obese and nonobese individuals. *Arch Intern Med* 2004; 164: 2135-2140.
7. Wang F, McDonald T, Champagne LJ, Edington DW. Relationship of body mass index and physical activity to health care costs among employees. *J Occup Environ Med* 2004; 46: 428-436.
8. Finkelstein MM. Obesity, cigarette smoking and the cost of physicians' services in Ontario. *Can J Pub Health* 2001; 92: 437-440.
9. Trogdon JG, Finkelstein EA, Hylands T, Dellea PS, Kamal-Bahl SJ. Indirect costs of obesity: a review of the current literature. *Obes Rev* 2008;9: 489-500.
10. Allison DB, Zannolli R, Narayan KM. The direct health care costs of obesity in the United States. *Am J Pub Health* 1999; 89: 1194-1199.
11. van Baal PH, Polder JJ, de Wit GA, et al. Lifetime medical costs of obesity: prevention no cure for increasing health expenditure. *PLoS Med* 2008; 5: e29.

Chapter 13

1. Lobstein T, Baur L, Uauy R. IASO International Obesity Task Force. Obesity in children and young people: a crisis in public health. *Obes Rev* 2004; 5: 4-104.
2. Wang Y, Lobstein T. Worldwide trends in childhood overweight and obesity. *Int J Pediatr Obes* 2006; 1: 11-25.
3. Dietz WH, Robinson TN. Use of the body mass index (BMI) as a measure of overweight in children and adolescents. *J Pediatr* 1998; 132: 191-193.
4. Cole TJ, Bellizzi MC, Flegal KM, Dietz WH. Establishing a standard definition for child overweight and obesity worldwide: international survey. *BMJ* 2000; 320: 1240-1243.
5. World Health Organization. Growth reference data for 5-19 years. World Health Organization: Geneva, 2007. www.who.int/growthref/. Accessed 4 April 2008.

Chapter 14

1. Rice DP, Hodgson TA, Kopstein AN. The economic costs of illness: a replication and update. *Health Care Financ Rev* 1985; 7: 61-80.
2. Wang F, McDonald T, Champagne LJ, Edington DW. Relationship of body mass index and physical activity to health care costs among employees. *J Occup Environ Med* 2004; 46: 428-436.
3. Wang F, McDonald T, Reffitt B, Edington DW. BMI, physical activity, and health care utilization/costs among Medicare retirees. *Obes Res* 2005; 13: 1450-1457.
4. Pratt M, Macera CA, Wang G. Higher direct medical costs associated with physical inactivity. *Physician Sportsmed* 2000; 28: 63-70.
5. Anderson LH, Martinson BC, Crain AL, et al. Health care charges associated with physical inactivity, overweight, and obesity. *Prev Chronic Dis* 2005; 2: A09.
6. Katzmarzyk PT, Janssen I, Ardern CI. Physical inactivity, excess adiposity and premature mortality. *Obes Rev* 2003; 4: 257-290.
7. Popkin BM, Kim S, Rusev ER, Du S, Zizza C. Measuring the full economic costs of diet, physical activity and obesity-related chronic diseases. *Obes Rev* 2006; 7: 271-293.
8. Katzmarzyk PT, Janssen I. The economic costs associated with physical inactivity and obesity in Canada: an update. *Can J Appl Physiol* 2004; 29: 90-115.
9. Allison DB, Zannolli R, Narayan KM. The direct health care costs of obesity in the United States. *Am J Pub Health* 1999; 89: 1194-1199.
10. van Baal PH, Polder JJ, de Wit GA, et al. Lifetime medical costs of obesity: prevention no cure for increasing health expenditure. *PLoS Med* 2008; 5: e29.

Chapter 15

1. Tjepkema M. Adult obesity. *Health Rep* 2006; 17: 9-25.
2. Colditz GA. Economic costs of obesity and inactivity. *Med Sci Sports Exerc* 1999; 31: S663-S667.
3. World Health Organization. *Preventing chronic diseases: a vital investment.* World Health Organization: Geneva, 2005.
4. Ganz ML. The economic evaluation of obesity interventions: its time has come. *Obes Res* 2003; 11: 1275-1277.
5. Roux L, Kuntz KM, Donaldson C, Goldie SJ. Economic evaluation of weight loss interventions in overweight and obese women. *Obesity (Silver Spring)* 2006; 14: 1093-1106.

6. Haby MM, Vos T, Carter R, Moodie M, Markwick A, Magnus A, et al. A new approach to assessing the health benefit from obesity interventions in children and adolescents: the assessing cost effectiveness in obesity project. *Int J Obes* 2006; 30: 1463-1475.

7. Roux L, Pratt M, Tengs TO, Yore MM, Yanagawa TL, Van Den Bos J, et al. Cost-effectiveness of community-based physical activity interventions. *Am J Prev Med* 2008; 35: 578-588.

8. Centers for Disease Control and Prevention. Guide to community preventive services (community guide). www.thecommunityguide.org/. Accessed 10 Sept 2008.

9. Centers for Disease Control and Prevention. Behavioral Risk Factor Surveillance System: 2003 survey data. www.cdc.gov/brfss/technical_infodata/surveydata/2003.htm.

Chapter 16

1. Dishman RK, Berthoud HR, Booth FW, Cotman CW, Edgerton VR, Fleshner MR, et al. Neurobiology of exercise. *Obesity (Silver Spring)* 2006; 14: 345-356.

2. Snitker S, Tataranni PA, Ravussin E. Spontaneous physical activity in a respiratory chamber is correlated to habitual physical activity. *Int J Obes Relat Metab Disord* 2001; 25: 1481-1486.

3. Levine JA, Eberhardt NL, Jensen MD. Role of nonexercise activity thermogenesis in resistance to fat gain in humans. *Science* 1999; 283: 212-214.

4. Kotz CM, Teske JA, Billington CJ. Neuroregulation of nonexercise activity thermogenesis and obesity resistance. *Am J Physiol Regul Integr Comp Physiol* 2008; 294: R699-710.

5. Levine JA, Kotz CM. NEAT—non-exercise activity thermogenesis—egocentric & geocentric environmental factors vs. biological regulation. *Acta Physiol Scand* 2005; 184: 309-318.

6. Tou JC, Wade CE. Determinants affecting physical activity levels in animal models. *Exp Biol Med (Maywood)* 2002; 227: 587-600.

7. Seburn KL. Measurements taken using the Comprehensive Lab Animal Monitoring System. MPD:92. Mouse Phenome Database Web Site, The Jackson Laboratory, Bar Harbor, Maine USA. World Wide Web (URL: http://www.jax.org/phenome).

8. Nogueiras R, Lopez M, Lage R, Perez-Tilve D, Pfluger P, Mendieta-Zeron H, et al. Bsx, a novel hypothalamic factor linking feeding with locomotor activity, is regulated by energy availability. *Endocrinology* 2008; 149: 3009-3015.

9. Schoeller DA. The importance of clinical research: the role of thermogenesis in human obesity. *Am J Clin Nutr* 2001; 73: 511-516.

10. Speakman J, Hambly C, Mitchell S, Krol E. Animal models of obesity. *Obes Rev* 2007; 8: 55-61.

Chapter 17

1. Rankinen T, Bouchard C. Genetics of physical activity. In: Clement K, Sorensen TIA (eds.), *Obesity genomics and postgenomics.* Informa Healthcare: New York, 2008, pp. 277-286.

2. Stubbe JH, Boomsma DI, Vink JM, Cornes BK, Martin NG, Skytthe A, et al. Genetic influences on exercise participation in 37,051 twin pairs from seven countries. *PLoS ONE* 2006; 20: 1: e22.

3. Cai G, Cole SA, Butte N, Bacino C, Diego V, Tan K, et al. A quantitative trait locus on chromosome 18q for physical activity and dietary intake in Hispanic children. *Obesity* 2006 Sept; 14: 1596-1604.

4. Bray MS, Hagberg JM, Perusse L, Rankinen T, Roth SM, Wolfarth B, et al. The human gene map for performance and health-related fitness phenotypes: the 2005 update. *Med Sci Sports Exerc* 2009; 41: 35-73.

5. Simonen RL, Rankinen T, Perusse L, Leon AS, Skinner JS, Wilmore JH, et al. A dopamine D2 receptor gene polymorphism and physical activity in two family studies. *Physiol Behav* 2003; 78: 751-757.

6. Loos RJ, Rankinen T, Tremblay A, Perusse L, Chagnon Y, Bouchard C. Melanocortin-4 receptor gene and physical activity in the Quebec Family Study. *Int J Obes (London)* 2005; 29: 420-428.

7. Stefan N, Vozarova B, Del Parigi A, Ossowski V, Thompson DB, Hanson RL, et al. The Gln223Arg polymorphism of the leptin receptor in Pima Indians: influence on energy expenditure, physical activity and lipid metabolism. *Int J Obes Relat Metab Disord* 2002; 26: 1629-1632.

8. Winnicki M, Accurso V, Hoffmann M, Pawlowski R, Dorigatti F, Santonastaso M, et al. Physical activity and angiotensin-converting enzyme gene polymorphism in mild hypertensives. *Am J Med Genet A* 2004; 125: 38-44.

9. Simonen RL, Rankinen T, Perusse L, Rice T, Rao DC, Chagnon Y, et al. Genome-wide linkage scan for physical activity levels in the Quebec Family Study. *Med Sci Sports Exerc* 2003; 35: 1355-1359.

10. De Moor MH, Spector TD, Cherkas LF, Falchi M, Hottenga JJ, Boomsma DI, et al. Genome-wide linkage scan for athlete status in 700 British female DZ twin pairs. *Twin Res Hum Genet* 2007; 10: 812-820.

11. De Moor MH, Posthuma D, Hottenga JJ, Willemsen G, Boomsma DI, De Geus EJ. Genome-wide linkage scan for exercise participation in Dutch sibling pairs. *Eur J Hum Genet* 2007; 15: 1252-1259.

12. De Moor MH, Stubbe JH, Boomsma DI, De Geus EJ. Exercise participation and self-rated health: do common genes explain the association? *Eur J Epidemiol* 2007; 22: 27-32.

Chapter 18

1. Gluckman PD, Hanson MA. Living with the past: evolution, development, and patterns of disease. *Science* 2004; 305: 1733-1736.

2. Vickers MH, Breier BH, McCarthy D, Gluckman PD. Sedentary behavior during postnatal life is determined by the prenatal environment and exacerbated by postnatal hypercaloric nutrition. *Am J Physiol Regul Integr Comp Physiol* 2003; 285: R271-273.

3. Watkins AJ, Wilkins A, Cunningham C, Perry VH, Seet MJ, Osmond C, et al. Low protein diet fed exclusively during mouse oocyte maturation leads to behavioural and cardiovascular abnormalities in offspring. *J Physiol* 2008; 586: 2231-2244.

4. Owen D, Matthews SG. Repeated maternal glucocorticoid treatment affects activity and hippocampal NMDA receptor expression in juvenile guinea pigs. *J Physiol* 2007; 578: 249-257.

5. Dahlgren J, Nilsson C, Jennische E, Ho HP, Eriksson E, Niklasson A, et al. Prenatal cytokine exposure results in obesity and gender-specific programming. *Am J Physiol Endocrinol Metab* 2001; 281: E326-334.

6. Meaney MJ, Szyf M. Environmental programming of stress responses through DNA methylation: life at the interface between a dynamic environment and a fixed genome. *Dialog Clin Neurosci* 2005; 7: 103-123.

7. Vickers MH, Gluckman PD, Coveny AH, Hofman PL, Cutfield WS, Gertler A, et al. Neonatal leptin treatment reverses developmental programming. *Endocrinology* 2005; 146: 4211-4216.
8. Waterland RA, Jirtle RL. Early nutrition, epigenetic changes at transposons and imprinted genes, and enhanced susceptibility to adult chronic diseases. *Nutrition* 2004; 20: 63-68.
9. Gluckman PD, Lillycrop KA, Vickers MH, Pleasants AB, Phillips ES, Beedle AS, et al. Metabolic plasticity during mammalian development is directionally dependent on early nutritional status. *Proc Natl Acad Sci USA* 2007; 104: 12796-12800.
10. Lillycrop KA, Phillips ES, Jackson AA, Hanson MA, Burdge GC. Dietary protein restriction of pregnant rats induces and folic acid supplementation prevents epigenetic modification of hepatic gene expression in the offspring. *J Nutr* 2005; 135: 1382-1386.

Chapter 19

1. Evans R. *Interpreting and addressing inequalities in health: from Black to Acheson to Blair to . . . ?* Office of Health Economics: London, UK, 2002.
2. Galobardes B, Lynch J, Davey Smith G. Measuring socioeconomic position in health research. *Br Med Bull* 2007; 81-82: 21-37.
3. Marmot M. Health in an unequal world. *Lancet* 2006; 368: 2081-2094.
4. McLaren L. Socioeconomic status and obesity. *Epidemiol Rev* 2007; 29: 29-48.
5. Batty GD, Leon DA. Socio-economic position and coronary heart disease risk factors in children and young people. *Eur J Pub Health* 2002; 12: 263-272.
6. Gidlow C, Johnston LH, Crone D, Ellis N, James D. A systematic review of the relationship between socio-economic position and physical activity. *Health Educ J* 2006; 65: 338-367.
7. Chaudhary N, Krieger N. Nutrition and physical activity interventions for low-income populations. *Can J Diet Pract Res* 2007; 68: 201-206.
8. Krieger N. A glossary for social epidemiology. *J Epidemiol Comm Health* 2001; 55: 693-700.
9. Kuh D, Ben-Shlomo Y, Lynch J, Hallqvist J, Power C. Life course epidemiology. *J Epidemiol Comm Health* 2003; 57: 778-783.
10. Pabayo R, Gauvin L. Proportions of students who use various modes of transportation to and from school in a representative population-based sample of children and adolescents, 1999. *Prev Med* 2008 46: 63-66.
11. Lopez AD, Mathers CD, Ezzati M, Jamison DT, Murray CJL. *Global burden of disease and risk factors.* World Bank and Oxford University Press: New York, 2006.

Chapter 20

1. U.S. Department of Health and Human Services. *Physical activity and health: a report of the surgeon general.* Centers for Disease Control and Prevention, National Centers for Chronic Disease Prevention and Health Promotion: Atlanta, 1996.
2. Crespo CJ, Ainsworth BE, Keteyian SJ, Heath GW, Ellen M. Prevalence of physical inactivity and its relation to social class in U.S. adults: results from the Third National Health and Nutrition Examination Survey, 1988-1994. *Med Sci Sports Exerc* 1996; 31: 1821-1827.
3. Macera CA, Ham SA, Yore MM, Jones DA, Ainsworth BE, Kimsey CD, et al. Prevalence of physical activity in United States: Behavioral Risk Factor Surveillance System. *Prev Chronic Dis* 2005; 2: A17.
4. Troiano RP, Berrigan D, Dodd KW, Masse LC, Tilert T, McDowell M. Physical activity in the United States measured by accelerometer. *Med Sci Sports Exerc* 2008; 40: 181-188.
5. Marshall SJ, Jones DA, Ainsworth BE, Reis JP, Levy SS, Macera CA. Race/ethnicity, social class, and leisure-time physical inactivity. *Med Sci Sports Exerc* 2007; 39: 44-51.
6. Bowman SA. Television-viewing characteristics of adults: correlations to eating practices and overweight and health status. *Prev Chronic Dis* 2006; 3: A38.
7. Matthews CE, Chen KY, Freedson PS, Buchowski MS, Beech BM, Pate RR, et al. Amount of time spent in sedentary behavior in the United States, 2003-2004. *Am J Epidemiol* 2008; 167: 875-881.
8. Andersen RE, Crespo CJ, Bartlett SJ, Cheskin LJ, Pratt M. Relationship of physical activity and television watching with body weight and level of fatness among children. Results from the third National Health and Nutrition Examination Survey. *JAMA* 1998; 279: 938-942.
9. Duke J, Heitzler C. Physical activity levels among children aged 9-13 years: United States, 2002. *MMWR* 2003; 52: 785-788.
10. Gordon-Larsen P, Adair LS, Popkin BM. Ethnic differences in physical activity and inactivity patterns and overweight status. *Obes Res* 2002; 10(3): 141-149.

Chapter 21

1. Dishman RK, Motl RW, Saunders R, Felton G, Ward DS, Dowda M, Pate RR. Self-efficacy partially mediates the effect of a school-based physical-activity intervention among adolescent girls. *Prev Med* 2004; 38: 628-636.
2. Rhodes RE, Smith NE. Personality correlates of physical activity: a review and meta-analysis. *Br J Sports Med* 2006; 40: 958-965.
3. Bandura A. Health promotion by social cognitive means. *Health Educ Behav* 2004; 31: 143-164.
4. Dishman RK, Saunders R, Dowda M, Felton G, Ward D, Pate RR. Goals and intentions mediate efficacy beliefs and declining physical activity in girls. *Am J Prev Med* 2006; 31: 475-483.
5. Ferreira I, van der Horst K, Wendel-Vos W, Kremers S, van Lenthe FJ, Brug J. Environmental correlates of physical activity in youth—a review and update. *Obes Rev* 2007; 8: 129-154.
6. Humpel N, Owen N, Leslie E. Environmental factors associated with adults' participation in physical activity: a review. *Am J Prev Med* 2002; 22: 188-199.
7. Dowda M, Dishman RK, Pfeiffer KA, Pate RR. Family support for physical activity in girls from 8th to 12th grade in South Carolina. *Prev Med* 2007; 44: 153-159.
8. Motl RW, Dishman RK, Saunders RP, Dowda M, Pate RR. Perceptions of physical and social environment variables and self-efficacy as correlates of self-reported physical activity among adolescent girls. *J Pediatr Psychol* 2007; 32: 6-12.
9. Dishman RK, Berthoud HR, Booth FW, Cotman CW, Edgerton VR, Fleshner MR, et al. Neurobiology of exercise. *Obesity (Silver Spring)* 2006; 14: 345-356.
10. Werme M, Messer C, Olson L, Gilden L, Thorén P, Nestler EJ, et al. Delta FosB regulates wheel running. *J Neurosci* 2002; 22: 8133-8138.

11. Kotz CM, Wang C, Teske JA, Thorpe AJ, Novak CM, Kiwaki K, et al. Orexin A mediation of time spent moving in rats: neural mechanisms. *Neuroscience* 2006; 142: 29-36.
12. Dishman RK. Gene-physical activity interactions in the etiology of obesity: behavioral considerations. *Obesity (Silver Spring)* 2008; 16: 560-65.

Chapter 22

1. Sallis J, Cervero RB, Ascher W, Henderson KA, Kraft MK, Kerr J. An ecological approach to creating active living communities. *Annu Rev Pub Health* 2006; 27: 297-322.
2. Kraft MK, Sallis JF, Vernez Moudon A, Linton LS (guest eds.). Active Living Research Supplement. *J Phys Act Health* 2006; 3(1): S1-S272.
3. Gebel K, Bauman AE, Petticrew M. The physical environment and physical activity: a critical appraisal of review articles. *Am J Prev Med* 2007; 32: 361-369.
4. Heath GW, Brownson RC, Kruger J, Miles R, Powell KE, Ramsey LT, and the Task Force on Community Preventive Services. The effectiveness of urban design and land use and transport policies and practices to increase physical activity: a systematic review. *J Phys Act Health* 2006; 3: S55-S76.
5. Cervero R, Kockelman KM. Travel demand and the 3Ds: density, diversity, and design. *Transport Res-D* 1997; 2: 199-219.
6. Sallis JF, Kerr J. Physical activity and the built environment. *Pres Counc Phys Fit Sports Res Dig* 2006; series 7, no. 4: 1-8.
7. Kerr J, Rosenberg D, Sallis JF, Saelens BE, Frank LD, Conway TL. Active commuting to school: associations with built environment and parental concerns. *Med Sci Sports Exerc* 2006; 38: 787-794.
8. Sallis JF, Conway TL, Prochaska JJ, McKenzie TL, Marshall SP, Brown M. The association of school environments with youth physical activity. *Am J Pub Health* 2001; 91: 618-620.
9. Frank LD, Saelens BE, Powell KE, Chapman JE. Stepping towards causation: do built environments or neighborhood and travel preferences explain physical activity, driving, and obesity? *Soc Sci Med* 2007; 65: 1898-1914.
10. Active Living Research. Designing for active living among adults. 2007. www.activelivingresearch.org.
11. Active Living Research. Designing for active living among children. 2008. www.activelivingresearch.org.

Chapter 23

1. Bauman A, lman-Farinelli M, Huxley R, James WP. Leisure-time physical activity alone may not be a sufficient public health approach to prevent obesity—a focus on China. *Obes Rev* 2008; 9: 119-126.
2. Blanck HM, McCullough ML, Patel AV, Gillespie C, Calle EE, Cokkinides VE, et al. Sedentary behavior, recreational physical activity, and 7-year weight gain among postmenopausal U.S. women. *Obesity (Silver Spring)* 2007; 15: 1578-1588.
3. Owen N, Leslie E, Salmon J, Fotheringham MJ. Environmental determinants of physical activity and sedentary behavior. *Exerc Sport Sci Rev* 2000; 28: 153-158.
4. Matthews CE, Chen KY, Freedson PS, Buchowski MS, Beech BM, Pate RR, et al. Amount of time spent in sedentary behaviors in the United States, 2003-2004. *Am J Epidemiol* 2008; 167: 875-881.
5. Shields M, Tremblay MS. Sedentary behaviour and obesity. *Health Rep* 2008; 19: 19-30.
6. Andersen RE, Franckowiak SC, Zuzak KB, Cummings ES, Bartlett SJ, Crespo CJ. Effects of a culturally sensitive sign on the use of stairs in African American commuters. *Soz Praventiv Med* 2006; 51: 373-380.
7. United States Census Bureau. Most of us still drive to work–alone. 2008. www.census.gov/Press-Release/www/releases/archives/american_community_survey_acs/010230.html.
8. Jakes RW, Day NE, Khaw KT, Luben R, Oakes S, Welch A, et al. Television viewing and low participation in vigorous recreation are independently associated with obesity and markers of cardiovascular disease risk: EPIC-Norfolk population-based study. *Eur J Clin Nutr* 2003; 57: 1089-1096.
9. Andersen RE, Wadden TA, Bartlett SJ, Zemel B, Verde TJ, Franckowiak SC. Effects of lifestyle activity vs structured aerobic exercise in obese women. *JAMA* 1999; 281: 335-340.
10. Weiss EC, Galuska DA, Kettel KL, Gillespie C, Serdula MK. Weight regain in U.S. adults who experienced substantial weight loss, 1999-2002. *Am J Prev Med* 2007; 33: 34-40.
11. Raynor DA, Phelan S, Hill JO, Wing RR. Television viewing and long-term weight maintenance: results from the National Weight Control Registry. *Obesity (Silver Spring)* 2006; 14: 1816-1824.
12. The Nielsen Company. Nielsen's three screen report, May 2008. 1-4.

Chapter 24

1. Matthews CE, Chen KY, Freedson PS, Buchowski MS, Beech BM, Pate RR, et al. Amount of time spent in sedentary behaviors in the United States, 2003-2004. *Am J Epidemiol* 2008; 167: 875-881.
2. Jordan AB, Robinson TN. Children, television viewing, and weight status: summary and recommendations from an expert panel meeting. *Ann Am Acad Pol Soc Sci* 2008; 615: 119-132.
3. Epstein LH, Roemmich JN, Robinson JL, Paluch RA, Winiewicz DD, Fuerch JH, et al. A randomized trial of the effects of reducing television viewing and computer use on body mass index in young children. *Arch Pediatr Adolesc Med* 2008; 162: 239-245.
4. Wiecha JL, Peterson KE, Ludwig DS, Kim J, Sobol A, Gortmaker SL. When children eat what they watch: impact of television viewing on dietary intake in youth. *Arch Pediatr Adolesc Med* 2006; 160: 436-442.
5. Institute of Medicine, Committee on Food Marketing and the Diets of Children and Youth, Food and Nutrition Board. *Food marketing to children and youth: threat or opportunity?* National Academies Press: Washington, DC, 2006.
6. Westerterp KR. Pattern and intensity of physical activity. *Nature* 2001; 410: 539-539.
7. Dietz WH, Gortmaker SL. Do we fatten our children at the television set? Obesity and television viewing in children and adolescents. *Pediatrics* 1985; 75: 807-812.
8. Robinson TN. Reducing children's television viewing to prevent obesity: a randomized controlled trial. *JAMA* 1999; 282: 1561-1567.
10. Roberts DF, Foehr UG, Rideout V. *Generation M: media in the lives of 8-18 year olds.* Henry J. Kaiser Family Foundation: Menlo Park, CA, 2005.

Chapter 25

1. Brien SE, Katzmarzyk PT. Physical activity and metabolic syndrome in Canada. *Appl Physiol Nutr Metab* 2006; 31: 40-47.
2. Van Pelt RE, Davy KP, Stevenson ET, Wilson TM, Jones PP, Desouza CA, et al. Smaller differences in total and regional adiposity with age in women who regularly perform endurance exercise. *Am J Physiol* 1998; 275: E626-E624.
3. Ball K, Owen N, Salmon J, Bauman A, Gore CJ. Associations of physical activity with body weight and fat in men and women. *Int J Obes* 2001; 25: 914-919.
4. Lee I-M, Paffenbarger R. Associations of light, moderate, and vigorous intensity physical activity with longevity: the Harvard Alumni Health Study. *Am J Epidemiol* 2000; 151: 293-299.
5. French SA, Jeffery RW, Forster JL, Kelder SH, Baxter JE. Predictors of weight change over two years among a population of working adults: the Healthy Worker Project. *Int J Obes* 1994; 18: 145-154.
6. DiPietro L, Dziura J, Blair SN. Estimated change in physical activity level (PAL) and prediction of 5-year weight change in men: the Aerobics Center Longitudinal Study. *Int J Obes* 2004; 28: 1541-1547.
7. DiPietro L, Kohl HW, Barlow CE, Blair SN. Improvements in cardiorespiratory fitness attenuate age-related weight gain in healthy men and women: the Aerobics Center Longitudinal Study. *Int J Obes* 1998; 22: 55-62.
8. Foster JA, Gore SA, West DS. Altering TV viewing habits: an unexplored strategy for adult obesity intervention? *Am J Health Behav* 2006; 30: 3-14.
9. Ball K, Brown W, Crawford D. Who does not gain weight? Prevalence and predictors of weight maintenance in young women. *Int J Obes* 2002; 26: 1570-1578.
10. Hu FB, Li TY, Colditz GA, Willett WC, Manson JE. Television watching and other sedentary behaviors in relation to risk of obesity and type 2 diabetes mellitus in women. *JAMA* 2003; 289: 1785-1791.

Chapter 26

1. National Association for Sport and Physical Education, Council on Physical Education for Children. *Physical activity for children: a statement of guidelines for children ages 5-12.* 2nd ed. 2004.
2. Institute of Medicine, National Academy of Science. *Dietary reference intakes for energy, carbohydrate, fiber, fat, fatty acids, cholesterol, protein, and amino acids (macronutrients).* National Academy Press: Washington, DC, 2002.
3. Torun B, Davies PSW, Livingstone MBE, Paolisso M, Sackett R, Spurr GB. Energy requirements and dietary recommendations for children and adolescents 1 to 18 years old. *Eur J Clin Nutr* 1998; 50: S37-S80.
4. Butte NF, Christiansen E, Sorensen TIA. Energy imbalance underlying the development of childhood obesity. *Obesity* 2007; 15: 3056-3066.
5. Moore LL, Gao D, Bradlee ML, Cupples LA, Sundarajan-Ramamurti A, Proctor MH, et al. Does early physical activity predict body fat change throughout childhood? *Prev Med* 2003; 37: 10-17.
6. Must A, Tybor DJ. Physical activity and sedentary behavior: a review of longitudinal studies of weight and adiposity in youth. *Int J Obes* 2005; 29: S84-S96.
7. Goran MI, Shewchuk R, Gower BA, Nagy TR, Carpenter WH, Johnson RK. Longitudinal changes in fatness in white children: no effect of childhood energy expenditure. *Am J Clin Nutr* 1998; 67: 309-316.
8. Johnson MS, Figueroa-Colon R, Herd SL, Fields DA, Sun M, Hunter GR, et al. Aerobic fitness, not energy expenditure predicts increasing adiposity in African-American and Caucasian children. *Pediatrics* 2000; 106: e50-e56.
9. Figueroa-Colon R, Arani RB, Goran MI, Weinsier RL. Paternal body fat is a longitudinal predictor of changes in body fat in premenarcheal girls. *Am J Clin Nutr* 2000; 71: 829-834.
10. Salbe AD, Weyer C, Harper I, Lindsay RS, Ravussin E, Tataranni PA. Assessing risk factors for obesity between childhood and adolescence: II. Energy metabolism and physical activity. *Pediatrics* 2002; 110: 307-314.
11. Treuth MS, Butte NF, Sorkin JD. Predictors of body fat gain in non-obese girls with a familial predisposition to obesity. *Am J Clin Nutr* 2003; 78: 1212-1218.
12. Bandini LG, Must A, Phillips SM, Naumova EN, Dietz WH. Relation of body mass index and body fatness to energy expenditure: longitudinal changes from pre-adolescence through adolescence. *Am J Clin Nutr* 2004; 80: 1262-1269.

Chapter 27

1. Malina RM, Bouchard C, Bar-Or O. *Growth, maturation, and physical activity.* 2nd ed. Human Kinetics: Champaign, IL, 2004.
2. Malina RM, Howley E, Gutin B. Body mass and composition. *Report prepared for the Youth Health subcommittee, Physical Activity Guidelines Advisory Committee,* 2007.
3. Moore LL, Gao D, Bradlee ML, Cupples LA, Sundarajan-Ramamurti A, Proctor MH, et al. Does early physical activity predict body fat change throughout childhood? *Prev Med* 2003; 37: 10-17.
4. Mundt CA, Baxter-Jones ADG, Whiting SJ, Bailey DA, Faulkner RA, Mirwald RL. Relationships of activity and sugar drink intake on fat mass development in youths. *Med Sci Sports Exerc* 2006; 38: 1245-1254.
5. Twisk JWR, Van Mechelen W, Kemper HCG, Post GB. The relation between "long term exposure" to lifestyle during youth and young adulthood and risk factors for cardiovascular disease at adult age. *J Adol Health* 1997; 20: 309-319.
6. Raitakari OT, Porkka KV, Taimela S, Telama R, Rasanen L, Viikari JS. Effects of persistent physical activity and inactivity on coronary risk factors in children and young adults: the Cardiovascular Risk in Young Finns Study. *Am J Epidemiol* 1994; 140: 195-205.
7. Viner RM, Cole TJ. Who changes body mass between adolescence and adulthood? Factors predicting change in BMI between 16 years and 30 years in the 1970 British birth cohort. *Int J Obes* 2006; 30: 1368-1374.
8. Sherar LB, Mirwald RL, Erlandson MC, Baxter-Jones ADG. Is boys' physical activity in childhood associated with being overweight in mid-adulthood? A longitudinal study spanning 35 years. *Can Stud Pop* 2007; 34: 85-99.
9. Lefevre J, Philippaerts R, Delvaux K, Thomis M, Claessens AL, Lysens R, et al. Relation between cardiovascular risk factors at adult age, and physical activity during youth and adulthood: the Leuven Longitudinal Study of Lifestyle, Fitness and Health. *Int J Sports Med* 2002; 23: S32-S38.
10. Hancox RJ, Milne BJ, Poulton R. Association between child and adolescent television viewing and adult health: a longitudinal birth cohort study. *Lancet* 2004; 364: 257-262.

11. Tammelin T, Laitinen J, Näyhä S. Change in the level of physical activity from adolescence into adulthood and obesity at the age of 31 years. *Int J Obes* 2004; 28: 775-782.

Chapter 28

1. Nader PR, O'Brien M, Houts R, Bradley R, Belsky J, Crosnoe R, et al. Identifying risk for obesity in early childhood. *Pediatrics* 2006; 118: 594-601.
2. Beunen G, Lefèvre J, Claessens AL, Lysens R, Maes H, Renson R, et al. Age-specific correlation analysis of longitudinal physical fitness levels in men. *Eur J Appl Physiol* 1992; 64: 538-545.
3. Twisk JWR, Kemper HCG, van Mechelen W, Post GB. Tracking of risk factors for coronary heart disease over a 14-year period: a comparison between lifestyle and biologic risk factors with data from the Amsterdam Growth and Health Study. *Am J Epidemiol* 1997; 145: 888-898.
4. Bloom BS. *Stability and change in human characteristics.* Wiley: New York, 1964.
5. Casey VA, Dwyer JT, Coleman KA, Valadian I. Body mass index from childhood to middle age: a 50-year follow-up. *Am J Clin Nutr* 1992; 56: 14-18.
6. Katzmarzyk PT, Perusse L, Malina RM, Bouchard C. Seven-year stability of indicators of obesity and adipose tissue distribution in the Canadian population. *Am J Clin Nutr* 1999; 69: 1123-1129.
7. Conner Gorber SC, Tremblay M, Moher D, Gorber B. A comparison of direct versus self-report measures for assessing height, weight and body mass index: a systematic review. *Obes Rev* 2007; 8: 307-326.
8. Whitaker RC, Wright JA, Pepe MS, Seidel KD, Dietz WH. Predicting obesity in young adulthood from childhood and parental obesity. *N Engl J Med* 1997; 337: 869-873.
9. Trudeau F, Shephard RJ, Arsenault F, Laurencelle L. Changes in adiposity and body mass index from late childhood to adult life in the Trois-Rivières study. *Am J Hum Biol* 2001; 13: 349-355.
10. Singh AS, Mulder C, Twisk JW, van Mechelen W, Chinapaw MJ. Tracking of childhood overweight into adulthood: a systematic review of the literature. *Obes Rev* 2008; 9: 474-488.
11. Clarke WR, Lauer RM. Does childhood obesity track into adulthood? *Crit Rev Food Sci Nutr* 1993; 33: 423-430.
12. Freedman DS, Khan LK, Serdula MK, Dietz WH, Srinivasan SR, Berenson GS. Racial differences in the tracking of childhood BMI to adulthood. *Obes Res* 2005; 13: 928-935.
13. Garn SM, LaVelle M. Two-decade follow-up of fatness in early childhood. *Am J Dis Child* 1985; 139: 181-185.
14. Sun SS, Liang R, Huang TT, Daniels SR, Arslanian S, Liu K, et al. Childhood obesity predicts adult metabolic syndrome: the Fels Longitudinal Study. *J Pediatr* 2008; 152: 191-200.
15. Magarey AM, Daniels LA, Boulton TJ, Cockington RA. Predicting obesity in early adulthood from childhood and parental obesity. *Int J Obes Relat Metab Disord* 2003; 27: 505-513.

Chapter 29

1. Ogden CL, Carroll MD, Curtin LR, McDowell MA, Tabak CJ, Flegal KM. Prevalence of overweight and obesity in the United States, 1999-2004. *JAMA* 2006; 295: 1549-1555.
2. Chumlea WC, Guo SS, Kuczmarski RJ, Flegal KM, Johnson CL, Heymsfield SB, et al. Body composition estimates from NHANES III bioelectrical impedance data. *Int J Obes Relat Metab Disord* 2002; 26: 1596-1609.
3. Kyle UG, Genton L, Gremion G, Slosman DO, Pichard C. Aging, physical activity and height-normalized body composition parameters. *Clin Nutr* 2004; 23: 79-88.
4. Koster A, Patel KV, Visser M, van Eijk JT, Kanaya AM, de Rekeneire N, et al. Joint effects of adiposity and physical activity on incident mobility limitation in older adults. *J Am Geriatr Soc* 2008; 56: 636-643.
5. Wang YC, Colditz GA, Kuntz KM. Forecasting the obesity epidemic in the aging U.S. population. *Obesity (Silver Spring)* 2007; 15: 2855-2865.
6. Centers for Disease Control and Prevention. Prevalence of regular physical activity among adults—United States, 2001 and 2005. *MMWR* 2007; 56: 1209-1212.
7. Pleis JR, Lethbridge-Cejku M. Summary health statistics for U.S. adults: National Health Interview Survey, 2006. *Vital Health Stat 10* 2007; 1-153.
8. Marcell TJ, Hawkins SA, Tarpenning KM, Hyslop DM, Wiswell RA. Longitudinal analysis of lactate threshold in male and female master athletes. *Med Sci Sports Exerc* 2003; 35: 810-817.
9. Villareal DT, Apovian CM, Kushner RF, Klein S. Obesity in older adults: technical review and position statement of the American Society for Nutrition and NAASO, The Obesity Society. *Obes Res* 2005; 13: 1849-1863.
10. Lang IA, Guralnik JM, Melzer D. Physical activity in middle-aged adults reduces risks of functional impairment independent of its effect on weight. *J Am Geriatr Soc* 2007; 55: 1836-1841.

Chapter 30

1. Chen LJ, Fox KR, Haase A, Wang JM. Obesity, fitness and health in Taiwanese children and adolescents 2006. *Eur J Clin Nutr* 2006; 60(12): 1367-1375.
2. Lee SJ, Arslanian SA. Cardiorespiratory fitness and abdominal adiposity in youth. *Eur J Clin Nutr* 2007; 61(4): 561-565.
3. Johnson MS, Figueroa-Colon R, Herd SL, Fields DA, Sun M, Hunter GR, et al. Aerobic fitness, not energy expenditure, influences subsequent increase in adiposity in black and white children. *Pediatrics* 2000; 106(4): E50.
4. Byrd-Williams CE, Shaibi GQ, Sun P, Lane CJ, Ventura EE, Davis JN, et al. Cardiorespiratory fitness predicts changes in adiposity in overweight hispanic boys. *Obesity (Silver Spring)* 2008; 16: 1072-1077.
5. Ross R, Katzmarzyk PT. Cardiorespiratory fitness is associated with diminished total and abdominal obesity independent of body mass index. *Int J Obes Relat Metab Disord* 2003; 27(2): 204-210.
6. Wong SL, Katzmarzyk P, Nichaman MZ, Church TS, Blair SN, Ross R. Cardiorespiratory fitness is associated with lower abdominal fat independent of body mass index. *Med Sci Sports Exerc* 2004; 36(2): 286-291.
7. DiPietro L, Kohl HW 3rd, Barlow CE, Blair SN. Improvements in cardiorespiratory fitness attenuate age-related weight gain in healthy men and women: the Aerobics Center Longitudinal Study. *Int J Obes Relat Metab Disord* 1998; 22(1): 55-62.
8. Sidney S, Sternfeld B, Haskell WL, Quesenberry CP Jr., Crow RS, Thomas RJ. Seven-year change in graded exercise treadmill test performance in young adults in the CARDIA study. Cardiovascular Risk Factors in Young Adults. *Med Sci Sports Exerc* 1998; 30(3): 427-433.

9. Mason C, Brien SE, Craig CL, Gauvin L, Katzmarzyk P. Musculoskeletal fitness and weight gain in Canada. *Med Sci Sports Exerc* 2007; 39(1): 38-43.

10. Brien SE, Katzmarzyk PT, Craig CL, Gauvin L. Physical activity, cardiorespiratory fitness and body mass index as predictors of substantial weight gain and obesity: the Canadian physical activity longitudinal study. *Can J Pub Health* 2007; 98(2): 121-124.

Chapter 31

1. Eaton CB, McPhillips JB, Gans KM, Garber CE, Assaf A, Lasater TM, et al. Cross-sectional relationship between diet and physical activity in two southeastern New England communities. *Am J Prev Med* 1995; 11: 238-244.

2. Rodrigo CP, Van Praagh E, Gibney M, Sjostrom M. ILSI Europe workshop on Diet and Physical Activity–Interactions for Health: summary and conclusions, 22-24 March, 1999 in Chamonix, France. International Life Sciences Institute. *Pub Health Nut* 1999; 2: 321-325.

3. Matthews CE, Hebert JR, Ockene IS, Saperia G, Merriam PA. Relationship between leisure time physical activity and selected dietary variables in the Worcester Area Trial for Counseling in Hyperlipidemia. *Med Sci Sports Exerc* 1997; 29: 1199-1207.

4. Brodney S, McPherson RS, Carpenter RS, Welten D, Blair SN. Nutrient intake of physically fit and unfit men and women. *Med Sci Sports Exerc* 2001; 33: 459-467.

5. Jago R, Nicklas T, Yang SJ, Baranowski T, Zakeri I, Berenson GS. Physical activity and health enhancing dietary behaviors in young adults: Bogalusa Heart Study. *Prev Med* 2005; 41: 194-202.

6. Gillman MW, Pinto BM, Tennstedt S, Glanz K, Marcus B, Friedman RH. Relationships of physical activity with dietary behaviors among adults. *Prev Med* 2001; 32: 295-301.

7. Jago R, Baranowski T, Yoo S, Cullen KW, Zakeri I, Watson K, et al. Relationship between physical activity and diet among African-American girls. *Obes Res* 2004; 12: 55S-63S.

8. Elder SJ, Roberts SB. The effects of exercise on food intake and body fatness: a summary of published studies. *Nutr Rev* 2007; 65: 1-19.

9. Thompson D, Jago R, Baranowski T, Watson K, Zakeri I, Cullen KW, et al. Covariability in diet and physical activity in African-American girls. *Obes Res* 2004; 12: 46S-54S.

10. Lioret S, Touvier M, Lafay L, Volatier JL, Maire B. Dietary and physical activity patterns in French children are related to overweight and socioeconomic status. *J Nutr* 2008; 138: 101-107.

Chapter 32

1. Tremblay A, Fontaine E, Poehlman ET, Mitchell D, Perron L, Bouchard C. The effect of exercise-training on resting metabolic rate in lean and moderately obese individuals. *Int J Obes* 1986; 10: 511-517.

2. Tremblay A, Fontaine E, Nadeau A. Contribution of post-exercise increment in glucose storage to variations in glucose-induced thermogenesis in endurance athletes. *Can J Physiol Pharmacol* 1985; 63: 1165-1169.

3. Tremblay A, Coveney S, Després JP, Nadeau A, Prud'homme D. Increased resting metabolic rate and lipid oxidation in exercise-trained individuals: evidence for a role of beta adrenergic stimulation. *Can J Physiol Pharmacol* 1992; 70: 1342-1347.

4. Plourde G, Rousseau-Migneron S, Nadeau A. Effect of endurance training on beta-adrenergic system in three different skeletal muscles. *J Appl Physiol* 1993; 74: 164-166.

5. Plourde G, Rousseau-Migneron S, Nadeau A. β-Adrenoreceptor adenylate cyclase system adaptation to physical training in rat ventricular tissue. *J Appl Physiol* 1991; 70: 1633-1638.

6. Tremblay A, Nadeau A, Fournier G, Bouchard C. Effect of a three-day interruption of exercise-training on resting metabolic rate and glucose-induced thermogenesis in trained individuals. *Int J Obes* 1988; 12: 163-168.

7. Bielinski R, Schutz Y, Jequier E. Energy metabolism during the post-exercise recovery in man. *Am J Clin Nutr* 1985; 42: 69-82.

8. Yoshioka M, Doucet E, St-Pierre S, Alméras N, Richard D, Labrie A, et al. Impact of high-intensity exercise on energy expenditure, lipid oxidation and body fatness. *Int J Obes* 2001; 25: 332-339.

9. Poehlman ET, Tremblay A, Nadeau A, Dussault J, Thériault G, Bouchard C. Heredity and changes in hormones and metabolic rates with short-term training. *Am J Physiol* 1986; 250: E711-E717.

10. Tremblay A, Poehlman ET, Després JP, Thériault G, Danforth E, Bouchard C. Endurance training with constant energy intake in identical twins: changes over time in energy expenditure and related hormones. *Metabolism* 1997; 46: 499-503.

11. Doucet E, St-Pierre S, Alméras N, Després JP, Bouchard C, Tremblay A. Evidence for the existence of adaptative thermogenesis during weight loss. *Br J Nutr* 2001; 85: 715-723.

12. Leibel RL, Rosenbaum M, Hirsch J. Changes in energy expenditure resulting from altered body weight. *New Engl J Med* 1995; 332: 621-628.

13. Pelletier C, Doucet E, Imbeault P, Tremblay A. Associations between weight-loss induced changes in plasma organochlorines, serum T_3 concentration and resting metabolic rate. *Toxicol Sci* 2002; 67: 46-51.

14. Tremblay A, Pelletier C, Doucet E, Imbeault P. Thermogenesis and weight loss in obese individuals: a primary association with organochlorine pollution. *Int J Obes* 2004; 28: 936-939.

Chapter 33

1. Jéquier E, Schutz Y. Resting energy expenditure, thermic effect of food, and total energy expenditure. In: Bray GA, Bouchard C (eds.), *Handbook of obesity, etiology and pathophysiology.* 2nd ed. Marcel Dekker: New York, 2004, pp. 615-629.

2. de Jonge L, Bray G. The thermic effect of food and obesity: a critical review. *Obes Res* 1997; 5: 622-631.

3. Granata GP, Brandon LJ. The thermic effect of food and obesity: discrepant results and methodological variations *Nutr Rev* 2002; 60: 295-297.

4. Levine JA. Non-exercise activity thermogenesis (NEAT). *Nutr Rev* 2004; 62: S82-97.

5. Segal KR, Gutin B, Albu J, Pi-Sunyer FX. Thermic effects of food and exercise in lean and obese men of similar lean body mass. *Am J Physiol* 1987; 252: E110-117.

6. Segal KR, Edaño A, Blando L, Pi-Sunyer FX. Comparison of thermic effects of constant and relative caloric loads in lean and obese men. *Am J Clin Nutr* 1990; 51: 14-21.

7. Segal KR, Blando L, Ginsberg-Fellner F, Edaño A. Postprandial thermogenesis at rest and postexercise before and after physical training in lean, obese, and mildly diabetic men. *Metabolism* 1992; 41: 868-878.

8. Broeder CE, Brenner M, Hofman Z, Paijmans IJ, Thomas EL, Wilmore JH. The metabolic consequences of low and moderate intensity exercise with or without feeding in lean and borderline obese males. *Int J Obes* 1991; 15: 95-104.
9. Davis JM, Sadri S, Sargent RG, Ward D. Weight control and calorie expenditure: thermogenic effects of pre-prandial and post-prandial exercise. *Addict Behav* 1989;14: 347-451
10. Nichols J, Ross S, Patterson P. Thermic effect of food at rest and following swim exercise in trained college men and women. *Ann Nutr Metab* 1988;32: 215-219

Chapter 34

1. Kelley DE, Mandarino LJ. Fuel selection in human skeletal muscle in insulin resistance: a reexamination. *Diabetes* 2000; 49: 677-683.
2. Ukropcova B, Sereda O, de Jonge L, Bogacka I, Nguyen T, Xie H, et al. Family history of diabetes links impaired substrate switching and reduced mitochondrial content in skeletal muscle. *Diabetes* 2007; 56: 720-727.
3. Simoneau JA, Colberg SR, Thaete FL, Kelley DE. Skeletal muscle glycolytic and oxidative enzyme capacities are determinants of insulin sensitivity and muscle composition in obese women. *FASEB J* 1995; 9: 273-278.
4. Ukropcova B, McNeil M, Sereda O, de Jonge L, Xie H, Bray GA, et al. Dynamic changes in fat oxidation in human primary myocytes mirror metabolic characteristics of the donor. *J Clin Invest* 2005; 115: 1934-1941.
5. Smith SR, de Jonge L, Zachwieja JJ, Roy H, Nguyen T, Rood J, et al. Concurrent physical activity increases fat oxidation during the shift to a high-fat diet. *Am J Clin Nutr* 2000; 72: 131-138.
6. Prats C, Donsmark M, Qvortrup K, Londos C, Sztalryd C, Holm C, et al. Decrease in intramuscular lipid droplets and translocation of HSL in response to muscle contraction and epinephrine. *J Lipid Res* 2006; 47: 2392-2399.
7. van Aggel-Leijssen DP, Saris WH, Wagenmakers AJ, Senden JM, van Baak MA. Effect of exercise training at different intensities on fat metabolism of obese men. *J Appl Physiol* 2002; 92: 1300-1309.
8. Houmard JA, Tanner CJ, Slentz CA, Duscha BD, McCartney JS, Kraus WE. Effect of the volume and intensity of exercise training on insulin sensitivity. *J Appl Physiol* 2004; 96: 101-106.
9. Venables MC, Jeukendrup AE. Endurance training and obesity: effect on substrate metabolism and insulin sensitivity. *Med Sci Sports Exerc* 2008; 40: 495-502.
10. Toledo FG, Menshikova EV, Azuma K, Radiková Z, Kelley CA, Ritov VB, et al. Mitochondrial capacity in skeletal muscle is not stimulated by weight loss despite increases in insulin action and decreases in intramyocellular lipid content. *Diabetes* 2008; 57: 987-994.

Chapter 35

1. Astrup A, Macdonald IA. Sympathoadrenal system and metabolism. In: Bray GA, Bouchard C, James WPT (eds.), *Handbook of obesity.* Marcel Dekker: New York, 1998, pp. 491-511.
2. Wallin BG, Charkoudian N. Sympathetic neural control of integrated cardiovascular function: insights from measurement of human sympathetic nerve activity. *Muscle Nerve* 2007; 36: 595-614.
3. McMurray RG, Hackney AC. Interactions of metabolic hormones, adipose tissue and exercise. *Sports Med* 2005; 35: 393-412.
4. Bell C, Day DS, Jones PJ, Christou DD, Petitt DS, Osterberg KO, et al. High energy flux mediates the tonically augmented beta-adrenergic support of resting metabolic rate in habitually exercising older adults. *J Clin Endocrinol Metab* 2004; 89: 3573-3578.
5. Alvarez GE, Halliwill JR, Ballard TP, Beske SD, Davy KP. Sympathetic neural regulation in endurance-trained humans: fitness vs. fatness. *J Appl Physiol* 2005; 98: 498-502.
6. Christin L, O'Connell M, Bogardus C, Danforth E Jr., Ravussin E. Norepinephrine turnover and energy expenditure in Pima Indian and white men. *Metabolism* 1993; 42: 723-729.
7. Cornelissen VA, Fagard RH. Effects of endurance training on blood pressure, blood-pressure regulating mechanisms, and cardiovascular risk factors. *Hypertension* 2005; 46: 667-675.
8. Grassi G, Seravalle G, Calhoun DA, Mancia G. Physical training and baroreceptor control of sympathetic nerve activity in humans. *Hypertension* 1994; 23: 294-301.
9. Sheldahl LM, Ebert TJ, Cox B, Tristani FE. Effect of aerobic training on baroreflex regulation of cardiac and sympathetic function. *J Appl Physiol* 1994; 76: 158-165.
10. Iwane M, Arita M, Tomimoto S, Satani O, Matsumoto M, Miyashita K, et al. Walking 10,000 steps/day or more reduces blood pressure and sympathetic nerve activity in mild essential hypertension. *Hypertens Res* 2000; 23: 573-580.

Chapter 36

1. Langin D, Dicker A, Tavernier G, Hoffstedt J, Mairal A, Rydén M, et al. Adipocyte lipases and defect of lipolysis in human obesity. *Diabetes* 2005; 54: 3190-3197.
2. Schiffelers SL, Akkermans JA, Saris WH, Blaak EE. Lipolytic and nutritive blood flow response to beta-adrenoceptor stimulation in situ in subcutaneous abdominal adipose tissue in obese men. *Int J Obes Relat Metab Disord* 2003; 27: 227-231.
3. Horowitz JF, Klein S. Whole body and abdominal lipolytic sensitivity to epinephrine is suppressed in upper body obese women. *Am J Physiol* 2000; 278: E1144-E1152.
4. Stich V, De Glisezinski I, Crampes F, Hejnova J, Cottet-Emard JM, Galitzky J, et al. Activation of alpha2-adrenergic receptors impairs exercise-induced lipolysis in subcutaneous adipose tissue of obese subjects. *Am J Physiol* 2000; 279: R499-R504.
5. Moro C, Pillard F, De Glisezinski I, Crampes F, Thalamas C, Harant I, et al. Sex differences in lipolysis-regulating mechanisms in overweight subjects: effect of exercise intensity. *Obesity* 2007; 15: 2245-2255.
6. De Glisezinski I, Crampes F, Harant I, Berlan M, Hejnova J, Langin D, et al. Endurance training changes in lipolytic responsiveness of obese adipose tissue. *Am J Physiol* 1998; 275: E951-E956.
7. Moro C, Pillard F, De Glisezinski I, Harant I, Riviere D, Stich V, et al. Training enhances ANP lipid-mobilizing action in adipose tissue of overweight men. *Med Sci Sport Exerc* 2005; 37: 1126-1132.
8. Horowitz JF, Braudy RJ, Martin III WH, Klein S. Endurance exercise training does not alter lipolytic or adipose tissue blood flow sensitivity to epinephrine. *Am J Physiol* 1999; 277: E325-E331.
9. Richterova B, Stich V, Moro C, Polak J, Klimcakova E, Majercik M, et al. Effect of endurance training on adrenergic control of lipolysis in adipose tissue of obese women. *J Clin Endocrinol Metab* 2004; 89: 1325-1331.

10. De Glisezinski I, Moro C, Pillard F, Marion-Latard F, Harant I, Meste M, et al. Aerobic training improves exercise-induced lipolysis in SCAT and lipid utilization in overweight men. *Am J Physiol* 2003; 285: E984-E990.
11. Polak J, Moro C, Klimcakova E, Hejnova J, Majercik M, Viguerie N, et al. Dynamic strength training improves insulin sensitivity and functional balance between adrenergic alpha 2A and beta pathways in subcutaneous adipose tissue of obese subjects. *Diabetologia* 2005; 48: 2631-2640.

Chapter 37

1. Yaspelkis BB 3rd, Singh MK, Krisan AD, Collins DE, Kwong CC, Bernard JR, et al. Chronic leptin treatment enhances insulin-stimulated glucose disposal in skeletal muscle of high-fat fed rodents. *Life Sci* 2004; 74: 1801-1816.
2. Steinberg GR, Dyck DJ. Development of leptin resistance in rat soleus muscle in response to high-fat diets. *Am J Physiol Endocrinol Metab* 2000; 279: E1374-1382.
3. Steinberg GR, Parolin ML, Heigenhauser GJ, Dyck DJ. Leptin increases FA oxidation in lean but not obese human skeletal muscle: evidence of peripheral leptin resistance. *Am J Physiol Endocrinol Metab* 2002; 283: E187-192.
4. Kraemer RR, Chu H, Castracane VD. Leptin and exercise. *Exp Biol Med* 2002; 227: 701-708.
5. Berggren JR, Hulver MW, Houmard JA. Fat as an endocrine organ: influence of exercise. *J Appl Physiol* 2005; 99: 757-764.
6. Hilton LK, Loucks AB. Low energy availability, not exercise stress, suppresses the diurnal rhythm of leptin in healthy young women. *Am J Physiol Endocrinol Metab* 2000; 278: E43-49.
7. Thong FS, Hudson R, Ross R, Janssen I, Graham TE. Plasma leptin in moderately obese men: independent effects of weight loss and aerobic exercise. *Am J Physiol Endocrinol Metab* 2000; 279: E307-313.
8. Steinberg GR, Smith AC, Wormald S, Malenfant P, Collier C, Dyck DJ. Endurance training partially reverses dietary-induced leptin resistance in rodent skeletal muscle. *Am J Physiol Endocrinol Metab* 2004; 286: E57-63.

Chapter 38

1. Berthoud HR, Morrison C. The brain, appetite, and obesity. *Annu Rev Psychol* 2008; 59: 55-92.
2. Saper CB, Scammell TE, Lu J. Hypothalamic regulation of sleep and circadian rhythms. *Nature* 2005; 437: 1257-1263.
3. Teske JA, Billington CJ, Kotz CM. Neuropeptidergic mediators of spontaneous physical activity and non-exercise activity thermogenesis. *Neuroendocrinology* 2008; 87: 71-90.
4. Kotz CM. Integration of feeding and spontaneous physical activity: role for orexin. *Physiol Behav* 2006; 88: 294-301.
5. Dishman RK, Berthoud HR, Booth FW, Cotman CW, Edgerton VR, Fleshner MR, et al. Neurobiology of exercise. *Obesity (Silver Spring)* 2006; 14: 345-356.
6. Bi S, Scott KA, Hyun J, Ladenheim EE, Moran TH. Running wheel activity prevents hyperphagia and obesity in Otsuka Long-Evans Tokushima fatty rats: role of hypothalamic signaling. *Endocrinology* 2005; 146: 1676-1685.

Chapter 39

1. Blundell JE, King NA. Physical activity and regulation of food intake: current evidence. *Med Sci Sports Exerc* 1999; 31: S573-583.
2. Drazen DL, Vahl TP, D'Alessio DA, Seeley RJ, Woods SC. Effects of a fixed meal pattern on ghrelin secretion: evidence for a learned response independent of nutrient status. *Endocrinology* 2006; 147: 23-30.
3. Martins C, Morgan LM, Bloom SR, Robertson MD. Effects of exercise on gut peptides, energy intake and appetite. *J Endocrinol* 2007; 193: 251-258.
4. Sodersten P, Bergh C, Zandian M. Understanding eating disorders. *Horm Behav* 2006; 50: 572-578.
5. Kotz CM, Teske JA, Billington CJ. Neuroregulation of non-exercise activity thermogenesis and obesity resistance. *Am J Physiol Regul Integr Comp Physiol* 2008; 294: R699-710.
6. Teske JA, Billington CJ, Kotz CM. Neuropeptidergic mediators of spontaneous physical activity and non-exercise activity thermogenesis. *Neuroendocrinology* 2008; 87: 71-90.
7. Kim HJ, Lee S, Kim TW, Kim HH, Jeon TY, Yoon YS, et al. Effects of exercise-induced weight loss on acylated and unacylated ghrelin in overweight children. *Clin Endocrinol (Oxford)* 2008; 68: 416-422.
8. Toshinai K, Kawagoe T, Shimbara T, Tobina T, Nishida Y, Mondal MS, et al. Acute incremental exercise decreases plasma ghrelin level in healthy men. *Horm Metab Res* 2007; 39: 849-851.
9. Hansotia T, Maida A, Flock G, Yamada Y, Tsukiyama K, Seino Y, et al. Extrapancreatic incretin receptors modulate glucose homeostasis, body weight, and energy expenditure. *J Clin Invest* 2007; 117: 143-152.
10. O'Connor AM, Pola S, Ward BM, Fillmore D, Buchanan KD, Kirwan JP. The gastroenteroinsular response to glucose ingestion during postexercise recovery. *Am J Physiol Endocrinol Metab* 2006; 290: E1155-1161.

Chapter 40

1. Griffin JE. The thyroid. In: Griffin JE, Odjeda SR (eds.), *Textbook of endocrine physiology.* 3rd ed. Oxford University: New York, 1996, pp. 260-283.
2. McMurray RG, Hackney AC. The endocrine system and exercise. In: Garrett W, Kirkendahl D (eds.), *Exercise and sports science.* Williams & Wilkins: New York, 2000, pp. 135-162.
3. McMurray RG, Hackney AC. Interactions of metabolic hormones, adipose tissue and exercise. *Sports Med* 2005; 35: 393-412.
4. Galbo H. The hormonal response to exercise. *Diab Metab Rev* 1986; 1: 385-408.
5. Hackney AC, Gulledge TP. Thyroid responses during an 8 hour period following aerobic and anaerobic exercise. *Physiol Res* 1994; 43: 1-5.
6. Berchtold P, Berger M, Cuppers HJ, et al. Non-glucoregulatory hormones (T4, T3, rT3 and testosterone) during physical exercise in juvenile type diabetes. *Horm Metab Res* 1978; 10: 269-273.
7. McMurray RG, Eubanks TE, Hackney AC. Nocturnal hormonal responses to weight training exercise. *Eur J Appl Physiol* 1995; 72: 121-126.
8. Balsam A, Leppo LE. Effect of physical training on the metabolism of thyroid hormones in man. *J Appl Physiol* 1975; 38: 212-215.
9. Baylor LS, Hackney AC. Resting thyroid and leptin hormone changes in women following intense, prolonged exercise training. *Eur J Appl Physiol* 2003; 88: 480-484.

10. Douyon L, Schteingart DE. Effect of obesity and starvation on thyroid hormone, growth hormone, and cortisol secretion. *Endocrinol Metab Clin North Am* 2002; 31: 173-189.

Chapter 41

1. Kyrou I, Tsigos C. Stress mechanisms and metabolic complications. *Horm Metab Res* 2007; 39: 430-438.
2. de Kloet ER, Karst H, Joels M. Corticosteroid hormones in the central stress response: quick-and-slow. *Neuroendocrinology* 2008; 29: 268-272.
3. Bale TL, Vale WW. CRF and CRF receptors: role in stress responsivity and other behaviors. *Annu Rev Pharmacol Toxicol* 2004; 44: 525-557.
4. Timofeeva E, Huang Q, Richard D. Effects of treadmill running on brain activation and the corticotropin-releasing hormone system. *Neuroendocrinology* 2003; 77: 388-405.
5. Duclos M, Guinot M, Le Bouc Y. Cortisol and GH: odd and controversial ideas. *Appl Physiol Nutr Metab* 2007; 32: 895-903.
6. Mastorakos G, Pavlatou M, Diamanti-Kandarakis E, Chrousos GP. Exercise and the stress system. *Hormones (Athens)* 2005; 4: 73-89.
7. Malendowicz LK, Rucinski M, Belloni AS, Ziolkowska A, Nussdorfer GG. Leptin and the regulation of the hypothalamic-pituitary-adrenal axis. *Int Rev Cytol* 2007; 263: 63-102.
8. Timofeeva E, Richard D. Activation of the central nervous system in obese Zucker rats during food deprivation. *J Comp Neurol* 2001; 441: 71-89.
9. Arvaniti K, Huang Q, Richard D. Effects of leptin and corticosterone on the expression of corticotropin-releasing hormone, agouti-related protein, and proopiomelanocortin in the brain of ob/ob mouse. *Neuroendocrinology* 2001; 73: 227-236.
10. Bjorntorp P. Stress and cardiovascular disease. *Acta Physiol Scand Suppl* 1997; 640: 144-148.
11. Barat P, Duclos M, Moisan MP, Mormède P. Involvement of hypothalamopituitary adrenal axis in abdominal obesity. *Arch Pediatr* 2008; 15: 170-178.
12. Therrien F, Drapeau V, Lalonde J, Lupien SJ, Beaulieu S, Tremblay A, et al. Awakening cortisol response in lean, obese, and reduced obese individuals: effect of gender and fat distribution. *Obesity (Silver Spring)* 2007; 15: 377-385.

Chapter 42

1. Pette D, Staron RS. Myosin isoforms, muscle fiber types, and transitions. *Microsc Res Tech* 2000; 50: 500-509.
2. Tanner CJ, Barakat HA, Dohm GL, Pories WJ, MacDonald KG, Cunningham PR, et al. Muscle fiber type is associated with obesity and weight loss. *Am J Physiol Endocrinol Metab* 2002; 282: E1191-E1196.
3. Bassett DR Jr. Skeletal muscle characteristics: relationships to cardiovascular risk factors. *Med Sci Sports Exerc* 1994; 26: 957-966.
4. Daugaard JR, Nielsen JN, Kristiansen S, Andersen JL, Hargreaves M, Richter EA. Fiber type-specific expression of GLUT4 in human skeletal muscle: influence of exercise training. *Diabetes* 2000; 49: 1092-1095.
5. Renganathan M, Messi ML, Delbono O. Dihydropyridine receptor-ryanodine receptor uncoupling in aged skeletal muscle. *J Membr Biol* 1997; 157: 247-253.
6. Ohlendieck K, Fromming GR, Murray BE, Maguire PB, Leisner E, Traub I, et al. Effects of chronic low-frequency stimulation on Ca2+-regulatory membrane proteins in rabbit fast muscle. *Pflugers Arch* 1999; 438: 700-708.
7. Rall JA. Energetic aspects of skeletal muscle contraction: implications of fiber types. *Exerc Sport Sci Rev* 1985; 13: 33-74.
8. Li JL, Wang XN, Fraser SF, Carey MF, Wrigley TV, McKenna MJ. Effects of fatigue and training on sarcoplasmic reticulum Ca(2+) regulation in human skeletal muscle. *J Appl Physiol* 2002; 92: 912-922.
9. Hood DA. Invited Review: contractile activity-induced mitochondrial biogenesis in skeletal muscle. *J Appl Physiol* 2001; 90: 1137-1157.
10. Koves TR, Li P, An J, Akimoto T, Slentz D, Ilkayeva O, et al. Peroxisome proliferator-activated receptor-gamma co-activator 1alpha-mediated metabolic remodeling of skeletal myocytes mimics exercise training and reverses lipid-induced mitochondrial inefficiency. *J Biol Chem* 2005; 280: 33588-33598.
11. Schenk S, Horowitz JF. Acute exercise increases triglyceride synthesis in skeletal muscle and prevents fatty acid-induced insulin resistance. *J Clin Invest* 2007; 117: 1690-1698.
12. Holloszy JO. Exercise-induced increase in muscle insulin sensitivity. *J Appl Physiol* 2005; 99: 338-343.

Chapter 43

1. Gaesser GA, Brooks GA. Metabolic bases of excess post-exercise oxygen consumption: a review. *Med Sci Sports Exerc* 1984; 16: 29-43.
2. Benedict FG, Carpenter TM. *The metabolism and energy transformations of healthy man during rest*. Carnegie Institution of Washington, publication no. 126. Lord Baltimore Press: Baltimore, 1910.
3. Bahr R. Excess postexercise oxygen consumption—magnitude, mechanisms and practical implications. *Acta Physiol Scand* 1992; suppl. 605: 1-70.
4. Børsheim E, Bahr R. Effect of exercise intensity, duration and mode on post-exercise oxygen consumption. *Sports Med* 2003; 33: 1037-1060.
5. Laforgia J, Withers RT, Gore CJ. Effects of exercise intensity and duration on the excess post-exercise oxygen consumption. *J Sports Sci* 2006; 24: 1247-1264.
6. Gore CJ, Withers RT. Effect of exercise intensity and duration on postexercise metabolism. *J Appl Physiol* 1990; 68: 2362-2368.
7. Barnard RJ, Foss ML. Oxygen debt: effect of beta-adrenergic blockade on the lactacid and alactacid components. *J Appl Physiol* 1969; 27: 813-816.
8. Wong T, Harber V. Lower excess postexercise oxygen consumption and altered growth hormone and cortisol responses to exercise in obese men. *J Clin Endocrinol Metab* 2006; 91: 678-686.
9. Harms CA, Cordain L, Stager JM, Sockler JM, Harris M. Body fat mass affects postexercise metabolism in males of similar lean body mass. *Med Exerc Nutr Health* 1995; 4: 33-39.
10. Børsheim E, Kien CL, Pearl WM. Differential effects of dietary intake of palmitic acid and oleic acid on oxygen consumption during and after exercise. *Metabolism* 2006; 55: 1215-1221.

Chapter 44

1. Maes HH, Neale MC, Eaves LJ. Genetic and environmental factors in relative body weight and human obesity. *Behav Genet* 1997; 27: 325-351.

2. Rankinen T, Zuberi A, Chagnon YC, Weisnagel SJ, Argyropoulos G, Walts B, et al. The human obesity gene map: the 2005 update. *Obes Res* 2006; 14: 529-644.
3. Young EH, Wareham NJ, Farooqi S, Hinney A, Hebebrand J, Scherag A, et al. The V103I polymorphism of the MC4R gene and obesity: population based studies and meta-analysis of 29,563 individuals. *Int J Obes* 2007; 31: 1437-1441.
4. Stutzmann F, Vatin V, Cauchi S, Morandi A, Jouret B, Landt O, et al. Non-synonymous polymorphisms in melanocortin-4 receptor protect against obesity: the two facets of a Janus obesity gene. *Hum Mol Genet* 2007; 16: 1837-1844.
5. Benzinou M, Creemers JWM, Choquet H, Lobbens S, Dina C, Durand E, et al. Common nonsynonymous variants in PCSK1 confer risk of obesity. *Nat Genet* 2008; 40: 943-945.
6. Kurokawa N, Young EH, Oka Y, Satoh H, Wareham NJ, Sandhu MS, et al. The ADRB3 Trp64Arg variant and BMI: a meta-analysis of 44,833 individuals. *Int J Obes* 2008; 32: 1240-1249.
7. Li S, Loos RJ. Progress in the genetics of common obesity: size matters. *Curr Opin Lipidol* 2008; 19: 113-121.
8. Saunders CL, Chiodini BD, Sham P, Lewis CM, Abkevich V, Adeyemo AA, et al. Meta-analysis of genome-wide linkage studies in BMI and obesity. *Obesity* 2007; 15: 2263-2275.
9. The Wellcome Trust Case Control Consortium. Genome-wide association study of 14,000 cases of seven common diseases and 3,000 shared controls. *Nature* 2007; 447: 661-678.
10. Frayling TM, Timpson NJ, Weedon MN, Zeggini E, Freathy RM, Lindgren CM, et al. The Wellcome Trust Case Control Consortium, Hattersley AT, McCarthy MI. A common variant in the FTO gene is associated with body mass index and predisposes to childhood and adult obesity. *Science* 2007; 316: 889-894.
11. Scuteri A, Sanna S, Chen W-M, Uda M, Albai G, Strait J, et al. Genome-wide association scan shows genetic variants in the FTO gene are associated with obesity-related traits. *PLoS Genetics* 2007; 3: e115.
12. Loos RJ, Lindgren CM, Li S, Wheeler E, Zhao JH, Prokopenko I, et al. Common variants near MC4R are associated with fat mass, weight and risk of obesity. *Nat Genet* 2008; 40: 768-775.
13. Andreasen CH, Stender-Petersen KL, Mogensen MS, Torekov SS, Wegner L, Andersen G, et al. Low physical activity accentuates the effect of the FTO rs9939609 polymorphism on body fat accumulation. *Diabetes* 2008; 57: 264-268.
14. Loos RJF, Bouchard C. Obesity—is it a genetic disorder? *J Intern Med* 2003; 254: 401-425.

Chapter 45

1. Nathanielsz PW. *Life in the womb: the origin of health and disease.* Promethean Press: Ithaca, NY, 1999.
2. Barraclough CA. Production of anovulatory, sterile rats by single injections of testosterone propionate. *Endocrinology* 1961; 68: 62-67.
3. Molnar J, Nijland MJM, Howe DC, Nathanielsz PW. Evidence for microvascular dysfunction after prenatal dexamethasone at 0.7, 0.75, and 0.8 gestation in sheep. *Am J Physiol Regul Integr Comp Physiol* 2002; 283: R561-R567.
4. Molnar J, Howe DC, Nijland MJM, Nathanielsz PW. Prenatal dexamethasone leads to both endothelial dysfunction and vasodilatory compensation in sheep. *J Physiol (London)* 2003; 547: 61-66.
5. Vickers MH, Breier BH, McCarthy D, Gluckman PD. Sedentary behavior during postnatal life is determined by the prenatal environment and exacerbated by postnatal hypercaloric nutrition. *Am J Physiol Regul Integr Comp Physiol* 2003; 285: R271-R273.
6. Khan IY, Taylor PD, Dekou V, Seed PT, Lakasing L, Graham D, et al. Gender-linked hypertension in offspring of lard-fed pregnant rats. *Hypertension* 2003; 41: 168-175.
7. Raygada M, Cho E, Hilakivi-Clarke L. High maternal intake of polyunsaturated fatty acids during pregnancy in mice alters offsprings' aggressive behavior, immobility in the swim test, locomotor activity and brain protein kinase C activity. *J Nutr* 1998; 128: 2505-2511.
8. Bayol SA, Simbi BH, Stickland NC. A maternal cafeteria diet during gestation and lactation promotes adiposity and impairs skeletal muscle development and metabolism in rat offspring at weaning. *J Physiol (London)* 2005; 567: 951-961.
9. Stannard SR, Johnson NA. Insulin resistance and elevated triglyceride in muscle: more important for survival than "thrifty" genes? *J Physiol (London)* 2004; 554: 595-607.
10. Zambrano E, Martinez-Samayoa PM, Bautista CJ, Deas M, Guillen L, Rodriguez-Gonzalez GL, et al. Sex differences in transgenerational alterations of growth and metabolism in progeny (F2) of female offspring (F1) of rats fed a low protein diet during pregnancy and lactation. *J Physiol* 2005; 566: 225-236.
11. Aagaard-Tillery KM, Grove K, Bishop J, Ke X, Fu Q, McKnight R, et al. Developmental origins of disease and determinants of chromatin structure: maternal diet modifies the primate fetal epigenome. *J Mol Endocrinol* 2008; 41: 91-102.
12. Samuelsson AM, Matthews PA, Argenton M, Christie MR, McConnell JM, Jansen EHJ, et al. Diet-induced obesity in female mice leads to offspring hyperphagia, adiposity, hypertension, and insulin resistance: a novel murine model of developmental programming. *Hypertension* 2008; 51: 383-392.

Chapter 46

1. Patel SR, Hu FB. Short sleep duration and weight gain: a systematic review. *Obesity* 2008; 16: 643-653.
2. Chen X, Beydoun MA, Wang Y. Is sleep duration associated with childhood obesity? *Obesity* 2008; 16: 265-274.
3. Agras WS, Hammer LD, McNicholas F, Kraemer HC. Risk factors for childhood overweight: a prospective study from birth to 9.5 years. *J Pediatr* 2004; 145: 20-25.
4. Chaput JP, Despres JP, Bouchard C, Tremblay A. The association between sleep duration and weight gain in adults: a 6-year prospective study from the Quebec Family Study. *Sleep* 2008; 31: 517-523.
5. Horne J. Short sleep is a questionable risk factor for obesity and related disorders: statistical versus clinical significance. *Biol Psychol* 2008; 77: 266-277.
6. Taheri S, Lin L, Austin D, Young T, Mignot E. Short sleep duration is associated with reduced leptin, elevated ghrelin, and increased body mass index. *PLoS Med* 2004; 1: 210-217.
7. Spiegel K, Tasali E, Penev P, Van Cauter E. Brief communication: sleep curtailment in healthy young men is associated with decreased leptin levels, elevated ghrelin levels, and increased hunger and appetite. *Ann Intern Med* 2004; 141: 846-850.
8. Youngstedt SD, Kripke DF. Long sleep and mortality: rationale for sleep restriction. *Sleep Med Rev* 2004; 8: 159-174.
9. Zielinski MR, Kline CE, Kripke DF, Bogan RK, Youngstedt SD. No effect of 8-week time-in-bed restriction on glucose tolerance in older long sleepers. *J Sleep Res* 2008; 17: 412-419.

10. Youngstedt SD, O'Connor PJ, Dishman RK. The effects of acute exercise on sleep: a quantitative synthesis. *Sleep* 1997; 20: 203-214.
11. Littman AJ, Vitiello MV, Foster-Schubert K, Ulrich CM, Tworoger SS, Potter JD, et al. Sleep, ghrelin, leptin and changes in body weight during a 1-year moderate-intensity physical activity intervention. *Int J Obes* 2007; 31: 466-475.

Chapter 47

1. Helmchen LA. Can structural change explain the rise in obesity? A look at the past 100 years. The Population Research Center at NORC and the University of Chicago. www.spc.uchicago.edu/prc/publications.php. Accessed 7 May 2008.
2. Brownson RC, Boehmer TK, Luke DA. Declining rates of physical activity in the United States: what are the contributors? *Annu Rev Pub Health* 2005; 26: 421-443.
3. Steffen LM, Arnett DK, Blackburn H, Shah G, Armstrong C, Luepker RV, et al. Population trends in leisure-time physical activity: Minnesota Heart Survey, 1980-2000. *Med Sci Sports Exerc* 2006; 38: 1716-1723.
4. Mayer J. Relation between caloric intake, body weight, and physical work: studies in an industrial male population in West Bengal. *Am J Clin Nutr* 1956; 4: 169-175.
5. Morris JN, Heady JA. Physique of London busmen: epidemiology of uniform size. *Lancet* 1956; 271: 569-570.
6. Taylor HL. Occupational factors in the study of coronary heart disease and physical activity. *Can Med Assoc J* 1967; 96: 825-831.
7. Oppert JM, Thomas F, Chalres MA, Benetos A, Basdevant A, Simon C. Leisure-time and occupational physical activity in relation to cardiovascular risk factors and eating habits in French adults. *Pub Health Nutr* 2006; 9: 746-754.
8. King GA, Fitzhugh EC, Bassett DR Jr., McLaughlin JE, Strath SJ, Swartz AM, et al. Relationship of leisure-time physical activity and occupational activity to the prevalence of obesity. *Int J Obes Relat Metab Disord* 2001; 25: 606-612.
9. Lakdawalla D, Phillipson T. Labor supply and weight. *J Health Hum Resour Adm* 2007; 42: 85-116.
10. Ainsworth BE, Haskell WL, Whitt MC, Irwin ML, Swartz AM, Strath SJ, et al. Compendium of physical activities: an update of activity codes and MET intensities. *Med Sci Sports Exerc* 2000; 32: S498-S516.
11. Mummery WK, Schofield GM, Steele R, Eakin EG, Brown WJ. Occupational sitting time and overweight and obesity in Australian workers. *Am J Prev Med* 2005; 29: 91-97.

Chapter 48

1. Tudor-Locke C, Bittman M, Merom D, Bauman A. Patterns of walking for transport and exercise: a novel application of time use data. *Int J Behav Nutr Phys Act* 2005; 2: 5.
2. Lindstrom M. Means of transportation to work and overweight and obesity: a population-based study in southern Sweden. *Prev Med* 2008; 46: 22-28.
3. Frank LD, Andresen MA, Schmid TL. Obesity relationships with community design, physical activity, and time spent in cars. *Am J Prev Med* 2004; 27: 87-96.
4. Pucher J, Dijkstra L. Promoting safe walking and cycling to improve public health: lessons from the Netherlands and Germany. *Am J Pub Health* 2003; 93: 1509-1516.
5. Sisson SB, Tudor-Locke C. Comparison of cyclists' and motorists' utilitarian physical activity at an urban university. *Prev Med* 2008; 46: 77-79.
6. Wen LM, Rissel C. Inverse associations between cycling to work, public transport, and overweight and obesity: findings from a population based study in Australia. *Prev Med* 2008; 46: 29-32.
7. Besser LM, Dannenberg AL. Walking to public transit: steps to help meet physical activity recommendations. *Am J Prev Med* 2005; 29: 273-280.
8. Wener RE, Evans GW. A morning stroll: levels of physical activity in car and mass transit commuting. *Environ Behav* 2007; 39: 62-74.
9. Kenworthy JR, Laube FB. Patterns of automobile dependence in cities: an international overview of key physical and economic dimensions with some implications for urban policy. *Trans Res* Part A 1999; 33: 691-793.
10. Wen LM, Orr N, Millett C, Rissel C. Driving to work and overweight and obesity: findings from the 2003 New South Wales Health Survey, Australia. *Int J Obes* 2006; 30: 782-786.

Chapter 49

1. Papas MA, Alberg AJ, Ewing R, Helzlsouer KJ, Gary TL, Klassen AC. The built environment and obesity. *Epidemiol Rev* 2007; 29: 129-143.
2. Owen N, Cerin E, Leslie E, duToit L, Coffee N, Frank L, et al. Neighborhood walkability and the walking behavior of Australian adults. *Am J Prev Med* 2007; 33: 387-395.
3. Brug J, van Lenthe FJ, Kremers SPJ. Revisiting Kurt Lewin—how to gain insight into environmental correlates of obesogenic behaviors. *Am J Prev Med* 2006; 31: 525-529.
4. Owen N, Humpel N, Leslie E, Bauman A, Sallis JF. Understanding environmental influences on walking: review and research agenda. *Am J Prev Med* 2004; 27: 67-76.
5. Giles-Corti B, Macintyre S, Clarkson JP, Pikora T, Donovan RJ. Environmental and lifestyle factors associated with overweight and obesity in Perth, Australia. *Am J Health Promot* 2003; 18: 93-102.
6. Rutt CD, Coleman KJ. Examining the relationships among built environment, physical activity, and body mass index in El Paso, TX. *Prev Med* 2005; 40: 831-841.
7. Frank LD, Saelens BE, Powell KE, Chapman JE. Stepping towards causation: do built environments or neighborhood and travel preferences explain physical activity, driving, and obesity? *Soc Sci Med* 2007; 65: 1898-1914.
8. Sugiyama T, Salmon J, Dunstan DW, Bauman AE, Owen N. Neighborhood walkability and TV viewing time among Australian adults. *Am J Prev Med* 2007; 33: 444-449.
9. Sallis JF, Owen N, Fisher EB. Ecological models of health behavior. In: Glanz K, Rimer BK, Viswanath K (eds.), *Health behavior and health education: theory, research, and practice.* 4th ed. Jossey-Bass: San Francisco 2008; 465-482.
10. Owen N, Leslie E, Salmon J, Fotheringham MJ. Environmental determinants of physical activity and sedentary behavior. *Exerc Sport Sci Rev* 2000; 28: 153-158.

Chapter 50

1. McLauren L. Socioeconomic status and obesity. *Epidemiol Rev* 2007; 29: 29-48.
2. Sobal J, Stunkard AJ. Socioeconomic status and obesity: a review of the literature. *Psychol Bull* 1989; 105: 260-275.
3. Ball K, Crawford D. Socioeconomic status and weight change in adults: a review. *Soc Sci Med* 2005; 60: 1987-2010.
4. Monteiro CA, Moura EC, Conde WL, Popkin BM. Socioeconomic status and obesity in adult populations of developing countries: a review. *Bull World Health Org* 2004; 82: 940-946.

5. Shrewsbury V, Wardle J. Socioeconomic status and adiposity in childhood: a systematic review of cross-sectional studies 1990-2005. *Obesity* 2008; 16: 275-284.
6. Wang Y, Beydoun MA. The obesity epidemic in the United States—gender, age, socioeconomic, racial/ethnic, and geographic characteristics: a systematic review and meta-regression analysis. *Epidemiol Rev* 2007; 29: 6-28.
7. Wang Y, Monteiro C, Popkin BM. Trends of obesity and underweight in older children and adolescents in the United States, Brazil, China and Russia. *Am J Clin Nutr* 2007; 75: 971-977.
8. Zhang Q, Wang Y. Trends in the association between obesity and socioeconomic status in U.S. adults: 1971 to 2000. *Obes Res* 2004; 12: 1622-1632.
9. Wang Y, Zhang Q. Are American children and adolescents of low socioeconomic status at increased risk of obesity? Changes in the association between overweight and family income between 1971 and 2002. *Am J Clin Nutr* 2006; 84: 707-716.
10. Zhang Q, Wang Y. Using concentration index to study changes in socioeconomic inequality of overweight among US adolescents between 1971 and 2002. *Int J Epidemiol* 2007; 36: 916-925.

Chapter 51

1. Curioni CC, Lourenco PM. Long-term weight loss after diet and exercise: a systematic review. *Int J Obes (London)* 2005; 29: 1168-1174.
2. King NA, Hopkins M, Caudwell P, Stubbs RJ, Blundell JE. Individual variability following 12 weeks of supervised exercise: identification and characterization of compensation for exercise-induced weight loss. *Int J Obes (London)* 2008; 32: 177-184.
3. Epstein LH, Wing RR. Aerobic exercise and weight. *Addict Behav* 1980; 5: 371-388.
4. King NA, Blundell JE. High-fat foods overcome the energy expenditure induced by high-intensity cycling or running. *Eur J Clin Nutr* 1995; 49: 114-123.
5. Tremblay A, Almeras N, Boer J, Kranenbarg EK, Despres JP. Diet composition and postexercise energy balance. *Am J Clin Nutr* 1994; 59: 975-979.
6. Murgatroyd PR, Goldberg GR, Leahy FE, Gilsenan MB, Prentice AM. Effects of inactivity and diet composition on human energy balance. *Int J Obes Relat Metab Disord* 1999; 23: 1269-1275.
7. Stubbs RJ, Sepp A, Hughes DA, Johnstone AM, King N, Horgan G, et al. The effect of graded levels of exercise on energy intake and balance in free-living women. *Int J Obes Relat Metab Disord* 2002; 26: 866-869.
8. Stubbs RJ, Sepp A, Hughes DA, Johnstone AM, Horgan G, King N, et al. The effect of graded levels of exercise on energy intake and balance in free-living men, consuming their normal diet. *Eur J Clin Nutr* 2002; 56: 129-140.
9. Skender ML, Goodrick GK, Del Junco DJ, et al. Comparison of 2-year weight loss trends in behavioral treatments of obesity: diet, exercise, and combination interventions. *J Am Diet Assoc* 1996; 96: 342-346.
10. Fogelholm M, Kukkonen-Harjula K, Nenonen A, Pasanen M. Effects of walking training on weight maintenance after a very-low-energy diet in premenopausal obese women: a randomized controlled trial. *Arch Intern Med* 2000; 160: 2177-2184.
11. Borg P, Kukkonen-Harjula K, Fogelholm M, Pasanen M. Effects of walking or resistance training on weight loss maintenance in obese, middle-aged men: a randomized trial. *Int J Obes Relat Metab Disord* 2002; 26: 676-683.
12. Church TS, Earnest CP, Skinner JS, Blair SN. Effects of different doses of physical activity on cardiorespiratory fitness among sedentary, overweight or obese postmenopausal women with elevated blood pressure: a randomized controlled trial. *JAMA* 2007 May 16; 297(19): 2081-2091.

Chapter 52

1. Fogelholm M, Männistö S, Vartiainen E, Pietinen P. Determinants of energy balance and overweight in Finland 1982 and 1992. *Int J Obes* 1996; 20: 1097-1104.
2. Martínez-González MA, Martínez JA, Hu FB, Gibney MJ, Kearney J. Physical inactivity, sedentary lifestyle and obesity in the European Union. *Int J Obes Relat Metab Disord* 1999; 23: 1192-1201.
3. Fogelholm M, Kukkonen-Harjula K. Does physical activity prevent weight gain—a systematic review. *Obes Rev* 2000; 1: 95-111.
4. Blanck HM, McCullough ML, Patel AV, Gillespie C, Calle EE, Cokkinides VE, et al. Sedentary behavior, recreational physical activity, and 7-year weight gain among postmenopausal U.S. women. *Obesity* 2007; 15: 1578-1588.
5. Droyvold WB, Holmen J, Midthjell K, Lydersen S. BMI change and leisure time physical activity (LTPA): an 11-y follow-up study in apparently healthy men aged 20-69 y with normal weight at baseline. *Int J Obes Relat Metab Disord* 2004; 28: 410-417.
6. Parsons TJ, Manor O, Power C. Physical activity and change in body mass index from adolescence to mid-adulthood in the 1958 British cohort. *Int J Epidemiol* 2006; 35: 197-204.
7. Hu FB, Li TY, Colditz GA, Willett WC, Manson JE. Television watching and other sedentary behaviors in relation to risk of obesity and type 2 diabetes mellitus in women. *JAMA* 2003; 289: 1785-1791.
8. Haskell WL, Lee IM, Pate RR, Powell KE, Blair SN, Franklin BA, et al. Physical activity and public health: updated recommendation for adults from the American College of Sports Medicine and the American Heart Association. *Med Sci Sports Exerc* 2007; 39: 1423-1434.
9. Fogelholm M, Lahti-Koski M. Community health-promotion interventions with physical activity: does this approach prevent obesity? *Scand J Nutr* 2002; 46: 173-177.
10. Brien SE, Katzmarzyk PT, Craig CL, Gauvin L. Physical activity, cardiorespiratory fitness and body mass index as predictors of substantial weight gain and obesity: the Canadian Physical Activity Longitudinal Study. *Can J Pub Health* 2007; 98: 121-124.

Chapter 53

1. Ross R, Janssen I. Physical activity, fitness and obesity. In: Bouchard C, Blair S, Haskell W (eds.), *Physical activity and health.* Human Kinetics: Champaign, IL, 2007, pp. 173-190.
2. Catenacci VA, Wyatt HR. The role of physical activity in producing and maintaining weight loss. *Nat Clin Pract Endocrinol Metab* 2007; 3: 518-529.
3. Wing RR. Physical activity in the treatment of the adulthood overweight and obesity: current evidence and research issues. *Med Sci Sports Exerc* 1999; 31: S547-S552.

4. Racette SB, Weiss EP, Villareal DT, Arif H, Steger-May K, Schechtman KB, et al. One year of caloric restriction in humans: feasibility and effects on body composition and abdominal adipose tissue. *J Gerontol A Biol Sci Med Sci* 2006; 61: 943-950.

5. Ross R, Dagnone D, Jones PJ, Smith H, Paddags A, Hudson R, et al. Reduction in obesity and related comorbid conditions after diet-induced weight loss or exercise-induced weight loss in men. A randomized, controlled trial. *Ann Intern Med* 2000; 133: 92-103.

6. Ross R, Janssen I, Dawson J, Kungl AM, Kuk JL, Wong SL, et al. Exercise-induced reduction in obesity and insulin resistance in women: a randomized controlled trial. *Obes Res* 2004; 12: 789-798.

7. Slentz CA, Duscha BD, Johnson JL, Ketchum K, Aiken LB, Samsa GP, et al. Effects of the amount of exercise on body weight, body composition, and measures of central obesity: STRRIDE—a randomized controlled study. *Arch Intern Med* 2004; 164: 31-39.

8. Donnelly JE, Hill JO, Jacobsen DJ, Potteiger J, Sullivan DK, Johnson SL, et al. Effects of a 16-month randomized controlled exercise trial on body weight and composition in young, overweight men and women: the Midwest Exercise Trial. *Arch Intern Med* 2003; 163: 1343-1350.

9. Irwin ML, Yasui Y, Ulrich CM, Bowen D, Rudolph RE, Schwartz RS, et al. Effect of exercise on total and intra-abdominal body fat in postmenopausal women: a randomized controlled trial. *JAMA* 2003; 289: 323-330.

10. Giannopoulou I, Ploutz-Snyder LL, Carhart R, Weinstock RS, Fernhall B, Goulopoulou S, et al. Exercise is required for visceral fat loss in postmenopausal women with type 2 diabetes. *J Clin Endocrinol Metab* 2005; 90: 1511-1518.

11. Lee CD, Blair SN, Jackson AS. Cardiorespiratory fitness, body composition, and all-cause and cardiovascular disease mortality in men. *Am J Clin Nutr* 1999; 69: 373-380.

Chapter 54

1. Serdula MK, Mokdad A, Williamson DF, Galuska D, Mendlein J, Heath G. Prevalence of attempting weight loss and strategies for controlling weight. *JAMA* 1999; 282: 1353-1358.

2. National Heart, Lung, and Blood Institute. *Clinical guidelines on the identification, evaluation, and treatment of overweight and obesity in adults; the evidence report.* National Institutes of Health, 1998, pp. 1-228. Bethesda, MD.

3. Ross R, Dagnone D, Jones PJ, Smith H, Paddags A, Hudson R, et al. Reduction in obesity and related comorbid conditions after diet-induced weight loss or exercise-induced weight loss in men. A randomized, controlled trial. *Ann Intern Med* 2000; 133: 92-103.

4. Donnelly JE, Hill JO, Jacobsen DJ, Potteiger J, Sullivan DK, Johnson SL, et al. Effects of a 16-month randomized controlled exercise trial on body weight and composition in young, overweight men and women: the Midwest Exercise Trial. *Arch Intern Med* 2003; 163: 1343-1350.

5. Pate RR, Pratt M, Blair SN, Haskell WL, Macera CA, Bouchard C, et al. Physical activity and public health: a recommendation from the Centers for Disease Control and Prevention and the American College of Sports Medicine. *JAMA* 1995; 273: 402-407.

6. Donnelly JE, Pronk NP, Jacobsen DJ, Pronk SJ, Jakicic JM. Effects of a very-low-calorie diet and physical-training regimens on body composition and resting metabolic rate in obese females. *Am J Clin Nutr* 1991; 54: 56-61.

7. Redman LM, Heilbronn LK, Martin CK, Alfonso A, Smith SR, Ravussin E. Effect of calorie restriction with or without exercise on body composition and fat distribution. *J Clin Endocrinol Metab* 2007; 92: 865-872.

8. Sparti A, DeLany JP, De la Bretonne JA, Sander GE, Bray GA. Relationship between resting metabolic rate and the composition of the fat-free mass. *Metabolism* 1997; 46: 1225-1230.

9. Donnelly JE, Jakicic JM, Pronk N, Smith BK, Kirk EP, Jacobsen DJ, et al. Is resistance training effective for weight management? *Evid Based Prev Med* 2003; 1: 21-29.

10. Jakicic JM, Clark K, Coleman E, Donnelly JE, Foreyt J, Melanson E, et al. Appropriate intervention strategies for weight loss and prevention of weight regain for adults. *Med Sci Sports Exerc* 2001; 33: 2145-2156.

Chapter 55

1. Li Z, Maglione M, Tu W, Mojica W, Arterburn D, Shugarman LR, et al. Meta-analysis: pharmacologic treatment of obesity. *Ann Intern Med* 2005; 142: 532-546.

2. Blackburn G. Effect of degree of weight loss on health benefits. *Obes Res* 1995; suppl. 2: 211s-216s.

3. Wadden TA, Berkowitz RI, Womble LG, Sarwer DB, Phelan S, Cato RK, et al. Randomized trial of lifestyle modification and pharmacotherapy for obesity. *N Engl J Med* 2005; 353: 2111-2120.

4. Pavlou KN, Krey S, Steffee WP. Exercise as an adjunct to weight loss and maintenance in moderately obese subjects. *Am J Clin Nutr* 1989; 49: 1115-1123.

5. Zachwieja JJ. Exercise as treatment for obesity. *Endocrinol Metab Clin North Am* 1996; 25: 965-988.

6. Jakicic JM, Winters C, Lang W, Wing RR. Effects of intermittent exercise and use of home exercise equipment on adherence, weight loss and fitness in overweight women: a randomized trial. *JAMA* 1999; 282: 1554-1560.

7. Weintraub M, Sundaresan PR, Schuster B, Averbuch M, Stein EC, Cox C, et al. Long-term weight control study. IV (weeks 156 to 190). The second double-blind phase. *Clin Pharmacol Ther* 1992; 51: 608-614.

8. Church T, Earnest C, Blair S. Dietary overcompensation across different doses of exercise. *Obesity* 2007; 15 suppl.: A17.

9. Gadde KM, Yonish GM, Foust MS, Tam PY, Najarian T. A 24-week randomized controlled trial of VI-0521, a combination weight loss therapy, in obese adults. *Obesity* 2006; 14 suppl.: A17.

10. Fujioka K, Greenway F, Cowley M, Guttadauria M, Robinson J, Landbloom R, et al. The 24-week experience with a combination sustained release product of zonisamide and bupropion: evidence of an encouraging benefit:risk profile. *Obesity* 2008; 15(suppl. 1): A85.

Chapter 56

1. Jakacic J, Clark K, Coleman E, Donnelly JE, Volek J, Volpe SL. Appropriate intervention strategies for weight loss and prevention of weight regain for adults. *Med Sci Sports Exerc* 2001; 33: 2145-2156.

2. Colles S, Dixon JB, O'Brien PE. Hunger control and regular physical activity facilitate weight loss after laparoscopic adjustable gastric banding. *Obes Surg* 2008; 18: 833-840.

3. Chaston T, Dixon JB, O'Brien PE. Changes in fat-free mass during significant weight loss: a systematic review. *Int J Obes (London)* 2007; 31: 743-750.

4. Metcalf B, Rabkin RA, Rabkin JM, Metcalf LJ, Lehman-Becker LB. Weight loss composition: the effects of exercise following obesity surgery as measured by bioelectrical impedance analysis. *Obes Surg* 2005; 15: 183-186.

5. Alvarez V, Dixon JB, Strauss BJG, Laurie CP, Chaston TB, O'Brien PE. Single frequency bioelectrical impedance is a poor method for determining fat mass in moderately obese women *Obes Surg* 2007; 16: 1-11.

6. Latner JD, Wetzler S, Goodman ER, Glinski J. Gastric bypass in a low-income, inner-city population: eating disturbances and weight loss. *Obes Res* 2004; 12: 956-961.

7. Pontiroli AE, Pizzocri P, Librenti MC, Vedani P, Marchi M, Cucchi E, et al. Laparoscopic adjustable gastric banding for the treatment of morbid (grade 3) obesity and its metabolic complications: a three-year study. *J Clin Endocrinol Metab* 2002; 87: 3555-3561.

8. Chevallier J, Paita M, Rodde-Dunet MH, Marty M, Nogues F, Slim K, et al. Predictive factors in outcome after gastric banding: a nationwide survey on the role of center activity and patients' behavior. *Ann Surg* 2007; 246: 1034-1039.

9. Larsen J, Geenen R, van Ramshorst B, Brand N, Hox JJ, Stroebe W, et al. Binge eating and exercise behavior after surgery for severe obesity: a structural equation model. *Int J Eat Disord* 2006; 39: 369-375.

10. Dunn AL, Marcus BH, Kampert JB, Garcia ME, Kohl HW 3rd, Blair SN. Comparison of lifestyle and structured interventions to increase physical activity and cardiorespiratory fitness: a randomized trial. *JAMA* 1999; 281: 327-334.

11. Tudor-Locke CE, Myers AM. Methodological considerations for researchers and practitioners using pedometers to measure physical (ambulatory) activity. *Res Q Exerc Sport* 2001; 72: 1-12.

12. Wing RR, Phelan S. Long-term weight loss maintenance. *Am J Clin Nutr* 2005; 82: 222S-225S.

Chapter 57

1. Chaston TB, Dixon JB. Factors associated with percent change in visceral versus subcutaneous abdominal fat during weight loss: findings from a systematic review. *Int J Obes (London)* 2008; 32: 619-628.

2. Checkly EA. *A material method of physical education making muscle and reducing flesh without diet or apparatus.* William C. Bryant: Brooklyn, NY, 1895, p. 88.

3. Olson AL, Edelstein E. Spot reduction of subcutaneous adipose tissue. *Res Q* 1968; 39: 647-652.

4. Noland M, Kearney JT. Anthropometric and densitometric response of women to specific and general exercise. *Res Q* 1978; 49: 322-328.

5. Ross R, Janssen I, Stalknecht B. Influence of endurance exercise on adipose tissue distribution. In: Nicklas BJ (ed.), *Endurance exercise and adipose tissue.* CRC Press: Boca Raton, FL, 2002, pp. 121-152.

6. Stallknecht B, Dela F, Helge JW. Are blood flow and lipolysis in subcutaneous adipose tissue influenced by contractions in adjacent muscles in humans? *Am J Physiol Endocrinol Metab* 2007; 292: E394-E399.

7. Gwinup G, Chelvam R, Steinberg T. Thickness of subcutaneous fat and activity of underlying muscles. *Ann Intern Med* 1971; 74: 408-411.

8. Roby FB. Effect of exercise on regional subcutaneous fat accumulations. *Res Q* 1962; 33: 273-278.

9. Schade M, Hellendrandt FA, Waterland JC, Carns ML. Spot reducing in overweight college women: its influence on fat distribution as determined by photography. *Res Q* 1962; 22: 461-470.

10. Katch FI, Clarkson PM, Kroll W, McBride T. Effect of sit up exercise training on adipose cell size and adiposity. *Res Q Exerc Sport* 1984; 55: 242-247.

11. Krotkiewski M, Aniansson A, Grimby G, Bjorntorp P, Sjostrom L. The effect of unilateral isokinetic strength training on local adipose and muscle tissue morphology, thickness, and enzymes. *Eur J Appl Physiol* 1979; 42: 271-281.

12. Kostek MA, Pescatello LS, Seip RL, Angelopoulos TJ, Clarkson PM, Gordon PM, et al. Subcutaneous fat alterations resulting from an upper-body resistance training program. *Med Sci Sports Exerc* 2007; 39: 1177-1185.

13. Katch FI, Katch VL. Measurement and prediction errors in body composition assessment and the search for the perfect prediction equation. *Res Q Exerc Sport* 1980; 51: 249-260.

Chapter 58

1. Goodpaster BH, Kelley DE, Wing RR, Meier A, Thaete FL. Effects of weight loss on regional fat distribution and insulin sensitivity in obesity. *Diabetes* 1999; 48: 839-847.

2. Ross R, Dagnone D, Jones PJ, Smith H, Paddags A, Hudson R, et al. Reduction in obesity and related comorbid conditions after diet-induced weight loss or exercise-induced weight loss in men. A randomized, controlled trial. *Ann Intern Med* 2000; 133: 92-103.

3. Mourier A, Gautier JF, De Kerviler E, Bigard AX, Villette JM, Garnier JP, et al. Mobilization of visceral adipose tissue related to the improvement in insulin sensitivity in response to physical training in NIDDM. Effects of branched-chain amino acid supplements. *Diab Care* 1997; 20: 385-391.

4. Goodpaster BH, Kelley DE. Skeletal muscle triglyceride: marker or mediator of obesity-induced insulin resistance in type 2 diabetes mellitus? *Curr Diab Rep* 2002; 2: 216-222.

5. Goodpaster BH, He J, Watkins S, Kelley DE. Skeletal muscle lipid content and insulin resistance: evidence for a paradox in endurance-trained athletes. *J Clin Endocrinol Metab* 2001; 86: 5755-5761.

6. Dube JJ, Amati F, Stefanovic-Racic M, Toledo FG, Sauers SE, Goodpaster BH. Exercise-induced alterations in intramyocellular lipids and insulin resistance: the athlete's paradox revisited. *Am J Physiol Endocrinol Metab* 2008; 294: E882-888.

7. Goodpaster BH, Thaete FL, Kelley DE. Thigh adipose tissue distribution is associated with insulin resistance in obesity and in type 2 diabetes mellitus. *Am J Clin Nutr* 2000; 71: 885-892.

8. Tamura Y, Tanaka Y, Sato F, Choi JB, Watada H, Niwa M, et al. Effects of diet and exercise on muscle and liver intracellular lipid contents and insulin sensitivity in type 2 diabetic patients. *J Clin Endocrinol Metab* 2005; 90: 3191-3196.

9. Devries MC, Samjoo IA, Hamadeh MJ, Tarnopolsky MA. Effect of endurance exercise on hepatic lipid content, enzymes, and adiposity in men and women. *Obesity (Silver Spring)* 2008; 16: 2281-2288.

10. Larson-Meyer DE, Heilbronn LK, Redman LM, Newcomer BR, Frisard MI, Anton S, et al. Effect of calorie restriction with or without exercise on insulin sensitivity, beta-cell function, fat cell size, and ectopic lipid in overweight subjects. *Diab Care* 2006; 29: 1337-1344.

Chapter 59

1. Forbes GB. Lean body mass-body fat interrelationships in humans. *Nutr Rev* 1987; 45: 225-231.
2. Forbes GB. Exercise and body composition. *J Appl Physiol* 1991; 70: 994-997.
3. Forbes GB. Exercise and lean weight: the influence of body weight. *Nutr Rev* 1992; 50: 157-161.
4. Hall KD. Body fat and fat-free mass inter-relationships: Forbes's theory revisited. *Br J Nutr* 2007; 97: 1059-1063.
5. Heymsfield SB, Harp JB, Rowell PN, Nguyen AM, Pietrobelli A. How much may I eat? Calorie estimates based upon energy expenditure prediction equations. *Obes Rev* 2006; 7: 361-370.
6. Ballor DL, Poehlman ET. Exercise-training enhances fat-free mass preservation during diet-induced weight loss: a meta-analytical finding. *Int J Obes* 1994; 18: 35-40.
7. Garrow JS, Summerbell CD. Meta-analysis: effect of exercise, with or without dieting, on the body composition of overweight subjects. *Eur J Clin Nutr* 1995; 49: 1-10.
8. Chaston TB, Dixon JB, O'Brien PE. Changes in fat-free mass during significant weight loss: a systematic review. *Int J Obes* 2007; 31: 743-750.
9. Redman LM, Heilbronn LK, Martin CK, Alfonso A, Smith SR, Ravussin E. Effect of calorie restriction with or without exercise on body composition and fat distribution. *J Clin Endocrinol Metab* 2007; 92: 865-872.

Chapter 60

1. Wood Baker C, Brownell KD. Physical activity and maintenance of weight loss: physiological and psychological mechanisms. In: Bouchard C (ed.), *Physical activity and obesity*. Human Kinetics: Champaign, IL, 2000, pp. 311-328.
2. Wing RR, Hamman RF, Bray GA, Delahanty L, Edelstein SL, Hill JO, et al. Achieving weight and activity goals among diabetes prevention program lifestyle participants. *Obes Res* 2004; 12: 1426-1434.
3. Jakicic J, Wing RR, Winters-Hart C. Relationship of physical activity to eating behaviors and weight loss in women. *Med Sci Sports Exerc* 2002; 34: 1653-1659.
4. Tate DF, Jeffery RW, Sherwood NE, Wing RR. Long-term weight losses associated with prescription of higher physical activity goals. Are higher levels of physical activity protective against weight regain? *Am J Clin Nutr* 2007; 85: 954-959.
5. Jakicic J, Wing R, Winters C. Effects of intermittent exercise and use of home exercise equipment on adherence, weight loss, and fitness in overweight women. *JAMA* 1999; 282: 1554-1560.
6. Jeffery RW, Wing RR, Sherwood NE, Tate DF. Physical activity and weight loss: does prescribing higher physical activity goals improve outcome? *Am J Clin Nutr* 2003; 78: 684-689.
7. Phelan S, Wyatt HR, O' Hill J, Wing RR. Are the eating and exercise habits of successful weight losers changing? *Obesity* 2006; 14: 710-716.
8. Catenacci VA, Ogden LG, Stuht J, Phelan S, Wing RR, Hill JO, et al. Physical activity patterns in the National Weight Control Registry. *Obesity (Silver Spring)* 2008; 16: 153-161.
9. Raynor DA, Phelan S, Hill JO, Wing RR. Television viewing and long-term weight maintenance: results from the National Weight Control Registry. *Obesity* 2006; 14: 1816-1824.
10. Phelan S, Roberts M, Lang W, Wing RR. Empirical evaluation of physical activity recommendations for weight control in women. *Med Sci Sports Exerc* 2007; 39: 1832-1836.

Chapter 61

1. Catalano PM, Ehrenberg HM. The short- and long-term implications of maternal obesity on the mother and her offspring. *BJOG* 2006; 113: 1126-1133.
2. Polley BA, Wing RR, Sims CJ. Randomized controlled trial to prevent excessive weight gain in pregnant women. *Int J Obes* 2002; 26: 1494-1502.
3. Gray-Donald K, Robinson E, Collier A, David K, Renaud L, Rodrigues S. Intervening to reduce weight gain in pregnancy and gestational diabetes mellitus in Cree communities: an evaluation. *CMAJ* 2000; 163: 1247-1251.
4. Olsen CM, Strawderman MS, Reed RG. Efficacy of an intervention to prevent excessive gestational weight gain. *Am J Obstet Gynecol* 2004; 191: 530-536.
5. Kinnunen T, Pasanen M, Aittasalo M, Fogelholm M, Hilakivi-Clarke L, Weiderpass E, et al. Preventing excessive weight gain during pregnancy—a controlled trial in primary health care. *Eur J Clin Nutr* 2007; 61: 884-891.
6. Claesson IM, Sydsjo G, Brynhildsen J, Cedergren M, Jeppsson A, Nystrom R, et al. Weight gain restriction for obese pregnant women: a case-control intervention study. *BJOG* 2008; 115: 44-50.
7. Artal R, Catanzaro R, Gavard J, Mostello D, Friganza JC. A lifestyle intervention of weight-gain restriction: diet and exercise in obese women with gestational diabetes mellitus. *Appl Physiol Nutr Metab* 2007; 32: 596-601.
8. Wolff S, Legarth J, Vangsgaard K, Toubro S, Astrup A. A randomized trial of the effects of dietary counseling on gestational weight gain and glucose metabolism in obese pregnant women. *Int J Obes* 2008; 32: 495-501.
9. Mottola MF, Giroux I, Gratton R, Hammond JA, Hanley A, McManus R, et al. Nutrition and exercise prevents excess weight gain in overweight pregnant women. *Med SciSports Exerc* 2010; Publish Ahead of Print: DOI: 10.1249/MSS.0b013e3181b5419a.
10. Wolfe LA, Mottola MF. PARmed-X for Pregnancy: physical activity readiness medical examination. www.csep.ca/communities/c574/files/hidden/pdfs/parmed-xpreg.pdf.
11. American College of Obstetricians and Gynecologists. Exercise during pregnancy and the postpartum period. Committee opinion no. 267. *Obstet Gynecol* 2002; 99: 171-173.
12. Mottola MF, Davenport M, Brun CR, Inglis SD, Charlesworth S, Sopper MM. VO_{2peak} prediction and exercise prescription for pregnant women. *Med Sci Sports Exerc* 2006; 38: 1389-1395.
13. Davenport MH, Charlesworth S, Vanderspank D, Sopper MM, Mottola MF. Development and validation of exercise target heart rate zones for overweight and obese pregnant women. *Appl Physiol Nutr Metab* 2008; 33: 984-989.

Chapter 62

1. Gunderson EP, Abrams B, Selvin S. The relative importance of gestational gain and maternal characteristics associated with the risk of becoming overweight after pregnancy. *Int J Obes* 2000; 24: 1660-1668.
2. Rooney BL, Schauberger CW. Excess pregnancy weight gain and long-term obesity: one decade later. *Obstet Gynecol* 2002; 100: 245-252.
3. Oken E, Taveras EM, Folasade AP, Rich-Edwards JW, Gillman MW. Television, walking, and diet: associations with postpartum weight retention. *Am J Prev Med* 2007; 32: 305-311.

4. Olson CM, Strawderman MS, Hinton PS, Pearson TA. Gestational weight gain and postpartum behaviors associated with weight change from early pregnancy to one year postpartum. *Int J Obes* 2003; 27: 117-127.

5. American Academy of Pediatrics. Breastfeeding and the use of human milk. *Pediatrics* 1997; 100: 1035-1039.

6. Lovelady CA, Lonnerdal B, Dewey KG. Lactation performance of exercising women. *Am J Clin Nutr* 1990; 52: 103-109.

7. Lovelady CA, Nommsen-Rivers LA, McCrory MA, Dewey KG. Effects of exercise on plasma lipids and metabolism of lactating women. *Med Sci Sports Exerc* 1995; 27: 22-28.

8. McCrory MA, Nommsen-Rivers LA, Mole PA, Lonnerdal B, Dewey KG. A randomized trial of the short-term effects of dieting vs dieting with aerobic exercise on lactation performance. *Am J Clin Nutr* 1999; 69: 959-967.

9. Lovelady CA, Garner KE, Moreno KL, Williams JP. The effect of weight loss in overweight, lactating women on the growth of their infants. *N Engl J Med* 2000; 342: 449-453.

10. Kinnunen TI, Pasanen M, Aittasalo M, Fogelholm M, Weiderpass E, Luoto R. Reducing postpartum weight retention—a pilot trial in primary health care. *Nutr J* 2007; 6: 21.

11. O'Toole ML, Sawicki MA, Artal R. Structured diet and physical activity prevent postpartum weight retention. *J Women's Health* 2003; 12: 991-998.

12. Leermakers EA, Anglin K, Wing RR. Reducing postpartum weight retention through a correspondence intervention. *Int J Obes Relat Metab Disord* 1998; 22: 1103-1109.

13. American College of Obstetricians and Gynecologists. Exercise during pregnancy and the postpartum period. Committee opinion no. 267. *Obstet Gynecol* 2002; 99: 171-173.

Chapter 63

1. Kramer MS. Aerobic exercise for women during pregnancy. *Coch Database Syst Rev* 2002; 2: CD000180. Update in *Coch Database Syst Rev* 2006; 3: CD000180.

2. Bonzini M, Coggon D, Palmer KT. Risk of prematurity, low birthweight and pre-eclampsia in relation to working hours and physical activities: a systematic review. *Occup Environ Med* 2007; 64: 228-243.

3. Dwarkanath P, Muthayya S, Vaz M, Thomas T, Mhaskar A, Mhaskar R, et al. The relationship between maternal physical activity during pregnancy and birthweight. *Asia Pac J Clin Nutr* 2007; 16: 704-710.

4. Barker DJ. The origins of the developmental origins theory. *J Intern Med* 2007; 261: 412-417.

5. Vickers MH, Breier BH, McCarthy D, Gluckman PD. Sedentary behavior during postnatal life is determined by the prenatal environment and exacerbated by postnatal hypercaloric nutrition. *Am J Physiol Regul Integr Comp Physiol* 2003; 285: R271-273.

6. Mattocks C, Ness A, Deere K, Tilling K, Leary S, Blair SN, Riddoch C. Early life determinants of physical activity in 11 to 12 year olds: cohort study. *BMJ* 2008; 336: 26-29.

7. Hallal PC, Wells JCK, Reichert FF, Anselmi L, Victora CG. Early determinants of physical activity in adolescence: prospective birth cohort study. *BMJ* 2006; 332: 1002-1007.

8. Boreham CA, Murray L, Dedman D, Davey Smith G, Savage JM, Strain JJ. Birthweight and aerobic fitness in adolescents: the Northern Ireland Young Hearts Project. *Pub Health* 2001; 115: 373-379.

9. Laaksonen DE, Lakka HM, Lynch J, Lakka TA, Niskanen L, Rauramaa R, et al. Cardiorespiratory fitness and vigorous leisure-time physical activity modify the association of small size at birth with the metabolic syndrome. *Diab Care* 2003; 26: 2156-2164.

10. Eriksson JG, Ylihärsilä H, Forsén T, Osmond C, Barker DJ. Exercise protects against glucose intolerance in individuals with a small body size at birth. *Prev Med* 2004; 39: 164-167.

Chapter 64

1. Bigaard J, Frederiksen K, Tjønneland A, Thomsen B, Overvad K, Heitmann B, et al. Body fat and fat-free mass and all-cause mortality. *Obes Res* 2004; 12: 1042-1049.

2. Gutin B, Johnson M, Humphries M, Hatfield-Laube J, Kapuku, Allison J, et al. Relations of visceral adiposity to cardiovascular disease risk factors in black and white teens. *Obesity* 2007; 15: 1029-1035.

3. Gutin B, Yin Z, Johnson M, Barbeau P. Preliminary findings of the effect of a 3-year after-school physical activity intervention on fitness and body fat. The Medical College of Georgia FitKid Project. *Int J Pediatr Obes* 2008; 3: 3-9.

4. Stallmann-Jorgensen I, Gutin B, Hatfield-Laube J, Humphries M, Johnson M, Barbeau P. General and visceral adiposity in black and white adolescents and their relation with reported physical activity and diet. *Int J Obes* 2007; 31: 622-629.

5. Ruiz JR, Rizzo NS, Hurtig-Wennlöf A, Ortega FB, Wärnberg J, Sjöström M. Relations of total physical activity and intensity to fitness and fatness in children: the European Youth Heart Study. *Am J Clin Nutr* 2006; 84: 299-303.

6. Gutin B. Child obesity can be reduced with vigorous activity rather than restriction of energy intake. *Obesity* 2008; 16: 2193-2196.

7. Gutin B, Barbeau P, Yin Z. Exercise interventions for prevention of obesity and related disorders in youths. *Quest* 2004; 56: 120-141.

8. Tolfrey K, Jones AM, Campbell IG. Lipid-lipoproteins in children: an exercise dose response study. *Med Sci Sports Exerc* 2004; 36: 418-427.

9. Barbeau P, Johnson M, Howe C, Allison J, Davis C. Gutin B, et al. Ten months of exercise improves general and visceral adiposity, bone, and fitness in black girls. *Obesity* 2007; 15: 2077-2085.

10. Connelly JB, Duaso MJ, Butler G. A systematic review of controlled trials of interventions to prevent childhood obesity and overweight: a realistic synthesis of the evidence. *Pub Health* 2007; 121: 510-517.

Chapter 65

1. Fontaine KR, Allison DB. Obesity and mortality rates. In: Bray G, Bouchard C (eds.), *Handbook of obesity*. 2nd ed. Marcel Dekker: New York, 2003, pp. 767-785.

2. LaMonte MJ, Blair SN. Physical activity, cardiorespiratory fitness, and adiposity: contributions to disease risk. *Cur Opin Clin Nutr Metab Care* 2006; 9: 540-546.

3. Lee IM, Skerrett PJ. Physical activity and all-cause mortality: what is the dose-response relation? *Med Sci Sport Exerc* 2001; 33(6 suppl.): S459-471.

4. Pedersen BK. Body mass index-independent effect of fitness and physical activity for all-cause mortality. *Scan J Med Sci Sports* 2007; 17: 196-204.

5. Hu FB, Willett WC, Li T, Stampfer MJ, Colditz GA, Manson JE. Adiposity as compared with physical activity in predicting mortality among women. *N Engl J Med* 2004; 351: 2694-2703.

6. Hu G, Tuomilehto J, Silventoinen K, Barengo NC, Peltonen M, Jousilahti P. The effects of physical activity and body mass index on cardiovascular, cancer and all-cause mortality among 47212 middle-aged Finnish men and women. *Int J Obes Relat Metab Disord* 2005; 29: 894-902.
7. Crespo CJ, Palmieri MR, Perdomo RP, McGee DL, Smit E, Sempos CT, et al. The relationship of physical activity and body weight with all-cause mortality: results from the Puerto Rico Heart Health Program. *Ann Epidemiol* 2002; 12: 543-552.
8. Sui X, LaMonte MJ, Laditka JN, Hardin JW, Chase N, Hooker SP, et al. Cardiorespiratory fitness and adiposity as mortality predictors in older adults. *JAMA* 2007; 298: 2507-2516.
9. Weinstein AR, Sesso HD. Joint effects of physical activity and body weight on diabetes and cardiovascular disease. *Exerc Sport Sci Rev* 2006; 34: 10-15.
10. Stevens J, Evenson KR, Thomas O, Cai J, Thomas R. Associations of fitness and fatness with mortality in Russian and American men in the Lipids Research Clinics Study. *Int J Obes Relat Metab Disord* 2004; 28: 1463-1470.

Chapter 66

1. Laakso M. Prevention of diabetes. *Curr Mol Med* 2005; 5: 365-374.
2. Pan XR, Li GW, Hu YH, Wang JX, Yang WY, An ZX, et al. Effects of diet and exercise in preventing NIDDM in people with impaired glucose tolerance. The Da Qing IGT and diabetes study. *Diab Care* 1997; 20: 537-544.
3. Tuomilehto J, Lindström J, Eriksson JG, Valle TT, Hämäläinen H, Ilanne-Parikka P, et al. Prevention of type 2 diabetes mellitus by changes in lifestyle among subjects with impaired glucose tolerance. *N Engl J Med* 2001; 344: 1343-1350.
4. Laaksonen DE, Lindström J, Lakka TA, Eriksson JG, Niskanen L, et al. Physical activity in the prevention of type 2 diabetes. The Finnish Diabetes Prevention Study. *Diabetes* 2005; 54: 158-165.
5. Knowler WC, Barrett-Connor E, Fowler SE, Hamman RF, Lachin JM, Walker EA, et al. Reduction in the incidence of type 2 diabetes with lifestyle intervention or metformin. *N Engl J Med* 2002; 346: 393-403.
6. Kosaka K, Noda M, Kuzuya T. Prevention of type 2 diabetes by lifestyle intervention: a Japanese trial in IGT males. *Diab Res Clin Pract* 2005; 67: 152-162.
7. Ramachandran A, Snehalatha C, Mary S, Mukesh B, Bhaskar AD, et al. The Indian diabetes prevention programme shows that lifestyle modification and metformin prevent type 2 diabetes in Asian Indian subjects with impaired glucose tolerance (IDPP-1). *Diabetologia* 2006; 49: 289-297.
8. Mensink M, Blaak EE, Corpeleijn E, Saris WH, de Bruin TW, Feskens EJ. Lifestyle intervention according to general recommendations improves glucose tolerance. *Obes Res* 2003; 11: 1588-1596.
9. Carr DB, Utzschneider KM, Boyko EJ, Asberry PJ, Hull RL, Kodama K, et al. A reduced-fat diet and aerobic exercise in Japanese Americans with impaired glucose tolerance decreases intra-abdominal fat and improves insulin sensitivity but not beta-cell function. *Diabetes* 2005; 54: 340-347.
10. Oldroyd JC, Unwin NC, White M, Mathers JC, Alberti KG. Randomised controlled trial evaluating lifestyle interventions in people with impaired glucose tolerance. *Diab Res Clin Pract* 2006; 72: 117-127.

Chapter 67

1. Centers for Disease Control and Prevention. www.cdc.gov/diseases.
2. Ford ES, Giles WH, Dietz WH. Prevalence of metabolic syndrome among US adults. *JAMA* 2002; 287: 356-359.
3. Steppel JH, Horton ES. Exercise in patients with diabetes mellitus. In: Kahn CR, Weir GC, King GL, Jacobson AM, Moses AC, Smith RJ (eds.), *Joslin's diabetes mellitis.* 14th ed. Lippincott Williams & Wilkins: Philadelphia, 2005, pp. 649-657.
4. Sigal RJ, Wasserman DH, Kenny GP, Castaneda-Sceppa C. Physical activity/exercise and type 2 diabetes (ADA technical review). *Diab Care* 2004; 27: 2518-2539.
5. Rockl KSC, Witzak CA, Goodyear LJ. Signaling mechanisms in skeletal muscle: acute response and chronic adaptations to exercise. *IUBM Life* 2008; 60: 145-153.
6. Boule NG, Haddad E, Kenny GP, Wells GA, Sigal RJ. Effects of exercise on glycemic control and body mass in type 2 diabetes mellitus: a meta analysis of controlled clinical trials. *JAMA* 2001; 286: 1218-1227.
7. Wing RR. Exercise and weight control. In: Ruderman N, Devlin JT, Schneider SH, Kriska A (eds.), *Handbook of exercise in diabetes.* 2nd ed. American Diabetes Association: Alexandria, VA, 2002, pp. 355-364.
8. Church TS, Cheng YJ, Earnest CP, Barlow CE, Gibbons LW, Priest EL, et al. Exercise capacity and body composition as predictors of mortality among men with diabetes. *Diab Care* 2004; 27: 83-88.
9. Look AHEAD Research Group, Pi-Sunyer X, Blackburn G, Brancati FL, Bray GA, Bright R, et al. Reduction in weight and cardiovascular disease risk factors in individuals with type 2 diabetes of the Look AHEAD trial. *Diab Care* 2007; 30: 1374-1383.
10. Sigal RJ, Kenny GP, Wasserman DH, Castenada-Sceppa C, White RD. Physical activity/exercise and type 2 diabetes: a consensus statement from the American Diabetes Association. *Diab Care* 2006; 29: 1433-1438.

Chapter 68

1. Stamler R, Stamler J, Riedlinger WF, Algera G, Roberts RH. Weight and blood pressure: findings in hypertension screening of 1 million Americans. *JAMA* 1978; 240: 1607-1610.
2. Poirer P, Giles TD, Bray GA, Hong Y, Stern JS, Pi-Sunyer X, Eckel RH. Obesity and cardiovascular disease: pathophysiology, evaluation, and effect of weight loss. An update of the 1997 AHA scientific statement on obesity and heart disease from the Obesity Committee of the Council on Nutrition, Physical Activity, and Metabolism. *Circulation* 2006; 113: 898-918.
3. Seals DR, Hagberg JM. The effect of exercise training on human hypertension: a review. *Med Sci Sports Exerc* 1984; 16: 207-225.
4. Fagard RH. Physical activity in the prevention and treatment of hypertension in the obese. *Med Sci Sports Exerc* 1999; 31:(suppl.): S624-S630.
5. Hagberg JM, Brown MD, Park JJ. The role of exercise training in the treatment of hypertension: an update. *Sports Med* 2000; 30: 193-206.
6. Whelton SP, Chin A, Xin X, He J. Effect of aerobic exercise on blood pressure: a meta-analysis of randomized controlled trials. *Ann Intern Med* 2002; 136: 493-503.

7. Wallace JP. Exercise in hypertension: a clinical review. *Sports Med* 2003; 33: 585-598.
8. Blair SN, Kohl HW III, Paffenbarger RS, Clark DG, Cooper KH, Gibbons LW. Physical fitness and all-cause mortality: a prospective study of healthy men and women. *JAMA* 1989; 262: 2395-2401.
9. Gordon NF, Scott CB, Levine BD. Comparison of single versus multiple lifestyle interventions: are the antihypertensive effects of exercise training and diet-induced weight loss additive? *Am J Cardiol* 1997; 79: 763-767.
10. Dengel DR, Galecki AT, Hagberg JM, Pratley RE. The independent and combined effects of weight loss and aerobic exercise on blood pressure and oral glucose tolerance in older men. *Am J Hypertension* 1998; 11: 1405-1412.

Chapter 69

1. Morris JN, Heady JA, Raffle PA, Roberts CG, Parks JW. Coronary heart-disease and physical activity of work. *Lancet* 1953; 265: 1053-1057.
2. Heady JA, Morris JN, Raffle PA. Physique of London busmen; epidemiology of uniforms. *Lancet* 1956; 271: 569-570.
3. Paffenbarger RS Jr., Laughlin ME, Gima AS, Black RA. Work activity of longshoremen as related to death from coronary heart disease and stroke. *N Engl J Med* 1970; 282: 1109-1114.
4. Wei M, Kampert JB, Barlow CE, Nichaman MZ, Gibbons LW, Paffenbarger RSJ, et al. Relationship between low cardiorespiratory fitness and mortality in normal-weight, overweight, and obese men. *JAMA* 1999; 282: 1547-1553.
5. Hu FB, Willett WC, Li T, Stampfer MJ, Colditz GA, Manson JE. Adiposity as compared with physical activity in predicting mortality among women. *N Engl J Med* 2004; 351: 2694-2703.
6. Blair SN, Church TS. The fitness, obesity, and health equation: is physical activity the common denominator? *JAMA* 2004; 292: 1232-1234.
7. Ardern C.I., Katzmarzyk PT, Janssen I, Church TS and Blair SN. Revised Adult Treatment Panel III Guidelines and Cardiovascular Disease Mortality in Men Attending a Preventive Medical Clinic. *Circulation* 2005;112:1478-1485.
8. Church TS, LaMonte MJ, Barlow CE, Blair SN. Cardiorespiratory fitness and body mass index as predictors of cardiovascular disease mortality among men with diabetes. *Arch Intern Med* 2005; 165: 2114-2120.

Chapter 70

1. Hu G, Tuomilehto J, Silventoinen K, Sarti C, Männistö S, Jousilahti P. Body mass index, waist circumference, and waist-hip ratio on the risk of total and type-specific stroke. *Arch Intern Med* 2007; 167: 1420-1427.
2. Lee CD, Folsom AR, Blair SN. Physical activity and stroke risk: a meta-analysis. *Stroke* 2003; 34: 2475-2481.
3. Sacco RL, Adams R, Albers G, Alberts MJ, Benavente O, Furie K, et al. Guidelines for prevention of stroke in patients with ischemic stroke or transient ischemic attack: a statement for healthcare professionals from the American Heart Association/American Stroke Association Council on Stroke. *Stroke* 2006; 37: 577-617.
4. Kalichman L, Rodrigues B, Gurvich D, Israelov Z, Spivak E. Impact of patient's weight on stroke rehabilitation results. *Am J Phys Med Rehabil* 2007; 86: 650-655.
5. MacKay-Lyons MJ, Makrides L. Cardiovascular stress during a contemporary stroke rehabilitation program: is the intensity adequate to induce a training effect? *Arch Phys Med Rehabil* 2002; 83: 1378-1383.
6. Ashe MC, Miller WC, Eng JJ, Noreau L and the PACC Research Group. Older adults, chronic disease and leisure-time physical activity. *Gerontology* 2008; 1: 64-72.
7. Rand D, Eng JJ, Tang PF, Jeng JS, Hung C. How active are people with stroke? Use of accelerometers to assess physical activity. *Stroke* 2009; 40: 163-168.
8. Pang MY, Eng JJ, Dawson AS, Gylfadóttir S. The use of aerobic exercise training in improving aerobic capacity in individuals with stroke: a meta-analysis. *Clin Rehabil* 2006; 20: 97-111.
9. Ivey FM, Ryan AS, Hafer-Macko CE, Goldberg AP, Macko RF. Treadmill aerobic training improves glucose tolerance and indices of insulin sensitivity in disabled stroke survivors: a preliminary report. *Stroke* 2007; 38: 2752-2758.
10. Greenlund KJ, Giles WH, Keenan NL, Croft JB, Mensah GA. Physician advice, patient actions, and health-related quality of life in secondary prevention of stroke through diet and exercise. *Stroke* 2002; 33: 565-570.
11. van der Ploeg HP, Streppel KR, van der Beek AJ, van der Woude LH, Vollenbroek-Hutten MM, van Harten WH, et al. Successfully improving physical activity behavior after rehabilitation. *Am J Health Promot* 2007; 21: 153-159.
12. Rimmer JH, Braunschweig C, Silverman K, Riley B, Creviston T, Nicola T. Effects of a short-term health promotion intervention for a predominantly African-American group of stroke survivors. *Am J Prev Med* 2000; 18: 332-338.

Chapter 71

1. Anderson T, Gerhard M, Meredith I, Charbonneau F, Delagrange D, Creager M, et al. Systemic nature of endothelial dysfunction in atherosclerosis. *Am J Cardiol* 1995; 75: 71B-74B.
2. Quyyumi A. Endothelial function in health and disease: new insights into the genesis of cardiovascular disease. *Am J Med* 1998; 105: 32S-39S.
3. Watts K, Beye P, Siafarikas A, O'Driscoll G, Jones TW, Davis EA, et al. Effects of exercise training on vascular function in obese children. *J Pediatr* 2004; 144: 620-625.
4. Van Guilder GP, Stauffer BL, Greiner JJ, Desouza CA. Impaired endothelium-dependent vasodilation in overweight and obese adult humans is not limited to muscarinic receptor agonists. *Am J Physiol Heart Circ Physiol* 2008; 294: H1685-H1692.
5. Dengel DR, Kelly AS, Steinberger J, Sinaiko AR. Effect of oral glucose loading on endothelial function in normal-weight and overweight children. *Clin Sci (London)* 2007; 112: 493-498.
6. Schjerve IE, Tyldum GA, Tjonna AE, Stolen TO, Loennechen JP, Hansen HE, et al. Both aerobic endurance and strength training programs improve cardiovascular health in obese adults. *Clin Sci (London)* 2008; 115: 283-293.
7. Van Guilder GP, Westby CM, Greiner JJ, Stauffer BL, DeSouza CA. Endothelin-1 vasoconstrictor tone increases with age in healthy men but can be reduced by regular aerobic exercise. *Hypertension* 2007; 50: 403-409.
8. Oliver JJ, Webb DJ, Newby DE. Stimulated tissue plasminogen activator release as a marker of endothelial function in humans. *Arterioscler Thromb Vasc Biol* 2005; 25: 2470-2479.
9. Van Guilder GP, Hoetzer GL, Smith DT, Irmiger HM, Greiner JJ, Stauffer BL, et al. Endothelial t-PA release is impaired in overweight and obese adults but can be improved with regular aerobic exercise. *Am J Physiol Endocrinol Metabol* 2005; 289: E807-E813.

10. Sudi KM, Gallistl S, Weinhandl G, Muntean W, Borkenstein MH. Relationship between plasminogen activator inhibitor-1 antigen, leptin, and fat mass in obese children and adolescents. *Metabol: Clin Exper* 2000; 49: 890-895.

Chapter 72

1. Wannamethee SG, Lowe GD, Whincup PH, Rumley A, Walker M, Lennon L. Physical activity and hemostatic and inflammatory variables in elderly men. *Circulation* 2002; 105: 1785-1790.
2. Mora S, Cook N, Buring JE, Ridker PM, Lee IM. Physical activity and reduced risk of cardiovascular events: potential mediating mechanisms. *Circulation* 2007; 116: 2110-2118.
3. Hammett CJ, Prapavessis H, Baldi JC, Varo N, Schoenbeck U, Ameratunga R, et al. Effects of exercise training on 5 inflammatory markers associated with cardiovascular risk. *Am Heart J* 2006; 151: 367.e7-367.e16.
4. Huffman KM, Samsa GP, Slentz CA, Duscha BD, Johnson JL, Bales CW, et al. Response of high-sensitivity C-reactive protein to exercise training in an at-risk population. *Am Heart J* 2006; 152: 793-800.
5. Kohut ML, McCann DA, Russell DW, Konopka DN, Cunnick JE, Franke WD, et al. Aerobic exercise, but not flexibility/resistance exercise, reduces serum IL-18, CRP, and IL-6 independent of beta-blockers, BMI, and psychosocial factors in older adults. *Brain Behav Immun* 2006; 20: 201-209.
6. Stewart LK, Flynn MG, Campbell WW, Craig BA, Robinson JP, McFarlin BK, et al. Influence of exercise training and age on CD14+ cell-surface expression of toll-like receptor 2 and 4. *Brain Behav Immun* 2005; 19: 389-397.
7. Nicklas BJ, Ambrosius W, Messier SP, Miller GD, Penninx BW, Loeser RF, et al. Diet-induced weight loss, exercise, and chronic inflammation in older, obese adults: a randomized controlled clinical trial. *Am J Clin Nutr* 2004; 79: 544-551.
8. Hamer M, Steptoe A. Prospective study of physical fitness, adiposity, and inflammatory markers in healthy middle-aged men and women. *Am J Clin Nutr* 2009; 89: 85-89.
9. Arsenault BJ, Lachance D, Lemieux I, Almeras N, Tremblay A, Bouchard C, et al. Visceral adipose tissue accumulation, cardiorespiratory fitness, and features of the metabolic syndrome. *Arch Intern Med* 2007; 167: 1518-1525.
10. Radak Z, Chung HY, Naito H, Takahashi R, Jung KJ, Kim HJ, et al. Age-associated increase in oxidative stress and nuclear factor kappaB activation are attenuated in rat liver by regular exercise. *FASEB J* 2004; 18: 749-750.
11. Danesh J, Wheeler JG, Hirschfield GM, Eda S, Eiriksdottir G, Rumley A, et al. C-reactive protein and other circulating markers of inflammation in the prediction of coronary heart disease. *N Engl J Med* 2004; 350: 1387-1397.

Chapter 73

1. Katzmarzyk PT, Baur LA, Blair SN, Lambert EV, Oppert JM, Riddoch C, et al. International Conference on Physical Activity and Obesity in Children: summary statement and recommendations. *Int J Pediatr Obes* 2008; 3: 3-21.
2. Simon GE, Ludman EJ, Linde JA, Operskalski BH, Ichikawa L, Rohde P, et al. Association between obesity and depression in middle-aged women. *Gen Hosp Psychiatry* 2008; 30: 32-39.
3. Jorm AF, Korten AE, Christensen H, Jacomb PA, Rodgers B, Parslow RA. Association of obesity with anxiety, depression and emotional well-being: a community survey. *Aust NZ J Pub Health* 2003; 27: 434-440.
4. Taylor AH, Faulkner G. Inaugural editorial. *Mental Health Phys Act;* 2008 1(1), 1-8.
5. Dunn AL, Trivedi MH, Kampert JB, Clark CG, Chambliss HO. Exercise treatment for depression: efficacy and dose response. *Am J Prev Med* 2005; 28: 1-8.
6. Taylor AH, Fox KR. Changes in physical self-perceptions: findings from a randomised controlled study of a GP exercise referral scheme. *Health Psychology* 2005; 24: 11-21.
7. Ekkekakis P, Lind E. Exercise does not feel the same when you are overweight: the impact of self-selected and imposed intensity on affect and exertion. *Int J Obes* 2006; 30: 652-660.
8. Thayer RE. *Calm energy: how people regulate mood with food and exercise.* Oxford University Press: New York, 2001.
9. Duman RS. Neurotrophic factors and regulation of mood: role of exercise, diet and metabolism. *Neurobiol Aging* 2005; 26: 88-93.
10. Wang GJ, Volkow ND, Fowler JS. The role of dopamine in motivation for food in humans: implications for obesity. *Expert Opin Ther Targets* 2002; 6: 601-609.
11. Taylor AH, Oliver A. Acute effects of brisk walking on urges to eat chocolate, affect, and responses to a stressor and chocolate cue: An experimental study. *Appetite* 2009; 52: 155-160.

Chapter 74

1. American Cancer Society. *Cancer facts & figures 2008.* American Cancer Society: Atlanta, 2008.
2. Friedenreich CM, Cust AE. Physical activity and breast cancer risk: impact of timing, type and dose of activity and population subgroup effects. *Br J Sports Med* 2008; 42: 636-647.
3. Carmichael AR. Obesity as a risk factor for development and poor prognosis of breast cancer. *BJOG* 2006; 113: 1160-1166.
4. Holmes MD, Chen WY, Feskanich D, Kroenke CH, Colditz GA. Physical activity and survival after breast cancer diagnosis. *JAMA* 2005; 293: 2479-2486.
5. Holick CN, Newcomb PA, Trentham-Dietz A, Titus-Ernstoff L, Bersch AJ, Stampfer MJ, et al. Physical activity and survival after diagnosis of invasive breast cancer. *Cancer Epidemiol Biomarkers Prev* 2008; 17: 379-386.
6. McNeely ML, Campbell KL, Rowe BH, Klassen TP, Mackey JR, Courneya KS. Effects of exercise on breast cancer patients and survivors: a systematic review and meta-analysis. *CMAJ* 2006; 175: 34-41.
7. Courneya KS, Katzmarzyk PT, Bacon E. Physical activity and obesity in Canadian cancer survivors: population-based estimates from the 2005 Canadian Community Health Survey. *Cancer* 2008; 112: 2475-2582.
8. Vallance JK, Courneya KS, Plotnikoff R, Mackay JR. Randomized controlled trial of the effects of print materials and step pedometers on physical activity and quality of life in breast cancer survivors. *J Clin Oncol* 2007; 25: 2352-2359.
9. Mefferd K, Nichols JF, Pakiz B, Rock CL. A cognitive behavioral therapy intervention to promote weight loss improves body composition and blood lipid profiles among overweight breast cancer survivors. *Breast Cancer Res Treat* 2007; 104: 145-152.
10. Irwin ML. Randomized controlled trials of physical activity and breast cancer prevention. *Exerc Sport Sci Rev* 2006; 34: 182-193.

Chapter 75

1. Giovannucci E, Wu K. Cancers of the colon and rectum. In: Schottenfeld D, Fraumeni Jr. JF (eds.), *Cancer epidemiology and prevention.* Oxford University Press: New York, 2006, pp. 809-829.
2. Flood DM, Weiss NS, Cook LS, Emerson JC, Schwartz SM, Potter JD. Colorectal cancer incidence in Asian migrants to

the United States and their descendants. *Cancer Causes Control* 2000; 11: 403-411.

3. Platz EA, Willett WC, Colditz GA, Rimm EB, Spiegelman D, Giovannucci E. Proportion of colon cancer risk that might be preventable in a cohort of middle-aged US men. *Cancer Causes Control* 2000; 11: 579-588.
4. Lee I-M, Oguma Y. Physical activity. In: Schottenfeld D, Fraumeni Jr. JF (eds.), *Cancer epidemiology and prevention.* Oxford University Press: New York, 2006, pp. 449-467.
5. World Cancer Research Fund/American Institute for Cancer Research. *Food, nutrition, physical activity, and the prevention of cancer: a global perspective.* AICR: Washington, D.C., 2007, pp. 198-228.
6. Physical Activity Guidelines Advisory Committee Report. 2008. Washington, D.C. U.S. Department of Health and Human Services.
7. Renehan AG, Tyson M, Egger M, Heller RF, Zwahlen M. Body-mass index and incidence of cancer: a systematic review and meta-analysis of prospective observational studies. *Lancet* 2008; 371: 569-578.
8. Larsson SC, Rutegard J, Bergkvist L, Wolk A. Physical activity, obesity, and risk of colon and rectal cancer in a cohort of Swedish men. *Eur J Cancer* 2006; 42: 2590-2597.
9. Lee IM, Paffenbarger RS Jr. Quetelet's index and risk of colon cancer in college alumni. *J Natl Cancer Inst* 1992; 84: 1326-1331.
10. Hou L, Ji BT, Blair A, Dai Q, Gao YT, Chow WH. Commuting physical activity and risk of colon cancer in Shanghai, China. *Am J Epidemiol* 2004; 160: 860-867.

Chapter 76

1. Flegal KM, Graubard B, Williamson DF, Gall MH. Case-specific excess deaths associated with underweight, overweight and obesity. *JAMA* 2007; 298: 2028-2037.
2. Lucenteforte E, Bosetti C, Talamini R, Montella M, Zucchetto A, Pelucci C, et al. Diabetes and endometrial cancer: effect modification by body weight, physical activity and hypertension. *Br J Cancer* 2007; 97: 995-998.
3. Friberg E, Mantzoros CS, Wolk A. Diabetes and the risk of endometrial cancer: a population-based prospective study. *Cancer Epidemiol Biomark Prev* 2007; 16: 276-280.
4. Machova N, Cizek L, Horakova D, Koutna J, Lorenc J, Janoutova G, et al. Association between obesity and cancer incidence in the population of the District Sumperk, Czech Republic. *Onkol* 2007; 30: 538-542.
5. Brinton LA, Sakoda LC, Frederiksen K, Sherman ME, Kjaer SK, Graubard BI, et al. Relationships of uterine and ovarian tumors to pre-existing chronic conditions. *Gynecol Oncol* 2007; 107: 487-494.
6. Reeves GK, Pirie K, Beral V, Green J, Spencer E, Bull D, et al. Cancer incidence and mortality in relation to body mass index in the million women cohort study. *BMJ* 2007; 335: 1134.
7. Friedenreich C, Cust A, Lahmann PH, Steindorf K, Butron-Ruault MC, Clavel-Chapelon F, et al. Anthropometric factors and risk of endometrial cancer: the European prospective investigation into cancer and nutrition. *Cancer Causes Control* 2007; 18: 399-413.
8. Luo J, Margolis KL, Adami HO, Lopez AM, Lessin L, Ye W, et al. Body size, weight cycling and risk of renal cell carcinoma among postmenopausal women: the Women's Health Initiative (United States). *Am J Epidemiol* 2007; 166: 752-759.
9. Setlawan VW, Stram DO, Nomura DO, Kolonel AMY, Henderson BE. Risk factors for renal cancer: the multiethnic cohort. *Am J Epidemiol* 2007; 166: 932-940.
10. Lin Y, Kikuchi S, Tamakoshi A, Yagyu K, Obata Y, Inaba Y, et al. Obesity, physical activity and the risk of pancreatic cancer in a large Japanese cohort. *Int J Cancer* 2007; 120: 2665-2671.
11. Merry AH, Schouten LJ, Goldbohm RA, van den Brandt PA. Body mass index, height, and risk of adenocarcinoma of the esophagus and gastric cardia: a prospective cohort study. *Gut* 2007; 56: 1503-1511.
12. Fowke JH, Motley SS, Wills M, Cookson MS, Concepcion RS, Eckstein CW, et al. Prostate volume modifies the association between obesity and prostate cancer or high-grade prostate intraepithelial neoplasia. *Cancer Causes Control* 2007; 18: 375-384.
13. Lamarre NS, Ruggieri MR, Braverman AS, Gerstein AS, Matthew I, Mydlo JH. Effect of obese and lean Zucker rat sera on human and rat cancer cells: implications in obesity-related prostate tumor biology. *Urology* 2007; 69: 191-195.
14. Olsen CM, Green AC, Whiteman DC, Sadeghi S, Kolahdooz F, Webb PM. Obesity and the risk of ovarian epithelial cancer: a systematic review and meta-analysis. *Eur J Cancer* 2007; 43: 690-709.
15. Larsson SC, Wolk A. Overweight, obesity and risk of liver cancer: a meta-analysis of cohort studies. *Br J Cancer* 2007; 97: 1005-1008.
16. Ahrens W, Timmer A, Vyberg M, Fletcher T, Guenel P, Merler E, et al. Risk factors for extrahepatic biliary tract carcinoma in men: medical conditions and lifestyle: results from a European multicentre case-control study. *Eur J Gastroenterol Hepatol* 2007; 19: 623-630.
17. Larsson SC, Wolk A. Obesity and the risk of gall-bladder disease. *Br J Cancer* 2007; 96: 1457-1461.
18. Larsson SC, Wolk A. Body mass index and risk of multiple myeloma: a meta-analysis. *Int J Cancer* 2007; 121: 2512-2516.
19. Skibola CF. Obesity, diet, and the risk of non-Hodgkin's lymphoma. *Cancer Epidemiol Biomark Prev* 2007; 16: 392-395.
20. Chiu BC, Soni L, Gapstur SM, Fought AJ, Evens AM, Weisenburger DD. Obesity and risk of non-Hodgkin's lymphoma (United States). *Cancer Causes Control* 2007; 18: 677-685.
21. Larsson SC, Wolk A. Obesity and risk of non-Hodgkin's lymphoma: a meta-analysis. *Int J Cancer* 2007; 121: 1564-1570.

Chapter 77

1. Reaven GM. Banting lecture 1988. Role of insulin resistance in human disease. *Diabetes* 1988; 37: 1595-1607.
2. Anonymous. Executive summary of the third report of the National Cholesterol Education Program (NCEP) Expert Panel on Detection, Evaluation, and Treatment of High Blood Cholesterol in Adults (Adult Treatment Panel III). *JAMA* 2001; 285: 2486-2497.
3. Park YW, Zhu S, Palaniappan L, Heshka S, Carnethon MR, Heymsfield SB. The metabolic syndrome: prevalence and associated risk factor findings in the US population from the third National Health and Nutrition Examination Survey, 1988-1994. *Arch Intern Med* 2003; 163: 427-436.
4. Grundy SM, Hansen B, Smith SC Jr., Cleeman JI, Kahn RA, for Conference Participants. Clinical management of metabolic syndrome: report of the American Heart Association/National Heart, Lung, and Blood Institute/American Diabetes Association conference on scientific issues related to management. *Circulation* 2004; 109: 551-556.
5. Ford ES, Li C. Physical activity or fitness and the metabolic syndrome. *Expert Rev Cardiovasc Ther* 2006; 4: 897-915.
6. Holme I, Tonstad S, Sogaard AJ, Larsen PG, Haheim LL. Leisure time physical activity in middle age predicts the metabolic syndrome in old age: results of a 28-year follow-up of men in the Oslo Study. *BMC Pub Health* 2007; 7: 154.

7. Wilsgaard T, Jacobsen BK. Lifestyle factors and incident metabolic syndrome: the Tromsø Study 1979-2001. *Diab Res Clin Pract* 2007; 78: 217-224.
8. Orchard TJ, Temprosa M, Goldberg R, Haffner S, Ratner R, Marcovina S, et al. The effect of metformin and intensive lifestyle intervention on the metabolic syndrome: the Diabetes Prevention Program randomized trial. *Ann Intern Med* 2005; 142: 611-619.
9. Katzmarzyk PT, Leon AS, Wilmore JH, Skinner JS, Rao DC, Rankinen T, et al. Targeting the metabolic syndrome with exercise: evidence from the HERITAGE Family Study. *Med Sci Sports Exerc* 2003; 35: 1703-1709.
10. Church TS, Earnest CP, Skinner JS, Blair SN. Effects of different doses of physical activity on cardiorespiratory fitness among sedentary, overweight or obese postmenopausal women with elevated blood pressure: a randomized controlled trial. *JAMA* 2007; 297: 2081-2091.
11. L. Grundy SM, Brewer HB Jr., Cleeman JI, Smith SC Jr., Lenfant C, for the Conference Participants. Definition of metabolic syndrome: report of the National Heart, Lung, and Blood Institute/American Heart Association Conference on Scientific Issues Related to Definition. *Circulation* 2004 Jan 27; 109(3): 433-438.

Chapter 78

1. Hootman JM. "These old bones"—a growing public health problem. *J Athl Train* 2007; 42: 325-326.
2. Whiting WC, Zernicke RF. *Biomechanics of musculoskeletal injury.* Human Kinetics: Champaign, IL, 1998, pp. 41-85.
3. Hagen KB, Tambs K, Bjerkedal T. A prospective cohort study of risk factors for disability retirement because of back pain in the general working population. *Spine* 2002; 27: 1790-1796.
4. Felson DT, Lawrence RC, Dieppe PA, Hirsch R, Helmick CG, Jordan JM, et al. Osteoarthritis: new insights. Part I: The disease and its risk factors. *Ann Intern Med* 2000; 133: 635-646.
5. Felson DT, Zhang Y. An update on the epidemiology of knee and hip osteoarthritis with a view to prevention. *Arthritis Rheum* 1998; 41: 1343-1355.
6. Ryb GE, Dischinger PC. Injury severity and outcome of overweight and obese patients after vehicular trauma: a crash injury research and engineering network (CIREN) study. *J Trauma* 2008; 64: 406-411.
7. Kohrt WM, Nelson ME, Fielding RA, Hootman JM, Lane NE. Musculoskeletal Subcommittee Report, Physical Activity Guidelines Advisory Committee. www.health.gov/paguidelines.
8. Wearing SC, Hennig EM, Byrne NM, Steele JR, Hills AP. The biomechanics of restricted movement in adult obesity. *Obes Rev* 2006; 7: 13-24.
9. Christensen R, Bartels EM, Astrup A, Bliddal H. Effect of weight reduction in obese patients diagnosed with knee osteoarthritis: a systematic review and meta-analysis. *Ann Rheum Dis* 2007; 66: 433-439.
10. Melissas J, Colakakis E, Hadjipavlou A. Low-back pain in morbidly obese patients and the effect of weight loss following surgery. *Obes Surg* 2003; 13: 389-393.
11. Messier SP, Loeser RF, Miller GD, Morgan TM, Rejeski WJ, Sevick MA, et al. Exercise and dietary weight loss in overweight and obese older adults with knee osteoarthritis: the Arthritis, Diet, and Activity Promotion Trial. *Arthritis Rheum* 2004; 50: 1501-1510.

Chapter 79

1. Corbeil P, Simoneau M, Rancourt D, Tremblay A, Teasdale N. Increased risk for falling associated with obesity: mathematical modeling of postural control. *IEEE Trans Neural Syst Rehabil Eng* 2001; 9: 126-136.
2. Schiller JS, Kramarow EA, Dey AN. Fall injury episodes among noninstitutionalized older adults: United States, 2001-2003. *Adv Data* 2007: 1-16.
3. Laurin D, Verreault R, Lindsay J, MacPherson K, Rockwood K. Physical activity and risk of cognitive impairment and dementia in elderly persons. *Arch Neurol* 2001; 58: 498-504.
4. Fjeldstad C, Fjeldstad AS, Acree LS, Nickel KJ, Gardner AW. The influence of obesity on falls and quality of life. *Dyn Med* 2008; 7: 4.
5. Finkelstein EA, Chen H, Prabhu M, Trogdon JG, Corso PS. The relationship between obesity and injuries among U.S. adults. *Am J Health Promot* 2007; 21: 460-468.
6. Owusu W, Willett W, Ascherio A, et al. Body anthropometry and the risk of hip and wrist fractures in men: results from a prospective study. *Obes Res* 1998; 6: 12-19.
7. Liu-Ambrose T, Khan KM, Eng JJ, Janssen PA, Lord SR, McKay HA. Resistance and agility training reduce fall risk in women aged 75 to 85 with low bone mass: a 6-month randomized, controlled trial. *J Am Geriatr Soc* 2004; 52: 657-665.
8. Gillespie L, Gillespie W, Robertson M, et, al. Interventions for preventing falls in elderly people. Cochrane Database Syst Rev. 2003; 4: CD000340.
9. Robertson MC, Campbell AJ, Gardner MM, Devlin N. Preventing injuries in older people by preventing falls: a meta-analysis of individual-level data. *J Am Geriatr Soc* 2002; 50: 905-911.
10. Maffiuletti NA, Agosti F, Proietti M, et al. Postural instability of extremely obese individuals improves after a body weight reduction program entailing specific balance training. *J Endocrinol Invest* 2005; 28: 2-7.
11. Hemmingsson E, Page A, Fox K, Rossner S. Influencing adherence to physical activity behaviour change in obese adults. *Scand J Nutr* 2001; 45: 114-119.

Chapter 80

1. Carnethon MR, Gulati M, Greenland P. Prevalence and cardiovascular disease correlates of low cardiorespiratory fitness in adolescents and adults. *JAMA* 2005; 294: 2981-2988.
2. Physical Activity Guidelines Advisory Committee. *Physical Activity Guidelines Advisory Committee Report, 2008.* U.S. Department of Health and Human Services: Washington, DC: 2008. Part G, section 10: Adverse events, pp. G10-1–G10-58. Accessed 16 Oct 2008. www.health.gov/paguidelines/Report/pdf/CommitteeReport.pdf.
3. Browning RC, Kram R. Effects of obesity on the biomechanics of walking at different speeds. *Med Sci Sports Exerc* 2007; 39: 1632-1641.
4. Wearing SC, Hennig EM, Byrne NM, Steele JR, Hills AP. Musculoskeletal disorders associated with obesity: a biomechanical perspective. *Obes Rev* 2006; 7: 239-250.
5. Carlson SA, Hootman JM, Powell KE, Macera CA, Heath GW, Gilchrist J, et al. Self-reported injury and physical activity levels: United States 2000 to 2002. *Ann Epidemiol* 2006; 16: 712-719.
6. Finkelstein EA, Chen H, Prabhu M, Trogdon JG, Corso PS. The relationship between obesity and injuries among U.S. adults. *Am J Health Promot* 2007; 21: 460-468.

7. Institute of Medicine. *Adequacy of evidence for physical activity guidelines development: workshop summary.* National Academies Press: Washington, DC, 2007.

8. Hootman JM, Powell KE. Physical activity, fitness, and musculoskeletal injury. In: Lee I-M (ed.), *Epidemiologic methods in physical activity studies.* Oxford University Press: New York, 2008.

9. American College of Sports Medicine. American College of Sports Medicine position stand on exertional heat illness during training and competition. *Med Sci Sports Exerc* 2007; 39: 556-572.

10. Gardner JW, Kark JA, Karnei K, Sanborn JS, Gastaldo E, Burr P, et al. Risk factors for predicting exertional heat illness in male Marine Corps recruits. *Med Sci Sports Exerc* 1996; 28: 939-944.

Chapter 81

1. Andersen LB, Harro M, Sardinha LB, Froberg K, Ekelund U, Brage S, et al. Physical activity and clustered cardiovascular risk in children: a cross-sectional study (The European Youth Heart Study). *Lancet* 2006; 368: 299-304.

2. Sundberg S. Maximal oxygen uptake in relation to age in blind and normal boys and girls. *Acta Pædiatr Scand* 1982; 71: 603-608.

3. Dela F, Larsen JJ, Mikines KJ, Ploug T, Petersen LN, Galbo H. Insulin-stimulated muscle glucose clearance in patients with NIDDM—effects of one-legged physical-training. *Diabetes* 1995; 44: 1010-1020.

4. Nielsen GA, Andersen LB. The association between high blood pressure, physical fitness, and body mass index. *Prev Med* 2003; 36: 229-234.

5. Gutin B, Yin Z, Humphries MC, Bassali R, Le NA, Daniels S, et al. Relations of body fatness and cardiovascular fitness to lipid profile in black and white adolescents. *Pediatr Res* 2005; 58: 78-82.

6. Gutin B, Owens S. Role of exercise intervention in improving body fat distribution and risk profile in children. *Am J Hum Biol* 1999; 11: 237-247.

7. Anderssen SA, Cooper AR, Riddoch C, Sardinha LB, Harro M, Brage S, et al. Low cardiorespiratory fitness is a strong predictor for clustering of cardiovascular disease risk factors in children independent of country, age and sex. *Eur J Cardiovasc Prev Rehabil* 2007; 14: 526-531.

8. Andersen LB, Sardinha LB, Froberg K, Riddoch CJ, Page AS, Anderssen SA. Fitness, fatness and clustering of cardiovascular risk factors in children from Denmark, Estonia and Portugal: the European Youth Heart Study. *Int J Pediatr Obes* 2008; 3 Suppl 1: 58-66.

9. Reinehr T, Stoffel-Wagner B, Roth CL, Andler W. High-sensitive C-reactive protein, tumor necrosis factor alpha, and cardiovascular risk factors before and after weight loss in obese children. *Metabolism* 2005; 54: 1155-1161.

10. Hotamisligil GS, Shargill NS, Spiegelman BM. Adipose expression of tumor necrosis factor-alpha: direct role in obesity-linked insulin resistance. *Science* 1993; 259: 87-91.

Chapter 82

1. Weiss R, Taksali SE, Tamborlane WV, Burger TS, Savoye M, Caprio S. Predictors of changes in glucose tolerance status in obese youth. *Diab Care* 2005; 28: 902-909.

2. Shaibi GQ, Roberts CK, Goran MI. Exercise and insulin resistance in youth. *Exerc Sports Sci Rev* 2008; 36: 5-11.

3. Schmitz KH, Jacobs Jr. DR, Steinberger J, Moran A, Sinaiko AR. Association of physical activity with insulin sensitivity in children. *Int J Obes* 2002; 26: 1310-1316.

4. Andersen LB, Haaro M, Sardinha LB, Froberg K, Ekelund U, Brage S, et al. Physical activity and clustered cardiovascular risk in children: a cross-sectional study (The European Youth Heart Study). *Lancet* 2005; 368: 299-304.

5. Imperatore G, Cheng YJ, Williams DE, Fulton J, Gregg EW. Physical activity, cardiovascular fitness, and insulin sensitivity among U.S. adolescents. *Diab Care* 2006; 29: 1567-1572.

6. Ferguson MA, Gutin B, Le N-A, Karp W, Litaker M, Humphries M, et al. Effects of exercise training and its cessation on components of the insulin resistance syndrome in obese children. *Int J Obes* 1999; 22: 889-895.

7. Nassis GP, Papantakou K, Skenderi K, Triandaffilopoulou M, Kavrouras SA, Yannakoulia M, et al. Aerobic exercise training improves insulin sensitivity without changes in body weight, body fat, adiponectin, and inflammatory markers in overweight and obese girls. *Metabolism* 2005; 54: 1472-1479.

8. Benson AC, Torode ME, Fiatarone Singh MA. Effects of resistance training on metabolic fitness in children and adolescents: a systematic review. *Obes Rev* 2008; 9: 43-66.

9. Shaibi GQ, Cruz ML, Ball GDC, Weigensberg MJ, Salem GJ, Crespo NC, et al. Effects of resistance training on insulin sensitivity in overweight Latino adolescent males. *Med Sci Sports Exerc* 2006; 38: 1208-1215.

10. Bell LM, Watts K, Siarfararikas A, Thompson A, Ratnam N, Bulsara M, et al. Exercise alone reduces insulin resistance in obese children independently of changes in body composition. *J Clin Endocrinol Metab* 2007; 92: 4230-4235.

Chapter 83

1. World Health Organization. *Preventing chronic diseases: a vital investment.* World Health Organization: Geneva, 2005.

2. World Health Organization. *The world health report 2002: reducing risks, promoting healthy life.* World Health Organization: Geneva, 2002.

3. World Health Organization. *Obesity and overweight.* Fact sheet no. 311. World Health Organization: Geneva, 2006.

4. World Health Organization. *Diet, nutrition and the prevention of chronic diseases. Report of a joint WHO/FAO expert consultation.* Technical report series no. 916. World Health Organization: Geneva, 2003.

5. World Health Organization. Resolution WHA57.17. *Global health strategy on diet, physical activity and health.* In: *Fifty-seventh World Health Assembly, Geneva, 22 May 2004.* Vol. 1. Resolutions and decisions. World Health Organization: Geneva, 2004. www.who.int/gb/ebwha/pdf_files/WHA57/A57_R17-en.pdf. Accessed 10 April 2008.

6. World Health Organization. *A guide for population-based approaches to increasing levels of physical activity: implementation of the WHO global strategy on diet, physical activity and health.* World Health Organization: Geneva, 2007. www.who.int/dietphysicalactivity/physical-activity-promotion-2007.pdf. Accessed 20 Feb 2008.

7. World Health Organization. *Preventing noncommunicable diseases in the workplace through diet and physical activity. WHO/World Economic Forum report of a joint event.* World Health Organization: Geneva, 2008.

8. World Health Organization. *School policy framework: implementation of the WHO global strategy on diet, physical activity and health.* World Health Organization: Geneva, 2008.

9. World Health Organization. *Global Strategy on diet, physical activity and health: a framework to monitor and evaluate implementation.* World Health Organization: Geneva, 2006. www.who.int/dietphysicalactivity/DPASindicators/en/. Accessed 20 Feb 2008.

10. World Health Organization. *STEPwise approach to surveillance (STEPS).* World Health Organization: Geneva. www.who.int/chp/steps/en/. Accessed 28 Feb 2008.

Chapter 84

1. DeBakey ME. The role of government in health care: a societal issue. *Am J Surg* 2006; 191: 145-157.
2. Nova Scotia Health Promotion and Protection: Public health 101: an introduction to public health. Nov 2007. www.gov.ns.ca/hpp/publications/PH-101.pdf.
3. Resolution WHA55.23. Diet, physical activity and health. In: *Fifty-fifth World Health Assembly, Geneva, 15-22 May 2004. Resolutions and decisions, annexes.* World Health Organization: Geneva, 2004.
4. www.phac-aspc.gc.ca/hl-vs-strat/index.html.
5. www.dh.gov.uk/en/Publicationsandstatistics/Publications/PublicationsPolicyAndGuidance/DH_082378.
6. Mason JO, Powell KE. Physical activity, behavioural epidemiology, and public health. *Pub Health Rep* 1985; 100: 113-115.
7. http://cmte.parl.gc.ca/Content/HOC/committee/391/hesa/reports/rp2795145/hesarp07/hesarp07-e.pdf.
8. Bauman A, Madill J, Craig C, Salmon A. ParticipACTION: the mouse that roared, but did it get the cheese? *Can J Pub Health* 2004; 95: S14-S19.
9. Edwards P, Tsouros A. *Promoting physical activity and active living in urban environments—the role of local governments.* World Health Organization: Geneva, 2006.

Chapter 85

1. Brown WJ, Eakin EG, Mummery WK, Trost SG. 10,000 Steps Rockhampton: establishing a multi-strategy physical activity promotion in a community. *Health Promot J Aust* 2003; 14: 95-100.
2. Brown W, Mummery K, Eakin EG, Schofield G. 10,000 Steps Rockhampton: evaluation of a whole community approach to improving population levels of physical activity. *J Phys Act Health* 2006; 3: 1-14.
3. Mummery WK, Schofield G, Hincliffe A, Joyner K, Brown W. Dissemination of a community-based physical activity project: the case of 10,000 steps. *J Sci Med Sport* 2006; 9: 424-430.
4. Rose G. *The strategy of preventive medicine.* Oxford: Oxford University Press; 1992.
5. Wareham NJ, Van Sluijs EMF, Ekelund U. Physical activity and obesity prevention: a review of the current evidence. *Proc Nutr Soc* 2005; 64: 229-247.
6. Strachan D, Rose G. Strategies of prevention revisited: effects of imprecise measurement of risk factors on the evaluation of high-risk and population-based approaches to prevention of cardiovascular disease. *J Clin Epidemiol* 1991; 44: 1187-1196.
7. Blair SN, LaMonte MJ, Nichaman MZ. The evolution of physical activity recommendations: how much is enough? *Am J Clin Nutr* 2004; 79: 913S-920S.
8. Laverack G, Labonte R. A planning framework for community empowerment goals within health promotion. *Health Policy Plan* 2000; 15: 255-262.
9. Zakocs RC, Edwards EM. What explains community coalition effectiveness? A review of the literature. *Am J Prev Med* 2006; 30: 351-361.
10. Kahn EB, Ramsey LT, Brownson RC, Heath GW, Howze EH, Powell KE, et al. The effectiveness of interventions to increase physical activity. A systematic review. *Am J Prev Med* 2002; 22: 73-107.

Chapter 86

1. Pate RR, Davis MG, Robinson TN, Stone EJ, McKenzie TL, Young JC. Promoting physical activity in children and youth: a leadership role for schools: a scientific statement from the American Heart Association Council on Nutrition, Physical Activity, and Metabolism (Physical Activity Committee) in collaboration with the Councils on Cardiovascular Disease in the Young and Cardiovascular Nursing. *Circulation* 2006; 114: 1214-1224.
2. Doak CM, Visscher TL, Renders CM, Seidell JC. The prevention of overweight and obesity in children and adolescents: a review of interventions and programmes. *Obes Rev* 2006; 7: 111-136.
3. Summerbell CD, Waters E, Edmunds LD, Kelly S, Brown T, Campbell KJ. Interventions for preventing obesity in children. *Coch Database Syst Rev* 2005: CD001871.
4. van Sluijs EM, McMinn AM, Griffin SJ. Effectiveness of interventions to promote physical activity in children and adolescents: systematic review of controlled trials. *BMJ* 2007; 335: 703.
5. DeMattia L, Lemont L, Meurer L. Do interventions to limit sedentary behaviours change behaviour and reduce childhood obesity? A critical review of the literature. *Obes Rev* 2007; 8: 69-81.
6. Simon C, Schweitzer B, Oujaa M, Wagner A, Arveiler D, Triby E, et al. Successful overweight prevention in adolescents by increasing physical activity: a 4-year randomized controlled intervention. *Int J Obes (London)* 2008; 32: 1489-1498.
7. Cleland V, Dwyer T, Blizzard L, Venn A. The provision of compulsory school physical activity: associations with physical activity, fitness and overweight in childhood and twenty years later. *Int J Behav Nutr Phys Act* 2008; 5: 14.
8. Gutin B, Yin Z, Johnson M, Barbeau P. Preliminary findings of the effect of a 3-year after-school physical activity intervention on fitness and body fat: the Medical College of Georgia Fitkid Project. *Int J Pediatr Obes* 2008; 3 suppl. 1: 3-9.
9. McKee R, Mutrie N, Crawford F, Green B. Promoting walking to school: results of a quasi-experimental trial. *J Epidemiol Comm Health* 2007; 61: 818-823.
10. Graf C, Koch B, Falkowski G, Jouck S, Christ H, Staudenmaier K, et al. School-based prevention: effects on obesity and physical performance after 4 years. *J Sports Sci* 2008; 26: 987-994.
11. Sollerhed AC, Ejlertsson G. Physical benefits of expanded physical education in primary school: findings from a 3-year intervention study in Sweden. *Scand J Med Sci Sports* 2008; 18: 102-107.
12. Martinez Vizcaino V, Salcedo Aguilar F, Franquelo Gutierrez R, et al. Assessment of an after-school physical activity program to prevent obesity among 9- to 10-year-old children: a cluster randomized trial. *Int J Obes (London)* 2008; 32: 12-22.

Chapter 87

1. Bureau of Labor Statistics. Employees on nonfarm payrolls by major industry sector and selected industry detail, seasonally adjusted. 2008. ftp://ftp.bls.gov/pub/suppl/empsit.ceseeb3.txt. Accessed 31 March 2008.
2. Anderson LH, Martinson BC, Crain AL, Pronk NP, Whitebird RR, Fine LJ, et al. Health care charges associated with physical inactivity, overweight and obesity. *Prev Chronic Dis* 2005; 2: A09.
3. Wang F, McDonald T, Champagne LJ, Edington DW. Relationship of body mass index and physical activity to health care costs among employees. *J Occup Environ Med* 2004; 45: 428-236.
4. Chapman LS. Meta-evaluation of worksite health promotion economic return studies: 2005 update. *Am J Health Promot* 2005; 19: 1-11.
5. Martinson BC, Crain AL, Pronk NP, O'Connor PJ, Maciosek MV. Changes in physical activity and short-term changes in health care charges: a prospective cohort study of older adults. *Prev Med* 2003; 37: 319-326.
6. Proper KI, Staal BJ, Hildebrand VH, van der Beek AJ, van Mechelen W. Effectiveness of physical activity programs at worksites with respect to work-related outcomes. *Scand J Work Environ Health* 2002; 28: 75-84.
7. Goetzel RZ, Ozminkowski RJ. The health and cost benefits of work site health-promotion programs. *Annu Rev Pub Health* 2008; 29: 303-323.
8. Dishman R. Worksite physical activity interventions. *Am J Prev Med* 2003; 15: 344-361.
9. Werbach A. The birth of blue. Speech delivered to the Commonwealth Club, San Francisco, 10 April 2008. www.saatchis.com/birthofblue/. Accessed 16 April 2008.

Chapter 88

1. World Health Organization. *Obesity: preventing and managing the global epidemic. Report on a WHO consultation.* Technical report series no. 894. World Health Organization: Geneva, 2000.
2. Wammes B, Oenema A, Brug J. The evaluation of a mass media campaign aimed at weight gain prevention among young Dutch adults. *Obesity* 2007; 15: 2780-2789.
3. Cavill N, Bauman A. Changing the way people think about health-enhancing physical activity: do mass media campaigns have a role? *J Sports Sci* 2004; 22: 771-790.
4. Bauman A, Bowles HR, Huhman M, Heitzler CD, Owen N, Smith BJ, et al. Testing a hierarchy of effects model—pathways from awareness to outcomes in the VERB™ campaign 2002-2003. *Am J Prev Med* 2008; 34: 249-256.
5. Rothschild ML. Carrots, sticks, and promises: a conceptual framework for the management of public health and social issue behaviors. *J Marketing* 1999; 63: 24-37.
6. Bauman A, Allman-Farinelli M, Huxley R, James WPT. Leisure-time physical activity alone may not be a sufficient public health approach to prevent obesity—a focus on China. *Obes Rev* 2008; 9 suppl. 1: 119-126.
7. Morabia A, Costanza M. Does walking 15 minutes per day keep the obesity epidemic away? Simulation of the efficacy of a population wide obesity prevention campaign. *Am J Pub Health* 2004; 94: 437-440.
8. Wardle J, Rapoport L, Miles A, Afuape T, Duman M. Mass education for obesity prevention: the penetration of the BBC's "Fighting Fat, Fighting Fit" campaign. *Health Educ Res* 2001; 16: 343-355.
9. Werder O. Battle of the bulge: an analysis of the obesity prevention campaigns in the United States and Germany. *Obes Rev* 2007; 8: 451-457.
10. Beaudoin CE, Fernandez C, Wall JL, Farley TA. Promoting healthy eating and physical activity: short-term effects of a mass media campaign. *Am J Prev Med* 2007; 32: 217-223.

Chapter 89

1. U.S. Department of Health and Human Services. *2008 Physical activity guidelines for Americans.* U.S. Department of Health and Human Services: Washington, DC, 2008.
2. World Health Organization. *The world health report 2002. Reducing risks, promoting healthy life.* World Health Organization: Geneva, 2002.
3. World Health Organization. *Preventing chronic diseases: a vital investment.* World Health Organization: Geneva, 2005.
4. Cole TJ, Bellizzi MC, Flegal KM, Dietz WH. Establishing a standard definition for child overweight and obesity worldwide: international survey. *BMJ* 2000; 320: 1240-1243.
5. World Health Organization. *Obesity: preventing and managing the global epidemic.* World Health Organization: Geneva, 1998.
6. Sisson SB, Katzmarzyk PT. International prevalence of physical activity in youth and adults. *Obes Rev* 2008; 9: 606-614.
7. Bouchard C, An P, Rice T, Skinner JS, Wilmore JH, Gagnon J, et al. Familial aggregation of VO_{2max} response to exercise training: results from the HERITAGE Family Study. *J Appl Physiol* 1999; 87: 1003-1008.

Index

Note: The italicized *f* and *t* following page numbers refer to figures and tables, respectively.

P

T

U

About the Editors

Photo courtesy of Alan R. Pesch. Pennington Biomedical Research Center.

Claude Bouchard, PhD, is executive director of the Pennington Biomedical Research Center in Baton Rouge, Louisiana. For more than 40 years he has researched the role of physical activity on physiology, metabolism, and health indicators, taking into account genetic uniqueness. He also has authored or coauthored more than 1,000 scientific publications, and he served as president of the International Society for the Study of Obesity from 2002 to 2006. He also has served as president of the Canadian Society for Applied Physiology and has directed the Physical Activity Sciences Laboratory at Laval University, Quebec City, Canada, for over 20 years.

Dr. Bouchard has received numerous awards over the years, including the TOPS award from the North American Association for the Study of Obesity in 1998 and the Albert Creff Award in Nutrition of the National Academy of Medicine of France in 1997.

Photo courtesy of Alan R. Pesch. Pennington Biomedical Research Center.

Peter T. Katzmarzyk, PhD, is a professor and the associate executive director for Population Science at the Pennington Biomedical Research Center in Baton Rouge, Louisiana. He holds the Louisiana Public Facilities Authority endowed chair in nutrition. His main research interest is the epidemiology and public health impact of obesity and physical inactivity, and determining the relationships between physical activity, physical fitness, obesity, and related disorders such as metabolic syndrome, cardiovascular disease, and diabetes.

Dr. Katzmarzyk has published his research findings in more than 190 scholarly journals and books, and he regularly participates in the scientific meetings of several national and international organizations. He is currently an editorial board member for the *International Journal of Pediatric Obesity, Journal of Physical Activity and Health,* and *Metabolic Syndrome and Related Disorders.*